Principle 11

A sound marketing and sales plan enables a new firm to identify the target customer, set its marketing objectives, and implement the steps necessary to sell the product and build solid customer relationships.

Principle 12

Effective leaders coupled with a good organizational plan, a collaborative performance-based culture, and a sound compensation scheme can help align every participant with the goals and objectives of the new firm.

Principle 13

Effective new ventures use their persuasion skills, credibility, and location advantages to secure the required resources for their firm in order to build a well-coordinated mix of outsourced and internal functions.

Principle 14

The design and management of an efficient, real-time set of production, logistical, and business processes can become a sustainable competitive advantage for a new enterprise.

Principle 15

All new technology business ventures should formulate a clear acquisition and global strategy.

Principle 16

A new firm with a powerful revenue and profit engine can achieve strong but manageable growth leading to a favorable harvest of wealth for the owners.

Principle 17

A sound financial plan demonstrates the potential for growth and profitability for a new venture and is based on the most accurate and reliable assumptions available.

Principle 18

Many kinds of sources for investment capital for a new and growing enterprise exist and should be compared and managed carefully.

Principle 19

The presentation of a compelling story about a venture and the resulting skillful negotiations to close a deal with investors are critical to all new enterprises.

Principle 20

The ability to continuously and ethically execute a business plan and adapt to changing conditions provides long-term success.

Technology Ventures
From Idea to Enterprise

Technology Ventures
From Idea to Enterprise

Thomas H. Byers
Stanford University

Richard C. Dorf
University of California, Davis

Andrew J. Nelson
University of Oregon

Connect
Learn
Succeed™

TECHNOLOGY VENTURES: FROM IDEA TO ENTERPRISE, THIRD EDITION

Published by McGraw-Hill, a business unit of The McGraw-Hill Companies, Inc., 1221 Avenue of the Americas, New York, NY 10020. Copyright © 2011 by The McGraw-Hill Companies, Inc. All rights reserved. Previous editions © 2008 and 2005. No part of this publication may be reproduced or distributed in any form or by any means, or stored in a database or retrieval system, without the prior written consent of The McGraw-Hill Companies, Inc., including, but not limited to, in any network or other electronic storage or transmission, or broadcast for distance learning.

Some ancillaries, including electronic and print components, may not be available to customers outside the United States.

This book is printed on acid-free paper.

1 2 3 4 5 6 7 8 9 0 DOC/DOC 1 0 9 8 7 6 5 4 3 2 1 0

ISBN 978–0–07–338018–6
MHID 0–07–338018–0

Vice President & Editor-in-Chief: *Martin Lange*
Vice President EDP & Central Publishing Services: *Kimberly Meriwether David*
Global Publisher: *Raghothaman Srinivasan*
Sponsoring Editor: *Debra B. Hash*
Developmental Editor: *Lora Neyens*
Senior Marketing Manager: *Curt Reynolds*
Project Manager: *Melissa M. Leick*
Senior Production Supervisor: *Laura Fuller*
Design Coordinator: *Brenda A. Rolwes*
Cover Designer: *Studio Montage, St. Louis, Missouri*
(USE) Cover Image: © *Punchstock/ Getty Images RF*
Compositor: *Laserwords Private Limited*
Typeface: *10.5/12 Times Roman*
Printer: *R. R. Donnelley*

All credits appearing on page or at the end of the book are considered to be an extension of the copyright page.

Library of Congress Cataloging-in-Publication Data

Byers, Thomas (Thomas H.)
  Technology ventures : from idea to enterprise / Thomas H. Byers, Richard C. Dorf,
Andrew J. Nelson. — 3rd ed.
      p. cm.
  Dorf's name appears first on the earlier editions.
  Includes bibliographical references and index.
  ISBN 978-0-07-338018-6 (alk. paper)
  1. Information technology. 2. Entrepreneurship. 3. New business enterprises. I. Dorf, Richard C.
II. Nelson, Andrew J. III. Dorf, Richard C. Technology ventures. IV. Title.
  HC79.I55D674 2010
  658.1'1—dc22
                                                        2009043315

www.mhhe.com

DEDICATION

For our spouses: Michele Mandell, Joy Dorf, and Ann Carney Nelson. We warmly recognize their love and commitment to this publication that will help others create important enterprises for the benefit of all.

THOMAS H. BYERS, RICHARD C. DORF, ANDREW J. NELSON

ABOUT THE AUTHORS

Thomas H. Byers is professor of management science and engineering at Stanford University and founder of the Stanford Technology Ventures Program, which is dedicated to accelerating technology entrepreneurship education around the globe. After receiving his B.S., MBA, and Ph.D. from the University of California, Berkeley, Dr. Byers held leadership positions in technology ventures including Symantec Corporation. His teaching awards include Stanford University's highest honor (Gores Award) and the Gordon Prize from the National Academy of Engineering.

Richard C. Dorf is professor of electrical and computer engineering and professor of management at the University of California, Davis. He is a Fellow of the American Society for Engineering Education (ASEE) in recognition of his outstanding contributions to the society, as well as a Fellow of the Institute of Electrical and Electronic Engineering (IEEE). The best-selling author of *Introduction to Electric Circuits* (8th ed.), *Modern Control Systems* (11th ed.), *Handbook of Electrical Engineering* (3rd ed.), *Handbook of Engineering* (2nd ed.), and *Handbook of Technology Management,* Dr. Dorf is cofounder of seven technology firms.

Andrew J. Nelson is assistant professor of management at the University of Oregon's Lundquist College of Business. Dr. Nelson holds a Ph.D. in management science and engineering from Stanford University, an M.S. from Oxford University, and a dual B.A. from Stanford. He is a Kauffman Foundation Junior Faculty Fellow in Entrepreneurship and has received research awards from the Institute for Operations Research and the Management Sciences (INFORMS) and the T&J Meyer Foundation. He is well known for his studies on technology diffusion networks and the commercialization of university research.

BRIEF CONTENTS

CONTENTS

FOREWORD

by John L. Hennessy, President of Stanford University

I am delighted to introduce this book on technology entrepreneurship by Byers, Dorf, and Nelson. Technology and similar high-growth enterprises are both an important part of our world's economic growth story as well as the place where many young entrepreneurs realize their dreams.

Unfortunately, there have been relatively few complete and analytical books on technology entrepreneurship. Byers, Dorf, and Nelson bring their years of experience in teaching to this book, and it shows. Their personal experiences as entrepreneurs are also clear throughout the book. Their connections and involvement with start-ups—ranging from established companies like Google and Genentech to new ventures just delivering their first products—add a tremendous amount of real-world insight and relevance.

One of the most impressive aspects of this book is its broad coverage of the challenges involved in technology entrepreneurship. Part I talks about the core issues involved when deciding to pursue an entrepreneurial vision and what characteristics are vital to success from the very beginning. I am pleased to see that key topics include building and maintaining a competitive advantage and market timing. During the Internet boom of 2000, several great new companies were built, but too many entrepreneurs and investors forgot several key principles: have a sustainable advantage, create a significant barrier to entry, and be a leader when the market and the technology are both ready. Hopefully, the material in these chapters will help prevent future irrational behavior by both entrepreneurs and investors.

Part II examines the major strategic decisions with which any group of entrepreneurs must grapple: how to balance risk and return, what entrepreneurial structure to pursue, how to find and cultivate the best employees and help make them productive, and the critical issue of intellectual property. Indeed, these are problems faced by every company, and ones that must be continuously examined by the leadership in any organization.

Part III discusses the operational and organizational challenges that all entrepreneurs must tackle. Virtually every start-up led by a technologist that I have been close to inevitably wonders whether it needs sales and marketing. Sometimes in such companies, you hear a remark like: "We have great technology and that will bring us customers; nothing else matters!" I remind them that without sales, there is no revenue, and without marketing, sales will be diminished. It is important to understand how to approach these vital aspects of any successful business. The related topics of building the organization, thinking about acquisitions, and managing operations are also important. If you fail to address these aspects of your company, it will not matter how good your technology is.

Finally, Part IV talks about putting together a solid financial plan for the company, including exit and funding strategies. Of course, such topics are crucial, and they are often the sole or dominant topics of "how-to" books on entrepreneurship. Certainly, the financing and the choice of investors are key, but unless the challenges discussed in the preceding sections are overcome, it is unlikely that a new venture, even if well financed, will be successful.

In looking through this sage and comprehensive treatment, my overwhelming reaction was, "I wish I had read a book like this before I started my first company (MIPS Technologies in 1984)." Unfortunately, I had to learn many of the topics covered here in real-time and often by making a mistake on the first attempt. In my experience, it is the challenges discussed in the earlier sections that really prove to be the minefields. Yes, it is helpful to know how to negotiate a good deal and to structure the right mix of financing sources, especially so that employees can retain as much equity as possible. If, however, you fail to create a sustainable advantage or lack a solid sales and marketing plan, the employees' equity will not be worth much.

Those of us who work at Stanford and live near Silicon Valley are in the heart of the land of technology entrepreneurship. With this book, many others will get to share the extensive and deep insights of Byers, Dorf, and Nelson on this wonderful process that builds tomorrow's enterprises and business leaders.

Entrepreneurship is a vital source of change in all facets of society, empowering individuals to seek opportunities where others see insurmountable problems. For the past century, entrepreneurs have created many great enterprises that subsequently led to job creation, improved productivity, increased prosperity, and a higher quality of life. Entrepreneurship is now playing a vital role in finding solutions to the huge challenges facing civilization, including energy, environment, health, security, and education.

Many books have been written to help educate others about entrepreneurship. Our textbook was the first to thoroughly examine a global phenomenon known as "technology entrepreneurship." Technology entrepreneurship is a style of business leadership that involves identifying high-potential, technology-intensive commercial opportunities, gathering resources such as talent and capital, and managing rapid growth and significant risks using principled decision-making skills. Technology ventures exploit breakthrough advancements in science and engineering to develop better products and services for customers. The leaders of technology ventures demonstrate focus, passion, and an unrelenting will to succeed.

Why is technology so important? The technology sector represents a significant portion of the economy of every industrialized nation. In the United States, more than one-third of the gross national product and about half of private-sector spending on capital goods are related to technology. It is clear that national and global economic growth depends on the health and contributions of technology businesses.

Technology has also become ubiquitous in modern society. Note the proliferation of cell phones, personal computers, and the Internet in the past decade and their subsequent integration into everyday commerce and our personal lives. When we refer to "high-technology" ventures, we include information technology enterprises, biotechnology and medical businesses, energy and sustainability companies, and those service firms where technology is critical to their missions. At the beginning of the twenty first century, many technologies show tremendous promise, including photonics and Internet advancements, medical devices and drug discovery, nanotechnology, and materials technologies related to energy and the environment. The intersection of these technologies may indeed enable the most promising opportunities.

The drive to understand technology venturing has frequently been associated with boom times. Certainly, the often-dramatic fluctuations of economic cycles can foster periods of extreme optimism as well as fear with respect to entrepreneurship. However, some of the most important technology companies have been founded during recessions (e.g., Intel, Cisco, and Amgen). This book's principles endure regardless of the state of the economy.

APPROACH

Just as entrepreneurs innovate by recombining existing ideas and concepts, we integrate the most valuable entrepreneurship and technology management theories from the world's leading scholars to create a fresh look at entrepreneurship. We also provide an action-oriented approach to the subject through the use of examples, exercises, and lists. By striking a balance between theory and practice, our readers gain from both perspectives.

Our comprehensive collection of concepts and applications provides the tools necessary for success in starting and growing a technology enterprise. We show the critical differences between scientific ideas and true business opportunities. Readers benefit from the book's integrated set of cases, examples, business plans, and recommended sources for more information.

We illustrate the book's concepts with examples from the early stages of high-technology firms (e.g., Intel, Google, and Genentech) and traditional firms that use technology strategically (e.g., FedEx and Wal-Mart). How did they develop enterprises that have had such positive impact, sustainable performance, and longevity? In fact, the book's major principles are applicable to any growth-oriented, high-potential venture, including high-impact nonprofit enterprises such as Conservation International and the Gates Foundation.

AUDIENCE

This book is designed for students in colleges and universities, as well as others in industry and public service, who seek to learn the essentials of technology and high-growth entrepreneurship. No prerequisite knowledge is necessary, although an understanding of basic accounting principles will prove useful.

Entrepreneurship was traditionally taught only to business majors. Because entrepreneurship education opportunities now span the entire campus, we wrote this book to be approachable by students of all majors and levels, including undergraduate, graduate, and executive education. Our primary focus is on science and engineering majors enrolled in entrepreneurship and innovation courses, but the book is also valuable to business students and others with a particular interest in high-growth ventures.

For example, our courses at Stanford University, the University of Oregon, and the University of California, Davis, based on this textbook regularly attract students from majors as diverse as computer science, product design, political science, economics, pre-med, electrical engineering, history, biology, and business. Although the focus is on technology entrepreneurship, these students find this material applicable to the pursuit of a wide variety of endeavors. Entrepreneurship education is a wonderful way to teach universal leadership skills, which include being comfortable with constant change, contributing to an innovative team, and demonstrating passion in any effort. Anyone can learn entrepreneurial thinking and leadership. We particularly encourage instructors to design courses in which the students form study teams early in the term and learn to work together effectively on group assignments.

WHAT'S NEW

Based upon feedback from readers and new developments in the field of technology entrepreneurship, numerous enhancements appear in this third edition. Recent compelling academic theories and practitioner insights in entrepreneurship are included in the text. Special attention is given to technology transfer and commercialization processes, open source innovation, and social entrepreneurship. All examples and exercises were reviewed to place even more emphasis on exciting technology ventures around the globe involved in energy and environmental technology applications, often referred to as clean or green tech.

Chapters 1 and 2 are now better organized to introduce the art and science of venturing. Chapter 4 on strategy development now contains important sections regarding alliances and social responsibility. The discussion in Chapter 5 on creativity and sources of innovation has a smoother flow. The concept story and business plan development materials and tools are expanded and summarized in Chapter 7. Chapter 13 now contains a section on clusters and regions of entrepreneurship. New sections on cost drivers and grants as a source of capital were added to Chapters 16 and 18, respectively. Three new full-length cases are included in the appendix including two from the famous Harvard Business School archives. Some reordering of sections within chapters streamlines the remaining content.

FEATURES

The book is organized in a modular format to allow for both systematic learning and random access of the material to suit the needs of any reader seeking to learn how to grow successful technology ventures. Readers focused on business plan development should consider placing a higher priority on Chapters 7, 10, 12, 17, 18, and 19. Regardless of the immediate learning goals, the book is a handy reference and companion tool for future use. We deploy the following wide variety of methods and features to achieve this goal, and we welcome feedback and comments.

Principles and Chapter Previews—A set of 20 fundamental principles is developed and defined throughout the book. They are listed in the inside front cover as well. Each chapter opens with a key question and outlines its content and objectives.

Examples and Exercises—Examples of cutting-edge technologies illustrate concepts in a shaded-box format. Information technology is chosen for many examples because students are familiar with its products and services. Exercises are offered at the end of each chapter to test comprehension of the concepts.

Sequential Exercise and Case—A special exercise called the "venture challenge" guides readers through a chapter-by-chapter formation of a new

TABLE P1 Overview of cases.

Cases in appendix B	Synopsis	Issues
Trexel	A university spin-out struggles to commercialize a novel environmental material	Opportunity identification and evaluation, product development, and innovation strategy
Biodiesel	Three founders consider an opportunity in the energy industry	Opportunity identification and evaluation, business model
Yahoo! 1995	Two founders face a decision on financing that forces them to confront their vision	Vision and business model, sources of capital, business plan
Barbara's Options	A soon-to-be graduate weighs two job offers	Stock options, finance
SolidWorks	A founder hits a snag while raising money for his venture	Team, finance, negotiations
Artemis Images	A promising image management company runs into trouble	Competitive strategy, business model, team, finance
Sirtris Pharmaceuticals	A life sciences firm faces major decisions about its future	Alliances, licensing, market strategy
Cooliris	A young entrepreneur struggles to hire a team	Hiring process, scaling issues

enterprise. In addition, a case study about an actual biotechnology firm, AgraQuest, runs from one chapter to the next.

Business Plans—Methods and tools for the development of a business plan are gathered into one special chapter, which includes a thoroughly annotated table of contents. A sample business plan is provided in appendix A.

Cases—Eight comprehensive cases are included in appendix B. A short description of each case is provided in Table P1. Additional cases from Harvard and ECCH are recommended on this textbook's websites.

References—References are indicated in brackets [Smith, 2001] and are listed as a complete set in the back of the book. This is followed by a list of entrepreneurship-related websites in appendix C and a comprehensive glossary.

Chapter Sequence—The chapter sequence represents our best effort to organize the material in a format that can be used in various types of entrepreneurship courses. The chapters follow the four-part layout shown in Figure P1. Courses focused on creating business plans can reorder the chapters with emphasis on Chapters 7, 10, 12, 17, 18, and 19.

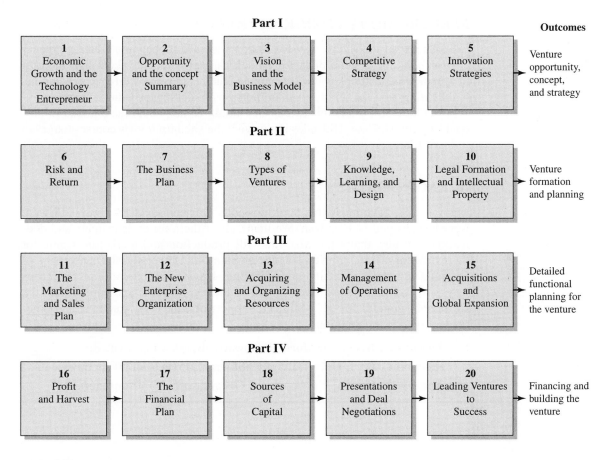

FIGURE P1 Chapter sequence.

Video Clips—A collection of suggested videos from world-class entrepreneurs, investors, and teachers is listed at the end of each chapter and provided on this textbook's websites. More free videos clips and podcasts are available at Stanford's Entrepreurship Corner website (see http://ecorner.stanford.edu).

Websites and Social Networking—Please visit websites for this book at both McGraw-Hill Higher Education (http://www.mhhe.com/byersdorf) and Stanford University (http://techventures.stanford.edu) for supplemental information applicable to educators, students, and professionals. For example, a complete syllabus for an introductory course on technology entrepreneurship and a sample presentation for each chapter are provided for instructors. Visitors to either website can link to the authors' blog and social networking sites for the latest news and advancements in technology venturing and entrepreneurship education.

ELECTRONIC TEXTBOOK OPTIONS

E-books are an innovative way for students to save money and create a greener environment at the same time. An e-book can save students about half the cost of a traditional textbook and offers unique features like a powerful search engine, highlighting, and the ability to share notes with classmates using e-books.

McGraw-Hill offers this text as an e-book. To talk about the e-book options, contact your McGraw-Hill sales rep or visit the site http://www.coursesmart.com to learn more.

ACKNOWLEDGMENTS

Many people have made this book possible. Our editors at McGraw-Hill were Lora Neyens and Debra Hash. We thank all of them for their insights and dedication. We also thank the McGraw-Hill production and marketing teams for their diligent efforts. Our colleagues at Stanford University, the University of Oregon, and the University of California, Davis, were helpful in numerous ways. We are indebted to them for all of their great ideas and support. Lastly, we remain grateful for the continued support of educators, students, and other readers of previous editions.

Thomas H. Byers, Stanford University, tbyers@stanford.edu

Richard C. Dorf, University of California, Davis, rcdorf@ucdavis.com

Andrew J. Nelson, University of Oregon, ajnelson@uoregon.edu

MEDIA SUPPLEMENTS FOR STUDENTS AND INSTRUCTORS

The 3rd edition is supplemented by two websites, collectively bringing students and instructors the most extensive resources available for technology and high-growth entrepreneurship courses. Visitors to either website can link to the authors' blog and social networking sites in order to interact with the authors and other readers.

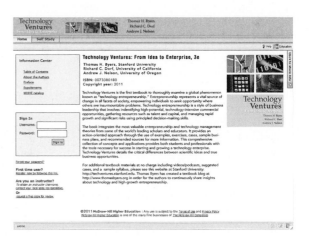

McGraw-Hill Website
www.mhhe.com/byersdorf

Accessed with a password, the McGraw-Hill website for instructors features:

■ Answers to end-of-chapter exercises

■ Teaching notes in Word and PDF format for the cases in appendix B

■ Extensive sample presentations based on the text

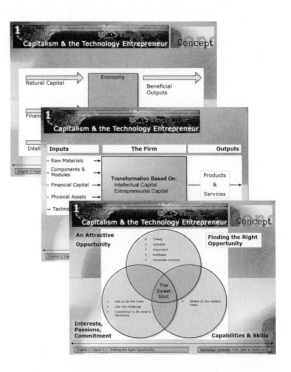

Sample presentations provide instructors with a framework for organizing their lectures, and reference topic-related videos on the textbook's websites.

MEDIA SUPPLEMENTS FOR STUDENTS AND INSTRUCTORS

Stanford University Website
http://techventures.stanford.edu

Rich with content, the author-created Stanford website provides relevant media for each chapter in *Technology Ventures,* including:

- Video clips and podcasts of entrepreneurial leaders including founders, CEOs, venture capitalists, authors, educators, and policy makers.

- Suggested case studies from Harvard Business School and other universities around the globe.

- Resources on how to best integrate the book's business plans and case studies into entrepreneurship courses.

- Links to compelling resources on entrepreneurship (appendix C).

- Additional sample business plans to augment the plan in appendix A.

- A sample syllabus, derived from an actual Stanford University course for students of all majors, includes all sessions with related content and links.

- A collection of the videos listed in the "Video Resources" section at the end of each chapter in this textbook.

Venture Opportunity, Concept, and Strategy

E ntrepreneurs have important roles in creating new businesses that fuel progress in societies worldwide. The entrepreneur uses innovation and technology to foster positive impact and activity in all facets of life. The capable entrepreneur learns to identify, select, describe, and communicate the essence of an opportunity that has attractive potential to become a successful venture. The entrepreneur is able to describe the valuable contributions of a venture and create the design of a business model that can be sustained by a competitive advantage. The venture team creates a road map (strategy) that can, with good chance, effectively lead to the commercialization of the new product or service in the marketplace with a sustainable competitive advantage. ∎

Economic Growth and the Technology Entrepreneur

There are risks and costs to a program of action. But they are far less than the long-range risks and costs of comfortable inaction.
John F. Kennedy

What drives global entrepreneurship?

Entrepreneurs strive to make a difference in our world and to contribute to its betterment. They identify opportunities, mobilize resources, and relentlessly execute on their visions. In this chapter, we describe the characteristics of the people called entrepreneurs and the process they use to create new enterprises. We identify firms as key structures in the economy and the role of entrepreneurship as the engine of economic growth. New technologies form the basis of many important ventures where scientists and engineers combine their technical knowledge with sound business practices to foster innovation. ■

1.1 The Entrepreneur's Challenge

The needs and problems of the world's population are immense. From environmental sustainability to security, from organizational inefficiencies to corruption, from information overload to disease, from transportation to communication, the opportunities for people to create a positive impact are enormous. **Entrepreneurs** are people who identify and pursue solutions among problems, possibilities among needs, and opportunities among challenges.

Entrepreneurship is more than the creation of a business and the wealth associated with it. It is focused on the creation of a new enterprise that serves society and makes a positive change. Entrepreneurs can create great and reputable firms that exhibit performance, leadership, and longevity. In Table 1.1 look at the examples of successful entrepreneurs and the enterprises they created. What contributions have these people and organizations made? What organization would you add to the list? What organization do you wish you had created or been a part of during its formative years? What organization might you create in the future?

TABLE 1.1 Selected entrepreneurs and the enterprises they started.

Entrepreneur	Enterprise started	Age of entrepreneur at time of start	Year of start
Bezos, Jeff	Amazon.com (USA)	31	1995
Brin, Sergey	Google (USA)	27	1998
Dell, Michael	Dell Computer (USA)	19	1984
Gates, William	Microsoft (USA)	20	1976
Greene, Diane	VMWare (USA)	42	1998
Hewlett, William	Hewlett-Packard (USA)	27	1939
Ibrahim, Mo	Celtel (Africa)	42	1998
Lerner, Sandra	Cisco (USA)	29	1984
Li, Robin	Baidu (China)	32	2000
Ma, Jack	Alibaba.com (China)	35	1999
Plattner, Hasso	SAP (Germany)	28	1972
Rottenberg, Linda	Endeavor (Chile, Argentina)	28	1997
Sasaki, Koji	AdIn Research (Japan)	43	1986
Shwed, Gil	Check Point (Israel)	25	1993
Tanti, Tulsi	Suzlon Energy (India)	37	1995
Yunus, Muhammed	Grameen Bank (India)	36	1976
Zuckerberg, Mark	Facebook (USA)	20	2004

Entrepreneurs seek to achieve a certain goal by starting an organization that will address the needs of society and the marketplace. They are prepared to respond to a challenge to overcome obstacles and build a business. As Martin Luther King, Jr. (1963), said, "The ultimate measure of a man is not where he stands in moments of comfort and convenience, but where he stands at times of challenge and controversy."

For an entrepreneur, a **challenge** is a call to respond to a difficult task and the commitment to undertake the required enterprise. Richard Branson, the creator of Virgin Group, reported [Garrett, 1992]: "Ever since I was a teenager, if something was a challenge, I did it and learned it. That's what interests me about life—setting myself tests and trying to prove that I can do it."

Entrepreneurs are resilient people who pounce on challenging problems, determined to find a solution. They combine important capabilities and skills with interests, passions, and commitment. Over nearly a decade, Fred Smith worked on perfecting a solution to what he viewed as a growing problem of organizations to find ways to rapidly ship products to customers. To address this challenge, Smith saw an opportunity to build a freight-only airline that would fly packages to a huge airport and then sort, transfer, and fly them to their destinations overnight. He turned in his paper describing this plan to his Yale University professor, who gave it an average grade, said to be a C. After he graduated, Smith served four years as a U.S. Marine Corps officer and pilot. Following his military service, he spent a few years in the aviation industry building up his experience and knowledge of the industry. Then, he prepared a fully developed business plan for an overnight freight service. By 1972, he had secured financial backing, and Federal Express took to the air in 1973. Federal Express became a new way of shipping goods that revolutionized the cargo shipping business worldwide.

Smith and other entrepreneurs recognize a change in society and its needs, and then, based on their knowledge and skill, they respond with a new way of doing things. Typically, entrepreneurs create a novel response to an opportunity by recombining people, concepts, and technologies into an original solution. Smith saw that the combination of dedicated cargo airplanes, computer-assisted tracking systems, and overnight delivery would serve a new market that required just-in-time delivery of critically important parts, documents, and other valuable items. Smith adapted computer technology to manage the complex task of tracking and moving packages. More fundamentally, Smith matched his passions and skills as a person with a good opportunity.

An **opportunity** is a favorable juncture of circumstances with a good chance for success or progress. Attractive opportunities combine good timing with realistic solutions that address important problems in favorable contexts. It is the job of the entrepreneur to locate new ideas, to determine whether they are actual opportunities, and, if so, to put them into action. Thus, **entrepreneurship** may be described as the nexus of enterprising individuals and promising opportunities [Shane and Venkataraman, 2000]. As illustrated in Figure 1.1, the "sweet

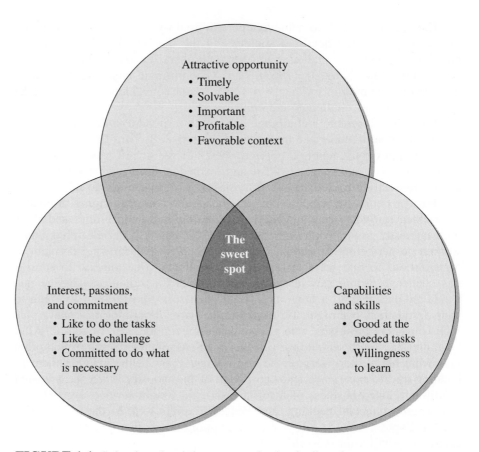

FIGURE 1.1 Selecting the right opportunity by finding the sweet spot.

spot" exists where an individual's or team's passions and capabilities intersect with an attractive opportunity.

Entrepreneurship is not easy. Only about one-third of new ventures survive their first three years. As change agents, entrepreneurs must be willing to accept failure as a potential outcome of their venture. But, regardless of whether the right opportunity has emerged, a person can learn to act as an entrepreneur by trying the activity in a low-cost manner. To avoid the realm of daydreams and fantasy, a person needs to start the practice of experimenting, testing, and learning about his or her entrepreneurial self [Ibarra, 2002]. The would-be entrepreneur should, therefore, engage in this sequence: do it, then reflect on it.

The first step is to craft small experiments in new activities with entrepreneurial teams or small ventures. Through these small experiments, the entrepreneur develops new contacts and mentors, while learning more about the process

TABLE 1.2 Four steps to starting a business.

1. The founding team or individual has the necessary skills or acquires them.

2. The team members identify the opportunity that attracts them and matches their skills. They create a solution to match the opportunity.

3. They acquire (or possess) the financial and physical resources necessary to launch the business by locating investors and partners.

4. They complete an arrangement or contract with their partners, with investors, and within the founder team to launch the business and share the ownership and wealth created.

of pursuing an opportunity. He or she may also find a challenge that serves as a catalyst for a new venture. If team members identify an opportunity that attracts them and matches their skills, they next obtain the resources necessary to implement their solution. Finally, they launch and grow an organization, which can grow to have a massive impact, like those enterprises listed in Table 1.1. These four steps to starting a business are outlined in Table 1.2.

Ultimately, entrepreneurship is centrally focused on the identification and exploitation of previously unexploited opportunities. Fortunately for the reader, successful entrepreneurs do not possess a rare entrepreneurial gene. Entrepreneurship is a systematic, organized, rigorous discipline that can be learned and mastered [Drucker, 2002]. This textbook will show you how to identify true business opportunities and how to start and grow a high-impact enterprise.

1.2 The Entrepreneur

The entrepreneur is a bold, imaginative deviator from established business methods and practices who constantly seeks the opportunity to commercialize new products, technologies, processes, and arrangements [Baumol, 2002]. Entrepreneurs thrive in response to challenges and look for unconventional solutions. They apply creativity, create visions, build stories that explain their visions, and then act to be part of the solution. They forge new paths and risk failure, but persistently seek success. Entrepreneurs distinguish themselves through their ability to accumulate and manage knowledge, as well as their ability to mobilize resources to achieve a specified business or social goal [Kuemmerle, 2002].

Entrepreneurs engage in eight key activities, as described in Table 1.3. They identify and select opportunities that match their skills and interests, they acquire resources, and they start organizations.

In order to successfully pursue these activities, entrepreneurs should possess several important capabilities, as noted in Table 1.4. Entrepreneurs are opportunity driven and work to find a strategy that can reasonably be expected to bring that opportunity to fruitful success. They seek new means

TABLE 1.3   Eight skills of entrepreneurship.

- Entrepreneurs initiate and operate a purposeful enterprise.

- Entrepreneurs operate within the context and industrial environment at the time of initiation.

- Entrepreneurs identify and screen timely opportunities.

- Entrepreneurs accumulate and manage knowledge and technology.

- Entrepreneurs mobilize resources— financial, physical, and human.

- Entrepreneurs assess and mitigate uncertainty and risk associated with the initiation of the enterprise.

- Entrepreneurs provide an innovative contribution or at least a contribution that encompasses novelty or originality.

- Entrepreneurs enable and encourage a collaborative team of people who have the capabilities and knowledge necessary for success.

or methods and are willing to commit to solving a social or business problem that will result in success. Entrepreneurs work toward needing shorter time periods to decide on an appropriate strategy and seize opportunities. Entrepreneurs have a passion to build an enterprise that will solve an important problem. They seek ways to express themselves and validate their ideas. They are creative, internally motivated, and attracted to new, big ideas or opportunities.

Entrepreneurs exhibit robust confidence, sometimes bordering on overconfidence [Hayward et al., 2006]. Entrepreneurial innovators tend to exhibit high self-efficacy—the belief that they can organize and effectively execute actions to produce desired attainments [Markman et al., 2002]. They believe they possess the capabilities and insights required for the entrepreneurial task. One or

TABLE 1.4   Required capabilities of the entrepreneurial team.

- Has talent, knowledge, and experience within the industry where the opportunity occurs

- Seeks important opportunities with sizable challenges and valuable potential returns

- Able to select an opportunity in a short period: timely

- Creatively explores a process that results in the concept of a valuable solution for the problem or need

- Able to convert an opportunity in to a workable and marketable enterprise

- Wants to succeed: achievement-oriented

- Able to accommodate uncertainty and ambiguity

- Flexibly adapts to changing circumstances and competitors

- Seeks to evaluate and mitigate the risks of the venture

- Creates a vision of the venture to communicate the opportunity of staff and allies

- Attracts, trains, and retains talented, educated people capable of multidisciplinary insights

- Skilled at selling ideas and have a wide network of potential partners

TABLE 1.5 Elements of the ability to overcome a challenge.

■ Able to deal with a series of tough issues	■ Resilient in the face of setbacks
■ Able to create solutions and work to perfect them	■ Willing to work hard and not expect easy solutions
■ Able to handle many tasks simultaneously	■ Well-developed problem-solving skills
	■ Able to learn and acquire the skills needed for the tasks at hand

more of the entrepreneur team usually have some experience in the industry in which the new venture will be operating.

Good entrepreneurs seek to be flexible so they can adapt to changing conditions and reduce the risks of the venture. They are resilient in the face of setbacks, able to multitask, and exercise well-developed problem-solving skills to overcome challenges. Table 1.5 lists some of the elements of this ability.

Finally, entrepreneurs create an overarching vision of the venture and use it to motivate employees, allies, and financiers. Perhaps the most important qualities or characteristics of an entrepreneur are the abilities to accomplish the necessary tasks, meet goals, and inspire others to help with these tasks. Successful entrepreneurial teams attract, train, and retain intellectually brilliant and educated people capable of multidisciplinary insights [van Praag, 2006].

Members of the entrepreneurial team must, therefore, exhibit leadership qualities. **Leadership** is the ability to create change or transform organizations. Leadership within an organization enables the organization to adapt and change as circumstances require. A real measure of leadership is the ability to acquire needed new skills as the situation changes.

Entrepreneurs vary widely in their backgrounds. Recall the list of entrepreneurs in Table 1.1. The age of these people when they launched their enterprises ranges from 19 to 43. The median age of all technology-based company founders is 39 and many founders are much older [Wadha et al., 2008]. Entrepreneurship is a lifelong pursuit that is accessible to people of all ages. Entrepreneurs are also well educated. Ninety-two percent of technology entrepreneurs surveyed by the Kauffman Foundation hold a bachelor's degree, 31 percent hold a master's degree, and 10 percent hold a Ph.D. At the same time, however, institutions such as the Grameen Bank, which lends primarily to women in the third world so that they can start businesses, have opened up entrepreneurship as a possibility for a wide range of people.

In general, entrepreneurs should have most of the qualities listed in Table 1.4 in order to participate in a new venture. But, not everyone will have the same blend of capabilities. In order to strengthen, diversify, and complement an organization's skills, insights, resources, and connections, most entrepreneurs work as part of a team.

Moreover, entrepreneurship is an attitude and capability that diffuses beyond the founding team to all members of an organization. Most growing

TABLE 1.6 Factors people use to determine whether to act as entrepreneurs.

Positive factors or benefits	
■ Independence: Freedom to adapt and use their own approach to work and flexibility of work, autonomy ■ Financial success: Income, financial security	■ Self-realization: Recognition, achievement, status ■ Innovation: Creating something new ■ Roles: Fulfilling family tradition, acting as leader
Negative factors	
■ Risk: Potential for loss of income and wealth	■ Work effort and stress: Level of work effort required, long hours, constant anxiety

firms strive to infuse the culture of the entire company with the entrepreneurial spirit. For example, Thomas Edison created an enterprise that became General Electric; Steve Jobs and Steve Wozniak founded Apple Computer; and Azim Premji started Wipro Technologies. These entrepreneurs combined their knowledge of valuable new technologies with sound business practices to build important new enterprises that continued to maintain their entrepreneurial spirit for years after founding.

Members of an entrepreneurial team decide whether to act as entrepreneurs based on the seven factors listed in Table 1.6 [Gatewood, 2001]. Good entrepreneurs tend to seek independence, financial success, self-realization, validation of achievement, and innovation, while fulfilling leadership roles. At the same time, potential entrepreneurs evaluate the risk and work efforts associated with an opportunity and balance them with the benefits. Successful entrepreneurs are able to answer positively the five questions listed in Table 1.7 [Kuemmerle, 2002].

Context can have an important effect on whether or not someone becomes an entrepreneur [Sørenson, 2007]. For example, people whose colleagues are entrepreneurial are more likely to become entrepreneurs themselves [Stuart and Ding, 2006]. Similarly, younger and smaller organizations are more likely to

TABLE 1.7 Five questions for the potential entrepreneur.

■ Are you comfortable stretching the rules and questioning conventional wisdom? ■ Are you prepared to take on powerful competitors? ■ Do you have the perseverance to start small and grow slowly?	■ Are you willing and able to shift strategies quickly? ■ Are you a good deal closer and decision maker?

spawn entrepreneurs [Dobrev and Barnett, 2005]. Environmental changes, such as an increase in the availability of venture capital financing, also affect the decision to become an entrepreneur [Hsu et al., 2007].

On an individual level, people act as self-employed entrepreneurs when that career path is felt to be better than employment by an existing firm. Consider the satisfaction (utility) derived from an employment arrangement. A utility function, U, is [Douglas and Shepherd, 1999]:

$$U = f\,(Y, I, W, R, O)$$

where Y = income, I = independence, W = work effort, R = risk, and O = other working conditions. It may be assumed that income depends in turn on ability. People will have an incentive to be entrepreneurs when the most satisfaction (utility) is obtained from the entrepreneurial activity. In other words, entrepreneurship pays off due to higher expected income and independence when reasonable levels of risk and work efforts are required.

For new entrepreneurial activities, the results of the venture are less known, and expected returns, independence, work effort, and risk can only be estimated. Potential entrepreneurs must be careful to do an honest assessment of their motivation and skills [Wasserman, 2008]. Regrettably, many entrepreneurs overweigh the benefits of independence and income, and underestimate the work effort required.

Based on the utility function above, we may postulate a utility index that we will call the Entrepreneurial Attractiveness (EA) index [Levesque et al., 2002]. For each factor (Y, I, W, and R), we use a scale of 1 to 5 with 1 = low, 3 = medium, and 5 = high.

$$EA = (Y + I) - (W + R) \tag{1.1}$$

As a simple example, consider the straightforward alternatives for a successful marketing manager in the electronics industry. She can earn $60,000 annually in her existing job (Y in equation 1.1). However, she values the independence of the new venture highly (I). The work effort for the new venture is estimated to be the same as for her current work (W). However, the risk is higher for the new independent venture (R). The potential entrepreneur estimates that she can obtain the same income over the next two years, although she will need a four-month period with a lower income at the start. The entrepreneur can compare the two options across these dimensions as shown in Table 1.8. In this case, over the first two years, the benefits of the new venture are Y + I = 8, and the costs of the venture are W + R = 7. The benefits of the existing job are equal to 5, and the costs are 6. Therefore,

$$\text{New venture: } (Y + I) - (W + R) = 8 - 7 = +1$$
$$\text{Existing job: } (Y + I) - (W + R) = 5 - 6 = -1$$

The new opportunity looks more favorable due to this entrepreneur's desire for independence. Thus, it warrants in-depth analysis.

**TABLE 1.8 Summary of the entrepreneur's analysis of a new opportunity
and the opportunity cost using a two-year period.**

Factor	New venture	Existing job
Income over	$120,000	$120,000
two years (Y)	Y = 3	Y = 3
Independence (I)	I = 5	I = 2
Work effort (W)	W = 4	W = 4
Risk (R)	R = 3	R = 2

In summary, entrepreneurs are multitalented individuals who leverage
their capabilities and interests to pursue a particular opportunity, almost
always with the help of a team.* The decision to pursue an entrepreneurial
path and a particular opportunity is determined by weighing the benefits of
independence and income against the work effort required and the risk of the
venture. In chapter 2, we learn how a potential entrepreneur can evaluate an
idea to determine if it is an actual opportunity.

1.3 Economics and the Firm

All entrepreneurs are workers in the world of economics and business. **Economics** is the study of the production, distribution, and consumption of
goods and services. Society, operating at its best, works through entrepreneurs to effectively manage its material, environmental, and human
resources to achieve widespread prosperity. An abundance of material and
social goods equitably distributed is the goal of most social systems. Entrepreneurs are the people who arrange novel organizations or solutions to
social and economic problems. They are the people who make our economic
system thrive [Baumol et al., 2007].

According to Global Entrepreneurship Monitor (GEM) researchers, the
United States maintained about a 10 percent entrepreneurial activity rate
between 1999 and 2007. This indicated that one in ten adults was engaged in
setting up or managing a new enterprise during that period, a rate 50 percent
higher than the average of all other participating high-income nations [Phinisee
et al., 2008]. New ventures have been the source of an estimated one-half to
two-thirds of the new jobs created in the United States over the past two
decades, meaning start-ups are a key to economic recovery and job growth
[Stangler, 2009]. The entrepreneur turns a social problem into an opportunity, a
productive organization, and new, well-paid jobs.

*Throughout this book, the word *entrepreneur* will refer to an individual or a team of individuals.

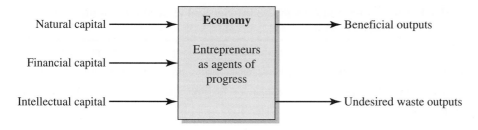

FIGURE 1.2 A model of the economy.

An economic system is a system for the production and distribution of goods and services. Given the limitations of nature and the unlimited desires of humans, economic systems are schemes for (1) administering scarcities and (2) improving the system to increase the abundance of goods and services. For a nation as a whole, its wealth is its food, housing, transportation, health care, and other goods and services. A nation is wealthier when it has more of these goods and services. Nations strive to secure more prosperity by organizing to achieve a more effective and efficient economic system. It is entrepreneurs who organize and initiate that change.

Almost all variation in living standards among countries is explained by **productivity,** which is the quantity of goods and services produced from the sum of all inputs, such as hours worked and fuels used. A model of the economy is shown in Figure 1.2. The inputs to the economy are natural capital, financial capital, and intellectual capital. The outputs are the desired benefits or outcomes and the undesired waste. An appropriate goal is to maximize the beneficial outputs and minimize the undesired waste [Dorf, 2001].

Natural capital refers to those features of nature, such as minerals, fuels, energy, biological yield, or pollution absorption capacity, that are directly or indirectly utilized or are potentially utilizable in human social and economic systems. Because of the nature of ecologies, natural capital may be subject to irreversible change at certain thresholds of use or impact. For example, global climate change poses a serious threat to sources of natural capital.

Financial capital refers to financial assets, such as money, bonds, securities, and land, which allow entrepreneurs to purchase what they need to produce goods and services. The **intellectual capital** of an organization includes the talents, knowledge and creativity of its people, the efficacy of its management systems, and the effectiveness of its customer and supplier relations. The sources of intellectual capital are threefold: human capital, organizational capital, and social capital. **Human capital** (HC) is the combined knowledge, skill, and ability of the company's employees. **Organizational capital** (OC) is the hardware, software, databases, methods, patents,

TABLE 1.9 Three elements of the intellectual capital (IC) of an organization.

Human capital (HC): The skills, capabilities, and knowledge of the firm's people

Organizational capital (OC): The patents, technologies, processes, databases, and networks

Social capital (SC): The quality of the relationships with customers, suppliers, and partners

IC = HC + OC + SC

and management methods of the organization that support the human capital. **Social capital** (SC) is the quality of relationships with a firm's suppliers, allies, partners, and customers. These elements of intellectual capital are summarized in Table 1.9.

The economy as portrayed in Figure 1.2 consists of the summation of all organizations, for-profit as well as nonprofit and governmental, that provide the beneficial outputs for society. These are the organizations that we study and will label as enterprises or firms*. Entrepreneurs constantly form new organizations or enterprises to meet social and economic needs.

The purpose of a firm is to establish an objective and mission and carry it out for the benefit of the customer. Thus, the purpose of Merck Corporation is to create pharmaceuticals that protect and enhance its customers' health. To do so, each individual firm transforms inputs into desirable outputs that serve the needs of customers.

A firm exists as a group of people because it can operate more effectively and efficiently than a set of individuals acting separately. Furthermore, a firm creates conditions under which people can work more effectively than they could on their own. Thus, firms exist to coordinate and motivate people's economic activity [Roberts, 2004]. A firm is more effective because (1) it has lower transaction costs and (2) the necessary skills and talent are gathered together in effective, collaborative work.

A model of the firm as a transformation entity is shown in Figure 1.3. The transformation of inputs into desired outputs is based primarily on the intellectual capital and the entrepreneurial capital of the firm. As an example, consider Microsoft, a powerful software firm. It creates and purchases technologies, develops new software, and builds a client base. The transformation of its inputs into outputs is based on its formidable stock of entrepreneurial capital and intellectual capital.

Entrepreneurial capital (EC) can be formulated as a combination of entrepreneurial competence and entrepreneurial commitment [Erikson, 2002]. **Entrepreneurial competence** is the ability (1) to recognize and envision taking advantage of opportunity and (2) to access and manage the

---

* Henceforth, we use firm to represent organizations, enterprises, and corporations.

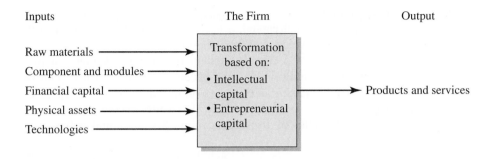

FIGURE 1.3 The firm as transforming available inputs into desired outputs.

necessary resources to actually take advantage of the opportunity. **Entrepreneurial commitment** is a dedication of the time and energy necessary to bring the enterprise to initiation and fruition. The presence of competence without any commitment creates little entrepreneurial capital. The presence of commitment without competence may waste both time and resources. Both commitment and competence are required to provide significant entrepreneurial capital. Thus, we can say that

Entrepreneurial Capital = entrepreneurial competence
$$\times \text{ entrepreneurial commitment}$$

or

$$EC = \text{Ecomp} \times \text{Ecomm} \tag{1.2}$$

where Ecomp is entrepreneurial competence and Ecomm is entrepreneurial commitment. Note that the symbol $\times$ is a multiplication sign, but it should be recognized that this equation is qualitative in nature.

The accretion of knowledge and experience over time leads to increased competence as people mature. However, commitment of energy and time may decline when people become less interested in or available for the necessary entrepreneurial competence activities. Both commitment and competence are qualities of the leadership team, and they may be complementary qualities shared among the team members.

To transform inputs into outputs, the firm also acts to develop, attract, and retain intellectual capital. The firm develops and uses intellectual capital to build the strengths of the firm and to provide the desired products.* The firm provides a place where people can collaborate, learn, and grow.

Intellectual capital can be thought of as the sum of knowledge assets of an organization. This knowledge is embodied in the talent, know-how, and skills of the members of an organization. Thus, a firm needs to attract and retain the best people for its requirements in the same way that it seeks the best technologies

---

* Henceforth, we use products to refer to products and services.

or physical assets. Knowledge is one of the few assets that grows when shared. By organizing around intellectual capital, a new firm strives to leverage it, usually through collaboration, development, and sharing.

The intellectual capital of a firm is used to transform raw material into something more valuable. Antinori succeeds because of the human capital of its grape growers and wine makers. KFC relies on the organizational capital of its recipes and processes. A local café where the waiter recognizes you and knows your favorite latté relies on its social capital. Social capital is based on strong, positive relationships.

The firm's actions are based on its knowledge of its customer, its product, and its markets. The firm must identify and understand its customers, its competitors, and their values and behavior. Knowledge of organizations, design, and technologies is filtered through a firm's strengths and weaknesses. The firm acts on all this knowledge.

First, a firm is clear about its mission and purpose. Second, the firm must know and understand its customers, suppliers, and competitors. Third, a firm's intellectual capital is understood, renewed, and enhanced as feasible. Finally, the firm must understand its environment or context, which is set by society, the market, and the technology available to it. We can call this the **theory of a firm's business,** or how it understands its total activities, resources, and relationships. Figure 1.4 depicts the business theory of

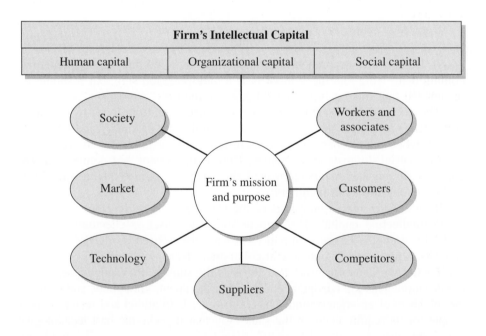

FIGURE 1.4 A firm's theory of business depicts how it understands and uses its total resources, activities, and relationships.

a firm. One hundred years ago, firms were hierarchical and bureaucratic with a theory of business that emphasized making long runs of standardized products. They regularly introduced "new and improved" varieties and provided lifetime employment. Today, firms compete globally with high-value, customized products. They use flattened organizations and base their future on intellectual capital. Firms look to brands and images to cut through the clutter of messages. In the future, a firm's human capital—talent—will become even more important.

One way to look at the future of a firm is as a competition among its stakeholders. Flexibility and leanness mostly benefit the firm's shareowners. Stakeholders include not only these shareholders, but also workers, customers, people in the community, and society in general. Placing a high valuation on talent gives more power to the workers. Customers stand to gain power as competitors vie for their attention. A good reputation means the firm needs to look after its community and society. The entrepreneur in the new firm strives to build a firm that serves all its stakeholders well.

1.4 Creative Destruction

One view of economic activity describes a world of routine in which little changes. In this static model, all decisions have been made, and all alternatives are known and explored. But clearly, no economy is static, and change appears to be certain.

Dynamic capitalism is the process of wealth creation characterized by the dynamics of new, creative firms forming and growing and old, large firms declining and failing. In this model, it is disequilibrium—the disruption of existing markets by new entries—that makes capitalism lead to wealth creation [Kirchhoff, 1994]. New firms are formed by entrepreneurs to exploit and commercialize new products or services, thus creating new demand and wealth. This renewal and revitalization of industry leads to a life cycle of formation, growth, and decline of firms.

The recorded music industry provides a good example of waves of change. Music lovers listened to their favorite music recorded on vinyl discs until about 1980, when cassette tapes grew in popularity. The compact size and recordability of the cassette tape caused a massive shift from vinyl records to tape. By the late 1980s, however, compact discs (CDs) overshadowed cassettes, due to the CD's better sound quality and instant access to tracks. In turn, the CD business peaked in 1995 just as the Internet was gaining momentum in society at large. A few years later, peer-to-peer file transfer began to allow piracy of music. By 2001, Apple had introduced the iPod and iTunes and eventually gained a commanding position in the music distribution and sales business. In a dynamic economy, companies need to reinvent their business arrangements or end up becoming irrelevant [Knopper, 2009].

Joseph Schumpeter (1883–1950) described this process of new entrepreneurial firms and waves of change as **creative destruction.** Born and educated in Austria, Schumpeter taught at Harvard University from 1932 until his death in 1950. His most famous book, *Capitalism, Socialism and Democracy,* which appeared in 1942 [Schumpeter, 1984], argued that the economy is in a perpetual state of **dynamic disequilibrium.** Entrepreneurs upend the established order, unleashing a gale of creative destruction that forces incumbents to adapt or die. Schumpeter argued that the concept of perfect competition is irrelevant because it focused entirely on market (price) competition, when the focus should be on technological competition. Creative destruction incessantly revolutionizes the economic structure from within, destroying the old structure and creating a new one. The average life span of a company in the Standard and Poors 500 declined from 35 years in 1975 to less than 20 years today. Less than 4 of the top 25 technology companies 30 years ago are leaders today—perhaps only IBM and Hewlett-Packard.

In a world of change, entrepreneurs seek to embrace it. Entrepreneurs match ideas for change with opportunity. These changes include the adoption of new and better (or cheaper) sources of input supplies, the opening of new markets, and the introduction of more profitable forms of business organization.

The profit of the new firm is the key to economic growth and progress. By introducing a new and valuable product, the innovator obtains temporary monopoly power until rivals figure out how to mimic the innovation. Lower costs may give the innovative firm profits higher than those of its rivals, which must continue to sell at higher prices to cover their higher expenses. Alternatively, a superior product may permit a price above that charged by other firms. The same concept clearly fits all forms of successful change. The business system works to drive out inefficiency and forces business process renewal.

Economic progress is reflected in productivity growth, which provides for increases in people's standard of living. Over the past half-century, the U.S. workforce (including immigration) has grown at about 1.7 percent annually, and productivity per worker has risen at 2.2 percent, generating real economic growth (excluding inflation) averaging 3.9 percent. This is an excellent record, due in great part to the impact of technology entrepreneurship.

Rising output per worker comes from two sources: (1) new technology and (2) smarter ways of doing work. Both paths have been followed throughout human history, and they became faster tracks with the coming of the Industrial Revolution. The twentieth century started with new techniques of management and many new inventions. The century ended with smarter management techniques and dramatic advances in electronic technology, which helped revive productivity growth after limited gains through much of the 1970s and 1980s.

The free spirit of entrepreneurs provides the vital energy that propels this capitalist system. During the past 30 years, the forces of entrepreneurship, competition, and globalization have encouraged new technologies and business methods that raise efficiency and efficacy. In recent years, due to competition, many of the benefits of strong productivity have flowed to consumers in the form of lower prices. Together, innovation, entrepreneurship, and competition are important sources of productivity growth.

1.5 Innovation and Technology

Little doubt now exists that the economy is driven by firms that capitalize on change, technology, and challenge. This book is focused on helping the reader to purposefully become an agent for creative destruction by creating his or her own firm. An example of an agent for creative destruction is Craig Venter, who founded Synthetic Genomics in order to use modified or synthetically produced microorganisms to create ethanol and hydrogen. The company is attempting to capitalize on the growing interest in alternative fuels and to design and synthesize specifically engineered cells to perform particular tasks.

New technologies such as these are often a source of disequilibrium or discontinuity, and Schumpeter's theory was based on disruptive, or "radical," innovations. **Technology** includes devices, artifacts, processes, tools, methods, and materials that can be applied to industrial and commercial purposes. For example, Intel was formed to apply semiconductor technology to the design and manufacture of semiconductor circuits. Microsoft was formed to create and distribute computer software products for applications in industry and the home. Apple has reshaped itself around mobile communications and mobile media technologies.

Modern entrepreneurial firms breed a constant flow of high-impact products that create value and stimulate economic growth by bringing new methods, technologies, and ideas to the global marketplace [Schramm, 2004]. Figure 1.5 illustrates "waves" of innovation based upon different technologies throughout history. Modern entrepreneurial firms are at the forefront of the sixth wave, which places a special emphasis on sustainability.

Population growth and a worldwide rising middle class, combined with tightening energy supplies and fears of climate changes, have prompted a move toward socially and environmentally responsible business. The goal is to provide housing, transportation, and energy systems that use less energy and emit less pollution and carbon dioxide. The concept is to use knowledge and innovation to create and implement sustainable energy systems and to increase resource productivity [Friedman, 2008].

A clean energy system would consist of a mixture of energy generation, transmission, and utilization in ways that best use natural resources and minimize environmental impacts. By clean and green we mean a system based on conservation, best uses of natural resources, and minimizing environmental impacts. Examples of green technology solutions include installing carbon capture systems at power plants, increasing the use of wind power systems, and developing high-efficiency biofuel systems. Improving the reliability and smart control of the electricity grid also offers a good opportunity for entrepreneurs.

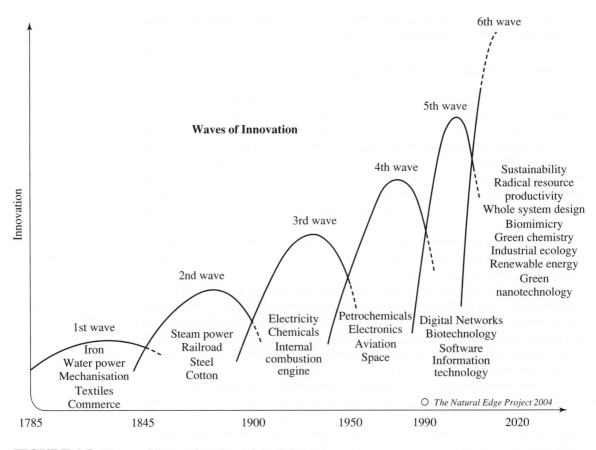

FIGURE 1.5 Waves of innovation throughout history.

As the green technology movement highlights, technology entrepreneurship is based upon intellectual capital. One hundred years ago, successful companies such as U.S. Steel were primarily managing physical assets. By contrast, today's successful firms, such as Microsoft and Genentech, manage knowledge and intellectual capital. In fact, for many, if not most, firms, intellectual capital is the organization's most important asset, more valuable than its other physical and financial assets. Many firms depend on their patents, copyrights, and software, and the capabilities and relationships of their people. This intellectual capital, appropriately applied, will determine success or failure. Thus, knowledge has become the most important factor of production.

While innovation and intellectual property are critical, however, a dynamic economy ultimately rests on the actions of entrepreneurs who assume and accept the benefits and risks of an initiative. It is people acting as leaders, organizers, and motivators who are the central figures of modern economic activity.

Three factors make up entrepreneurial action: (1) a person or group who is responsible for the enterprise, (2) the purposeful enterprise, and (3) initiation and growth of the enterprise. The individuals responsible for the organization were described in section 1.2. The purposeful enterprise may be a new firm organized for a suitable and attractive purpose or a new unit within or separated from an existing business corporation. Furthermore, the organization may be based on radical innovation, incremental changes, imitation, or rent-seeking behavior.

In the first type of enterprise, the entrepreneur engages in an innovative activity that results in novel methods, processes, and products. The second form emphasizes the founding and management of a business that builds upon and improves an existing product or service. The imitative venture is founded by an entrepreneur who is involved in the rapid dissemination of an innovative idea or process. This person or group finds a novel innovation and transfers it to another environment, region or country. The final means of entrepreneurship is called rent-seeking or profit-seeking and focuses on the use of regulation, standards, or laws to appropriate some of the value of a monopoly that is generated somewhere in the economy.

In this book, we emphasize the creation of the venture that capitalizes on technological changes and that will have a significant impact on a region, a nation, or the world. A new regulation or clever financial restructuring may afford the entrepreneur a new opportunity. But, a radical or transforming innovation may provide an entrepreneur an important opportunity to make a productive and very significant contribution to the world as we know it.

1.6 The Sequential Case: AgraQuest

The AgraQuest case illustrates and illuminates the issues raised in each chapter. It focuses on a real-life emerging firm in the life science industry that illustrates each factor described in a chapter. AgraQuest (www.agraquest.com) is an entrepreneurial firm that may significantly contribute to improved environmental and social conditions and agricultural industries around the world. Read the segment on the case at the end of each chapter and learn of a real-life effort that could make a big difference to the world.

Every seven years in the woodsy town of Killingworth, Connecticut, where she grew up, Pamela Marrone would feel the droppings of gypsy moth caterpillars raining down on her head as the cyclical pests gorged on maples and oaks. Desperate to save a heavily infested dogwood, her father once ignored his own organic gardening tenets and blasted the tree with a chemical called a carbamate.

By the next morning, every bee, every ladybird beetle, every lacewing— all the "good" bugs that fed on plant pests—lay dead on the ground. In her youth, Marrone knew that she wanted to keep the good bugs while deterring bad pests. She recognized a great opportunity that, if solved, could help farmers prosper while using natural pest control agents (not chemicals). Furthermore, as a youth, Marrone had tried, with her parents' encouragement, several modest entrepreneurial ventures at craft fairs and state fairs.

Marrone studied entomology (the study of the forms and behavior of insects) at Cornell University, going on to North Carolina State University, from which she received her doctorate in 1983. She then spent seven years as the leader of the new pest control unit at Monsanto in St. Louis, where she acted on her dedication to the natural control of pests. At Monsanto, Marrone built her technical and entrepreneurial skills. As a result, in 1990 she was recruited by Novo Nordisk, a Danish company, to create a biopesticide subsidiary called Entotech Inc. in Davis, California.

Entotech's goal was to hunt for natural products that can defeat plant scourges without wreaking havoc on human beings, animals, helpful insects, or soil. But in 1995, Entotech was sold to Abbott Laboratories, prompting Marrone to start her own firm to meet the challenge of building a successful company that would use a new search process for identifying natural products for pest control. Thus was born AgraQuest. Marrone possessed the interest and passion, the capabilities and skills, and saw an attractive opportunity in the sweet spot of Figure 1.1.

1.7 Summary

The entrepreneur is the creative force that allows free enterprise to flourish. Entrepreneurship is the process through which individuals and teams bring together the necessary resources to exploit opportunities and in doing so create wealth, social benefits, and prosperity.

The critical ideas of this chapter are:

■ The entrepreneur as creator of a great enterprise.
■ The entrepreneur responds to an attractive opportunity.
■ A person can learn to be an entrepreneur.
■ The entrepreneur knows how to use knowledge to create innovation and new firms.
■ Positive entrepreneurship activity flows from a combination of entrepreneurial capital and intellectual capital that leads to productivity and prosperity.
■ The entrepreneur uses an appropriate organizational structure to achieve his or her goals.

Principle 1
Entrepreneurs develop enterprises with the purpose of creating prosperity and wealth for all participants—investors, customers, suppliers, employees, and themselves—using a combination of intellectual capital and entrepreneurial processes.

Video Resources

Visit http://techventures.stanford.edu to view experts discussing content from this chapter.

Entrepreneurial Skills Learned	Mark Zuckerberg	Facebook
Do What You Like to Get Where You Want	John Melo	Amyris
Technology Cycles Start with a Breakthrough Innovation	Judy Estrin	JLabs
Broad Environmental Solutions Require Brawny Change	Vinod Khosla	Khosla Ventures

1.8 Exercises

1.1 What is the difference between an idea and an opportunity? Why is this difference important to entrepreneurs?

1.2 Consider opportunities that have occurred to you over the past month and list them in a column. Then, describe your strong interests and passions, and list them in a second column. Finally, create a list of your capabilities in a third column. Is there a natural match of opportunity, interests, and capabilities? If so, does this opportunity appear to offer a good chance to build an enterprise? What would you need to do to make this opportunity an attractive chance to build an enterprise business?

1.3 Name an entrepreneur that you personally admire. Why do you consider this person to be an entrepreneur? What sets him or her apart from other business leaders? What path did this person take to entrepreneurship? What personal sacrifices or investments did this person make in the journey? What people were important to this person's success?

1.4 Name a successful entrepreneurial team you personally admire. How would you classify it in the context of the entrepreneur capabilities shown in Table 1.4? Do these elements of entrepreneurship apply to it?

1.5 Research the number of companies that either had an IPO (initial public offering) or have been acquired in the last five years. What industries were these companies in? Where is the number of IPOs vs. M&As (mergers and acquisitions) trend leading? What implications does this have on the number of new ventures being started?

1.6 Given an understanding of the waves of innovation throughout history (Figure 1.5), explore opportunities that are created in a wave after the peak. For example, how can an entrepreneur take advantage of a mature or declining market?

VENTURE CHALLENGE

Select a high-potential opportunity that interests you and then use it for the venture challenge exercises at the end of each chapter. For example, you might consider one of these current trends in science and technology: mobile applications, Internet and services, nanotechnology, clean technologies, pandemic and biodefense treatments, and advancements in stem cell research.

1.   Describe the opportunity that attracts you and why you think it is a new venture opportunity.

2.   Describe the competencies and skills you and your team members possess.

3.   What important stakeholders will you need to be successful?

4.   Describe the passion and commitment you have for the opportunity.

5.   Is this a good opportunity for you?

CHAPTER 2

Opportunity and the Concept Summary

In the field of observation, chance only favors minds which are prepared.
Louis Pasteur

How can an entrepreneur identify and select a valuable opportunity?

The identification and evaluation of opportunities is one of the entrepreneur's most important tasks. Good opportunities address important market needs. Examining social, technological, and economic trends can lead to the identification of emerging needs. Entrepreneurs seek to build new ventures and to act on a good opportunity when it matches their capabilities and interests, exists in a favorable context, exhibits the potential for sustainable long-term growth, and facilitates the acquisition of required resources. Such opportunities offer a reasonable chance of success and require the entrepreneur to make a difficult decision to act or not act. The choice of an opportunity and the decision to act is a critical juncture in the life of an entrepreneur. With the decision to act, the entrepreneur prepares a business summary for the venture that is used to test the new venture with potential investors, employees, and customers. The six steps to action as an entrepreneur are shown in Figure 2.1, which summarizes the tasks described in this chapter. ■

2.1 Opportunity Identification

"Every problem is an opportunity." – Vinod Khosla

The first role of the entrepreneur—an individual or a group of people—is to identify and select an appropriate opportunity. An opportunity is a timely and favorable juncture of circumstances providing a good chance for a successful venture. Effective entrepreneurs often find that opportunity identification is a creative process that relates a need to the methods, means, or services that solve the problem. They recognize and pursue opportunities that are based on meeting a need in the marketplace, solving a problem, or filling a niche within a reasonable time. There is timeliness to every opportunity.

Entrepreneurship begins with an idea that upon reflection is a valuable opportunity. Ideas for new ventures are easy to find but difficult to evaluate. Often the idea will be reviewed by a team or group of creative individuals working together to select a good opportunity. A critical task of the entrepreneur is to distinguish between an idea and an opportunity.

Good opportunities are usually disguised, so most people don't easily recognize them. An entrepreneur's awareness of opportunities is shaped by the people she knows, the activities she pursues, the books and magazines she reads, and other such factors [Ozgen and Baron, 2007]. New opportunities open up because customers' needs change or new technologies lead to new ways of accomplishing tasks. Good opportunities also emerge from circumstances of employment or experience. Often they emerge from the personal experience of a need or problem that cries out for a solution. An example is the need for a pharmaceutical that can mitigate or cure the effects of AIDS. This type of opportunity can be called opportunity pull, since the size of the opportunity draws opportunity seekers to attempt to exploit it. The founders of new industries capitalize on opportunity pull to create disruptive innovations that lead to new products that solve significant problems.

Another type of opportunity occurs from the discovery of a capability or resource that can be applied to a problem or need. An example of this type of opportunity is the discovery of a new technology, such as HDTV in consumer electronics and stem cells in biotechnology. This type of opportunity can be called a capability push, since it flows from a capability or resource availability. New organizational firms and industries are founded by individuals who recognize big opportunities as a result of technological change.

Often, being in the right line of business at the right place and time is the source of good opportunity. For example, Cisco Systems was formed in 1984 to exploit the capabilities of the founders and their associates at Stanford University. The firm was founded by Sandra Lerner and Leonard Bosack, who discovered the capability to enable a router to transmit and translate data to and from disparate computers [Bunnell, 2000]. By 2009, Cisco had revenues of almost $40 billion.

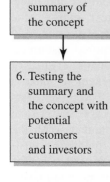

1. Determining the entrepreneur's capabilities and interests

2. Identifying the opportunity

3. Evaluating the opportunity

4. Deciding to act on the opportunity or look elsewhere

5. Writing a summary of the concept

6. Testing the summary and the concept with potential customers and investors

FIGURE 2.1 Six steps to acting as an entrepreneur.

> ### Opportunity Pull at ResMed
>
> Obstructive sleep apnea (OSA) was a widespread but underdiagnosed problem during the 1980s and early 1990s. OSA occurs when tissue at the back of the throat collapses during sleep, blocking the airway and preventing breathing. Oxygen levels drop in the bloodstream causing sharp fluctuations in heart rate and blood pressure. OSA is strongly correlated with other severe conditions—nearly half of all heart failure patients and 60 percent of type 2 diabetes patients suffer from OSA. It was estimated that 2 percent of the U.S. population suffered from OSA in some form. It was clearly a massive problem waiting for a solution.
>
> ResMed was founded in Australia to combat this problem. The company created a novel device that pressurized the airway during sleep to prevent the airway from blocking. The device was fantastically successful and as recognition of OSA expanded during the early 1990s, ResMed took off. ResMed correctly identified a huge unsolved problem and provided a solution that fit into both the patient's life and the health insurers' plans. As a result, the company has been incredibly successful. It is now public on the New York Stock Exchange with revenues of almost $1 billion in 2009.

A good opportunity has the potential to create significant value for the customer. Another way of describing a good opportunity is to describe the customer's pain, which represents the extent of need for the solution to a problem. The pain of need is the converse of value. Thus, a high-value solution is sought by a customer who feels significant pain of need. For example, today's airline customer often experiences a fear of flying. The solution to that problem should lead to the improved value of security of airline travel. Both the customer and the airline want a security solution.

Some would-be entrepreneurs have a new technology and often mistake it for a solution. Customers want a solution to their problem and usually do not care what technology is employed. Unfortunately, some believe that entrepreneurship is having a great technological idea. Entrepreneurship is really about creating a new business that solves a problem.

Successful new ventures are often initiated by people who have experienced significant painful problems. Sam Goldman, the founder of d.light, grew up in Mauritania, Pakistan, Peru, India, and Rwanda before becoming a Peace Corps volunteer in Benin. He then moved on to study biology and environmental studies in Canada before receiving his MBA from Stanford. While Goldman was living in Benin, his neighbor's son was badly burned by a kerosene lamp. This inspired him to create a new source of light, which could match kerosene lamps on price, but be safe for use around small children. d.light now creates extremely efficient LED lights that are 8 to 10 times brighter than a kerosene lamp and 50 percent more efficient than fluorescent lights.

Other new successful ventures occur due to shifts in regulatory policies. The opening of the wireless radio spectrum for mobile devices and cell phones is an example of a big shift in opportunity. We only need look around us to see the proliferation of the wireless revolution.

We can summarize the nine categories of opportunity as shown in Table 2.1. We use these nine categories of opportunity to describe a way of identifying opportunities. The first, and perhaps most common, is to increase the value of the product or service. This can include improved performance, better quality or experience, and improved accessibility or other values unique to the product. For example, Shokay is a for-profit social enterprise based in Tibet that manufacturers and distributes 100 percent yak down products including scarves and throws. The products are sourced from impoverished Tibetan yak herders with the broader mission of fostering economic development to remote areas of western China. Their products are luxurious, soft, functional, and have wide appeal.

The second category seeks new applications of existing means or technologies. Credit cards with magnetic stripes were available in the 1960s, but a thoughtful innovator recognized the application of this technology to hotel door cards and created a wholly new application and industry.

The third category concentrates on creating mass markets for existing products. A good example is the introduction of the disposable camera, which is often used at weddings or parties.

Customization of products for individuals, category 4, affords a new opportunity for an existing product or technology. Examples of customization can be found in the personal computer industry. Dell Computer is a good example of a company that offers customization.

Expanding geographic reach or online reach, category 5, allows a new venture to increase its number of customers. Founded in Scotland, Optos developed a novel eye exam technology and innovative pay-per-use business model. Backed by angel investors for many years, it carefully expanded its operations into the U.S. and Germany markets. It is now a viable public company listed on the London stock exchange.

TABLE 2.1 Nine categories of opportunity.

1. Increasing the value of a product or service

2. New applications of existing means or technologies

3. Creating mass markets

4. Customization for individuals

5. Increasing reach

6. Managing the supply chain

7. Convergence of industries

8. Process innovation

9. Increasing the scale of the firm

Managing the supply chain, category 6, is a powerful force for improvement. Wal-Mart used its distribution system and large stores connected to an inventory information system to reap the economic benefits of inventory management.

Convergence of industries, category 7, affords potential benefits to innovative teams. For example, genetic engineering is the convergence of electron microscopy, micromanipulation, and supercomputing.

Innovation of business and manufacturing processes, category 8, is another source of opportunity. For example, the shipping of goods has been greatly changed by the introduction of FedEx and other airborne shipping systems.

Finally, the ninth category of opportunity is the increasing scale or consolidation of industry. Historically, the railroad industry provides a powerful example of consolidation in the United States. Consolidation of the railroads began by the turn of the twentieth century. Today, there are five major railroad companies, down from the thousands of companies in the late 1890s. More recent consolidation examples include the personal computer and video rental industries. Through mergers and acquisitions, an industry can be consolidated with attendant cost savings and value for the customer.

Great opportunities are often disguised as difficult problems. For example, Scott Cook saw a problem experienced by individuals who wanted to easily and reliably keep their own home budget records, do their taxes, and pay the bills. He thought that problem could be solved by using a personal computer. The software program his new firm developed was intuitive and easy to use so that most people could use it without resorting to the manual—thus, the name of the firm: Intuit (www.intuit.com). Scott Cook solved a big problem with an easy-to-use solution. The identification of problems depends on preparation, experience, competence, and a keen sense of observation. Entrepreneurs like Cook leverage their curiosity and an observant nature.

Once a problem has been identified, a solution can be deduced by first asking, "How would an unconstrained person solve the problem?" Starting without constraints such as price and physical limits opens up many possibilities. Once a good unconstrained solution appears attractive, it can often be rearranged to accommodate reasonable constraints [Nalebuff and Ayres, 2003].

The power of *serendipity*—making useful discoveries by accident—can also lead to good opportunities. Working in a microwave lab, Percy Spencer observed a chocolate bar melting by microwave power—thus, leading to the microwave oven. Clarence Birdseye was a fur trader in Canada when he noticed a phenomenon while ice fishing. At 50 degrees below zero, fish froze rock-hard almost instantly, yet when thawed, they were fresh and tender. After some experimentation, he learned that the key was the speed at which foods were frozen. That observation led to the flash freezing process that created a multi-billion dollar industry and made Birdseye a success.

Another means of finding a good opportunity is to look for a discontinuity in culture, society, or markets. Table 2.2 describes examples of discontinuities that lead to a big opportunity. An example of a big opportunity is addressing

TABLE 2.2 Sources of discontinuities.

Society	Technology	Markets
■ Aging society	■ Innovation	■ Deregulation
■ Lifelong education	■ Disruptive technologies	■ Supply chain disruption
■ Food and population	■ New knowledge	■ Globalization
■ Regulation		

the need for creation of new pharmaceuticals to help prevent the increasing incidence of Parkinson's and Alzheimer's diseases among older people.

Any specific business opportunity may be portrayed in the three-dimensional cube of Figure 2.2. The entrepreneur identifies the customer, the required technology, and the application of this technology to create a solution. Several websites for identifying new ideas are listed in appendix C.

Good opportunities display the characteristics of a potential to solve important problems within economic constraints. Usually, they will look attractive because they can be profitable to the new venture as well as valuable to the

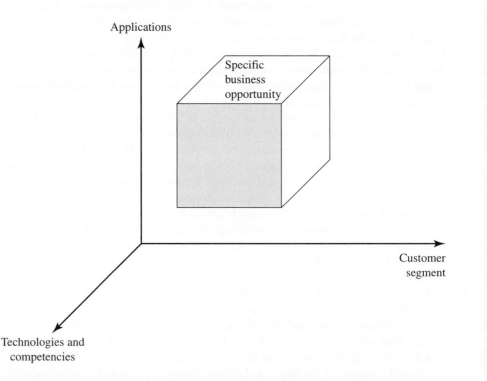

FIGURE 2.2 Finding a specific business opportunity with a combination of customer segment, technology and competencies, and applications.

TABLE 2.3 Five characteristics of an attractive opportunity.

- Timely—a current need or problem
- Solvable—a problem that can be solved in the near future with accessible resource
- Important—the customer deems the problem or need important
- Profitable—the customer will pay for the solution and allow the enterprise to profit
- Context—a favorable regulatory and industry situation

customers. An attractive opportunity displays the five characteristics listed in Table 2.3. The entrepreneur seeks a timely, solvable, important problem with a favorable context that can lead to profitability.

It is the entrepreneur who adds value to the opportunity by creating a response to a good opportunity. The opportunity, and a general response to it, is not unique—many recognize it but few possess the relevant passion to solve the problem as well as the capability to do so. For example, many people propose to exploit the new science of nanotechnology to solve various problems, but few of them will act. The true entrepreneur finds the best opportunity that matches his or her interests, skills, and knowledge—and acts to get it done. Thus, it is really the passion and capabilities that distinguish the entrepreneurial team. The selection process consists of looking for the best match of opportunity, capabilities, and interest (passion).

Jeremy Jaech is a good example of an entrepreneur who found this match. Jaech attended the University of Washington, receiving a BA in mathematics in 1977. He joined the computer science graduate program after graduation, completing his master's degree in 1980, while working at Boeing on computer graphics. In 1983, he joined Atex, a maker of computer systems for newspapers. After nine months, Atex closed the facility where he worked, and Jaech needed to find an opportunity for himself. His capabilities were in computer programming for graphics, and his interest was to achieve independence and success. His passion was for developing software for desktop computer graphics. His former boss at Atex suggested they form their own company that would create software for desktop computer graphics. Jaech was a good technical leader, and his boss was a good manager; together, they made a solid team. In 1984, the two men founded Seattle-based Aldus Corporation, which created the software called PageMaker, which launched desktop publishing on personal computers.

By 1989, although Aldus had grown, Jaech was faced with a new challenge. He wanted to broaden the product line, but his partner/CEO wanted to remain focused on desktop publishing. Jaech saw an opportunity to create a Windows-based software product for general-purpose drawing. He matched his capabilities with his interests and in 1990 started a new firm that was later called Visio Corporation. When the company's first product was shipped in 1992, it had 14 employees. It went public as a 200-person company in 1995 and was eventually purchased in January 2000 by

Microsoft Corporation for $1.5 billion in stock. Jaech had exploited two successive opportunities: Aldus and Visio both used his ability to design software while matching his capabilities and skills with his passions and interest to create two important companies. Jaech is now CEO of Verdium, which is an enterprise software venture that helps customers reduce energy consumption.

2.2 Trends and Convergence

Trends in technologies and demographics can lead to large opportunities. For example, just 30 years ago online shopping and mobile phones were far-off dreams. Today, hundreds of millions of people shop online and mobile phones are widespread on every continent. Opportunities abound in medicine, agriculture, materials, energy, transportation, housing, computers, and education, to mention a few industries.

The world's food supply and nutritional sources will get a big boost from biotechnology in the decades ahead. Biotech's benefits will include more environmentally friendly agriculture. Farmers will have more tools to combat pests, overcome difficult conditions, and grow more food from fewer acres and resources.

Crops may be designed with built-in resistance to diseases and pests, boosting yields worldwide. This is already being done with corn, cotton, and soybeans. Plants will also be endowed with new tolerance to weather, greatly expanding the land area where grains and vegetables can be profitably grown. Genetic engineering will also produce trees that grow faster or resist disease.

The trend toward globalization of business is based on the negation of time and distance with the emergence of the Internet and overnight shipping. With a billion new capitalists in Asia, "globalization is a mega-trend that will shape all other trends" [Prestowitz, 2005].

As prosperity grows and spreads worldwide, many opportunities occur from lifestyle changes. Starbucks, for example, offers a quality, customized coffee product and makes it broadly available. Other premium product segments such as wine, gifts, and flowers will grow with prosperity. Entrepreneurs should try to identify destabilizing influences. These come about through technological change, as well as through changes in taste. Entire industries can be made or broken by a shift in fashion. For example, the replacement of silk stockings by nylon ones led to the popularity of synthetic fabrics in clothing generally.

Demographic and cultural trends offer many examples of opportunity. Several social and cultural trends are listed in Table 2.4. The biggest current trend in America is the aging of the baby-boom generation—those born between 1946 and 1973. During those years, 107.5 million Americans were born, making up 50 percent of everyone alive in 1973 [Hoover, 2001]. Those born in the peak-birth year of 1961 will be 50 in 2011 and acting as wealthy consumers of goods and services such as new homes, furniture, travel, and retirement plans.

> **Sirtris and the Battle Against Age-Related Diseases**
>
> Sirtris was founded to address some of the diseases that are strongly associated with age. Many industrialized nations are seeing massive demographic shifts as the average age of their population consistently increases. Countries like Japan are confronting whole new problems as large portions of the population retire, but are still expected to live for another generation. Sirtris is hoping to address this problem by creating drugs to combat diseases like Alzheimer's, diabetes, and cancer. These drugs not only would improve quality of life for the elder portion of the population, but also would allow it to remain in the workforce longer. This could have massive implications for everything from health care costs to Medicare and Medicaid.

TABLE 2.4 Social and cultural trends that will create opportunities.

■ Aging of the baby-boom generation	■ Changing role of religious organizations
■ Increasing diversity of the people of the United States (e.g., Latino population)	■ Changing role of women in society
■ Two-working-parent families	■ Pervasive influence of media— television, DVDs, Internet
■ Rising middle class of developing nations	

Other critical trends include the rise of diversity as massive waves of immigrants arrive in the United States and as the role of women changes in many societies and nations in the world. Any new enterprise must fit the social and cultural context of its service area. One of the most promising areas of science and engineering is based on several breakthroughs that enable the manipulation of matter at the molecular level. Mass production of products with these molecular adjustments now offers a world of possibilities. Nanotechnology will make materials lighter, more durable, and more stain resistant. (One nanometer is one-billionth of a meter.) Soon we may also get a host of miniaturized products, from semiconductors to pumps that work more efficiently and accurately. The areas of application range from medical and industrial to the home [Ratner and Ratner, 2004].

Tiny robots acting in coordinated teams may be used in events such as fires, toxic spills, and bomb threats. This activity, like that depicted in the movie *Minority Report,* can be used for safety and reconnaissance activities [Grabowski et al., 2003]. Companies such as iRobot have recently marketed robots that have been used in dangerous military and firefighting situations.

Another example of a current trend that results in a big problem is the unsolicited e-mail (spam) that is received by all e-mail users. In 2009, 85 percent of

TABLE 2.5 Trends and opportunities.

- Life science: Genetic engineering, genomics, biometrics

- Information technology: Internet, wireless device, cloud computing

- Food preservation: Improved distribution of food

- Video gaming: Learning, entertainment

- Speech recognition: Interface between computers and people

- Security devices and systems: Identification devices, baggage checkers, protective clothes

- Nanotechnology: Devices 100 nanometers or less for drug delivery, biosensors

- Renewable energy: Solar cells and wind turbines

- Fuel cells: Electrochemical conversion of hydrogen or hydrocarbon fuels into electric current

- Superconductivity: Energy savings on utility power lines

- Designer enzymes: Protein catalysts that accelerate chemical reactions in living cells for consumers and health products

- Cell phones: Communications and computing

- Software security: Blocking unsolicited e-mail (spam), preventing "phishing"

- Robots: Teams of small coordinated robots for monitoring and safety functions

all e-mail messages sent over the Internet were spam. Any new firm that can sell a foolproof spam blocker would solve a problem for most e-mail users.

With the need for security and safety of personal information, the emergence of personal identification cards, or smart cards, may be the next trend in America. Cards for pay telephones and money transfer are one application. Another important use of smart cards would be a common approach for driver's licenses and personal information. A smart card is a plastic card incorporating an integrated circuit chip and memory that stores and transfers data such as personal data and identification information such as finger or palm print or facial scans. These cards have been adopted in several European and Asian countries and could spread worldwide. One form of smart card, the Octopus Card, is used in Hong Kong to pay for everything from subway rides to groceries.

Table 2.5 lists some important technology trends and opportunities for the future. Perhaps the most important advances will come from the energy and environment sectors.

Moreover, the boundaries between many once-distinct businesses, from agribusiness and chemicals to pharmaceuticals and health care to energy and computing, will continue to blur. The **convergence** of technologies or industries is the coming together or merging of several technologies or industries thought to be different or separate. Often they emerge from creative combinations that build on complementary technologies. One example of industry convergence is that of computing and communications, which merged into the field of networks. Another example is convergence of a handheld computer and a cell phone. The idea is to let users carry just one device instead of two or three and still stay connected via voice, e-mail, or data. Observers say the devices could be a boon to the wireless industry. Large innovations emerging from convergences can drive growth and create vistas of opportunity [Mandel, 2004].

An excellent example of the convergence of two technologies leading to an opportunity is the development of global positioning systems (GPS) and their wide use by hikers, travelers, surveyors, and farmers. Satellite imaging and data and the handheld computer converged into the GPS device, which is a widely used, inexpensive device that addresses the need for accurate locational data.

Think creatively about possible convergences. How about the convergence of scanners, computers, and security systems that enables shoppers to bag their own groceries in a self-checkout system? Or, consider a gene chip that uses semiconductor technology to speed up gene lab analysis. Another example is the new world of medicine driven by innovation and the needs of aging Americans for ever-more-intense levels of care. Already, the United States spends $2.3 trillion, or 16 percent of its gross domestic product, on medical care. The health care transformation could be as big as the computer revolution.

Another big trend is the convergence of the computer and communications and the trend toward wireless phones and devices, as illustrated in Figure 2.3. Cell phones become more like computers, and handheld computers transform to phones [Lohr, 2003a]. People are excited by the opportunities as wireless

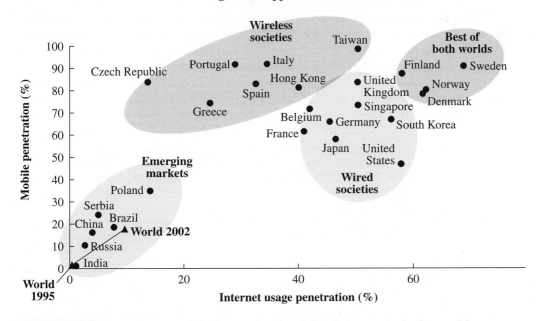

FIGURE 2.3 Convergence of the Internet and mobile phone usage in the world. (*Source:* Applied Materials Corporation.)

start-ups proliferate. Another trend is the convergence of medical and robotic technology. As robotic devices become more advanced, they are increasingly finding broader applications in the world of medicine. Intuitive Surgical manufactures robots that can help surgeons perform certain operations with greater precision and accuracy. Its da Vinci medical robot has four arms and flexible wrists mounted with tools and cameras that can be controlled by a surgeon. The robots are being used for prostate surgeries, hysterectomies, and more complicated surgeries like heart valve repair. Effective entrepreneurs look for a new technology in one industry that can be applied in another one—and they are savvy at identifying future trends as a result.

2.3 Opportunity Evaluation

Choosing the right opportunity is a difficult and important task. We select the opportunity that affords the best chance of success within the context of the marketplace. This choice is analogous to the selection of an equity investment in a company. Entrepreneurs will invest time, effort, and money in the venture they choose in a manner similar to how people invest in the stock of a company. Some sound investment principles that can be used for selecting opportunities are listed in Table 2.6.

The entrepreneur finds and thoroughly analyzes the best opportunities, since for many people, only one or two are needed to make a good life of entrepreneurial activity. One goal is to invest in a firm for which you pay less than it is worth; this provides some cushion for unforeseen challenges. Also, entrepreneurs try to find an opportunity with solid long-term potential in an industry they understand. They put together a good management team that can execute the strategy for this opportunity. And, they ensure that the customer will allow their firm to profit from the venture. Thus, they avoid industries that sell commodities where price is the only differentiation unless they have a new, innovative business process that enables their firm to be the low-cost provider.

TABLE 2.6 Guiding principles for selecting good opportunities.

■ Only one or two very good opportunities are needed in a lifetime.	■ If the opportunity is selected and turns out unfavorably, can you exit with minor losses?
■ Invest less time, money, and effort in the venture than it will be worth in one or two years. Calculate the probability of a large return in four years.	■ Does this opportunity provide a potential for a long-term success, or is it a fad? Go to where the potential future gains are significant.
■ Do not count on making a high-priced sale of your firm to the public or another company.	■ Can the management team execute the strategy selected for this opportunity?
■ Carry out a solid analysis of the current and expected conditions of the industry where the opportunity resides.	■ Will the customer enable your firm to profit from this venture?

Tom Stemberg, the founder of Staples, conceived the idea of a supermarket store for office supplies in the mid-1980s. He didn't like the politics of big companies and sought independence. He started with a single store in Brighton, Massachusetts, and built 1,500 outlets. He carried out a complete analysis of the opportunity and determined it was a $100 billion market growing at 15 percent per year with large profit margins.

The review of opportunities will always include the evaluation of alternatives. The **opportunity cost** of an action is the value (cost) of the forgone alternative action. Selecting one opportunity will involve rejecting others. Chapter 1 discussed some of the considerations that people should use when deciding whether to become an entrepreneur by pursuing a specific opportunity. A critical part of this decision hinges on the quality of the opportunity in terms of a market assessment, feasibility of implementation, and differentiation of the product. Much of this analysis requires additional information. Judgment regarding the qualities of the opportunity can be made by the entrepreneurial team as it considers all the aspects of the opportunity. A comprehensive analytical approach to evaluation of the opportunity does not suit most start-ups. Entrepreneurs often lack the time and money to interview a representative cross section of potential customers, analyze substitutes, reconstruct competitors' cost structures, or project alternative planning scenarios.

Most entrepreneurial teams instead follow a basic five-step process, as outlined in Table 2.7. The goal is to quickly weed out unpromising ventures and conserve energy and time for the promising ones. In general, it is best to reject ventures in industries or markets in which the entrepreneurs have little experience or knowledge. Standard checklists or approaches don't work for most entrepreneurs. The appropriate analytical effort and the issues that are most worthy of research and analysis depend on the characteristics of each venture. For example, the exploration process should be short for potential ventures with low degrees of novelty [Choi et al, 2008]. In general, however, an entrepreneur works through the five steps and eliminates the opportunities that do not pass muster. Those that do pass a quick review are worth looking into further.

TABLE 2.7 Basic five-step process of evaluating an opportunity.

1. **Capabilities:** Is the venture opportunity consistent with the capabilities, knowledge, and experience of the team members?

2. **Novelty:** Does the product or service have significant novel, proprietary, or differentiating qualities? Does it create significant value for the customer—enough so that the customer wants the product and will pay a premium for it?

3. **Resources:** Can the venture team attract the necessary financial, physical, and human resources consistent with the magnitude of the venture?

4. **Return:** Can the product be produced at a cost so that a profit can be obtained? Is the expected return of the venture consistent with the risk of the venture?

5. **Commitment:** Do the entrepreneurial team members feel compelled to commit to this venture? Are they passionate about the venture?

The iPod Opportunity

In the late 1990s, many people had been listening to music stored on their computer that was obtained through the Napster file-sharing service. The challenge was to design and sell a portable music storage and player device. Tony Fadell, who had worked at General Magic, started his own company, Fuse, to design consumer electronic products. He tried to secure financing for the design of a portable music player. Without financing, he went to Apple in February 2001 as a contractor and then in April 2002 joined Apple as an employee to lead the iPod project. Fadell and Apple together recognized that the iPod opportunity possessed all the characteristics of Table 2.7. The capabilities, resources, and commitment of Apple and Fadell enabled them to build a device, the iPod, that was truly novel and that could offer a significant return.

It is difficult for many to ascertain their dreams and goals. Often it helps to write them down. Keep revising them until you are sure of them. John James Audubon was a taxidermist who had a passion for painting the birds of America. By 1829, he published the first of his famous four volumes entitled *The Birds of America*. Most of the paintings were life-size. His opportunity became a life's work and a legacy that we value today.

The entrepreneur has to live with critical uncertainties, such as the relative competence of rivals or the preferences of customers, which are not easy to analyze. Who could have forecast, for example, that IBM would turn to Microsoft for an operating system for its personal computer, allow Microsoft to retain the rights to this operating system, and thus gain monopolistic dominance of the operating system marketplace? Entering a race requires faith in one's ability to finish ahead of whoever else might participate.

A new product has to offer customers exceptional value at an attractive price, and the company must be able to deliver it at a good profit. The initial opportunity review can be based on the five characteristics of the opportunity and its assessment by the team: capabilities, novelty, resources, return, and commitment, as depicted in Table 2.7.

When evaluating an opportunity, the entrepreneur considers whether it fits or matches the contextual conditions, the team's capabilities and characteristics, and the team's ability to secure the necessary resources to initiate a new venture based on the opportunity. Figure 2.4 shows a diagram of fit or congruence that can be used to review an opportunity. A big diamond with high grades of fit are best.

Consider an opportunity that has existed for over 100 years—the electric automobile. We will assume that a capable set of engineers is available and the entrepreneurial team has the attitudes and capabilities required. However, the team is insecure about the risky nature of the venture, given the numerous failures over the past century. We will rate the entrepreneurial team at 75 percent on the team scale. The characteristics of the context are very mixed since regulations and

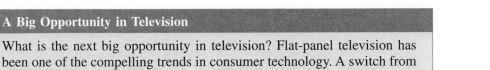

A Big Opportunity in Television

What is the next big opportunity in television? Flat-panel television has been one of the compelling trends in consumer technology. A switch from the bulky cathode-ray tube to flat panel displays is under way in the multi billion unit global market. Opportunities exist in the market for glass and the devices and chips for digital light processing. Another opportunity is making the sets and selling them. As nations switch to high-definition TV (HDTV), the market for flat panel displays has grown significantly. Which of these opportunities pass the evaluation process of Table 2.7?

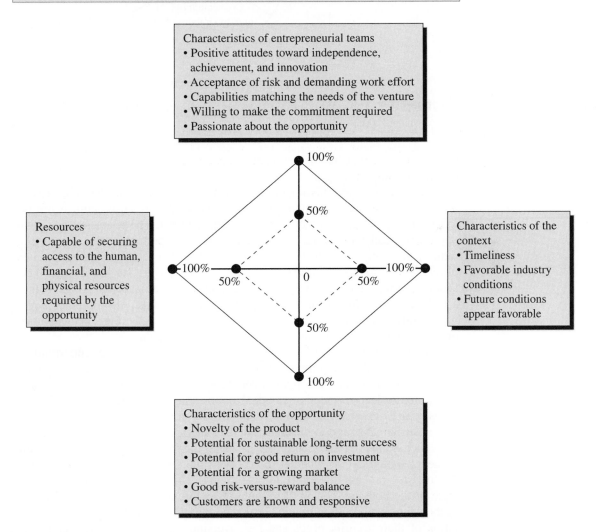

FIGURE 2.4 Diagram of the fit of an opportunity, the context, the entrepreneurial team, and the resources required. Rate each factor on a scale of 0 to 100 percent.

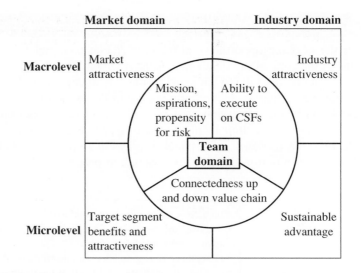

FIGURE 2.5 The seven domains of attractive opportunities.
(*Source:* Mullins, 2006.)

support for electric cars are continually changing as potential customers and government organizations adjust their assessment of the benefits and costs of these vehicles. We will rate this opportunity as only 50 percent on the context scale. Next we turn to the opportunity, which is challenged by costs, limited life batteries, and short ranges before a recharge is required. The characteristics of the opportunity call for a rating of 75 percent on the opportunity scale. Given these ratings, most teams would be severely limited in their ability to secure the tens of millions of dollars required to launch this venture. Thus, we rate it only 50 percent on the resource scale. Clearly, this opportunity is a challenging one. Without a technical breakthrough in battery performance and cost, electric autos have a risky future—valuable as an electric car might be to the environmental conditions of auto-impacted regions such as Los Angeles and Beijing.

Another way of envisioning the concept of a fit with an opportunity is shown graphically in Figure 2.5. Both markets and industries must be examined on the macrolevel and the microlevel. Moreover, the team must be evaluated across multiple dimensions. An ideal opportunity lies where the market and industry are attractive, customer benefits are compelling, the start-up's advantage is sustainable, and the team can deliver the results it seeks [Mullins, 2006].

After evaluating an opportunity by using the factors in Table 2.7, the entrepreneurs should decide whether to act. With the knowledge generated by using the five-step process in Table 2.7, the entrepreneurs will tend to act on their estimate of the potential benefits and gains, B, while accounting for the total costs of the venture, C. Within the total cost accounting, there will need to be a recognition of their security needs and loss aversion. An individual will tend to act if the ratio B/C is greater than 1. The lucrative opportunity (high benefits and low losses) will tend to cause higher intention to act [McMullen and Shepherd, 2002].

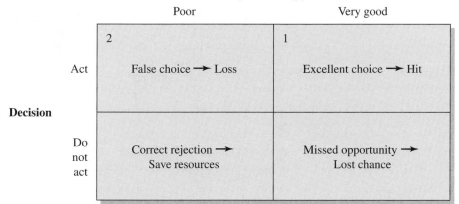

FIGURE 2.6 Decision matrix.

If one acts and it is a false choice, the cost of that choice is important. Opportunities that can be attempted with low initial financial and time commitment costs may offer the chance for lucrative returns at a low initial cost.

The matrix in Figure 2.6 shows the decision to act or not act. Then, the actual resulting quality of the opportunity is shown (this can be determined only after the decision). Life is about choices, and the best case is when we choose to act and it turns out we are right!

The entrepreneur attempts to make a rational decision based on (1) his or her current psychological and financial assets, and (2) the possible consequences of the choice [Hastie and Dawes, 2001]. The decision challenge is the task of turning incomplete knowledge of an opportunity into an action consistent with that knowledge. Competitive advantage comes from actually doing something that others cannot do. Analysis and reports cannot substitute for action. Reworking a plan is no substitute for acting to get things done. In the end, an opportunity can be evaluated only so much [Pfeffer and Sutton, 2000]. Ambiguity remains, and the entrepreneur needs to act on or reject the opportunity. Fear of failure may overwhelm all but the best opportunities.

Perhaps the best way to find a really good opportunity is to examine it by estimating fit in the diagram of Figure 2.4 and then act on the best opportunity, trying it out in the marketplace of ideas and investors. This can lead to a refinement of the opportunity. Tom Peters and Robert Waterman [1982] called this approach "ready, fire, aim". Banishing fear of failure and learning from a series of small failures can lead to a good new venture. The act-review-fix cycle, as shown in Figure 2.7, summarizes the critical ability to act, review, and learn from the results, and then, fix and adjust the business scheme as required. As John Stuart Mill stated: "There are many truths of which the full meaning cannot be realized until personal experience has brought it home."

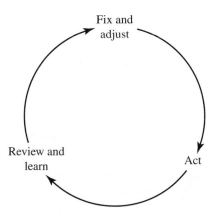

FIGURE 2.7 Act-learn-fix cycle
of building a new venture.

2.4 The Concept Summary

Once a business opportunity has been selected for action, it is important to
prepare a **concept summary** of the new venture. This can be a simple statement of the problem being addressed and how the venture will solve it. This
statement of the business concept is a short description of the new business.
The elements of a concept summary are given in Table 2.8. For example, the
original business concept for Amazon.com might be summarized as "an Internet-based retail service that allows customers to search for and purchase at a
discounted price books that will be delivered quickly."

A **story** is a narrative of factual or imagined events. The new business story
depicts a business problem responded to with a new means to solve the problem. The story tells the goal of the venture, the challenge, and the response of
the new firm. The creation of the story is used to communicate verbally the
business idea and the profitable solution of the problem. The investor or new
team member will be drawn to a good story. The three elements of a story
summarized in Table 2.9 are (1) background, (2) challenge, and (3) resolution.
The goal is to tell a compelling story in three acts: background and characters,
challenge, and a workable resolution. All good stories exhibit coherence and
flow. Coherence generates the listener's trust [Ibarra and Lineback, 2005]. By
the end of the story, the reader or listener should know how the new venture
will make a real difference to the customer [Kawasaki, 2004].

TABLE 2.8 Elements of a concept summary.

1. Explain the problem or need and identify the customer.

2. Explain the proposed solution and the uniqueness of the solution.

3. Tell why the customer will pay for the solution.

TABLE 2.9 Elements of the business story.

1. **Background:** Describe the current situation, characters, and problem.

2. **Challenge:** Describe the challenges and conflicts that impede a coherent plan to solve the problem.

3. **Resolution:** Portray a solution to the challenges and the problem and how the venture will succeed by resolving the problem.

As an illustration of an important story, consider the world's energy challenge. Energy is the lifeblood of industrial civilization and necessary for lifting the world's poor out of poverty. However, current methods of mobilizing energy are highly disruptive of local and global environmental conditions and processes. Thus, the challenge is to develop a new, more favorable energy system and its associated sources. The resolution of this challenge will, it is hoped, be the discovery of an energy technology that can economically convert solar energy to a locally useful form. One possibility is a solar conversion system yielding hydrogen to be used in fuel cells. Many technology ventures could exploit this opportunity favorably. It can become a great story with an important outcome.

In today's fast-paced, dynamic world, a business concept and associated story are all one needs to start working on building a business. For the first steps, the entrepreneur (1) builds a concept to solve the business challenge, (2) fashions a story that conveys the meaning of the new venture, and (3) prepares a presentation of a few slides that tell the story and explain the concept. The elements of a presentation are given in Table 2.10. After testing the concept summary and the story with potential investors and partners, the entrepreneur may go on to develop a complete business plan, which is a process described in Chapter 7.

The story can be told to all would-be investors or employees. The concept summary can be left with them for later review. The presentation is for more formal occasions with investors or allies. The story, concept summary, and presentation should be professional, novel, provocative, creative, and, where possible, customized. *Novelty* refers to newness and freshness. *Provocative* and *creative* mean it provokes interest and is creative in layout or format.

For many entrepreneurs, the executive summary of the business plan is what the investors and potential team members will want to review. If the entrepreneurs can piece together the initial elements of a business plan, they can write a reasonable summary without actually completing the plan. In a

TABLE 2.10 Elements of the presentation.

1. Explain the concept and give the story. Emphasize the customers and their pain.

2. Clearly explain the problem and the solution. Tell why the customer cares.

3. Describe the competencies of the team. Tell about the passion and skills of the team.

4. Provide a picture of the competition. Name a few competitors and tell how you are different and better.

TABLE 2.11 Elements of an executive summary.

1. Business concept: The problem and the solution

2. Market, customer, and industry

3. Marketing and sales strategy

4. Organization and key leaders

5. Financial plan: Four years of summary results

6. Financing and key allies required

sense, the executive summary is the essence of the business plan. As such, for many new ventures, it stands alone as a short business plan. The executive summary succeeds by capturing the readers' attention and imagination, causing them to want to learn more. When readers finish the executive summary, they should have a good sense of what the entrepreneurs are trying to do in their business. The executive summary should be no longer than three pages. Most professional investors will ask you to e-mail it to them. An executive summary states the problem, the solution, the customer, the competitive advantage, and who will lead the effort. This summary is intended to convey the core of the business and draw the reader into a follow-up conversation.

The executive summary portrays the content and purpose of your business. The elements of an executive summary are listed in Table 2.11. Not all these elements will be necessary for all ventures. A fictitious example is provided in Table 2.12. Also see appendix A for another executive summary. The executive summary for AgraQuest is provided in Table 2.13.

TABLE 2.12 Example of a business summary.

Security Robots Inc. (SRI) was formed in 2009 to design and build mobile robots for clearing and cleaning up facilities that have or may have experienced security breaches. Office buildings, factories, schools, and laboratories may be subject to intrusion by terrorists who plant biological, chemical, or explosive devices. The SRI robots are capable of remote operation by security and police organizations and can be used to examine and clear or destroy terrorists' weapons.

SRI is a Subchapter S corporation seeking an initial set of investors to bring its new products to market. Founded in 2009 by Dr. Henry Morgan and Ms. Angela Wolfe, the firm has designed a mobile robot platform that can be customized for many high danger security tasks. SRI has filed for a design patent on the robot platform system.

Dr. Morgan holds a MS in electrical engineering and a PhD in mechanical engineering from the University of Texas. Dr. Morgan served as chief technology officer of FMA Corporation of Dallas, Texas, from 1997–2008. Ms. Wolfe, CPA, holds an MBA from Duke University and was formerly CFO of Moore Systems, Austin, Texas.

SRI has secured 3000 sq. ft. of industrial space in Austin Technology Park. The current staff of six has an operating robot under review and certification by the U.S. Department of Homeland Security and the Texas State Police. Manufacture of the robot product line is to be provided by SelectTech Systems, a national contractor, at its Huntsville, Alabama, facility. The marketing plan calls for a regional strategy in the first year, 2010; with expansion throughout the southern and eastern United States in 2011. A highly trained direct sales force will sell the SRI robots to police and security organizations.

TABLE 2.12 (continued)

The funds requested for commencement of manufacturing and marketing operations are $400,000. The co-founders of the firm have already invested $120,000 of their funds. The funds will be used for facilities, equipment, contracting, and marketing communications. The founders will not receive a salary until January 1, 2011, or at cash-flow breakeven, whichever occurs first.

Financial projections show revenues of $1.3 million in 2010 and $7.4 million in 2011. The company intends to go public (IPO) within five years of beginning sales operations. Investors may purchase units of 10,000 shares for $20,000. After issuing the 200,000 shares to the investors for $400,000, there will be total of 2 million shares issued. All individual investors should contact the firm for an investment prospectus and further information.

Contacts: Henry Morgan, CEO; Angela Wolfe, CFO

Security Robots Inc.   (512) 555-0121

Austin Technology Park   www.securityrobots.net

Austin, Texas 78712

TABLE 2.13 AgraQuest executive summary.

Mission: AgraQuest's mission is to be the best and most efficient at discovery and development of environmentally friendly natural products for pest management.

The business: AgraQuest discovers, develops, and markets environmentally friendly natural product pesticides from microorganisms. It has three sources of revenues: (1) sales of natural product pesticides to farmers and consumers, (2) sales of lead molecules that do not fit our development criteria to large pesticide companies, and (3) contract testing for pesticide companies.

Market need and market opportunity: 25 billion dollars are spent each year on chemical pesticides. Consumers have increasing expectations that their food is free of pesticide residues. Society has increasing concerns about how chemicals affect the environment, including fish, wildlife, groundwater, and air quality. The regulatory agencies are responding by establishing stricter criteria for registration of new chemical pesticide products and reregistration of older ones. The cost and time for registering a new chemical pesticide have ballooned to $40–70 million and 7–10 years. As a result, few new products are being registered, and many older products are being taken off the market or are so tightly regulated that their use is limited.

Technology: Natural products are substances produced by microbes, plants, and other organisms that can kill pests. Unlike natural products, currently marketed biopesticides, such as *Bacillus thuringiensis* (Bt), insect viruses, and insect-killing fungi, use living organisms as pesticides. As a result, they are negatively affected by heat, wind, rain, and sunlight. Therefore, they do not have efficacy as good as chemical pesticides and have not significantly penetrated chemical pesticide markets. Natural products can have efficacy against the targeted pest that is as good as chemical pesticides. This is not speculation. We have found them. We know they are there. Unlike many chemical pesticides, natural products are biodegradable and specific to the pest, without harmful effects on fish, wildlife, and beneficial insects.

Microbial natural products can be registered with the U.S. Environmental Protection Agency (EPA) as "biochemicals." This means that bringing a specific natural product to the market takes considerably less time and money (approximately 3–5 years and less than $5 million) than chemical pesticides.

Competition: If microbial natural product pesticides are so ideal, why aren't they the target of large companies? Pharmaceutical companies have the technical expertise for discovery of microbial natural products with pesticidal activity, but they are often not set up to assess agricultural applications of the molecules and lack the knowledge and experience to commercialize them. There is currently no independent company dedicated to screening for microbial natural product insecticides, fungicides, nematicides, and herbicides.

(concluded on next page)

TABLE 2.13 (continued)

Company's competitive advantage: AgraQuest can find a higher number of novel pesticidal natural products more quickly than anyone else. We have unique knowledge of the groups and sources of microorganisms that yield the highest number of novel pesticidal natural products. Our proprietary isolation and fermentation media generate higher numbers of "hits." We find more novel natural products because of our focus on difficult chemistry that very few in the industry attempt. We have proprietary automated, high-throughput in-vivo and in-vitro pesticidal and extraction assays. At a very early stage, we can rapidly recognize pesticidal molecules with product potential and activity as good as synthetic chemicals. We know how to develop and market bio-based pesticides in specialty markets; we have extensive and unique knowledge of the market and competition. We are experienced at creating a company culture that results in exceptional and sustained productivity, creativity, motivation, and commitment by employees.

Management team: AgraQuest has assembled a management team experienced in pesticide, biopesticide, and agricultural biotechnology business, research and development, marketing, management, and finance.

Pamela G. Marrone, PhD, President/CEO. Dr. Marrone left Novo Nordisk in January 1995 to start up AgraQuest. Under her tenure as president of Novo's subsidiary, Entotech, Inc., which she built from the ground up, the company extended its Bt product line into three new crop segments, brought a Bt product, two new Bt product formulations, and a new gypsy moth virus product formulation to the market. In addition, Entotech found six novel pesticidal natural products, including a novel Bt enhancer (now on the commercial track), and has filed or has pending 20 patent applications. Dr. Marrone wrote and implemented marketing plans and developed a new approach to generating revenue from biopesticides, which is now the flagship strategy of the division. Prior to Novo Nordisk, Dr. Marrone worked for Monsanto Agricultural Company (1983–1990). Her Insect Biology group led pioneering projects in natural product and genetically engineered microbial pesticides and Bt transgenic crops (to be on the market in 1996).

Ralph Sinibaldi, Vice President of Research and Development. Dr. Sinibaldi worked for Sandoz Agro, Inc., from 1982 to 1994 and was most recently Associate Research Director and Project Manager, where he coordinated two major products on crop transformation and regulation of gene expression. Dr. Sinibaldi has received or filed several patents, and he turned over three major pieces of technology to Sandoz Seeds for development.

Duane Ewing, Vice President of New Business and Product Development. Duane Ewing has 13 years of management experience and a total of 17 years of experience in agriculture-related industries. As one of the first employees of Pan-Ag Labs, Inc. (1981), Mr. Ewing played a crucial role in Pan-Ag's growth from $120,000 to almost $6 million annually in roles as Director of Field Research, Director of Business Development, Vice President, and de facto President during the owner's absence.

Bruce Holm CPA, Chief Financial Officer. Bruce Holm has over 30 years of accounting experience. For 16 years (1971–1987), he was Corporate Controller for Zoecon Corporation, where he was responsible for financial reporting in six SEC filings. Following Zoecon (1987–1991), he was employed by California Energy Company, Inc., as Corporate Controller and Joint Venture Controller.

Financial summary and amount and structure of proposed financing: AgraQuest requires start-up funding in the first full year of approximately $1.1 million for equipment, $2.5 million for operations, and $2.9 million for cash reserves. *First-round financing of $2.5 million will allow us to identify our first commercial product candidate from our own R and D and to develop an externally acquired product.* The following two years are expected to require approximately $11.4 million for operating expenses, $5.8 million funded by sales of research

(concluded on next page)

TABLE 2.13 (concluded)

services and molecules and product sales, leaving a net operations requirement of $5.1 million. Also, approximately $1.1 million is projected for equipment and improvements purchases, and $0.8 million is required for cash reserves.

A public offering is projected to occur early in year five, with a target of $20 million. We are confident that AgraQuest can, by year three, develop a pipeline that subsequently generates 5–10 new natural products per year. Our novel natural product portfolio will specifically include two for corn rootworm (to be sold), one for sucking insects, one fungicide, one nematicide, and one herbicide (to be sold).

The business projects a profit in year five and approximately $40 million in sales of molecules, services, and products in year seven.

Projected capital requirements:	($thousands)			
Year ended June 30:	1996	1997	1998	1999
Operating expenditures and interest	$2,500	$4,700	$6,700	$9,800
Equipment and furniture	1,100	500	600	400
Cash reserves buildup	2,900		1,300	
Total	$6,500	$5,200	$8,600	$10,200

Projected funding:				
Revenue from contract screening, molecule and product sales, and government grants	$200	$1,800	$4,000	$7,500
Equity financing	5,200	2,600	4,300	
Capital lease and/or bank financing of equipment, net of repayment	1,100	300	300	
Cash reserves usage		500		2,700
Total	$6,500	$5,200	$8,600	$10,200

Status of the company: AgraQuest was incorporated in the state of Delaware in January 1995. The company is in the process of completing seed financing (approximately $200,000), which is being used for starting the microbial library and pest colonies. Also, we expect to obtain one product candidate from outside the company and secure at least one corporate collaboration.

2.5 AgraQuest

Pam Marrone and her colleagues at Entotech were informed by Novo Nordisk that the firm was being sold to Abbott Laboratories. Marrone believed that natural biological controls could protect crops—an old idea that environmental enterprises are making fresh again. Driving the quest is pressure from government and consumer activists to reduce the use of synthetic chemicals on the nation's farms and ranches. The challenge for such companies is to develop reliable biopesticides at a price with chemicals.

Marrone contemplated leaving Entotech and starting a new venture based on developing an innovation (item 8 of Table 2.1) in the biotechnology industry: finding naturally occurring microorganisms that can serve as biopesticides and developing a process for producing them for reliable use on farms. With the growing trend toward natural, environmentally friendly products and processes, the opportunity looked favorable.

Marrone had already worked in corporate new ventures with Monsanto and Novo Nordisk, and she was confident of her technical and leadership competencies. She examined the opportunity using the principles of Table 2.6 and determined that this opportunity was very good. The agricultural pesticide industry showed a tendency to be slow to adopt risky innovations such as natural pesticides. However, she was convinced she could overcome the risk-averse nature of the farmer and the long regulatory review by government. The fit of her proposed company (see Figure 2.4) with the opportunity seemed to be high. Therefore, she decided to act by finding the key members of her team and founding the company, to be called AgraQuest. She was convinced that the opportunity was of very high quality (see Figure 2.6).

Pam Marrone and her fellow founders wrote an executive summary dated May 5, 1995, provided in edited form in Table 2.13.

2.6 Summary

The entrepreneur identifies numerous ideas and needs that may point to good opportunities that can be made into great companies. However, he or she searches for the one that best fits the capabilities of the team, the characteristics of the business context, the characteristics of the opportunity, and the team's capability to secure the necessary resources. Then, the entrepreneur decides whether to act or not act on that best-fit opportunity.

The important ideas of the chapter are:

- A great enterprise displays leadership in its industry, profitability, reputation, and longevity.
- Great opportunities are often disguised as problems that are difficult to describe.
- An important problem well stated is a problem on its way to solution.
- The entrepreneurial team should cumulatively possess all the necessary capabilities.
- Entrepreneurs should, if possible, act on favorable opportunities in a timely manner.
- Entrepreneurs should prepare a story and summary of the venture and use it to test the venture with potential customers, employees, and investors.

Principle 2
The capable entrepreneur knows how to identify, select, describe, and communicate an opportunity that has good potential to become a successful venture.

Video Resources

Visit http://techventures.stanford.edu to view experts discussing content from this chapter.

Disruptive Technologies	John Doerr	KPCB
Find a Wave and Ride It	Erik Straser	MDV
Problem-Solving Paradigm	Vinod Khosla	Khosla Ventures
The Founding of AgraQuest	Pam Marrone	AgraQuest

2.7 Exercises

2.1 One approach to classifying market entry is by (a) creating a new market, (b) attacking an existing market, or (c) resegmenting an existing market. Using Table 2.1, indicate how each of these categories of opportunities would be applicable to these market-entry approaches.

2.2 What were some of the key customer, technology, and market trends that drove entrepreneurship during the last decade? What factors do you predict will drive entrepreneurial challenges in the next decade?

2.3 The next big wave of innovation may be the convergence of bio-, info-, and nanotechnologies. Each holds promise in its own right, but together in combination, they could give rise to many important products. Describe one opportunity motivated by the convergence of these new areas, and develop a story about the opportunity.

2.4 Some imagine that within a few years it will be possible, through the use of stem cells, to create new cells and eventually new organs to replace those that fail. Summarize the potential opportunity for stem cell enterprises. How would you begin to estimate the size of this opportunity? Develop a story depicting the opportunity.

2.5 The convergence of biology with computers and nanotechnology may lead to safer and more effective medicines. Visit www.research.cornell.edu/anmt and examine the potential for nanomedical technologies. Write a brief concept summary for a nanotechnology start-up.

2.6 As energy costs rise and the impact on the environment becomes clearer, clean tech has become an area of significant new investment. Quantify the trends driving this renewed investment interest. How would you evaluate and market size the clean tech opportunity?

2.7 Great companies often create tools that solve people's everyday problems. People like to chat and say hello often. What innovations were motivated by this simple desire? What new opportunities in this space are being created by technology innovation?

2.8 Consider a software application you use regularly. What task(s) does it improve or enable? Suggest three ways the application could be improved. Would any of these improvements be considered an opportunity for a new venture? Why or why not?

2.9 Global sales of radio frequency identification tags (RFID) and related equipment have been forecasted to explode multiple times in the last decade. Describe the problems solved by RFID and the opportunities presented. What have been the barriers to commercialization of this technology? What types of opportunities will be created when RFID tags are widely adopted in products?

2.10 The trend of performance of two electronic technologies is given in Figure 2.8. Determine the performance trend of another technology. Prepare a chart of its performance over time.

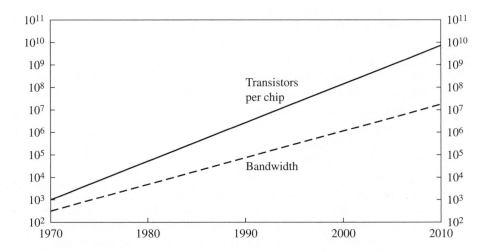

FIGURE 2.8 Technology trends: (a) transistors per chip; (b) bandwidth per household (bits/second). (*Source:* Dorf, 2004.)

VENTURE CHALLENGE

Consider the opportunity that you identified at the end of Chapter 1.

1. Evaluate it using Table 2.6's principles and Table 2.7's process. Write a concept summary using the format provided in Table 2.8.

2. Create a brief business story for the opportunity (venture) as summarized in Table 2.9 and present it to your team. Be sure to clearly describe the product or service, what problem it is solving, and who the customer is.

CHAPTER 3

Vision and the Business Model

Success in any enterprise requires the right product, methods, and workers, and each must complement the others.
Joseph Burger

How do successful entrepreneurs create a compelling business design for their new ventures?

A new business is defined by the wants or needs customers satisfy when they buy a product or service. To create a theory of a new business, the entrepreneur must cogently and clearly describe the customers and their needs and how the new venture will satisfy those needs. To describe the business, the entrepreneur prepares a series of statements and propositions that clearly outline the business. These are ultimately summarized in a model of the business activities and goals. Based on the core competencies of the organization coupled with the business model and the key resources available, the firm acts to attempt to create and retain a sustainable competitive advantage. The six steps of designing and creating a theory of the new business are summarized in Figure 3.1. ■

3.1 The Vision

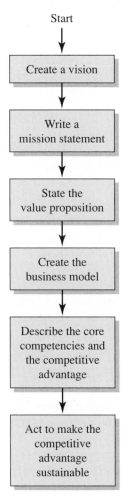

Start

Create a vision

Write a
mission statement

State the
value proposition

Create the
business model

Describe the core
competencies and
the competitive
advantage

Act to make the
competitive
advantage
sustainable

FIGURE 3.1
Creating a business
theory of a new
venture.

Once the entrepreneur identifies a good opportunity and decides to pursue it, the next step is to formulate a vision. A **vision** is an informed and forward-looking statement of purpose that defines the long-term destiny of the firm. Thus, if the entrepreneur recognizes a good opportunity to meet a real customer need, he or she describes a vision of a future venture that will respond effectively to that opportunity. The vision is a statement of insight, intention, ambition, and purpose. It reflects clearly the novelty of the solution and uniqueness of the entrepreneur's commitment. Successful entrepreneurs are able to communicate their vision and their enthusiasm about that vision to others. A vision often constitutes a novel idea about serving its market. McDonald's vision is: low-priced, fast food in a clean restaurant for people short on time. Google's vision is: online search that reliably provides fast and relevant results.

A solid vision provides direction and shows a path forward. A vision also motivates and influences decisions that are made by the team members. A clear vision can bind and inspire the entire community of a firm. A good vision is clear, consistent, unique, and purposeful [Hoover, 2001]. A clear vision is also easily understood. A consistent vision is one that does not change in response to daily challenges and fads. Any sound vision clearly explains the purpose of the firm. Remember, a good opportunity embodies a response to a big problem and calls forth a clear picture of your response. The 7 elements of a vision are summarized in Table 3.1.

Any business must understand what outcome the customer will really pay for. For example, doctors and their patients seek biomedical devices that really improve the lives of the patient. Therefore, a stent placed in an artery should keep the artery open. For Southwest Airlines, passengers want low prices, on-time arrivals, and the ability to fly between a chosen pair of cities [Chatterjee, 2005].

The purpose of the firm defines the enduring character of the organization, consistently held to and understood through the life of the firm. The purpose, or core ideology, of Hewlett-Packard has been a respect for the individual, a dedication to innovation, and a commitment to service to society. The purpose of Merck is to gain victory against disease and help mankind. This core ideology provides the glue that holds an organization together [Collins and Porras, 1996]. The vision provides a clear picture of the future for all concerned.

TABLE 3.1 Elements of a vision.

- **Clarity:** Easily understood and focused
- **Consistency:** Holds constant over a time period, but is adjustable as conditions warrant
- **Uniqueness:** Special to this enterprise
- **Purposeful:** Provides reason for being and others to care

TABLE 3.2 Example of a vision for an innovative firm.

We strive to preserve and improve people's lives through the innovation of biomedical devices while supporting, training, and inspiring our employees so that individual ability and creativity are released and rewarded. Our goal is to be the leader in our industry by 2012 and be widely known throughout the world for devices that save and extend lives.

The core ideology is based on the core values of the organization, such as respect for the individual.

A vision describes a specific desired outcome and promotes action and change. It serves as a picture of its destiny as the firm moves through challenge and change. It also provides the basis for a strategy. A vision is an imaginable picture of the future. It is like a rudder on a boat in a turbulent sea. An example of a simple, clear vision is given in Table 3.2. This vision statement provides the reader with a clear mental model of where the firm is going and how it will get there. Notice that the vision statement of Table 3.2 states the values and goals of the firm, and it inspires and motivates people.

Entrepreneurs need to create a shared vision or meaning for their venture. A dialogue of meaning and commitment will help bring a shared sense of urgency and importance for the venture. The vision can be written as a statement and verbally expressed as a story. The vision is used as a part of the business plan and described often to potential team members and investors. Stories play an important role in the processes that enable new businesses to emerge. A story is a narrative version of the vision, told in an engaging way. It helps to make the unfamiliar new enterprise more familiar, understandable, acceptable, and thus more legitimate to key constituencies [Lounsbury and Glynn, 2001]. By clarifying the core idea behind an enterprise, a story can also help an enterprise raise money and gather other resources [Martens et al., 2007].

Jim Clark started three companies: Silicon Graphics, Netscape, and Healtheon (now WebMD). As recounted in *The New New Thing* [Lewis, 2000], Clark stated:

> "The only thing I can do is start 'em." His role in the Valley was suddenly clear: he was the author of the story. He was the man with the nerve to invent the tale in which all the characters—the engineers, the VCs, the managers, the bankers—agreed to play the role he assigned them. And if he was going to retain the privilege of telling the stories, he had to make sure the stories had happy endings.

Clark had a vision for eliminating waste in the $300 billion costs of the U.S. health system by using the Internet to enable all the parties of any health transaction to connect via an online network—no paper forms, no hassle. Clark sketched a diamond depicting the players, as shown in Figure 3.2, and placed his proposed company, Healtheon, in the middle as intermediary. This was the way Clark told his ambitious stories—graphically, using sketches. Tales told by the entrepreneur aim to show plausibility and build confidence that the enterprise can

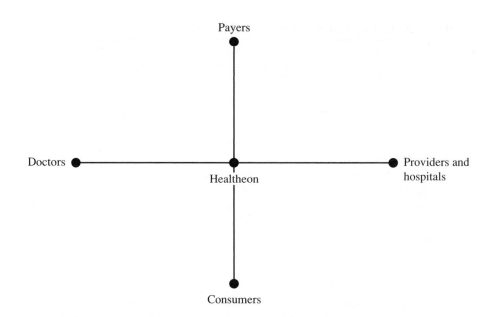

FIGURE 3.2 Vision for Healtheon (now WebMD). WebMD's three main businesses are providing electronic transaction services to doctors and hospitals, marketing software to help doctors run their practices, and providing online health information to doctors and consumers (see www.webMD.com).

succeed. To construct an identity that legitimates a new venture, entrepreneurial stories must have narrative clarity and resonate with the expectations, interests, and agendas of potential stakeholders [Lounsbury and Glynn, 2001].

Entrepreneurs need to learn how to tell their story about their team and venture, and to explain how their products will solve a problem. Their vision of the future can capture the interest of investors and team members.

Vinod Khosla is a prominent investor in green (clean) technologies. He believes that lifestyle changes and more conventional technology like hybrids will not be sufficient to solve the world's climate change crisis. According to Khosla, anything requiring people to spend more money or change their habits has a low probability of success. He is investing in revolutionary technology like cellulosic biofuels, which could fundamentally change the game by producing oil cheaper than any company today. In his view, any solution must make a difference at scale and cost less than conventional alternatives. This would drive people toward climate-friendly solutions for market reasons rather than ideological ones.

A vision told as a story helps people to see the situation and visualize the solution. The vision can also help people respond to the emotionally charged idea and want to help bring about useful change to the situation. The vision can be told as a story describing the potential outcome.

Henry Ford had a vision in 1910 of an automobile that could be available to all [Hounshell, 1985]:

> [The] greatest need today is a light, low-priced car with an up-to-date engine of ample horsepower, and built of the very best material. . . . It must be powerful enough for American roads and capable of carrying its passengers anywhere that a horse-drawn vehicle will go without the driver being afraid of ruining his car.

3.2 The Mission Statement

The mission statement for a new venture will more completely describe the company's goals and customers, while incorporating the basic tenets of the vision statement. A vision is an imaginative picture of the future, while a mission is a description of the course of action to implement the vision. The mission of an organization is lofty and audacious. It provides for a theory of change.

The potential elements of a mission statement are shown in Table 3.3. Most mission statements include only some of these elements. For example, the mission statement of eBay is given in Table 3.4.

Most mission statements are short—fewer than 100 words. The mission statement should be a concise, clear explanation of the purpose, values, product, and customer. The eBay statement clearly describes its mission. An example of a concise yet clear mission statement for an electronics firm is: "Our mission is to design and manufacture electronic devices that serve the needs of the aerospace industry on a timely basis and at reasonable prices."

A good mission statement can help align all the stakeholders and provide a rationale for allocating resources. If possible, the mission statement should be developed by the entrepreneurial team with other employees. The mission

TABLE 3.3 Possible elements of a mission statement.

■ Core values	■ Competitive advantage
■ Customers and/or stakeholders	■ Values provided to customer
■ Products	■ Markets or industry

TABLE 3.4 Mission statement of eBay.

We help people trade practically anything on earth. eBay was founded with the belief that people are basically good. We believe that each of our customers, whether a buyer or a seller, is an individual who deserves to be treated with respect.

We will continue to enhance the online trading experiences of all—collectors, hobbyists, dealers, small business, unique item seekers, bargain hunters, opportunistic sellers, and browsers. The growth of the eBay community comes from meeting and exceeding the expectations of these special people.

Our mission is to be the leading biotechnology company. using human genetic information to discover, develop, manufacture, and commercialize biotherapeutics that address significant unmet medical needs. We commit ourselves to high standards of integrity in contributing to the best interests of patients, the medical profession, our employees, and our communities, and to seeking significant returns to our stockholders, based on the continual pursuit of scientific and operational excellence.

FIGURE 3.3 Genentech mission statement.

TABLE 3.5 Mission statement of Symantec Corporation.

At Symantec, we know what happens when people have the confidence to achieve their best; we help make it possible.

We're the global leader in Internet security—solely dedicated to making the connected world a safer place. The more connected the world becomes, the more pivotal our role in making it an environment where commerce, culture, and ideas can flourish.

Symantec breeds confidence.

statement for Genentech is shown in Figure 3.3. This statement is very complete and describes its commitment to all its stakeholders—customers, employees, and community. The mission statement of Symantec Corporation is provided in Table 3.5. This statement speaks clearly to the customers—Symantec breeds confidence.

3.3 The Value Proposition

Value delivered to the customer results in a satisfied customer who will pay a reasonable price in return for the product or service. **Value** is the worth, importance, or usefulness to the customer. In business terms, value is the worth in monetary terms of the social and economic benefits a customer receives from paying for a product or service. To be successful, firms must offer products that meet the needs and values of the customer. The needs of the customer often include ease of locating or accessing the product as well as its qualities and features.

The five key values held by a customer can be summarized as product, price, access, service, and experience. These five values are listed in Table 3.6, along with specific descriptors for each value. Price, for example, can have high value to the customer when it is fair, visible, and consistent. A product may have value if it has high performance and quality, and is easy to find and use. Most technology-based products are initially focused on performance and functionality [Markides and Geroski, 2005].

TABLE 3.6 **Five values offered to a customer.**

1.	**Product:**	Performance, quality, features, brand, selection, search, easy to use, safe
2.	**Price:**	Fair, visible, consistent, reasonable
3.	**Access:**	Convenient, location, nearby, at-hand, easy to find, in a reasonable time
4.	**Service:**	Ordering, delivery, return, check-out
5.	**Experience:**	Emotional, respect, ambiance, fun, intimacy, relationships, community

The value proposition defines the company to the customer. Most value propositions can be described using the five key values. Crawford and Matthews have shown that one value is selected to dominate the value proposition offered to the customer. A second value differentiates the offering, and the remaining three values must meet the industry norm [Crawford and Matthews, 2001]. Consider a performance rating on a 1-to-5 scale where 5 is world-class, 1 is unacceptable, and 3 is industry par. Crawford states that a venture should plan a good product offering to have a value score of 5, 4, 3, 3, 3 for its five value proposition attributes in the following order: dominant, differentiating, norm, norm, norm.

Consider Wal-Mart, where price is the dominant value of its offering. Wal-Mart differentiates itself on product in terms of selection and quality. By contrast, the values offered by Target are dominated by product and differentiated by price. Many firms focus on good service, which is about human interaction. For example, Honda has great service as its dominant value, and its secondary, differentiating value is product.

Access can be described by ease of locating, connecting to, and then navigating the physical or virtual facility of a business. Very good accessibility is offered by Amazon.com, and its website is relatively easy to navigate. Accessibility can also be described as convenience or expedience. For a customer with a high demand for time, convenience is very important. A readily accessible website can be very valuable to a time-starved customer.

Zappos.com sells shoes and other clothing and accessories through its website. The company is known for providing an excellent customer experience. According to Zappos, "Customer service isn't just a department, it is the entire company." As a result, the company has a 75 percent repeat business rate, and enjoys a very good reputation through word-of-mouth referrals.

Apple Computer realized that by opening retail stores, it could make buying its products more of a recreational experience. Apple Stores provide a place to gather casually and learn how to do interesting things with Apple products. Customers can do everything from buying a computer or phone to learning how to record their own music and interacting with other Apple aficionados. As important, these extra services that Apple Stores provide are complimentary to all customers.

Most customers seek a provider of a product or service who saves them time, charges a reasonable price, makes it easy to find exactly what they want, delivers where they ask, pays attention to them, lets them shop when they want to, and

TABLE 3.7 Primary and secondary values for leading firms.

		Primary Value			
		Product	Price	Access	Service	Experience
Secondary Value	**Product**	——	Wal-Mart	Amazon.com	Honda	Harley-Davidson Disney World
	Price	Target	——	Holiday Inn	Wal-Mart	Olive Garden
	Access	Google Barnes & Noble	Priceline Visa	——	Dell Computer	Starbucks
	Service	Toyota Home Depot Intel	Southwest Airlines	McDonald's	——	Carnival Cruise Line
	Experience	Mercedes	Virgin Atlantic Best Buy	AT&T	Nordstrom	——

makes it a pleasurable experience. Any firm that fashions a value proposition to that set of customer values and actually delivers on that promise should do well.

The product value is described by its performance, range of selection, ability to search for it, and quality. Volvo built its business on the idea of product safety. Volvo became the first car company to offer three-point, lap and shoulder seat belts. Home Depot focuses on providing a very wide selection of quality products. The differentiating (secondary) value for Home Depot is its service. The primary and secondary values for selected leading firms are shown in Table 3.7.

One value of product is range of selection or choice. Often to appeal to many different customers, a firm offers many versions of a product. However, too much choice is often debilitating [Schwartz, 2004]. If a firm offers extensive choice, it should help the customer search and select the right version. Amazon and TiVo offer such help to their customers.

Remember, a firm must meet at par the three remaining variables. Consider the plight of today's department stores. Their primary value is product selection. However, they are struggling to be accessible to today's shopper and just be at par on service, price, and experience.

Google's Value Proposition

What are the primary and secondary values for Google? It offers product as its primary value with fast, relevant results for the most ill-described inquiry. Its secondary value is access, which is embodied in the easy online connection right to the search page without annoying pages or advertisements obscuring the search box.

The **value proposition** states who the customer is and describes the values offered to this customer. The value proposition for Amazon.com could be described as:

> An easily accessible Internet site that is convenient all of the time to provide a wide selection of books, CDs, and videos at a fair price to the busy, computer-literate customer.

The value proposition for Starbucks could be described as:

> We provide a friendly, comfortable, well-located place offering a wide range of fresh, customized quality coffees, teas, and other beverages for the person who enjoys a good experience and a good beverage.

Home Depot and Lowe's stores are the two large home-improvement chains in the United States. Home Depot's dominant value is product selection, and its secondary value is service. Lowe's has a dominant value of accessibility and a secondary value of product selection. These value differences lead to clearly separate value propositions offered to the customers of these two competitors.

The **unique selling proposition** (USP) is a short version of a firm's value proposition and is often used as a slogan or summary phrase to explain the key benefits of the firm's offering versus that of a key competitor. For example, Hewlett-Packard uses a USP as follows:

> Excellent technical products with reliable service at a fair price.

The clear, simple USP for FedEx is:

> Positively, absolutely overnight.

USPs are useful for succinctly describing a new venture to would-be investors, customers, or team members. In the jargon of investors, it is often called "the elevator pitch." This is a short description of your venture that can be told during the brief ride on an elevator between getting on and getting off. The USP is widely used in Hollywood for screenwriters to "pitch" their movie idea in a single sentence. For example, the pitch for *Spiderman* is "After a chance encounter with a spider in a chemical lab, a teenage boy realizes that he has super powers that he must use to save the city and win the girl he loves."

New ventures can use their value proposition and unique selling proposition to clarify the business values offered to the customer. This will help all stakeholders understand the purpose of the firm's business concept.

3.4 The Business Model

The design of a business is the means for delivering value to customers and earning a profit from that activity. The **business design** incorporates the selection of customers, its offerings, the tasks it will do itself and those it will outsource, and how it will capture profits. Business design is often called business concept. A successful business design represents a better way than existing alternatives.

TABLE 3.8 Elements of a business model.

■ **Customer selection:**	Who is the customer?
	Is our offering relevant to this customer?
■ **Value proposition:**	What are the unique benefits?
■ **Differentiation and control:**	How do we protect our cash flow and relationships?
	Do we have a sustainable competitive advantage?
■ **Scope of product and activities:**	What is the scope of our product activities?
	What activities do we do, and what do we outsource?
■ **Organizational design:**	What is the organizational architecture of the firm?
■ **Value capture for profit:**	How does the firm capture some of the total value for profit?
	How does the firm protect this profitability?
■ **Value for talent:**	Why will good people choose to work here?
	How will we leverage their talent?

It is like writing a news story, which will be used to attract investors, customers, and team members. In other words, the business design itself is an opportunity to be innovative [Zott and Amit, 2007].

A good business design involves what your firm will and will not do and how the firm will create a sound value proposition. The business design answers three key questions: who is the customer, how are the needs of the customer satisfied, and how are the profits captured and profitability protected. The resulting outcome of the business design process is the **business model,** which is the description of the business and how it will work in economic terms. A business model is a set of planned assumptions about how a firm will create value for all its stakeholders [Magretta, 2002]. A business model is the framework that connects a technology to economic profits.

The business model answers questions about the customer, profit, and value. The elements of a business model are shown in Table 3.8. The business model for Dell Computer is given in Table 3.9 [Slywotzky, 2000]. The first critical element of a business model is the selection of the customer. The business design

TABLE 3.9 Dell Computer business model.

■ Customer selection: High relevance	Four segments: Corporate, government, education, consumer
■ Value proposition: Unique benefits	A customized computer at a good price with great service readily accessible via phone or Internet
■ Differentiation and control: Sustainable competitive advantage	Customized products via a direct sales channel via phone or Internet with strong service and customer relationships
■ Scope of product and activities	Desktops, laptops, and servers Strong supply-chain management
■ Organizational design	Divisional organization for each customer segment
■ Value capture for profit	Opportunities for cross-sell and up-sell. Avoid price as the key value and focus on service and accessibility.
■ Value for talent: Learn, grow, prosper	Training, learning, and career opportunities

aims to specify the customers with unmet or latent needs, which will then define the target market. Dell uses four market segments to describe its customers and then prepares separate offerings for each segment. It is important to choose customers who will permit you to profit and spurn customers who want great value but are difficult or unfairly demanding. Instead of making all customers very happy, focus on the right customer and create an offering that allows good value to the customer and a reasonable profit to your firm. If possible, the price of your offering should include a reasonable profit margin, as well as good value for the customer. Once you know who your selected customer is, start saying no to those who don't fit the model.

The second step is to clearly state a unique value proposition that will provide differentiation for your firm. Show how the value proposition will address the market segment you have identified. For example, Dell sells customized computers at a good price with great service. Dell differentiates itself by relying on a direct sales model via the phone, mail, or the Internet and providing offerings suitable for each separate market segment.

Next, explain the scope of product and activities and organizational design that will enable you to implement the value proposition. A clear path to profitability is critical. You should determine your company's actual and projected revenues and expenses, identifying the key factors that influence total revenues and costs. Then, plot cash flow versus time to determine your financing needs [Hamermesh et al., 2002]. More on this process is detailed in Chapter 17.

Furthermore, it is important that you can retain good profit margins so that you can invest for the future. In general, it is best to avoid competing solely on price and making price the dominant value of the value proposition. Those companies that do make price the dominant value, such as Wal-Mart, Costco, Dollar General, and Family Dollar, are careful to differentiate themselves along other dimensions, too. Dollar General and Family Dollar use smaller stores in well-located strip malls so that accessibility is their differentiating value. Wal-Mart and Costco compete on the product quality and selection as their secondary value. The business model of Wal-Mart is successful because of its use of technology to achieve strong supply-chain management and store inventory control.

Southwest Airlines is another example of a business with price as the dominant value in its value proposition. Its secondary value is service: on-time arrival and departure, online ticket ordering, and a customer-friendly attitude. It captures profit from the valued service by controlling costs. It uses one type of aircraft, which keeps its costs of maintenance and training lower than its competitors'. It also heavily promotes the online sales of tickets. As a result, Southwest has been profitable every year since 1973 [Freiberg and Freiberg, 1997]. The business model of Southwest Airlines is compared with the business model of American Airlines in Table 3.10.

Customers influence changes in sound business models as their priorities change. Many business models fit a context that eventually evolves and necessitates changes in the models. The obsolescence of an outmoded

TABLE 3.10 Business model of two airlines.

	American Airlines	Southwest Airlines
Customer	Traveler who needs to fly many places throughout the world	Traveler who desires to fly routes served point to point in the U.S.
Value proposition: Dominant value Differentiating value	Product Accessibility	Price Service
Differentiation	Wide scope of product: goes almost anywhere	Limited point-to-point flights Easy maintenance and training for low cost
Scope of products and activities	Very broad: connects everywhere	Narrow: only flies to selected cities (point to point)
Organizational design and implementation	Hub-and-spoke High fixed cost	Point-to-point Lower, flexible costs Control costs
Value capture for profit	Dominate hub city Requires high occupancy	Requires high occupancy
Value for talent	High pilot salaries Good career	Participation in stock options and camaraderie

business model and the necessity for a redesign of the business model is called value migration [Slywotzky, 1996]. For example, the appropriate business model for Hewlett-Packard Corporation has changed significantly since the company's founding in 1938. In recent years, many of the firm's manufacturing activities have been outsourced as the company migrated toward a computer company competing on the primary value of product and the secondary value of price.

The business design process is summarized in Figure 3.4. The dynamic firm continuously tests for changing conditions and redesigns its value proposition to meet the values of its customers.

Almost every aspiring entrepreneur assumes that his or her first business plan (plan A) will be successful. What should an entrepreneur do when it falters or investors are not attracted? Determining what projected customers will actually pay for is difficult. A few focus groups and surveys is a start, but will likely come up short of what is needed. Good entrepreneurial teams try plan A on customers, but then are ready to adjust and move to plan B. They test hypotheses against reality and then are ready to adjust plans.

One solid approach is to use a dashboard to track key information for the team and key investors. The results (numbers) displayed on the dashboard give key indicators, which can be shown to the rest of your team and investors. The key elements of your business are described in the business model as shown in Table 3.8. In most cases, investors will want quantitative evidence of the soundness of this business model [Mullins and Komisar, 2009].

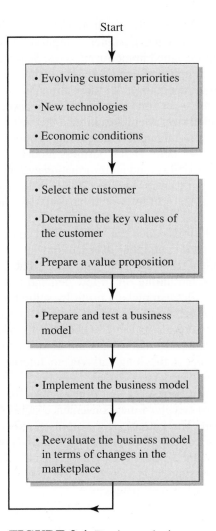

Start

- Evolving customer priorities

- New technologies

- Economic conditions

- Select the customer

- Determine the key values of the customer

- Prepare a value proposition

- Prepare and test a business model

- Implement the business model

- Reevaluate the business model in terms of changes in the marketplace

FIGURE 3.4 Business design process.

3.5 Business Model Innovation in Challenging Markets

The collapse of stability in the marketplace challenges any business team to keep its place in the list of successful companies. Business model innovation is the capacity to reconceive existing business models in new ways that create new value for customers [Hamel, 2000]. For example, Hewlett-Packard changed its model to become America's premier brand for computer printers. The key to Hewlett-Packard's business model is the consistency of product and service. Ryanair has taken the lead for profit and return on investment in the European airline industry. IKEA has designed a high-volume business model for selling affordable, well-designed home furnishings. Searching for a new job in another state can be accomplished using Monster.com, which lists hundreds of thousands of jobs across the country. All these firms have effectively reconceived their business models over time in response to market change.

While competitors will always attempt to imitate the best practices of a market leader, a unique, difficult-to-imitate business model is often based on a unique competency or technology—or both. Dell's business model is based on a direct sales, customized product capability, and an information system that enables Dell to manage its supply chain efficiently. As a result, Dell has an average inventory of four days, while a typical competitor would have 40 days' inventory.

Markets are dynamic, and companies and nations respond slowly. An economic model for many industries called the "long tail" is emerging that is based on providing access to a massive number of selections to suit a wide range of customer tastes. Many industries show a demand curve that shows high appeal for the most popular items, but tailing off to low demand for the others. Companies like Amazon and Netflix can make a profit by selling low-demand products to people who access their website. More than 50 percent of Amazon's book sales come from titles outside its top 130,000. Netflix offers over 100,000 movie titles, while the average Blockbuster store can offer only 1,000. New ventures can potentially access the long tail via an Internet site and attract sales for less-popular items [Anderson, 2006].

Amazon.com started as a bookseller, but it now looks like the Wal-Mart of the Internet. Gas stations have evolved into convenience stores selling beverages, food, newspapers, and fuel. The entrepreneur within an existing company can help to build new value for customers and new profit for the company by reconfiguring the firm's business model before its competitors do theirs. One powerful way to find a new business model is to look for the customers' latent, unstated dissatisfactions with existing business practices.

Electric Cars in Israel

Better Place is an excellent example of an interesting business model. Founded in Israel in 2007 by Shai Agassi, Better Place seeks to create a complete infrastructure for electric cars in all of Israel. Israel is a country that has a particular interest in ending its dependence on foreign oil. It is also a relatively small country covering about 8,000 square miles. These market conditions have lent themselves to Better Place developing an innovative business model.

One of the major problems of electric cars is the long refueling time and the lack of infrastructure to support it. Better Place seeks to solve this

(continued on next page)

(continued from page 64)

problem by creating an electric car infrastructure from the ground up. The company plans to import its own electric cars and to provide the expensive batteries on a lease basis. The customer would pay a monthly fee not unlike a typical cell phone bill. Any time the electric car needs refueling, the customer can go to a Better Place station and have the battery replaced with a fully charged one without additional cost. Better Place is an example of a company with an unconventional business model that is uniquely well-suited to its market.

3.6 Core Competencies

The **core competencies** of a firm are its unique skills and capabilities. A capability is the capacity of the firm, or a team within the firm, to perform some task or activity. Firms with core competencies that match those necessary to effectively implement their business model have the best chance to succeed. It is very important that the core competencies of your firm match the requirements of your business. The core competency of Honda is the ability to design and build internal combustion engines of all sizes. The core competency of Intel is the ability to design and manufacture integrated circuits for computers and communication systems.

We care about competencies since they are the roots of competitive advantage. The real sources of advantage are found in the competencies of a firm. Core competencies include the collective learning in the organization, the skills of its people, and its capabilities to coordinate and integrate know-how and proprietary knowledge. Unlike physical assets, core competencies do not deteriorate as they are applied and shared. They can grow as a firm learns to build its competencies. Physical assets wear out, but intellectual assets such as core competencies can improve over time.

The core competencies of 3M are in designing and manufacturing materials, coatings, and adhesives, and devising various ways of combining them for new, valuable products. Honda's core competencies in engines and power trains have enabled it to provide distinctive products for lawnmowers, motorcycles, automobiles, and electric generators. Core competencies provide potential access to a wide variety of markets. Core competencies are the wellspring of new business ventures.

A successful firm's core competencies are valuable, unique capabilities that enable the firm to implement its business model and thus deliver a valuable product or service to its customers. These unique capabilities will be rare, difficult to imitate, and difficult to substitute.

Core competencies are dynamic by nature and an integral part of organizational learning and competence building. These distinctive capabilities are those activities that a firm does better than its competitors. These competencies are the critical asset of a technology venture.

The core competency of Google is the design and operation of massively scaled Web services. It is the dominant online search engine. After starting as a search tool for finding information on diverse subjects, it has also become the leader in the Internet advertising industry.

3.7 Sustainable Competitive Advantage

The **competitive advantage** of a firm is its distinctive factors that give it a superior or favorable position in relation to its competitors. Competitive advantage is measured relative to a firm's competitors. A **sustainable competitive advantage** is a competitive advantage that can be maintained over a period of time—hopefully, measured in years. The duration, D, of a competitive advantage, CA, leads to the estimate of the market value, MV, of a firm as

$$MV = CA \times D \qquad (3.1)$$

That is, the market value of the firm is proportional to the size or magnitude of the competitive advantage and dependent on the expected duration of that advantage. A pharmaceutical firm with a 20-year patent and a strong competitive advantage will be highly valued, indeed!

The competitive advantage of a firm is directly dependent on its core competencies, its assets, and its organization architecture. A firm such as General Electric is said to have a sustainable competitive advantage in the electric power industry. It has higher profit margins than all its competitors in this field.

In general, the more value, V, customers place on a firm's products, the higher price, P, the company can charge for these products. The cost of producing the products is C, and the profit margin is P − C. The company profits as long as P > C. The value created is V − C, and the net value to the customer is V − P. These relationships can be portrayed as shown in Figure 3.5. The profit margin, P − C, is vanishingly small in some very competitive industries where the competitive advantage is small or nonexistent for all the firms.

American Express invented the traveler's check as a means of getting money while traveling abroad. The value to the customer was high. The cost to issue the checks was low, and (V − P) remained high to the customer. All parties have been pleased with this business model for more than 100 years [Magretta, 2002].

Many profitable firms are built on differentiation: offering customers something they value that competitors don't have. This unique offering can be in the product, service, or sales, delivery, or installation of the product. While the basic product may be a commodity, the differentiation can be obtained somewhere in the various interactions or services for the customer. Firms selling personal computers attempt to differentiate themselves by offering high-quality service. Harley-Davidson lends money to people to buy its motorcycles.

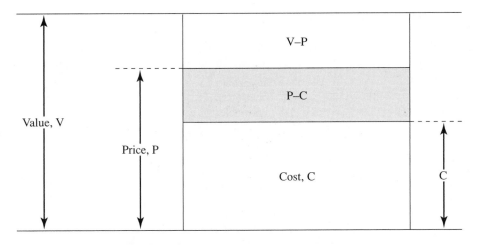

FIGURE 3.5 Value and return to the customer and the firm.

A competitive advantage is a significant difference in a product or service that meets a customer's key buying criteria. The sustainability of a firm's competitive advantage is a function of the competitors' difficulty in imitating or innovating around the incumbent's unique product or service attributes. One hospital service company successfully differentiates saline, a commodity product, by delivering it premeasured and frozen in plastic bags directly to hospital wards, thereby saving hospitals handling costs.

All firms seek to erode competitors' advantages by acting to *imitate* their product or service attributes or innovation.

Competitive advantage can be based on lower costs or differentiation of product or both. Most firms try to improve the efficiency of their operations to lower costs. They also strive to innovate or provide superior quality to outdo their competitors. Another point of differentiation can be in customer relationships. Examples of these approaches are shown in Table 3.11.

A firm can create new value and thus establish sustainable competitive advantage. The pyramid of value creation is shown in Figure 3.6. From a solid

TABLE 3.11 Potential sources of competitive advantage.

Source	Example
Efficiency, low costs	Alcoa
Product innovation	Intel
Quality, reliability	Mercedes
Customer responsiveness	Dell
Manufacturing innovation	Toyota

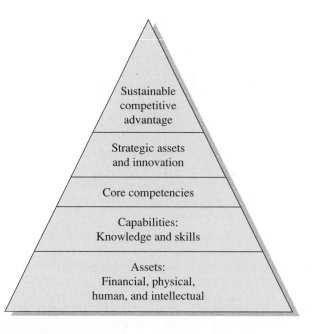

FIGURE 3.6 Pyramid of value creation.

base of assets, a firm builds its capabilities, which lead to its core competencies. With its core competencies and knowledge, it develops new products, processes, and other activities to build a competitive advantage. The sustainability of a firm's competitive advantage depends on its ability to continually innovate.

The duration, D, of a competitive advantage is longer when it is difficult to imitate. This difficulty is present when unique skills and assets are required and hard for a competitor to replicate or obtain.

In 1876, Sir Joseph Lister, a longtime advocate of improving sanitation conditions, was invited to speak at a medical conference in Philadelphia. In attendance was Robert Wood Johnson, who became inspired by Lister's speech. In 1886, Johnson, joined by two of his brothers, started Johnson and Johnson in order to manufacture a line of sterile surgical dressings. After experiencing some success, the company diversified into other segments of the medical industry. The company now manufactures pharmaceuticals and medical devices for physicians and consumers alike. Their products, including Tylenol, Band-Aids, and Listerine, are well-known and trusted in the United States. The core competency of Johnson and Johnson is the ability to select and market trusted, useful products. Their consistency in doing so allowed the company to retain a sustainable competitive advantage over the last century.

Selling the vision of the sustainable venture requires a passionate commitment to the venture. Candy Lightner started a nonprofit organization called

Mothers Against Drunk Driving (MADD) after her 13-year-old daughter was killed by a drunk driver in a hit-and-run accident. She communicated the vision of the organization with a passion born of loss and injustice.

The most powerful new venture provides a great sense of value at a reasonable price resulting in a high ratio of value to price for the customer. With an added sense of emotion or importance, the potential success of a venture can be seen as the ratio:

$$\text{Potential success} = \frac{\text{Value} + \text{Emotion}}{\text{Price}}$$

Clearly, organizations such as Doctors Without Borders and MADD incorporate the powerful emotion of a cause or importance.

A business model is the result of a firm's decision about how a business should be structured. The securities brokerage industry, in the past, operated on a theory of high commission fees and personal service. In the 1990s, the model changed to low commission fees and reduced personal service. Few business models are unchallenged.

Core competencies built by a firm that have the potential to generate value can be a source of competitive advantage in the marketplace. For airlines, one important capability is providing a memorably pleasant experience to passengers during flight. In the software services business, a dominant capability lies in the combination of high quality and low cost. Having capabilities that are distinctive and difficult for others to imitate can give your firm sustainable competitive advantage.

Ten types of sustainable competitive advantage are given in Table 3.12.

TABLE 3.12   Ten types of sustainable competitive advantage.

Type	Example
■ High quality	Hewlett-Packard
■ Customer service	Dell
■ Low-cost production or operation	Wal-Mart
■ Product design and functionality	Google
■ Market segmentation	Facebook
■ Product-line breadth	Amazon.com
■ Product innovation	Medtronic
■ Effective sales methods	Pfizer
■ Product selection	Oracle
■ Intellectual property	Genentech

3.8 AgraQuest

Like the hungry microbes in its natural products, AgraQuest is eating its way into the $28 billion global pesticide market—a field dominated by chemical giants Dow, DuPont, and Monsanto. Steering the biotech start-up into the fray is company president and CEO Pam Marrone, an international expert in agricultural biotechnology and biopesticide science. Marrone has led AgraQuest in its vision to research and develop safe and environmentally friendly alternatives for farm, home, and public health pest management.

Natural products are nonliving substances such as proteins and enzymes that are produced by organisms such as microbes and plants. Every day, scientists at AgraQuest make their rounds, carefully checking and rechecking various biological experiments and meticulously recording their findings. They check out new samples of soils, plant roots, or lichen arriving from across the globe, hoping that one will lead to the next breakthrough natural, environmentally safe pesticide or fungicide.

The vision statement for AgraQuest is given in Table 3.13. The vision of this company is to make a difference in the worldwide agricultural industry by providing pesticides and herbicides that do not cause environmental problems.

The mission statement, provided in Table 3.14, is clearly stated and motivational in character. The vision and mission statements are useful to provide information about the company to employees, investors, and other stakeholders.

The value proposition of AgraQuest must clearly state the key values for the firm, while also identifying those values that will match the competition. The five core values for the firm are provided in Table 3.15. The dominant value is to ensure that the efficacy of the product is as good as that for chemicals

TABLE 3.13 AgraQuest's vision statement.

Agriculture badly needs safer, biodegradable pesticides that fit well in pest management systems in order to create an environmentally sustainable agricultural system. The goal is to reduce the use of synthetic chemicals on the nation's farms and ranches. AgraQuest develops its own natural product pesticides that meet its criteria for in-house development and aggressively licenses or acquires natural products from outside the company to reduce the time line until market entry. AgraQuest discovers new pesticidal natural products from microorganisms and sells these natural compounds to agrochemical companies for non-core markets. AgraQuest plans to build a natural pesticide and herbicide business that will make a difference in world agricultural practices and environmental impacts. We will be the premier source of pest management knowledge and technology, and be accountable to our customers, our shareholders, our families, our community and ourselves.

TABLE 3.14 AgraQuest's mission statement.

AgraQuest discovers, develops, manufactures, and markets effective, safe, and environmentally friendly natural products for farm, home, and public health pest management.

TABLE 3.15 Five values for AgraQuest's products.

Dominant value: Product—The efficacy of the product is equivalent to that of chemicals, but it also can be used right up to harvest time. Furthermore, natural products are less susceptible to pest resistance buildup. It also is safer, reliable, and easy to use.

Differentiating value: Experience—A "green," natural product that is environmentally friendly and can lead to more sustainable agriculture and healthy conditions worldwide

Expected norm value: Price

Expected norm value: Service

Expected norm value: Access

while providing safer products that can be used right up to harvest time. The differentiating value is that of a "green," natural product than can lead to a sustainable agricultural system. AgraQuest matches its competitors on price, service, and access. The value proposition is provided in Table 3.16.

The unique selling proposition for AgraQuest is:

Innovative natural product solutions for pest management

The business model of AgraQuest is given in Table 3.17. AgraQuest could be well positioned to grow in the global pesticide market. Its challenge is to exploit its differentiation as a natural, "green," and safe product in a somewhat skeptical agricultural industry.

TABLE 3.16 AgraQuest's value proposition.

AgraQuest discovers, develops, manufactures, and markets effective, safe, and environmentally friendly natural products for pest management that serve worldwide agriculture and make it more environmentally sustainable.

TABLE 3.17 Business model for AgraQuest.

■	Customer selection	Farmers with higher-value products who want a safer "green," natural pesticide solution
■	Value proposition	"Green" products for herbicide and pesticide use at comparable efficacy and price
■	Differentiation	Natural products that can be used right up to harvest time
■	Scope of product	A moderate range of natural products
■	Organizational design	A flat organization with good communication
■	Value capture	Reasonable costs and growing revenues allowing for net positive income
■	Value for talent: Scientists and staff	Opportunity to work in an organization with a "green" mission

3.9 Summary

The theory of a business is a description of the elements required for the entrepreneur to act to build a business that satisfies the customers' needs. Coupled with the firm's core competencies and resources, the firm uses the elements of its business design to build a sustainable competitive advantage. The elements of a firm's theory of its business include: vision, mission, value proposition, business model, competitive advantage, and how it acts to retain a sustainable competitive advantage.

■ Great vision is a statement of purpose (or story) in response to an opportunity.
■ The mission describes the firm's goals, products, and customers, providing a theory of change for all to see.
■ The value proposition describes customer needs that will be satisfied.
■ The business model describes the economics and activities of the new enterprise.
■ The firm strives to create a competitive advantage and make it sustainable.

> **Principle 3**
> The vision, mission, value proposition, and business model embodied within the business design of a firm and powered by a sustainable competitive advantage can lead to compelling results.

Video Resources

Visit http://techventures.stanford.edu to view experts discussing content from this chapter.

Don't Write a Mission Statement Write a Mantra	Guy Kawasaki	Garage
Innovate in Technology and Business: The Founding of Google	Larry Page, Eric Schmidt	Google
Beyond Socially Responsible Business	William McDonough	McDonough + Partners
The Product Vision	Pam Marrone	AgraQuest

3.10 Exercises

3.1 How would you define Google's vision? Construct a mission statement for Google. After completing both of these tasks, go to Google's website and compare its actual corporate mission statement to your impression.

3.2 Social networking takes advantage of a compelling trend toward leveraging social connections to link people for viral marketing and affinity marketing. Compare and contrast the value propositions offered by these leading social networking sites: MySpace, Facebook, Linkedin, and Friendster.

3.3 Compare the business models for Yahoo and Google using Table 3.8. Make sure to identify how they are different. How do you see their business models evolving over the next five years?

3.4 Purchasing a used car is one of the least desirable experiences for most people. eBay Motors offers fraud protection, a warranty, and a title history (www.ebaymotors.com). What is the value proposition for eBay Motors? Would you buy a car using eBay?

3.5 Twitter.com continues to see explosive user growth. However, a business model has yet to materialize. Describe three business models Twitter could pursue to become a profitable business.

3.6 The branded and generic pharmaceutical industries have continued to grow rapidly over the past decade. Describe how branded pharmaceutical companies have innovated their business models to address the generic drug market. Generic drug companies have also experienced challenges as the sector has grown globally. How have these companies responded?

3.7 Apple has been successful in expanding its product and service portfolio from computers to MP3 players to mobile phones. What are Apple's business models? Describe the core competencies that have allowed Apple to make the moves from Mac to iPod, and from iPod to iPhone.

3.8 Woot.com is an online seller of mainly closeout products at a cheap price. This site provides low-priced sales and an online community to talk about the product of the day. Visit Woot.com and determine the business model of the firm. How does Woot generate a profit for this service?

VENTURE CHALLENGE

1. Create a brief vision statement for your venture.

2. State the value proposition for the venture.

3. Create a draft business model for the venture using the elements of Table 3.8.

4. What are your venture's core competencies and competitive advantage?

CHAPTER 4

Competitive Strategy

Praise competitors. Learn from them. There are times when you can cooperate with them to their advantage and to yours.
George Mathew Adams

How can a venture create a strategy to fit the new business opportunity?

Every new venture has a strategy or approach to achieve its goals. This strategy is in response to its plan to implement a solution to an important problem or opportunity. The process for creating a strategy for a new firm is shown in Table 4.1. Steps 1 and 2 were described in Chapter 3. With sound vision and mission statements and an initial business model, the entrepreneur examines the political and economic context of the industry, along with its growth rate and typical profit margins (step 3). Once the industry is understood, steps 4 and 5 are used to describe the firm's strengths and weaknesses and its opportunities and threats (SWOT). In step 6, the entrepreneur integrates his or her knowledge of the industry and competitors with his or her own SWOT to identify key success factors. Based on the information gathered in the preceding steps, the entrepreneur refines his or her vision, mission, and business model and creates a strategy to achieve a sustainable competitive advantage. The formation of cooperative alliances with other enterprises can be an important way for a new venture to position itself within an industry. Long-term success depends upon addressing the needs of all stakeholders and acting in a responsible manner. ■

TABLE 4.1 Management process for developing a strategy.

1. Develop the vision and mission statements, and the business model.

2. Describe the firm's core competencies, its customers, and its competitive advantage.

3. Describe the industry and context for the firm and its competitors.

4. Determine the firm's strengths and weaknesses in the context of the industry and environment.

5. Describe the opportunities and threats for the venture.

6. Identify the key factors for success using the six forces model.

7. Formulate strategic options and select the appropriate strategy.

8. Translate the strategy into action plans with suitable measures and controls.

4.1 Venture Strategy

A **strategy** is a plan or road map of the actions that a firm or organization will take to achieve its mission and goals, but it is not static. Imagine the difficulties of navigating through most towns with a map from 1900. In other words, a strategy is a firm's theory about how to compete successfully within the current realities of its industry. To be useful, the plan must be action-oriented and based on the firm's opportunities, strengths, and competencies. For example, the most efficient route for a cyclist to move from point A to point B may be different for a motorist. A corporate or organizational strategy is an integrated plan for the whole organization [Hill and Jones, 2001]. It is a firm's way of doing things and a theory of business [Drucker, 1995]. The desired outcome of a strategy is a sustainable competitive performance. Because of the dynamic nature of the competitive business world, a strategy has to be simple and clear. This allows everyone to work on a commonly understood plan.

Strategies help to set a firm on a course and then focus their efforts on it. Often, a strategy emerges as actions are taken and tested, eventually converging toward a pattern [Mintzberg et al., 1998]. With a strategy, the firm can differentiate its offerings and activities. For some, the essence of strategy is choosing what not to do [Magretta, 2002]. The process for developing a strategy is summarized in Table 4.1.

A strategy is a response to opportunity. The word *opportunity* is derived from the Latin expression "toward the port." The builder of value is like a merchant sea captain who secures the right payloads from the best customers, manages his crew, and adjusts his mix of established ports and new ports with high potential [McGrath et al., 2001]. The formulation of a sound strategy is based on deep knowledge of the opportunity, the industry, and its context. In describing the opportunity as a vision, a sense of drama and vitality emerges. With this vitality, the entrepreneur motivates the team and the investors to share the

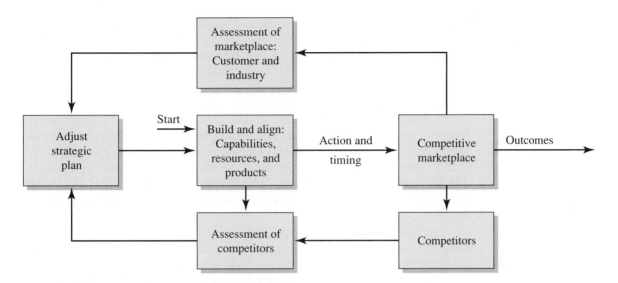

FIGURE 4.1 Framework for a firm operating in a dynamic marketplace.

vision, embrace the strategy, and act on it. In this case, the strategy emerges as the details unfold.

Long-term planning is very difficult due to the dynamic nature of the competitive marketplace. Industries are not in equilibrium, and industry analysis is difficult. It is hard to define where an industry begins and ends. Also, it is difficult to distinguish competitors from collaborators from suppliers. Thus, all strategies are subject to change and reemergence as conditions, alliances, and competition change.

Entrepreneurs start in the center of Figure 4.1 by building and aligning their capabilities, resources, and products. They then act on their initial strategy or business plan. Entry into the competitive marketplace will force a reassessment of the marketplace and industry as well as their competitor analysis. This leads strategic managers to redeploy and adjust the capabilities, resources, products, and actions to effectively compete in the dynamic market. These managers strive to attain a competitive advantage by securing and managing the assets of the firm. How the internal management responds to a changing customer, industry, and competition is crucial in the reestablishment of the strategic plan and the firm's assets to act competitively. Venture leaders strive to identify the fundamental forces for creating and capturing customer value. Those who focus on continuously adjusting and aligning a firm's strategy and capabilities will constantly evolve from one strategic position to the next strategic position in response to changing conditions.

GE Aircraft Engines (GEAE) provides an example of an adjustment in a strategic plan as a result of changes in the market. GEAE had a product strategy to develop engines with more power, efficiency, and better reliability.

Because of relentless competition and shorter product cycles, sustaining profitability was difficult. GEAE shifted to operating as an engine production and services provider, generating significant profits in the after-market services business [Demos et al., 2002]. Faced with a dynamic marketplace, the strategic leader develops a strategic response and adapts to the changes in the market.

To summarize Figure 4.1, the first step is to determine the basic driving forces in the industry: the economic, demographic, technological, or competitive factors that either constitute threats or create opportunities. The second step is to formulate a strategy that addresses the driving forces identified in step 1. The third step is to create a plan to implement the new strategy. Finally, the new strategy is implemented by building and realigning the firm's capabilities, resources, and products.

Entrepreneurs define their strategy within their perception of opportunity. They are not constrained by the present resources or capabilities but seek to acquire the necessary resources and capabilities. The theory of resource dependence states that a company's freedom of action is limited to satisfying the needs of customers and investors that give it the resources to survive [Christensen, 1999]. Investors and customers dictate how money will be spent because companies that do not satisfy them will be unable to survive.

A good strategy answers the questions asked by Kipling [1902]:
I keep six honest serving-men (They taught me all I knew);
Their names are What and Why and When and How and Where and Who.

The six questions for creating a sound, dynamic strategy are summarized in Figure 4.2. With solid, effective answers to these six questions, a firm will have formed a strategy that has the potential to lead to profitability.

Profitability		
Why are we pursuing this objective? • Vision • Mission	Where will we be active? • Customer • Market	How will we achieve our objective? • Innovation • Acquisitions
When will we act and at what speed? • Timing • Execution	What will differentiate our product? • Positioning • Competitor response	With whom will we compete and cooperate? • Competition • Alliances

FIGURE 4.2 The six questions for creating a dynamic strategy. Profitability rests on six solid answers to these questions.

A strategy can be viewed as a plan that integrates a firm's goals and actions into a cohesive whole that draws effectively on its resources and capabilities. The essence of strategy is choosing the priorities and deciding what to do and what not to do. The strategic priorities determine how a business is positioned relative to the alternatives. As the competitive conditions change, the new venture adjusts its strategy to meet the new conditions.

The development of a strategy often uses reasoning by analogy [Gavetti and Rivkin, 2005]. For example, Staples began by asking: "Could we be the Toys-R-Us of office supplies?" Analogical reasoning makes efficient use of information, but can be built on superficial similarities and inaccurate information. It is necessary to understand the source of the analogy and check the similarities.

4.2 The Industry and Context for a Firm

The eight steps for developing a strategic plan are outlined in Table 4.1. In the remaining sections of the chapter, we will discuss steps 3 through 8 since steps 1 and 2 were described in Chapter 3. In this section, we address step 3 of Table 4.1. Also, multiple methods exist for understanding the activities of Figure 4.1. We will highlight some of them in this and later sections.

A full description of the customer and the industry will help the entrepreneur build a sound strategic plan. The main elements of an industry analysis are given in Table 4.2. The first step is to accurately name and describe the industry in which the firm is or will be operating. The definition should be narrow and focused. An **industry** is a group of firms producing products that are close substitutes for each other and serve the same customers. Thus, selecting the telecommunications industry may be too broad. The definition of the industry should be more focused, such as "the Internet service provider industry serving homes and businesses in Ohio and Indiana." If data are not available for the targeted area of the market, the closest proxy should be used. For example, if statistics are not available for Ohio and Indiana, they may be available for the Midwest or the United States. Then, define this market and describe the customer. The second step is to describe the regulatory and legal issues within the industry. Both national, as well as state and local regulations, should be considered. Also, changes in regulations can influence both industry-funding trends and particular types of companies within an industry [Sine et al., 2005].

TABLE 4.2 Five elements of an industry analysis.

1. Name and describe the industry.

2. Describe the regulatory, political, and legal issues in this industry.

3. Describe the growth rate of the industry and the state of the evolution of the industry.

4. Describe the profit potential and the typical return on capital in the industry.

5. Describe the competitors in the industry and the rivalry among them.

TABLE 4.3 Four stages of an industry life cycle.

Stage	Examples
1. Emergence	Artificial organs
	Nanotechnology
	Genomics
2. Growth	Medical technology
	Software
	Smart phones
3. Maturation	Electric appliances
	Automobiles
	Personal computers
4. Decline	Steel
	Fax machines
	Car phones

The third step of Table 4.2 suggests describing the growth rate and state of evolution of the industry. Most industries tend to emerge through an initial period of slow growth with limited sales and few competitors. Then, they expand through a period of rapid growth as sales take off and many firms enter the industry. This is followed by a third period of maturation marked by slower growth and stability. Eventually, the number of firms in the industry declines [Low and Abrahamson, 1997]. We depict in Table 4.3 these four stages as (1) emergence, (2) growth, (3) maturation, and (4) decline. It is important to know where your industry is in the evolution cycle. In the emerging phase, significant product and market uncertainty exists. Producers are unsure of what features are required for the product. Customers may be unsure of the elements of the product they need. Many technology ventures begin in the emerging phase of an industry. For a technology venture, an emerging industry will not yet have a dominant design and will respond well to new firms with a wealth of knowledge that can be used to build a powerful new venture [Shane, 2005].

The growth stage emerges when the necessary features and performance become clear and a dominant design emerges. A **dominant design** is one whose major components and core concepts do not substantially vary from one product offering to another. With the emergence of a dominant design, the number of competitors stabilizes.

Eventually, an industry enters its mature phase as the number of competitors stabilizes and profit margins slowly decline as price becomes the primary competitive weapon. Finally, an industry enters a declining phase as the number of firms decline and profit margins erode. These four phases are described in Table 4.3.

The personal computer market began in 1978, with a number of small, emerging firms such as Apple Computer. IBM entered the personal computer market in 1982, and its PC quickly emerged as the dominant design. Many other firms entered after IBM made the design open to all, and the PC industry experienced a growth phase between 1984 and 1998. Eventually, the market reached a period of maturity, with only a few dominant firms having standardized or slightly differentiated products and relatively stable sales and market shares.

Table 4.2 shows that the next step in the industry analysis is a statement of the profit potential and the typical return on investment capital in the industry. The **return on capital** is defined as the ratio of profit to the total invested capital of a firm. The average return on capital in the computer software industry is about 16 percent, while the return on capital in the steel industry is about 6 percent. The steel industry is less attractive, while the computer software industry is attractive. One of the most effective ways to identify realistic profit opportunities for a new venture is to look at the Securities Exchange Commission filings of a young representative firm in the industry (www.sec.gov).

The **six forces model,** shown in Figure 4.3, is one popular method for evaluating the competitive forces in an industry. The six forces are: (1) firm rivalry, (2) threat of entry by new competitors, (3) threat of substitute products, (4) bargaining power of customers, (5) bargaining power of complementors, and (6) bargaining power of suppliers. This framework is an extension of the five forces model [Porter, 1998]. The six forces model enables the analyst to consider all the issues facing a new entrant by describing the key industry factors. The rivalry among the industry competitors may be intense or modest. In some industries, the bargaining power of the customer may be modest.

Consider the automobile industry, which has about 10 competitors. The rivalry is extremely intense. The bargaining power of customers regarding a new vehicle is very high since they have access to broad information on the relative performance and price of the products of the competitive companies and their dealers. The bargaining power of the suppliers in the industry is modest. Furthermore, the threat of a substitute product is small. The threat of new entrants is very small, due to the costs of developing a new product and dealer network. Thus, the auto industry experiences intense competition with the buyer wielding significant power.

Consider the online bookselling industry: Amazon.com and BarnesandNoble.com are the two large online booksellers in the United States, but there are many regional competitors such as Powells.com. The rivalry among these competitors is high. Their suppliers have low bargaining power, and the barriers to entry are moderate. The bargaining power of the customer is large, resulting in low prices, and profitability is modest. The threat of substitute products is low. However, e-books could undermine the printed book industry as more attractive devices emerge such as Amazon's Kindle.

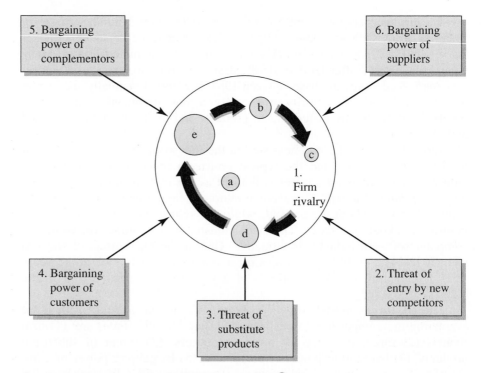

Note: Firms are represented by a circle; for example, ⓐ represents firm a. The size of the circle indicates the size of revenues of the firm. The six forces are numbered for clarity. The rivalry of the firms is shown as a vortex of competition illustrated by the solid arrows.

FIGURE 4.3 Six forces model.

By contrast, many new firms enter the computer software industry each year. The bargaining power of customers is moderate, and the threat of substitute products is low. As a result, profitability in the industry is high. However, the rivalry of the firms is intense.

A competitive analysis explains how you will do better than your rivals. And doing better, by definition, means being different. Organizations achieve superior performance when they are unique, when they do something no other business does in ways that no other business can duplicate. In military competition, strategy refers to the large-scale plan for how the generals intend to fight and win a war. The word *tactics,* in contrast, refers to small-scale operations, such as the conduct of a single battle [Clemons and Santamaria, 2002]. Very few strategic plans survive the first contact with competitors. Competitors respond and change the situation.

Complementors are companies that sell complements to the enterprise's own product offerings. A **complement** is a product that improves or perfects another product. For example, the complementors to Sony's PlayStation 3 and Nintendo's Wii are the companies that produce the video games that run on these consoles. Without an adequate supply of complementary products, demand for the player product would be modest. The complementary product to the automobile is the interstate road system that enables automobiles to safely and rapidly travel long distances. Without suitable, widely located electric recharge stations, the future of electric vehicles is very limited.

The entrepreneurial firm is likely to be a new entrant to the industry. Thus, the new venture should describe the barriers to entry, the threat of substitutes, and the bargaining power of the suppliers, customers, and complementors. One of the main factors that drives traditional analyses of the determinants of market structure involves comparing the size that a firm must be to compete efficiently to the overall size of the market in which it competes. If the industry has few firms, a new firm may be able to readily enter and gain market share. Using the six forces model, a new technology venture is likely to perform better when it operates in an industry with high barriers to entry, low rivalry, low threat of substitutes, low buyer power, low supplier power, and low bargaining power of complementors.

In Figure 4.3, examine the bargaining power of the suppliers. When the supplier industry is composed of many small companies and the buyers are few and large, the buyers tend to dominate the supply companies. An example is the automotive component supply industry in which the buyers are few and large and dominate the many small suppliers.

To complete the industry analysis, it will be necessary to name the competitors and describe the profitability of the industry. One method is to use *Standard and Poors Reports* or the *Value Line Investment Survey.* For example, if the new firm is entering the biomedical devices industry, the leading competitors are Medtronic and Boston Scientific. Using Value Line, we note that the average return on total invested capital for these companies is 15 percent. Value Line projects a 13 percent future growth rate of sales for this industry. With these attractive measures, the industry appears to be very attractive to new entrants with well-differentiated, fairly priced products.

4.3 Strengths and Opportunities—SWOT Analysis

Steps 4 and 5 of the management process for developing a strategic plan (Table 4.1) suggest that a strategy is based on the firm's strengths and opportunities, while avoiding or mitigating its weaknesses and managing threats. As discussed in chapters 2 and 3, a new firm is focused on securing the capabilities and resources necessary to succeed in its industry. Furthermore, the new firm concentrates on an attractive opportunity that was selected using Table 2.6. Thus, a strategy addresses the four aspects of the setting in which a firm operates: (1) a firm's strengths, (2) its weaknesses, (3) the opportunities, and (4) the threats in its competitive environment. This analysis is often called a SWOT analysis, which allows a firm to match its strengths and weaknesses with opportunities and threats and find the purpose for which it is best suited.

A firm's strengths are its resources and capabilities. Its weaknesses are its limitations of organization or lack of capabilities or resources. A firm's opportunities

TABLE 4.4 SWOT analysis for Amgen.

Organizational (internal)	Environmental (external)
1. Strengths: ■ Expertise in development and manufacturing of biologic drugs (e.g., proteins and antibodies) ■ High-margin products and limited competition	1. Opportunities: ■ Expansion of marketed products for new geographies, indications, and formulations ■ Allocation of resources to discover novel therapeutics to sustain growth
2. Weaknesses: ■ Inability to discover novel therapeutics to avoid declines in revenue	2. Threats: ■ Pharmaceutical companies entering the biologics arena ■ Competition from follow-on biologics and pricing pressures

are its chances for success in a new entry or product in its industry. The threats are actions or events outside its control in the competitive environment.

A basic SWOT analysis for Amgen is given in Table 4.4. The SWOT analysis provides the questions for a strategic response and helps a firm exploit its strengths, avoid or fix its weaknesses, seize its good opportunities, and mitigate its threats. Examples of threats are market shifts, regulatory changes, and delays in product development. Positive opportunities include increasing demand, repeated use, and willingness to pay.

We can examine opportunities in three dimensions, as shown in Figure 4.4. Perhaps the safest strategy is to take new products to existing customers via existing distribution channels using existing approaches. We can call the three

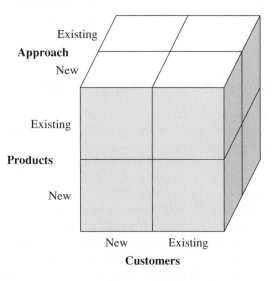

FIGURE 4.4 Three dimensions for examining opportunities.

dimensions: products, customers, and approach [Black and Gregersen, 2002]. Approach is the method or means of taking the product to the customer. The most risky strategy would be a new product taken to new customers via a new approach. Amazon.com started selling books (existing products) to book buyers (existing customers) via a new approach—online.

4.4 Barriers to Entry

Barriers to entry are factors that make it costly for companies to enter an industry. The greater the costs that potential competitors must bear to enter an industry, the greater are the barriers to entry. The six potential barriers to entry are listed in Table 4.5. Economies of scale can be a barrier in industries where the costs of production are low for a narrow range of volume or occur only for higher volumes. An example is the aircraft design and production industry. It is difficult to enter that industry since a low volume of production of aircraft is most likely uneconomic for the new entrant [Barney, 2002].

Cost advantages independent of scale may be held by existing companies and will deter a new company from entering. For example, incumbent firms may have proprietary technology, know-how, favorable geographic locations, and learning-curve advantages. These can all be barriers to a new entrant.

Product differentiation means that incumbent firms possess brand identification and customer loyalty that serve as barriers to new entrants. For example, Dell, Hewlett-Packard, and Apple have brand and customer loyalty, making it difficult for a new personal computer company to enter the industry on a large scale. Of course, this barrier may be less important to a specialty manufacturer that seeks a small niche in the personal computer market. A formidable barrier to entry is the reputation or brand equity of the incumbents. Providing ratings for bonds is an attractive industry since it is not asset-intensive and the profit margins are very good. If a new firm tries to enter this market, it would have to compete with Moody's and Standard and Poors, both competitors with strong reputations.

Contrived deterrence as a barrier occurs when incumbent firms strive to throw up unnatural barriers at a cost to them. They can use lower prices, newer

TABLE 4.5 Potential barriers to entry into an industry.

- Economies of scale
- Cost advantages independent of scale
- Product differentiation
- Contrived deterrence
- Government regulation
- Switching costs

products, or brand building to send a signal to potential entrants that intense responses will result if they try to enter. For example, a potential entrant to television broadcasting is deterred by government allocation of regular broadcast channels. A response to this limitation is for the new entrant to choose another means such as cable as the distribution channel—for example, the Fox Channel.

Two kinds of economic markets exist: substitutable and nonsubstitutable. Substitutable products are commodities such as groceries, cola drinks, and gasoline. In a nonsubstitutable market such as semiconductor manufacturing equipment, the required associated infrastructure means that once purchasers choose a system, they are not inclined to switch due to high switching costs.

Switching costs are the costs to the customer to switch from the product of an incumbent company to the product of the new entrant. When these costs are high, customers can be locked into the product of the incumbents even if new entrants offer a better product. An example is the cost of switching from Microsoft to the Apple computer operating system. Users would need to purchase a new set of software to use on the Apple computer as well as train their employees to use the new software.

Low Barriers to Entry in Web 2.0

Web 2.0 start-ups have been attractive to some entrepreneurs because the market seems relatively easy to enter. To them, it is a growth industry with low barriers to entry. The cost to set up a website is relatively low. Creating a website requires technical and programming knowledge, but modest capital investment. This industry offers an opportunity for entrepreneurs with little financial backing to create a product for a huge market quickly. Website services can easily be made accessible worldwide, without the need for physical distribution channels.

4.5 Achieving a Sustainable Competitive Advantage

Recall from Chapter 3 that a core competency is a matchless strength that a firm can use to achieve superior operating conditions that lead to a strong competitive advantage. A SWOT analysis helps the entrepreneur identify this unique competency. The unique competency of a firm arises from its capabilities and resources, as shown in Figure 4.5. Resources are financial, human, physical, and organizational, and include patents, brand, know-how, plants and equipment, and financial capital. The capabilities of a firm include skills, methods, and process management. It is the usefulness of both capabilities and resources in a coordinated way that leads to distinctive competencies. A firm must have: (1) a valuable set of resources and the capability to exploit those resources, or (2) a unique capability to manage common resources. Intel possesses unique

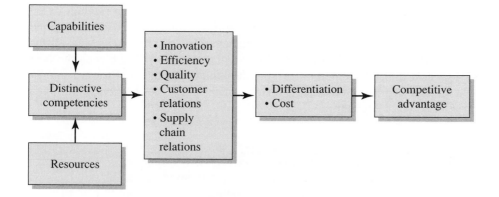

FIGURE 4.5 Distinctive competencies lead to a competitive advantage.

patent and know-how resources and the capabilities to exploit that knowledge and intellectual property. Ryanair and Southwest possess common resources—aircraft and aircraft equipment—but have unique capabilities to manage these resources. Disney has unique resources in its film library, brand, and theme parks but a mixed record of managing them well.

If a new technology venture possesses a particular valuable resource, then that firm can gain a competitive advantage and thus improve its efficiency and effectiveness in ways that competing firms cannot [Barney, 2001].

As shown in Figure 4.5, a firm uses its unique competencies to manage its innovation, efficiency, product quality, customer relations, and supplier relations to differentiate its product and manage its costs. A technology venture works to design and produce at a low cost the highest-quality product that possesses unique differentiating factors. Four common ways in which a firm will distinguish itself from its competitors are differentiation, cost, combined differentiation and cost, and niche, as summarized in Table 4.6.

TABLE 4.6 Four common types of strategies and their characteristics.

| Factor | Type of strategy | | | |
	Differentiation	Low cost	Differentiation-cost	Niche
Distinctive competencies	Innovation and relationships	Processes, logistics	Innovation and processes	Relationships
Product differentiation	High	Low	Medium	Medium
Market segmentation	Many segments	Mass market	Many segments	One or two segments
Examples	Intel	RadioShack	Dell	Getty Images
	Microsoft	Wal-Mart	Southwest Airlines	Incyte

The goal of a differentiation strategy is to create a unique product based on a firm's unique competencies. The low-cost strategy is based on unique competencies that enable the efficient management of processes. Many firms can achieve a combined differentiation–low-cost strategy that blends the best of low cost and differentiation. The niche strategy is directed toward one or two smaller segments of a larger market. This niche can be geographic or a product or price segment.

Intel's Competitive Advantage

Since the founding of Intel, its strategy was focused on technology leadership, first-mover advantage, and the dominance of important new markets. Intel emerged as the dominant supplier of microprocessors, which are used in 90 percent of personal computers. Intel is also a leading manufacturer of flash memory, embedded control chips, and communication chips. A unique competency is Intel's ability to build, manage, and exploit the world's best semiconductor manufacturing facilities. As an example of its technology leadership strategy, Intel announced a new material that will replace silicon, enabling Intel to build more density (transistors per area) while reducing heating and current leakage. For decades, Intel has had a successful differentiation strategy.

Niche ventures often require less capital and achieve financial success rather quickly. Typically, a niche business is too small for the mass-market supplier, and thus, competition is low. A niche can be geographic or a product or price segment. Niche businesses typically are started in one market segment and based on a focused core competency and good customer and supplier relationships.

Southwest Airlines is an example of an airline that started as a niche, low-cost business operating only in Texas. It served three cities—Dallas, Houston, and El Paso—and operated using standardized Boeing 737 aircraft. It used highly productive crews, frequent, reliable departures, and a no-frills (low-cost), short-haul, point-to-point system. Eventually, Southwest moved to other western states and many locations across the nation. Thus, its strategy evolved from a niche strategy to a differentiation-cost strategy.

Fastenal Company of Winona, Minnesota, is the largest distributor of nuts and bolts in the United States. It uses a low-cost strategy for its manufacturing and distribution business with about 2,200 warehouse stores achieving total sales exceeding $2 billion. Each store has at least one delivery truck. Customers talk to the local store and receive personal service. Fastenal sees itself as an inventory and delivery manager offering excellent customer service. It currently has a fleet of 4,100 pickup trucks that respond to customer orders on an expedited basis (www.fastenal.com).

A differentiation strategy is commonly based on an innovation or capability others do not possess. Led by Carlos Perea, Miox is a New Mexico-based venture that produces water purification systems. Traditionally, these systems used volatile and hazardous chlorine gas. Miox developed a technology that allows their products to function using only salt and water (www.miox.com).

Paychex is an example of a company with a differentiation-cost strategy. It provides payroll-processing services and began by targeting small- and medium-sized businesses that needed this service. The company offers customer service and payroll accuracy at a reasonable price, leading to wide acceptance. Once it has a satisfied customer, the switching costs for this customer are sizable. Paychex's revenues have grown to over $2 billion, and it serves more than 540,000 businesses. Paychex's annual revenue growth rate has been greater than 18 percent for over 20 years.

With good hardware and friendly software, Apple's portable player made a profitable business out of digital music—a business that eluded Sony, Microsoft, and Napster. What was the strategy that Apple adopted that led to success? The iPod was introduced in 2001, but it was the iTunes online-music store, introduced in 2003, that caused the iPod to take off. Apple sold more than 40 million iPods in 2008. Furthermore, iTunes sold 400 million songs in the first year and 5 billion songs through 2008. The iPod is easy to use, readily portable, and able to synchronize automatically with iTunes to download songs. Apple used the differentiation-cost strategy of Table 4.6 to achieve rapid success.

IKEA provides furniture to customers who are young, not wealthy, likely to have children, and work for a living. These customers are willing to forgo service to obtain low-cost furniture. IKEA designs its own low-cost, modular, and ready-to-assemble furniture. In large stores, it displays a wide range of products. While IKEA is a low-cost provider, it also offers several differentiated factors, such as extended hours and in-store childcare. Its strategy is a differentiation-cost strategy.

Some firms attempt to create new markets by breaking the existing value/cost trade-off [Kim and Mauborgne, 2005]. For example, it became possible for music, video, and videogames to be downloaded to your cell phone. This disruptive strategy created a new market for direct delivery of media.

4.6 Alliances

Many businesses use competitive strategies to shape their business strategies but often ignore *cooperative* strategies. Business is a complex mix of both competition and cooperation. A new venture possesses valuable novelty and innovation that will attract the attention of suppliers, customers, competitors, and complementors, acting as a value network, as shown in Figure 4.6. All the participants are connected and participate in this network of activity. Consider the value network for a university, shown in Figure 4.7

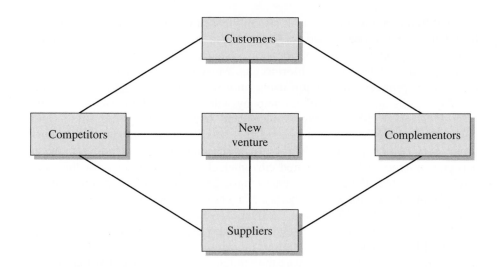

FIGURE 4.6 Value network.

[Brandenburger and Nalebuff, 1997]. The complementors to a university include kindergarten through grade-12 schools, local housing, community activities, and computing systems. All the members of the value network are connected together in the higher education value network. The university, to succeed, must cooperate with its suppliers, customers, competitors, and complementors. Competitors can be seen as rivals but also will be, in many instances, collaborators.

Many technology ventures offer products or services that require distinctive strategies because the products are parts of systems with complements provided by others. If a platform leader emerges and works with complementors, an ecosystem of innovation is formed [Gawer and Cusumano, 2008]. A platform strategy requires a compelling vision and strong leadership. A platform product or technology should provide a core function and be easy to connect for complementors. Examples of platform leaders are Google and Microsoft.

The value network is important to entrepreneurial ventures as they strive to accumulate the resources and capabilities required for success. The value of exploiting complementary resources can be significant [Hitt et al., 2001]. For example, a smaller, new biotechnology firm and a large pharmaceutical firm can both benefit from an alliance. The biotech firm provides new technologies and innovation, while the pharmaceutical firm provides the distribution networks and marketing capabilities to successfully commercialize the new products. The larger established pharmaceutical firm also gains value through access to its partner's innovation. Thus, firms usually search for partners with complementary assets or capabilities. An excellent example of complementary partners is Google and Firefox. The default start page for Firefox is the Google search engine, bringing Web traffic to Google.

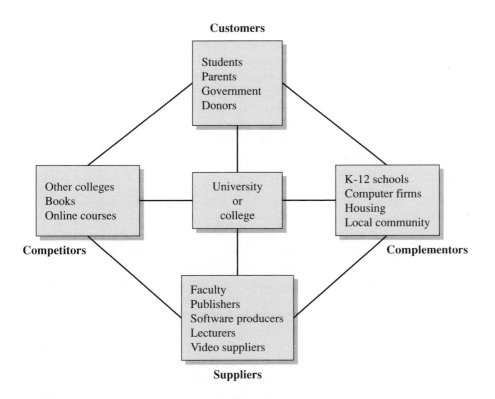

FIGURE 4.7 Value network for a university or college.

Google provides Firefox both monetary compensation and a popular search product for its users. This is a partnership that enhances the value of each participant.

A **partnership** or **alliance** is an association of two or more firms that agree to cooperate with one another to achieve mutually compatible goals that would be difficult for each to accomplish alone [Spekman and Isabella, 2000]. Proactive firms take the initiative rather than react to events. Proactive formation of strategic alliances is an important dimension of entrepreneurial activity that enables a new firm to acquire access to unowned but required strategic assets. All alliances are based on some exchange of knowledge in addition to a flow of products, capital, or technology. Alliances function best when mutual benefits and commitment are clear to all parties [Lee et al., 2005].

The configuration of alliances of a start-up impacts its early performance. External alliances can substitute for internal resources. A firm's decision to enter an alliance can be motivated by a desire to exploit an existing capability or technology, or to explore for new opportunities and new technologies [Rothaermel and Deeds, 2004].

The new firm should consider developing an alliance when it lacks the necessary assets that a complementor can provide. To select a partner, it must be clear which missing capabilities or resources are required. Then, it must determine which firms possess those assets and look at their characteristics. It will be necessary to build a relationship of trust with the potential partner and craft an agreement that will yield benefits for both partners [Doz and Hamel, 1998]. For example, the alliance that Alibaba.com and Yahoo! made in 2005 allows Yahoo! access to the Chinese market while Alibaba gains significant capital and expertise in scaling from Yahoo!.

Alliances have a variety of structures and are usually governed by a contract that delineates the roles and responsibilities of each partner. Complementor firms may also be potential competitors. Many a well-conceived alliance has fallen apart due to the tension between cooperative and competitive forces. These can be culture clashes, poor conflict management, and lack of effective coordination mechanisms. Furthermore, the entrepreneurial firm may be seeking access to needed assets but may, as a result, be exposed to the risk of losing its own vital internal knowledge. An example of this occurred during the development of the Apple Macintosh. Apple partnered with Microsoft to develop spreadsheet, database, and graphical applications for the Mac. As a result, Microsoft acquired critical knowledge about Apple's graphical user interface products, which eventually enabled its engineers to develop the Windows operating system [Norman, 2001]. Knowledge transfer occurs in conversation and association and is difficult to control. Starbucks wants to share its expertise with its partners. When it places its coffee shops in Barnes and Noble stores, it makes sure the bookstore employees at its counters are well versed in the Starbucks way of doing things.

Cell Phones and Gaming

As mobile phones became more and more popular, a significant opportunity arose for the development of software for these devices. In particular, significant potential was seen for sales of games for mobile phones. Jamdat developed a number of these games, including the popular Bejeweled. In order to distribute these games, it had to form an alliance with mobile phone service providers like Verizon Wireless and Cingular (now AT&T). The service providers controlled the only distribution channel for these games. Because of the partnerships it was able to establish, the company also served as a link between other game developers and service providers. Acting as a gaming publisher in this way is now one of the company's main functions. Jamdat was purchased by Electronic Arts in 2006 and is now called EA Mobile.

The benefits of alliances can be significant. Both firms learn and acquire new capabilities. Furthermore, they have access to complementary resources that they cannot easily duplicate. An entrepreneurial new venture wisely will consider the

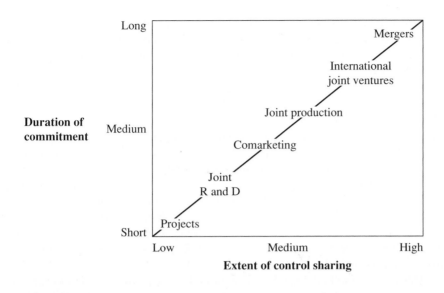

FIGURE 4.8 Range of alliances dependent on commitment and control sharing.

development of one or two partnerships consistent with its strategic goals. Going it alone can be a major liability for entrepreneurs. Innovators who get together in alliances can be more successful, especially where uncertainty prevails. Few start-up firms will have all the necessary capabilities and resources, and alliance networks can enable them to move forward effectively. The type of alliance can range from a joint short-term project to a merger, as shown in Figure 4.8.

Although a portfolio of alliances can be powerful, alliances also place significant demands on an organization's management capability. Different types of alliances also require different types of management. Thus, too many alliances can actually harm an organization's performance and it is critical to examine each potential alliance for both the benefits it brings and the time, resources, and attention it will require [Rothaermel and Deeds, 2006]. Table 4.7 outlines five simple rules for effectively managing alliances [Hughes and Weiss, 2007].

TABLE 4.7 Five simple rules for making alliances work.

1. **Develop the right working relationship** by specifying how you will work together
2. **Peg metrics to alliance progress**, not just progress toward alliance goals
3. **Leverage differences**, rather than trying to eliminate them
4. **Encourage collaboration** by moving beyond formal structures
5. **Manage internal stakeholders** to ensure that all involved players are committed to the success of the alliance

Source: Hughes and Weiss, 2007.

In particular, it is necessary to build real working relationships and establish effective collaborative behavior [Hughes and Weiss, 2007]. Failed alliances are usually due to a breakdown in trust and communication. True collaboration rests on a real relationship, not a contract.

4.7 Matching Tactics to Markets

A company can be said to be successful if it outperforms its competitors over time. Another view of how to formulate the best strategy for a venture is to match the firm's approach to the pace of the market. Table 4.8 summarizes three competitive approaches [Eisenhardt and Sull, 2001]. The first approach is based on establishing a *position* in an industry and defending it. The goal is to position the company so that its capabilities provide the best defense against the competitive forces of Figure 4.3 [Porter, 1998]. Furthermore, the positioning approach can be defended by anticipating shifts in the six forces of Figure 4.3 and responding to them.

The second method focuses on *resources,* such as patents and brand, and attempts to leverage those resources against the resources of the competitors. For example, the powerful brand of Southwest Airlines has enabled the firm to issue its own Visa card to many of its customers.

The third approach may be called *emergent* and is based on flexible and simple rules [Eisenhardt and Sull, 2001]. Firms using this method to develop a strategy select a few significant strategic processes and build simple rules to guide them through the ever-changing marketplace. The strategic processes could be innovation, alliances, or customer relationships. Dell, for example, has

TABLE 4.8 Three types of competitive tactics.

	Position	**Resources**	**Emergent**
Approach	Establish a position and defend it	Leverage resources such as brands, patents, or assets	Pursue emerging opportunities
Firm's basic question	Where should we be?	What should we be?	How should we take our next step?
Basic steps	Identify an attractive market Locate a defensible position, and fortify and defend it	Acquire unique, valuable resources	Choose one or two core strategic processes and use them to guide to the next step
Works best in	Slowly changing, well-understood markets	Moderately changing, well-understood market	Rapidly changing, uncharted markets
Duration of competitive advantage	Relatively long (3–6 years)	Relatively long (3–6 years)	Short period (1–3 years)
Risk or difficulty	Difficult to change position	Difficult to build new resources, if needed	Difficult to choose best opportunities
Performance goal	Profitability	Long-term dominance	Growth and profitability

chosen its customer relationships and customized products as its basic strategy. It then adjusts this strategy as conditions require.

Cisco Systems used an innovation strategy to guide it through emergent opportunities for its first years of operation. Later, it changed to a basic strategy of acquisitions to respond to rapidly changing markets. These basic tenets for guidance in emerging markets may be called simple rules and are summarized in Table 4.9 [Eisenhardt and Sull, 2001]. These rules allow a firm to compete in a fast-moving marketplace such as the emergent markets that many technology ventures start in.

A good way to understand strategic planning in emerging industries is to imagine an American football team trailing by a touchdown with only two minutes left to play, and it has the ball. The team refuses to panic. It has well-established rules of play for this situation. It switches to the "no-huddle" offense, with the quarterback calling the plays at the line of scrimmage as he surveys the defense.

Uncertainty is endemic in strategy formulation. Thus, the quality of a strategy cannot be fully assessed until it is tried. Strategy making can be thought of as an organizational capability, where different approaches are generated and considered, and where past successful approaches are just options for the future among many.

Sam Walton started with a strategy based on low-cost retail discount stores. He gained differentiation by locating many of these stores in relatively rural cities that were only large enough to support one large discount retail operation. His second differentiating factor was his organizational culture, which inspired his employees. As competition emerged, he developed one of the most cost-efficient distribution networks based on information technology systems. Walton's simple rules of strategy and operation were part of Wal-Mart's success.

In addition to matching the approach to the pace of the market, Table 4.10 highlights two key factors for determining a successful strategy: (1) specific industry-related competence, and (2) the existing level of competitive rivalry in the industry [Shepherd et al., 2000]. The venture capitalists who participated in Shepherd's study stated in summary: The most attractive strategy is led by a team that

TABLE 4.9 Simple rules for emergent markets.

Rules	Purpose	Example
Boundary	State which opportunities can be pursued	Cisco acquisition rule: No more than 75 employees in an acquired company
Priority	Rank the possible opportunities	Expected return on investment
Timing	Synchronize the selection of opportunities and the conditions of the firm	When product must be delivered
Exit	Know when to pull out of opportunities	Key team member leaves

TABLE 4.10 Factors for determining a successful strategy, in priority order.

1. Industry-related competencies: Distinctive competencies

2. Competitive rivalry: Low rivalry in the industry

3. Time of entry: Enters industry early at appropriate time

4. Educational capability: Able to obtain the skills, knowledge, and resources required to overcome market ignorance

5. Lead time: Significant time between the pioneer's entry and the appearance of the first follower

has strong competence in an industry that has not yet built up intense rivalries. The timing of entry may be favorable in these circumstances.

The success of a new venture arises, in part, from a fit between the distinctive competencies of the venture team and the major success factor requirements of the industry. The better the fit, the greater is the competitive advantage. A competitive advantage is sustainable if the competencies of the venture quickly track and match the changing requirements of the industry. In addition, it is important to build alliances with critical stakeholders, such as suppliers or distributors, thus erecting barriers to new entrants.

Every business strategy is unique since it is a unique mix of resources, context, goals, competencies, and organizational values. The potential for differentiation of a firm's strategy can also occur along a selected part of the consumption sequence shown in Figure 4.9 [McGrath et al., 2001]. Unique

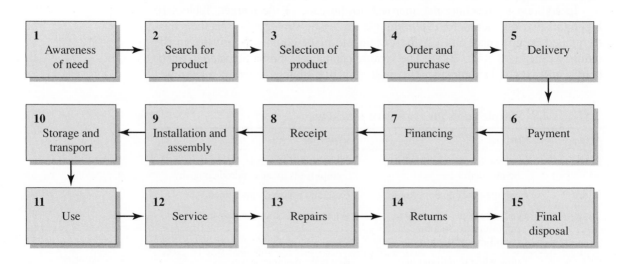

FIGURE 4.9 Consumption sequence.

methods, tools, or arrangements can be used at each step in the sequence. Every new technology venture should look at the consumption sequence and decide where it can differentiate its product or service.

The power of Dell Computer is its direct sales model offered to three different customer segments. The Dell direct sales model incorporates all 15 steps of the consumption sequence. On the other hand, CDW (www.cdw.com) acts as a middleman reseller for Hewlett-Packard and offers excellent customer service for the purchaser who needs help in choosing a computer. Its large sales force helps customers choose a total system that fits them, and a single salesperson is assigned to each customer for follow-up and later purchases. The CDW sales model incorporates steps 3 through 10 of the consumption sequence.

4.8 The Socially Responsible Firm

Any strategy adopted by a new venture firm inevitably affects the welfare of its stakeholders: customers, suppliers, stockholders, and the community. While a specific strategy may enhance the welfare of some stakeholders, it may harm others. The leaders of new ventures are challenged to build a strategy that attempts to meet the economic and social needs of stakeholders while protecting the social and environmental needs of its region. An explicit statement of a new firm's strategy for acting responsibly and ethically may be an appropriate part of a business plan [McCoy, 2007].

The quality of life on our planet depends on three factors, as illustrated in Figure 4.10. The quality of life in a society depends on equity of liberty, opportunity, and health, and the maintenance of community and households, which can be called **social capital,** or social assets. The growth of the economy and the standard of living are critical needs for all people; we call this **economic capital,** or economic assets. Finally, the environmental quality of a region or the world can be called **natural capital.** The interrelationship between these three factors adds up to the total quality of life. Quality of life includes such basic necessities as clothing, shelter, food, water, and safe sewage disposal. Beyond that, quality of life includes access to opportunity, liberty, and reasonable material and cultural well-being [Dorf, 2001].

Business, government, and environmental leaders need to build up capabilities for measuring and integrating these three factors and using them for decision making. We define the sum of these factors as the **triple bottom line.**

As they strive to treat nature and society respectfully while enhancing people's quality of life, corporations need to use nature only for what is necessary and in balance with what can be recycled and replenished.

Recognizing the interconnectedness and interdependence of all living things, corporate leaders can seek a balance using the triple bottom line concept. Economics, ecology, and society can be portrayed as a whole that depends on the person, the corporation, cultural values, and the community. Decisions

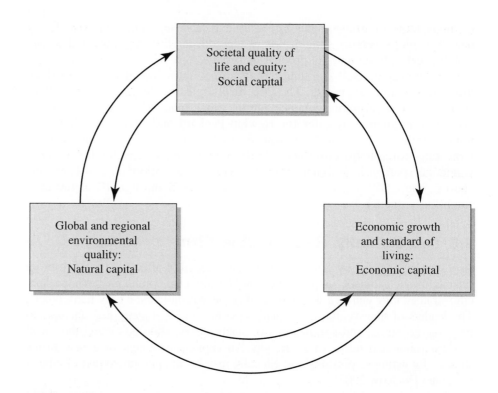

FIGURE 4.10 Three interrelated factors that determine the quality of life on our planet.

made by corporations or society need to account for all the three factors of the triple bottom line.

For many, there is a presumption that a company exists to enhance the welfare of society at large. For others, the only goal is the maximization of profits. We assert that the public welfare can be in the best interest of the corporation itself. One of the purposes of a firm is to make a profit—but service of society is also an implied expectation. In many ways, socially responsible behavior—remembering its obligations to its employees, its communities, and the environment, even as it pursues profits for shareholders—is in a firm's self-interest. A growing number of companies make corporate responsibility part of their value proposition. For example, Henry Ford believed he should pay his workers enough to afford to buy the cars they produced. His decision ultimately benefited Ford Motor Company by making it an attractive employer and stimulating demand for its products.

Some of the best companies in history have tended to pursue a mixture of objectives, of which making money is just one—and not necessarily the primary one. For Merck, a top priority is patient welfare. For Boeing, it is advancement of aviation technology. Profitability is a necessary condition for existence, but it is not the end in itself for many visionary companies. Consider Johnson & Johnson, whose credo, published in the early 1940s, was the basis for its response to the 1982 Tylenol crisis, when a cyanide tampering incident caused the deaths of seven people in the Chicago area. The company quickly removed all Tylenol capsules from the entire U.S. market at a cost of $100 million, though the deaths occurred only in Chicago.

BioFuelBox and GreenFuel: Fuel from Waste Products

BioFuelBox and GreenFuel Technologies are start-ups that have developed innovative processes for producing biofuels from waste products, avoiding the negative consequences of fossil fuels. GreenFuel produces a photosynthetic bioreactor that is fed exhaust from industrial reactors. The exhaust is converted into nutrients, which allow the algae in the bioreactors to multiply. The algae are harvested to yield a substance called algae oil. This oil is then used in the production of a variety of materials including plastics, ethanol, and biodiesel.

BioFuelBox's technology uses undesirable products like waste trap grease and wastewater sludge to produce biodiesel fuel from a small plant colocated next to this waste. Its technology turns what is currently trash into a highly valuable commodity. This approach of converting waste to profit not only is more environmentally friendly, but also has great economic benefits. Since many companies currently pay to get rid of their trap grease or process their exhaust, BioFuelBox and GreenFuel can get their source materials at low to no cost. New ventures such as GreenFuel and BioFuelBox often find that it is possible to strive for both profitability and social responsibility.

The social virtue matrix of Figure 4.11 illustrates the four possible responses to social responsibility challenges. The response of the lower-left quadrant (box 3) is conduct that corporations engage in by choice, in accordance with norms and customs. The lower-right quadrant (box 4) represents compliance—responsible conduct mandated by law or regulation [Martin, 2002]. These two lower quadrants represent the basic commitment of companies to society's values and laws. Actions in the two lower quadrants (boxes 3 and 4) of Figure 4.11 generate little credit since the public expects actions to be in compliance with its laws and norms. The most significant impediment to the growth of corporate virtue is limited vision for actions beyond compliance and allegiance to society's norms.

The two upper quadrants encompass activities that have high social virtue. The strategic benefits quadrant (box 1) includes activities that may add to shareholder value by generating positive reactions from customers, employees,

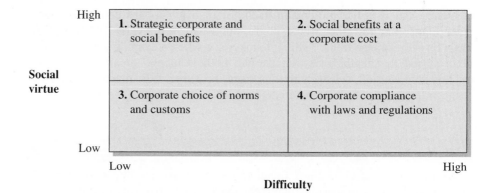

FIGURE 4.11 Social virtue matrix.

or legal authorities. These actions may ultimately benefit the firm by accruing customer goodwill and community support. The upper-right quadrant (box 2) encompasses activities that clearly benefit society or the environment, but at a cost to the corporation.

An example of a firm active in the upper-left quadrant (box 1) is Patagonia, founded by Yvon Chouinard in 1979 as a designer, marketer, and distributor of high-performance outdoor wear with a commitment to protect the natural environment. The firm sought the use of low-impact fibers and drifted to organic cotton by 2000. Patagonia considered three criteria during the design and development of a project: its quality, its impact on the environment, and its aesthetics. In support of their commitment to sustainability of the environment, the firm featured a speaker series of prominent environmentalists. Furthermore, it often shared information with other clothing firms regarding the use of environmentally favorable fabrics.

One great opportunity to enter business in the top-left quadrant (by offering strategic corporate and social benefits) is to stimulate commerce at the bottom of the economic pyramid. For example, an entrepreneur can help the world's poor by partnering with them to innovate new products and services that are valuable and profitable [Prahalad, 2005]. While individual incomes may be low, the aggregate buying power of poor communities is actually quite large, representing a substantial market in many countries. In these markets, entrepreneurs need to reconsider their focus on high gross margins and shift toward securing good returns on invested capital while delivering social and environmental benefits [Prahalad and Hammond, 2002].

Environmental challenges represent another opportunity to enter business in the top-left quadrant. Businesses that combat environmental degradation can be both profitable and socially beneficial [Dean and McMullen, 2007]. New technologies and business models can help build a sustainable world incorporating clean energy, drought-resistant crops, sound fish farming, biodiversity,

TABLE 4.11 Sampling of highly ranked, socially responsible companies.

■ Apple	■ General Electric	■ Procter & Gamble
■ BMW	■ Google	■ Steelcase
■ Cisco	■ Johnson and Johnson	■ Toyota
■ FedEx	■ Microsoft	■ UPS
■ Hewlett-Packard	■ Nordstrom	

Source: Fortune's Most Admired Companies, 2008.

and much more. Non-fossil fuel sources of energy such as wind, solar, hydro, geothermal, and biofuels will be developed over the next 10 years as we shift to lower-impact fuel systems. New technology ventures will emerge as entrepreneurs find new means for big opportunities [Sachs, 2008].

Actions in the upper-right quadrant (box 2) may ultimately engender benefit for shareholders. However, actions that provide benefits to society at a cost to a firm are difficult to defend to shareholders. For example, if only one automaker had decided to add air bags, it would lose some profits. When such an addition is mandated, all automakers can provide added social benefits at a competitive cost. Corporate coalitions, in which firms agree to provide benefits despite the costs, can also help firms take action in the upper-right quadrant.

The public wants information about a company's record on social and environmental responsibility to help decide which companies to buy from, invest in, and work for. As an example, see Starbucks website at http://www.starbucks.com/aboutus/gr.asp. Starbucks estimates that it saved about $36 million because the company's socially responsible actions increased employee loyalty and reduced turnover. The Mexican cement company Cemex helps low-income families construct concrete homes, helping to tackle housing problems while connecting the company with a large and untapped market [Austin et al., 2007]. Good deeds can redound to a company's credit. But, they can also backfire if the company fails to live up to the good-neighbor image it tries to project. Fifteen highly ranked, socially responsible companies are listed in Table 4.11.

4.9 AgraQuest

AgraQuest has a business model, as given in Table 3.17. The basis of AgraQuest's strategy is differentiation of its product. Its natural products have no environmental impacts, they can be used right up to harvest, and pests do not build up resistance to them as they do to chemicals. Thus, the differentiation is the efficacy of the product compared to chemical pesticides and herbicides.

The industry that AgraQuest participates in is the agricultural pesticide and herbicide industry. The goal of using pesticides is to increase the yield per acre of the crop. The industry is heavily regulated by the U.S. Environmental Protection Agency (EPA), as well as state agencies such as the California EPA. The global pesticide market (2001) is about $28 billion. The largest portion (26 percent) of the market consists of fruits, nuts, and vegetables. AgraQuest's target markets are grapes, tomatoes, peppers, bananas, lettuce, apples, cherries, and home gardens.

AgraQuest's biologically based products fight plant pests and diseases with as much success as synthetic chemical pesticides and compete favorably on cost, pest resistance, shelf life, ease of use, food and worker safety, and environmental impact.

The first useful microorganism was found by one of the AgraQuest scientists in a handful of dirt from a Fresno farmyard. Lab tests showed that the bacterium had an appetite for the fungus that causes bunch rot and mildew in grapes. At that point, AgraQuest went to work on finding a formula to grow the beneficial bug in industrial quantities.

Although the exact formula is proprietary and secret, this is how AgraQuest goes about creating it. A flask of bacteria is dumped into a 10,000-gallon tank filled with a special food source. Forty-eight hours later, the gooey slime in the tank is harvested. To create a usable product, the bacterial concentrate is dried so that it becomes something resembling powdered milk. The powder is put into 24-pound bags and shipped to the farmer, who dumps it into a spray tank, mixes it with water, and applies it just like a chemical fertilizer.

AgraQuest's competitors include many firms worldwide such as Valent Biosciences, Chicago; Dow Agrasciences, Indianapolis; BASF, Germany; Syngenta, Switzerland; and Bayer, Germany.

The barriers to entry are significant since an entrant must have the technical capabilities as well as recognition and reputation in the natural pesticide industry. The biggest strength of AgraQuest is its scientific capability to identify, develop, and manufacture microorganisms for agricultural pesticide control. This capability is the strength of AgraQuest and critical to the firm's success in the industry. The weakness of AgraQuest is its limited ability to build a large product line of products for various crops in a timely way.

The differentiated product strategy based on a strong scientific capability is sound, but its weakness is in the delay of creating new products. The economics of the natural pesticide market requires a product line in place that creates a positive cash flow.

4.10 Summary

The strategy of a new business venture is its plan to act to achieve its goals. Given the challenge of an important problem (opportunity), the strategy provides a road map for the new firm to act to achieve a profitable solution to the problem. The strategy is designed to solve the problem by creating a unique and sustainable way of acting that, it is hoped, will lead to a profitable and valuable outcome for the customer and the firm. A solid strategy is based on:

- Sound knowledge of the industry and the context for the venture.
- A deep understanding of the firm's strengths and weaknesses as well as its opportunities and threats.
- A solid competitor analysis and review of the six forces encountered by firms in a rival market.
- A strategic design that can lead to a sustainable competitive advantage.
- A choice of a differentiation, low cost, differentiation and low cost, or niche strategy that provides unique value to the customer.
- Formation of productive alliances with others and always acting in a socially responsible manner.

Principle 4

A clear road map or strategy for a new venture states how it will act to achieve its goals and attain a sustainable competitive advantage in a socially responsible manner.

Video Resources

Visit http://techventures.stanford.edu to view experts discussing content from this chapter.

Problem-Solving Paradigm	Vinod Khosla	Khosla Ventures
Alliances and Collaboration	Dr. Don Francis	Vaxgen
Honing in on the High Points of Clean Tech	Erik Straser	MDV
Social Responsibility from the Ground Up	Mitch Kapor	Foxmarks
Surviving Competition	Jeff Housenbold	Shutterfly
Understanding Your Customer	Pam Marrone	AgraQuest

4.11 Exercises

4.1   Zipcar offers a sophisticated form of car sharing (www.zipcar.com). The firm opened for business in Boston in late 2000. Describe the strategy of Zipcar using the six questions of Figure 4.2. Is the Zipcar strategy sustainable, and will it lead to profitability?

4.2   Podcasting, blogging, online photo sharing, online video, and twittering are five technologies that are enabling a much broader set of content publishers and content consumers. Describe the nature of these industries and analyze the competitive situation using all six forces in Figure 4.3.

4.3   Cypress Semiconductor is an integrated circuit chip company in a very competitive industry. Identify the firm's core industry and key customers. Complete a SWOT analysis for the firm following Table 4.4.

4.4   Nektar is an innovative drug delivery company creating differentiated products to allow for the inhalation of a number of medicines. Examine Nektar's website and publicly available information. Describe Nektar's strategy using Tables 4.6 and 4.8.

4.5   During the 1990s, DVD players became widely available and the rental DVD market took off. NetFlix (www.netflix.com) initiated an online DVD rental service creating a new market. Examine the Netflix website and determine the firm's basic strategy. What are the challenges to its strategy? Consider the timing of the initiation of NetFlix: was it too early or right on time? How have Blockbuster and Wal-Mart attempted to differentiate their online services from NetFlix?

4.6   eBay has modeled worldofgood.com after early green marketplaces, positioning its activities in an environmentally friendly niche. Visit the website, describe its social mission, and describe how this fits into eBay's broader corporate mission.

4.7   With the release of the iPhone 3G and OS V2.0 in 2008, Apple created a new ecosystem or market for mobile software developers to compete and succeed in. Many iPhone developers made loud proclamations of early success. By 2009, over 30,000 separate iPhone Apps became available for iPhone users, presenting unique challenges to new iPhone developers. Using the competitive concepts and frameworks in this chapter, describe (a) the industry and context for a new firm entering into the iPhone App market, (b) how a sustainable competitive advantage could be built given this market's very low entry costs, and (c) competitive tactics that could match well to the unique nature of this market.

4.8   Identify a technology company that incorporated more than 100 years ago. Describe the industry and context for the firm today. Describe a significant industry and context shift for the firm in its history. Has the firm maintained a sustainable, competitive advantage in the markets it competes in? If so, how?

4.9   Many online search competitors are moving to compete in the mobile local search market. Providing location tailored information to mobile phones is expected to be a large opportunity for both wireless carriers and local advertisers. Select one of these mobile local search companies and create a value network for this company (e.g., Figure 4.6).

VENTURE CHALLENGE

1. Develop a SWOT analysis using the format of Table 4.4.

2. Select your strategic approach from Table 4.6.

3. Create a partnership strategy as described in Section 4.6.

4. Describe your strategy in one or two sentences that could be circulated to your employees and allies.

5. Why and how will your venture be socially responsible?

CHAPTER 5

Innovation Strategies

There's a better way to do it. Find it!
Thomas Edison

How can an entrepreneur build an effective strategy based on innovation that will lead to a sound technology venture?

Many people believe that those who are quick to act will win the race while the slow and deliberate will trail behind. The decision to be the first mover needs to be addressed by all entrepreneurs. Using an idealized model of window of opportunity, the entrepreneur can decide when to act. The entrepreneur needs to maintain a sense of urgency but avoid being too early or too late to market. Entrepreneurs also seek to build an innovation strategy that involves new technologies, ideas, and creativity, which lead to invention and ultimately commercialization. An innovation strategy is part of most new firms' road map to success. A firm that encourages creativity and inventiveness can create the ingredients of sustained innovation. ■

5.1 First Movers Versus Followers

Many entrepreneurs believe that the quick survive while the slow struggle. The firm that leads the way with a new product or into a new market expects to lock in a competitive advantage that ensures superior profits over the long run. In this section, we consider the circumstances in which a pioneer may benefit from being a first mover. A **first-mover advantage** is the gain that a firm attains when it is first to market a new product or enter a new market.

We will describe the industries that a new venture enters as mature, growing, or emergent, as noted in Table 5.1. **Emergent industries** are newly created or newly recreated industries formed by product, customer, or context changes [Barney, 2002]. **Mature industries** have slow revenue growth, high stability, and intense competitiveness. **Growing industries** exhibit moderate revenue growth and have moderate stability and uncertainty. New technology ventures often start in uncertain, emergent industries.

The pioneering, first-mover firm has to bear the costs of promoting and establishing a product, including the potentially high costs of educating customers and suppliers. Furthermore, due to the high uncertainty of emergent markets, it is subject to potential mistakes in product, strategy, and execution. The follower firm can learn from the pioneer's mistakes and exploit the market potential created by the pioneer. Some firms successfully exploit a **follower strategy.**

Early entrants (second or third movers) into an emergent industry can also benefit from the additional time to develop, commercialize, and exploit new products if they possess the resources to wait for the opportunity to materialize [Agarwal et al., 2002]. Many examples exist of new start-ups that arrived early but didn't stay long. Pets.com, Helio, and Amp'd Mobile all burned through their investment capital before attracting enough customers to sustain a business. For most start-ups, it is more like a marathon, where how fast you get out of the starting block is irrelevant.

In many cases, pioneer entrants tend to make a large and lasting impression on customers, obtaining strong brand recognition, and buyers often face high switching costs in moving their business to a later entrant. The simplest reason in favor of a first-mover strategy is the ease of recalling the first brand name

TABLE 5.1 Three types of industries and their characteristics.

	Type of industry		
Characteristics	**Mature**	**Growing**	**Emergent**
Revenue growth	Slow	Moderate	Potentially fast
Stability	High	Moderate	Low
Uncertainty	Low	Moderate	High
Industry rules	Fixed	Fluid	Unestablished
Competitiveness	High	Moderate	Low or none

in a category. However, one study found that pioneers gained significant sales advantages but incurred large cost disadvantages relative to a fast follower entrant [Boulding and Christen, 2001]. The return on investment for pioneers was less than that for followers.

Of course, many conditions exist in which a first-mover advantage may be clear and compelling. Consider a mature industry such as restaurants or grocery stores. The attainment of a strategic resource such as a superior location may warrant acting as a first mover in a geographic market segment. Starbucks, for example, wants a store on the busiest corner in a city and acts when it finds an available site. First movers with the right set of competencies and organizational practices can reap the returns from being in the right place at the right time.

If a market is insufficiently ordered or unstable, the first entry may be too early. A market is said to be stable if the requirements necessary for success will not change substantially during the period of industry development. Amazon.com entered the online bookstore market and created intellectual property and the standard for this market. However, it incurred high development costs and had its advantages challenged by a later entry, BarnesandNoble.com. Nevertheless, Amazon.com became the leader in the race by continuous innovation.

Pioneers are often said to gain a low-cost advantage from having a head start down the experience curve, which describes improvements in productivity as workers gain experience. Often these lower costs are an advantage over later entrants [Shepherd and Shanley, 1998]. New technology ventures often act as pioneers in a new or emerging industry to gain brand, cost, and switching cost advantages. The potential advantages and disadvantages of first-mover action are summarized in Table 5.2.

When both technological innovation and consumer acceptance advance rapidly, first movers may be left behind [Suarez and Lanzolla, 2005]. However, first movers may gain advantage in an evolving market if they involve customers and

TABLE 5.2 First-mover potential advantages and disadvantages.

Possible advantages	Possible disadvantages
■ Create the standard and the rules	■ Short-lived advantages disappear with competition
■ Low-cost position	■ Higher development costs
■ Create and protect intellectual property	■ Established firms circumvent or violate patents and intellectual property rights
■ Tie up strategic resources	■ Cost of attaining the resources
■ Increase switching costs for the producer	■ High uncertainty of designing the right product. If vision is wrong, then costs to switch are large
■ Increase switching costs for the customer	■ Customer is reluctant to buy when a large cost to switch may be incurred

suppliers in the innovation process [Langerak and Haltink, 2005]. New technology ventures can exploit their nimbleness and competencies to build a competitive advantage. Amazon.com built a large business in a new market (e-commerce) in spite of the existence of large retailers such as Wal-Mart and Target.

Numerous examples exist of later entrants overtaking first movers and eventually bypassing them in profitability. Superior performance comes from distinctive competencies combined with an appropriate strategy leading to a competitive advantage (see Figure 4.5). Unfortunately, the first mover can develop a strategy based on uncertain or inaccurate assumptions about the six forces (see Figure 4.3). A follower who learns from the first mover's mistakes can move quickly to catch up or pass the first mover. The first mover also suffers from uncertainty about the customer, the organizational capabilities needed, and the industry context.

However, pioneering ventures can use their lead time to build relationships among suppliers, customers, and even competitors. These relationships can build trust and brand that a follower may not easily reproduce. A first-mover advantage can usually be attained under conditions of low market and internal firm uncertainty. Regrettably, most new ventures encounter large measures of uncertainty and must weigh carefully when to enter the market [Kessler and Bierly, 2002]. Entrepreneurs should emphasize speed to market in predictable markets. In an uncertain market, the new venture can probe or test the market by trying product tests, focus groups, and other means of market probes.

A commitment is an action taken in the present that binds an organization to a future course of action. A decision to act as a first mover is usually binding and should not be taken lightly. Preemptive actions can deter potential rivals from entering but may also result in heavy, irreversible investments [Sull, 2005].

The entrepreneur considers entering a market during an estimated period of opportunity often called a window of opportunity. The first mover envisions a greater cash flow as a result of early entry, as shown in Figure 5.1. Uncertainty about the period of opportunity can erode the actual results. If the first mover misestimates the timing of the window, a less attractive cash flow curve will result.

An entrepreneur's objective is to decide when to stop searching for additional information and enter the new market so as to maximize the expected profit. With insufficient information, a firm can enter too early and incur a large cost. However, if it takes too long to gather sufficient information, the firm may lose the first-mover advantage. Entrepreneurs should stop searching for information and enter a market when they estimate that the marginal benefit of additional knowledge is less than the payoff of entry [Lévesque and Shepherd, 2004].

Being a first mover means recognizing what direction existing technologies and industries are heading before competitors do. Today most personal computers (PCs) are built to support intense multimedia applications. In the early 1980s, most people did not recognize the role that gaming would play in the development of the PC. Chong-Moon Lee founded Diamond Multimedia Systems in 1982 to supply PC products such as color graphic and acceleration add-on boards. They significantly enhanced consumers' ability to play high-quality games on their computers. Eventually, Diamond's first-mover strategy

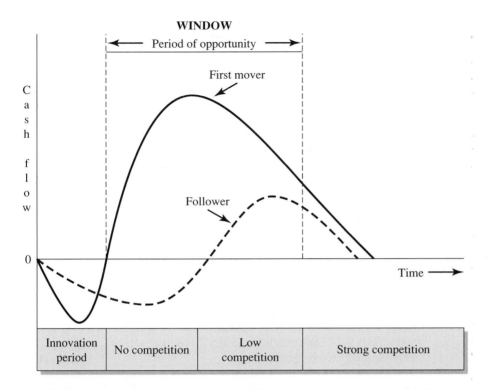

WINDOW

← Period of opportunity →

FIGURE 5.1 Expected first mover advantage and the concept of a window of opportunity.

and superior products captured the attention of IBM and Tandy (then the number 1 and number 2 PC makers in the world, respectively). As these companies increased the standard multimedia features on their PCs, they continued to buy Diamond products to provide a better customer experience. In 1995, Diamond went public and raised $126 million by selling 30 percent of the company. Moon had successfully identified the direction that personal computing would be going. His first-mover position and perseverance over time allowed him to leave a lasting impression on both the computer and gaming markets.

History is replete with companies that were first movers that did not succeed. The CPM operating system preceded Apple, which preceded DOS, which eventually became the early dominant operating system for the PC. Safety razors were introduced a decade before Gillette introduced its successful safety razor. The product must have the right mix of attributes and features, and must be understood as well as demanded by the customer. Early movers don't always have all the requisite characteristics. Prodigy was the first commercial e-mail system, but it received poor acceptance. The second entrant, CompuServe, was equally unsuccessful. Only later did AOL and MSN put together the right mix of attributes to succeed.

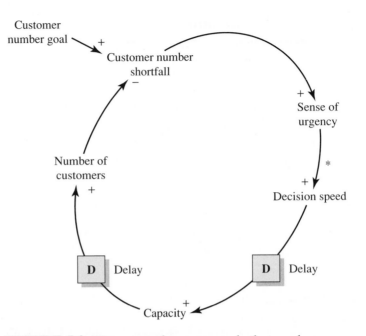

FIGURE 5.2 The sense-of-urgency cycle that can be experienced by new enterprises.

Many new ventures set a fast pace as they and their competitors enter a window of opportunity. Many start-ups exhibit a torrid pace due to a high sense of urgency, as illustrated by the causal diagram in Figure 5.2. A causal diagram can help protray causal links in a system. Variables are related by causal links, shown by arrows. For example, the link denoted with an ∗ implies that if "sense of urgency" increases, then "decision speed" increases. As the firm experiences a sense of urgency due to a shortfall of customers, it acts to build capacity to design, build, and sell its products. However, inevitable delays, D, slow down the buildup of capacity. As capacity increases, the firm expects customers to buy, but again it may experience delay as customers consider the purchase carefully. A slowdown in the growth of customer buildup results in a sales shortfall and an increasing sense of urgency [Perlow et al., 2002]. One way to decrease this unfortunate urgency cycle is to reduce the delay in capacity building and the time delay to customer purchase.

An encouraging case of good timing and entry into a marketplace is that of Google. Google entered the Internet search engine market in 1998, well after other search engines were firmly established. Larry Page and Sergey Brin, the founders of Google, met in 1995 as Ph.D. candidates at Stanford University. Over the next 18 months they collaborated to build a new search engine that ranked search query page results based not only on keywords, but also on popularity. Popularity was measured, in part, by the number of sites linked to each

Web page. The vision of the firm was "to organize the world's information." They allowed limited, less-intrusive advertising on the their site.

During late 1998, they wrote a business plan and raised $1 million in funding from family, friends, and angel investors. Working out of a garage in 1998, Google was answering 10,000 queries a day. In late 1999, it was answering three million queries each day. Google received $25 million in venture capital funding. In August 2004, Google sold about 20 million shares in an initial public offering at $85 per share. By 2009, the company had a market capitalization of $150 billion. Google's competitive advantages include its search technologies and its technical competencies. There is always a place for a new entry in a rapidly growing marketplace with a great technological innovation.

An example of identifying a window of opportunity both in a geographic market and for a technology is the founding of Baidu by Robin Li and Eric Xu. After spending several years working in the search industry, Li recognized that there was a need for a Chinese language Internet search engine in China. After Baidu was founded in 1999, the next four years were spent developing the best technology for China. With over a billion Chinese citizens, Li recognized that it was important to develop a search technology that would best serve everyone in China. In 2004, once Li felt Baidu had created the best search engine for the Chinese market, the firm shifted its focus to increasing brand awareness in China. 2005 saw an increased focus on revenue generation. Baidu's commitment to both search technology and the Chinese market helped it to become the second largest independent search engine in the world.

Machiavelli wrote in *The Prince* (XVII): "The prince ought to be slow to believe and to act, nor should he himself show fear, but proceed in a temperate manner with prudence and humanity, so that too much confidence may not make him incautious." An entrepreneur will be temperate and patient to move. On the other hand, an entrepreneur has a propensity to act. If a window of opportunity appears to be in the distant future, the entrepreneur may be wise to abandon the distant opportunity and seek one that is available and active now. If a window is about to "open", action may be prudent.

Silicon Valley Bank (SVB): Founding at Optimal Time

In the early 1980s, the deregulation of the banking industry led to a need for new innovative banks. In the same period, the Bank of America, which served the San Francisco area, was discontinuing its lending to high-tech companies. At that time, one of the founders of SVB had a series of meetings with bankers who were interested in participating in this opportunity. These factors converged, and the lead founder acted on his intuition, all of which led to the formation of Silicon Valley Bank in 1983 (see www.svb.com). Since its founding, SVB has played an important role in the early days of such successful ventures as Cisco Systems, Electronic Arts, Intuit, JDS Uniphase, KLA Tencor, and Veritas. The launch of Silicon Valley Bank is a sound example of good timing.

5.2 Imitation

Imitation is said to be the greatest form of flattery. Many important new ventures have been based on the replication or modification of an existing business that the entrepreneur encounters through previous employment or by chance [Bhide, 2000]. Entrepreneurs start a business because they believe they can manage the business as well or better than the example they are copying. Sam Walton opened his first Wal-Mart in Rogers, Arkansas, after making numerous trips to study discount retailers in other regions of the United States. Walton once said: "Most everything I've done, I've copied from someone else." Technologists attend trade shows and conferences and often notice competitors' new products that their firm may readily produce.

Unfortunately, most attempts to replicate excellent businesses fail [Szulanski and Winter, 2002]. The difficulty of imitation springs from the lack of deep understanding of the excellent business example. Furthermore, the transfer of the best business practices from one setting to another can be fraught with unforeseen uncertainty. Imitation by independent entrepreneurs can be difficult because when they look at the existing business, they cannot fully understand what makes it work. Thus, the best approach is to copy it in detail but recognize that quick response to customer comments will be necessary.

In 1986, Howard Schultz started his first independent effort as Il Giornale, a store modeled on his experience of Italian espresso bars. He played Italian opera in the Seattle store, and servers wore bowties. Il Giornale was set up as a stand-up bar, as is common in Italy, and it did not offer nonfat milk. Schultz had transferred the Italian coffee bar to Seattle with mixed success. People wanted chairs, and servers did not want ties. Nonfat milk quickly found its way onto the menu [Schultz, 1997].

Close copying may be the best method of imitation. It is important, however, to recognize the value of management and leadership, which is difficult to clone. A talented leader of an excellent business possesses some skills and capabilities that may be difficult to readily understand or copy.

Once the new business is up and running, customer comments can be used to adjust the business procedures to local conditions. Schultz loved the experience of Italian coffee bars but eventually adjusted his coffee café to fit Seattle's desires. Schultz was successful in adapting his Il Giornale store and ultimately created a successful system. He opened a second Seattle store after six months and a third store in Vancouver in 1987. By August 1987, Schultz arranged for his investor group to purchase all the Starbucks stores and its coffee roasting facility. He then merged his Il Giornale and Starbucks under the Starbucks name. By 2009, Starbucks had about 9,000 stores worldwide and revenues of over $9 billion.

JetBlue Airways is a good imitation of Southwest Airlines. Based on a low-cost, all-coach, point-to-point business model, JetBlue started in February 2000 with two aircraft serving New York City and Fort Lauderdale, Florida. JetBlue's initial public offering in 2002 raised almost $150 million for expansion. JetBlue was an excellent example of sound imitation.

5.3 Creativity and Invention

Creativity leads to invention and thus to innovation. **Creativity** is the ability to use the imagination to develop new ideas, new things, or new solutions. Creative ideas flow to invention, and invention flows to innovation. Creative thinking is a core competency of most new ventures, and entrepreneurs strive to have creative people on their team. Creative ideas often arise when creative people look at established solutions, practices, or products and think of something new or different. The successful company generates cost-effective surprises [Schrage, 2001]. This firm is committed to making innovation the underlying focus of its business.

The creative enterprise is based on six resources, as shown in Table 5.3 [Sternberg et al., 1997]. To create something new, one needs knowledge of the field and of the domain of knowledge required. Domains are areas such as science, engineering, or marketing. Fields within a domain might be circuit design or market research. Wise, knowledgeable, creative people avoid being blinded or limited by their knowledge.

The intellectual ability required is the ability to see linkages between things, redefine problems, and envision and analyze possible practical solutions. Creative people use intuitive thinking that reflects in novel ways on a problem. A creative thinker is motivated to make something happen. Creative people are open to taking reasonable risks and acting when advised otherwise. Finally, the creative person understands the context of the problem and is willing to take a risk and advocate change. The person who has most of these skills is often called *intuitive;* that is, he or she has an instinctive ability to perceive or learn relationships, ideas, and solutions.

The intuitive person suspends critical and conventional thinking long enough to consider the possibility of new solutions. One method of creative discovery includes the following steps: (1) slow down to explore different ideas, (2) read about the field, but not too much, (3) look at the available

TABLE 5.3 Six resources for a creative enterprise.

■ Knowledge in the required domain and fields: and knowing what is new.	■ Motivation toward action.
■ Intellectual abilities to recognize connections, redefine problems, and envision and analyze possible practical ideas and solutions.	■ Opportunity-oriented personality and openness to change.
■ Inventive thinking about the problem in novel ways.	■ Contextual understanding that supports creativity and mitigates risks.

raw data, and (4) cultivate smart friends who have good intellectual skills [Paydarfar and Schwartz, 2001].

Creative thinking involves divergent thinking, which is the ability to see the differences among various data and events. Creativity involves the ability to synthesize, working through information to come up with combinations that are new and useful [Florida, 2002]. Incubation of the issues and time to reflect are important steps to creativity. One process of creative thinking is shown in Figure 5.3. It starts with a description of a problem and rests for a period of incubation. Then, intuitive brainstorming leads to good insights and ideas that can be evaluated and tested. Finally, a prototype is built and shown to the potential customer. This may lead to a reframing of the question or problem and a second cycle through the process. An iterative process around the loop is followed until the prototype product solves the problem. Often, the creative process is collaborative as shown in Figure 5.3. The customer-firm interaction is the locus of value co-creation [Prahalad and Ramaswamy, 2004].

It is also useful to think of the innovation process as involving multiple personas, each with their own skills and points of view. The first three personas occupy learning roles: the *anthropologist* observes behaviors and develops a deep understanding of how people interact with products, services, and each other; the *experimenter* prototypes new ideas continuously; and the

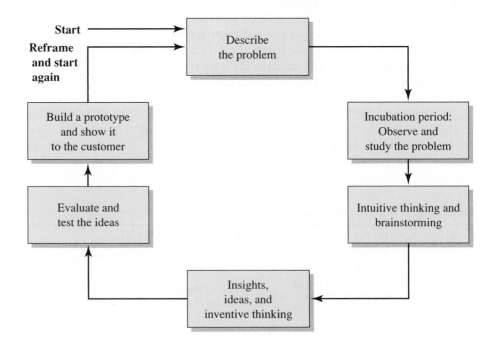

FIGURE 5.3 Creativity process.

cross-pollinator explores other industries and settings and borrows relevant ideas from them. The next three personas occupy organizing roles: the *hurdler* develops a knack for overcoming and outsmarting potential obstacles; the *collaborator* helps to bring diverse groups together; and the *director* gathers and inspires the team. The last four personas occupy building roles: the *experience architect* designs compelling experiences that go beyond mere functionality; the *set designer* transforms physical environments to facilitate the work of innovation team members; the *caregiver* anticipates and attends to customer needs; and the *storyteller* conveys a compelling narrative about the project [Kelley and Littman, 2005].

Managing for creativity can clash with rational management [Sutton, 2001]. Organizing for creativity can be different than organizing for routine work [Freeman and Engel, 2007]. One way to spur creativity is to find new uses for old materials, products, or concepts. In 1954, Kay Zufall was looking for new things for children to do. She didn't like the modeling clay sold for children because it was too stiff. However, her brother-in-law made a doughy mixture for cleaning wallpaper. Zufall tried it as a modeling medium and discovered it was soft and easy to mold and cut up. She and her brother-in-law reformulated it as a safe and colorful product for children, and they came up with the name: PlayDoh [Sutton, 2002].

All firms need a culture that sustains a creative process that enables the team members to engage and interact with ideas and new solutions. Apple was the first to develop the Newton personal digital assistant (PDA) with handwriting recognition software. In practice, few people were willing to wait for the Newton to slowly learn to recognize their handwriting. The Palm used a Graffiti interface and succeeded in capturing the market. Palm recognized that it was much easier to let humans learn to standardize their script than it was to develop software that enabled a computer to recognize all possible script.

A natural conflict exists between creatively generating ideas and inventions and implementing them. Creativity leads to new inventions and ideas. Bringing these inventions to market, however, takes routine processes. The forces of creativity and process can conflict or interact, depending on the firm's culture. Creativity flourishes when companies hire creative people, invest resources in risky projects, and get their workers to critique the ideas. These unconventional practices work because they make companies vary their thinking, see old things in new ways, and break from the past.

Rules and policies stifle creativity, and undisciplined thinking undermines routine manufacturing processes. The conflict is between managing for replication and managing for creativity. A small, emerging firm can accommodate both tendencies within it. As a firm grows, it needs to build a culture that reinforces the best qualities of creativity as well as efficient execution of its business processes [Brown, 2001].

Creativity can be seen as the ability to link together two seemingly unrelated ideas or concepts. Many ideas ignite in a free-form environment where people have capabilities and self-confidence. Mixing creative people in unexpected ways to unleash new ideas pays off. Creating new methods, products, or business models requires a powerful vision showing people what the problems are and how to resolve them. Compelling dramatic portrayals or visualizations can help people to see and feel new opportunities [Kotter and Cohen, 2002].

5.4 Types and Sources of Innovation

Innovation is based on teamwork and creativity, and is defined as invention that has produced economic value in the marketplace. Innovation is based on the commercialization of new technology. An innovation can include new products, new processes, new services, and new ways of doing business.

There are several different types of innovation, as illustrated in Figure 5.4. **Incremental innovation** is characterized by faster, better, and/or cheaper versions of existing products. Thus, they take an existing idea and creatively expand on it. To be successful, the incremental innovator must understand specific customer needs that are unmet by current offerings. For example, portable, battery-driven radios have been used since the 1950s. But, Trevor Bayles saw an opportunity to bring information to remote Africa by creating a windup spring- and dynamo-powered radio. Twenty-five seconds of winding gives the user one hour of listening. Bay Gen in Cape Town, South Africa, now manufactures more than 60,000 of these radios a month [Handy, 1999].

Like incremental innovation, architectural innovation leaves core design concepts untouched. But, **architectural innovation** changes the way in which components of a product are linked together. Thus, the components remain unchanged, but the architecture of module connection is the innovation. (The overall architecture of the product describes how the components will work together.) The essence of an architectural innovation is the reconfiguration of an established system to link together existing components in

		Basic design concepts	
		Reinforced	Overturned
Linkages between modules	Unchanged	Incremental innovation ("faster, better, cheaper")	Component or modular innovation
	Changed	Architectural innovation	Radical or disruptive innovation ("brave new world")

FIGURE 5.4 Four types of innovation.

a new way [Henderson and Clark, 1990]. By contrast, **modular innovation** is focused on the innovation of new components and modules. But, it does not disrupt the linkages between modules.

Finally, **radical innovation** or **disruptive innovation** uses new modules and new architecture to create new products. The Internet is an example of a network system with new modules and new architecture – a radical or disruptive innovation. Disruptive innovation transforms the relationship between customer and supplier, restructures markets, displaces current products, and often creates new product categories [Leifer et al., 2000]. Disruptive products also introduce a new value proposition [Christensen et al., 2004]. For example, e-mail is an application on the Internet that is a disruptive application (often called a "killer app"). It is e-mail that makes the Internet so widely used.

The iPod—A Disruptive Innovation

The iPod was introduced in 2001 with an overall performance below that required, as illustrated in Figure 5.7. Eventually, the introduction of the iTunes store enabled the iPod to succeed. The iPod became a music business, not a computer business. The online music business took off on the positive cycle illustrated in Figure 5.5. The size, charm, and elegance of the iPod device facilitated a rapid growth of the business. Excellent portability and easy-to-use controls also were a positive factor for success. In a way, the iPod became a modern, personal jukebox. The iPod was a significant disruptive innovation.

An example of a disruptive application for the personal computer is spreadsheet software. VisiCalc, created in 1979, was the first electronic spreadsheet that helped make the personal computers widely useful. Other good examples include the invention of the "800" number toll-free telephone call by AT&T and the development of the CT scanner for medical imaging, which combined X-ray technology with computer technology [Adner and Levinthal, 2002]. These types of disruptive applications bring significant value to a product and cause an industry to grow exponentially. Entrepreneurs need to discern a possible disruptive application for their start-up firm's product.

In the search for disruptive applications, people often look for attributes that will attract users to a new product. Another approach is to recognize that customers "hire" a product or service to get a job done. Home Depot and Lowe's are organized around jobs to be done [Christensen and Raynor, 2003]. Customers become aware of needing to get something done or fix some problem and set out to hire or engage a product or service that can meet their need. Customers will pay a significant premium for products that do a job well.

Geothermal Power — A Disruptive Application

A technology called engineered geothermal systems (EGS) offers a way to harness the heat trapped in the ground. EGS involves drilling two parallel wells into the earth. Water is forced down one well and steam emerges from the other after flowing through the hot rocks underground. The steam is used to power a generator that makes electricity. Unlike solar or wind power, which depend on climate conditions, geothermal hotspots are able to provide consistent power generation. The Phillipines is a world leader in ESG. The country gets nearly a quarter of its electricity from underground heat. At current levels of energy consumption, the earth under the United States could power the country for 2,000 years [Economist, 2008]. With rising fossil fuel costs and improved technology, geothermal power is growing in potential and could become a disruptive innovation.

Sources of innovation for new ventures include research laboratories, independent inventors, and universities [Branscomb et al., 1999]. In many areas of science and technology, universities can be an especially important source of innovation. Professors and other university researchers are responsible for much cutting-edge research. Since academic research typically is not driven by direct market needs, however, a major challenge in the commercialization of university breakthroughs lies in the need to move this research from lab prototypes and concepts into full working models that can be manufactured reliably at a reasonable cost [Jensen and Thursby, 2001]. To get the most benefit from a relationship with a university, a new venture should take a long-term view and imagine a partnership focused on both technical and strategic issues. When companies take a transactional approach to the relationship, attempting to pick technologies, sign a contract, and quickly commercialize them, they are likely to fail [Wright, 2008]. By contrast, when an inventor stays involved with product development as it moves from the university to a start-up, the chances of success increase dramatically [Thursby and Thursby, 2004]. University-sourced innovations also present special legal challenges, which we discuss in Chapter 10.

Another source of innovation is the ultimate customer. But, unfortunately, customers cannot always say what they want. For example, many students will say they want more detailed and complete information in their textbooks, when they actually want worked examples that will help them pass an exam. Akio Marita, the founder of Sony, asked his engineers to design a small portable radio and cassette player that would provide good audio quality and be attached to a person's head. No customer was asking for this product, yet eventually, the Sony Walkman became one of the twentieth century's disruptive applications of miniaturized electronics. In a capitalist economy, success is the ability to anticipate and meet the difficult-to-anticipate needs and wants of customers, and the most successful entrepreneurs are those who do this best.

Moreover, most customers have limited skills in predicting new products. They are best at reacting to a potential product and describing the outcomes they desire [Ulwick, 2002]. Customers are best at describing their experiences but limited in describing future needs. Who knew in advance that people wanted the Internet or electric-hybrid cars? For this reason, it is critical to observe potential customers to learn firsthand about their problems and needs. In fact, customers can be engaged to co-create valuable products and services, enabling a firm to create personalized experiences and customized solutions [Prahalad and Krishnan, 2008].

IDEO's Philosophy on Innovation

Successful innovation may require helping customers understand what it is they would like to see in a new product. As Tom Kelley puts it in *The Art of Innovation* [2001]:

> Your customers may lack the vocabulary or the palate to explain what's wrong and especially what's missing. Companies shouldn't ask them to. This is particularly true of new-to-the-world products or services. A user of a new type of remote control may not be able to recognize that it has too many buttons. Inexperienced computer users may not be able to explain that your Website lacks navigational clues. And they shouldn't have to. We saw this firsthand when a software company asked us to find out how users would react to one of their new applications. We set up a few computers and observed people struggling with the program. More than a couple were having a terrible time, grimacing and sighing audibly as they fumbled with the keyboard and mouse. But in exit interviews, the software company was given a different story. Those same people swore that they'd had no trouble with the new application and couldn't imagine a single improvement.

Another danger is the common practice of listening to the recommendations of a group of customers called **lead users** (sometimes called early adopters), who have an advanced understanding of a product and are experts in its use. Lead users can offer product ideas, but since they are not average users, their recommendations may have limited appeal. A good approach is to ask users what results or outcomes they want to see in doing their job using your product. Lead users can be very helpful at identifying valuable solutions. The process of innovation begins with identifying the outcomes customers want to achieve; it ends in the creation of items they will buy. One method is to ask potential customers to describe their typical day. They may reveal an important gap and a potential need. Many firms find new innovation from their customers (users) particularly in the software industry and with physical products. Lead users who are ahead on marketplace trends can provide innovations to their suppliers [von Hippel, 2005].

Lead users often are part of technical communities. These technical communities play a crucial role in helping new enterprises to develop and deploy innovations [West and O'Mahony, 2008]. Firms benefit from participating in open communities by gathering information on potential alliances, identifying new opportunities, and sharing work and risk. One example is the open source software development community. Individual members of such a community share common goals but not a common employer. Firms sponsoring an open source community are said to have an open source strategy. Examples are the Mozilla and Linux communities. An **open source innovation** community may be defined as a collection of many firms and individuals collaborating to develop and deploy an innovation. These communities share a common goal and an agreed-to governance system.

Benefits of open source community projects include shared goals, skills, resources, and ideas. These communities may be autonomous or sponsored by a firm. Furthermore, such communities offer the benefit of transparency of developments and ready accessibility to shared knowledge. One goal is to build modular products that can be readily accessed and utilized by all members. Governance of open source communities is established in several forms to give members rights and responsibilities. Firms include nonprofit, member, and sponsored structures.

Effective open source organizations enable new ventures to build on the ideas of others to create new innovations Murray and O'Mahony, 2008. The sharing process enables knowledge to be reused, recombined, and accumulated. This process supports the flow of innovation and progress. The process needs to provide rewards and incentives to community members in order to flourish. Cumulative innovation is enabled by ready access, disclosure, and incentives associated with community activities.

One example of a good open source effort is the response of states of the United States in monitoring and tracking infectious diseases [Kingsbury, 2008]. A new venture, called Collaborative Software Initiative (CSI), based in Portland, Oregon, was created to solve this tracking problem. A collaborative solution is better than 50 independent solutions. The CSI software is shared across all states. Another example is Wikipedia, a free online encyclopedia that anyone can edit. It is a good case of the mass collaboration of peers. Another example is YouTube [Tapscott and Williams, 2008].

5.5 Technology and Innovation Strategy

Most inventions are never actually commercialized. Only about 6 percent of inventions developed by independent inventors actually reach a market [Astebro, 1998]. In established firms, the success rate is about four times as high, which still means that three-quarters of these inventions are never commercialized. Given this low success rate, it is critical for entrepreneurs to have a sound innovation strategy.

It is often a long road from invention to commercialization. Chester Carlson developed the photocopier process—converting an image into a pattern of electrostatic charges that attract a powdered ink—in his kitchen. He patented the process in 1942. After years of little interest from established companies, he obtained help from the Battelle Institute. Then the Haloid Company purchased a license in 1946. Haloid, which became Xerox, successfully demonstrated a working product in 1949. In 1960, Haloid Xerox introduced the first successful office copier, the Xerox 914.

Schumpeter asserted that the process by which independent entrepreneurs created inventions to produce new goods, services, raw materials, and organizing methods is central to understanding business organization, the process of technical change, and economic growth. An innovation strategy rests on the competencies and knowledge of the new firm. Continual product and process innovation can enable the firm to maintain a strategic advantage. Figure 5.5 illustrates the innovation and competition cycle between firms. Competitors create innovations and offer new value to customers, fueling demand and sales and increasing the innovator's market share. The struggle is for each competitor to keep up in this innovation cycle.

While inventors can license or sell technological opportunities to others, the creation of new firms is an important mechanism through which entrepreneurs use technology to bring new products, processes, and ways of organizing into existence. Three factors influence the decision to exploit an independent invention through firm creation: the interests of the entrepreneurial team, the characteristics of the industry in which the invention would

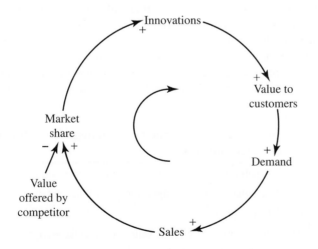

FIGURE 5.5 Innovation and competition cycle for market share.

be exploited, and the characteristics of the invention itself. We have discussed the first two of these factors in Chapters 1 through 4. Of course, the entrepreneurial team must be interested in the opportunity to be solved by the invention, and it must be satisfied that the industry will welcome and support the commercialized invention. In this section, we will consider the characteristics of the invention itself.

Three dimensions of technological inventions impact the probability that they will be commercialized through a new firm formation: importance, radicalness, and patent scope [Shane, 2001]. *Importance* reflects the magnitude of the economic value of an invention. The importance of an invention should increase the likelihood that a new firm will be founded to commercialize it because more important inventions have higher economic value and thus payoff to the entrepreneurs. Many inventions have limited commercial value and thus are not attractive to the entrepreneur. A critical determinant of an invention's importance is whether or not it addresses a real need. For example, if an invention makes it easier to do something that customers were not trying to do in the first place, it will fail [Christensen, 2002]. Thus, a "build it and they will come" innovation strategy will most likely fail.

Radicalness measures the degree to which an invention, regardless of economic value, differs from previous inventions in the field. Radical inventions have the potential, therefore, to be disruptive innovations. The radicalness of an invention is a reflection of the potential market effect of the commercialized invention. Radical technologies destroy the capabilities of existing firms because they depend on new capabilities and resources. Finally, *patent scope* describes the breadth of intellectual property protection for the invention. These three dimensions of likelihood of commercialization are listed in Table 5.4.

Dean Kamen, the holder of more than 440 patents, is one of the well-known inventors of the past three decades. He invented devices for infant care, for insulin delivery to diabetics, and for replacing the wheelchair [Brown, 2002]. Kamen invented the Segway Human Transporter in 2001. It is an electric scooterlike device. Gyroscopes inside the base platform make the scooter highly stable and self-balancing. There is no brake handle,

TABLE 5.4 Factors that influence the entrepreneur to exploit an independent invention.

1. Business interests, capabilities, and experiences of the entrepreneurial team

2. Characteristics of the industry in which the invention will be exploited

3. Characteristics of the invention:

    a. Importance of the invention: Economic value and potential payoff

    b. Radicalness of the invention: Differentiation of the invention from its predecessors

    c. Breadth of patent protection of the intellectual property

engine, throttle, or gearshift. Users lean forward to go forward and backward to reverse direction. The inventor says the Segway can traverse ice, snow, or even large rocks. Only time will tell us the extent of this invention's importance.

We can portray the new business formation process for an invention as shown in Figure 5.6. Using this process to review the potential of the Segway Human Transporter, one can obtain different conclusions for the various proposed uses: postal service, warehouse workers, or urban dwellers. Perhaps the best application for this device is not yet named.

The difficulty with deciding whether to proceed to commercialize an invention can depend, especially, on the radicalness of the invention. Disruptive or radical innovations introduce a set of attributes to a marketplace different from the ones that mainstream customers historically have valued, and the products often initially perform unfavorably along one or two dimensions

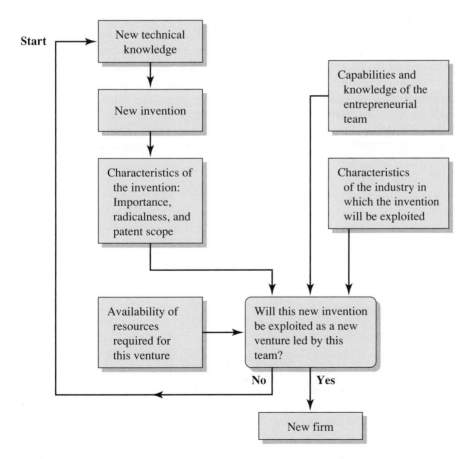

FIGURE 5.6 New business formation process for an invention.

of performance that are particularly important to those customers. As a result, mainstream customers are unwilling or unable to use disruptive products in applications they know and understand. At first, therefore, disruptive innovations tend to be used and valued only in new uncertain markets or applications.

Often the disruptive technology will not immediately serve a mainstream market, as shown in Figure 5.7. It will initially serve a niche market but will eventually enter the low end of the range of the mainstream market, as shown in Figure 5.7. For example, consider voice recognition software. The current performance of computer software for voice recognition is not always adequate for high-accuracy speaking (dictating) of documents for typing them; this might require 95 percent accuracy. Undoubtedly, there are many less-demanding uses for voice recognition software, such as voice-generated e-mails, customer service by telephone, or chat room messages. Thus, this innovation has entered the low end of the range of required performance and is progressing upward toward wider application.

Consider the disruptive innovation of U.S. discount stores in the 1960s. The increased mobility of shoppers enabled discount stores such as Kmart to select locations at the edge of town, reducing department stores' competitive advantage of prime city-center locations. The discount store had a new innovative business model: low-cost, high-unit volume and turnover provided at convenient suburban locations. It executed a trajectory from low-cost hard goods to low-cost hard and soft goods, and entered the mass markets in the 1970s and 1980s. Today, Target and Wal-Mart are in the center of the mass

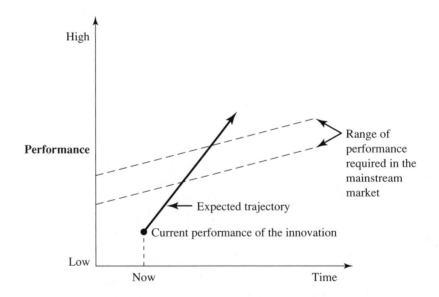

FIGURE 5.7 Expected trajectory of a disruptive innovation.

market. A recent disruptive challenge in the retail industry was Amazon.com, which appeared originally as an online bookstore but rapidly migrated toward becoming an online department store. Disruptive innovations, therefore, begin by addressing niche markets. With the right resources and capabilities, a new firm can satisfy initial needs in these markets, leveraging this early success toward mainstream dominance.

5.6 New Technology Ventures

Often, a new technology becomes available to an entrepreneur, but an economic application of the technology is not obvious. This new technology usually becomes available due to scientific discoveries or a new invention. Entrepreneurs find that this new technology may offer myriad opportunities for new ventures. However, it is unclear what, if any, application will be economically viable. Often, a new technology can be characterized as a solution looking for a problem. Neither the first companies to use the technology nor the companies with the best technology necessarily win. Instead, the firms that find the right application for the technology succeed [Balachandra et al., 2004].

The elements of an attractive innovation strategy are provided in Table 5.5. Any new venture should have a defined customer, one or two key benefits, a short period to payback, and a proprietary advantage. Finally, the new venture team must possess the necessary core competencies to exploit the new technology.

One way of describing the potential applications is to use the model shown in Table 5.6. The new technology is described briefly, and the key assumptions are listed. Then, the core competencies required for the venture are described. Finally, the possible applications market challenges are noted. Table 5.6 shows the summary of two fictional ventures. The Rotary Engine Inc. example illustrates an attractive new technology venture for vehicle engines, marine engines, appliances, and recreational vehicles. The market challenges are listed, and an attempt is made to fund the best application that will satisfy the required elements of an attractive innovation strategy.

A second example of a new technology is Fuel Cell Inc. Fuel cell technologies have been of great interest over the past decade. However, an

TABLE 5.5 Elements of an attractive innovation strategy.

- Well-defined customer
- Key customer benefit that is measurable in dollars
- Short period until economic payback and positive cash flow
- High benefit-to-price ratio for the customer
- Proprietary advantage that can be maintained or defended
- Core competencies required to exploit the new technology present or available to the new venture
- Access to the necessary resources

TABLE 5.6 Two potential new technology ventures.

Potential venture	Rotary Engine Inc.	Fuel Cell Inc.
Technology	Advanced rotary gasoline engine technology	Hydrogen fuel cell technology
Key assumptions and benefits	Improved engine efficiency and reduced pollution	Pollution reduced to near zero
Core competencies required	Engine design and manufacture	Fuel cell design and manufacture
Potential applications	1. Automobiles 2. Marine (ships) 3. Small appliances such as lawn mowers 4. Snowmobiles and off-road vehicles	1. Automobiles 2. Small, local electric generators 3. Battery replacements 4. Marine (ships)
Market challenges	1. Limited acceptance of rotary engines by customer 2. Lack of service knowledge for rotary engines 3. Benefits may be unclear to the customer	1. Limited infrastructure for hydrogen fuel cells 2. Benefits unclear to customer 3. Reliability of fuel cells is unproven

economic application is not yet proven. With the lack of supporting infrastructure, fuel cells have limited automobile applications. On the other hand, fuel cells as energy storage devices serving as battery replacements may be viable soon.

Examples of new technologies that eventually found attractive economic applications include semiconductors, genomics, stents, and wireless telephony. All these technologies eventually traversed the four steps necessary for a favorable technology innovation, as shown in Figure 5.8.

Any new attractive technology has to be feasible and manufacturable, and provide valued performance. With a sound business model and strategy, the new technology venture strives to achieve profitability in a reasonably short period.

Perhaps the greatest challenge is to develop an innovation that replaces fossil fuels [Carr, 2008]. The challenge is to find a low-cost, high-energy renewable fuel. Examples include wind, geothermal, waves, biomass, and solar technologies. These innovations must be effective, consistent, sustainable, and low cost. Solving global warming and creating green technologies will challenge innovators to develop economic innovations that can be brought to market and enable significant growth of use and scale [Krupp and Horn, 2008].

A model of a technology innovation process is shown in Figure 5.9 using the introduction of electric refrigeration as an illustration. By the late nineteenth

FIGURE 5.8 Four steps to achieve a favorable technology innovation.

century, electric power, electric motors, and refrigeration science were available. With the creation of the electric refrigerator [widely available by 1915], a totally new industry was created by a discontinuous innovation. The innovation model shown in Figure 5.9 can be used to illustrate new technology applications introduced today.

Technology entrepreneurs bring together the technical world and the business world in a profitable way. Entrepreneurship is a fundamental driver of the technological innovation process [Burgelman et al., 2004]. In summary, technology entrepreneurship is about the creation of a new business enterprise that generates benefits (wealth, jobs, value, progress) for participating parties by creating unique, new arrangements of resources, including technology, to meet the needs of customers and society.

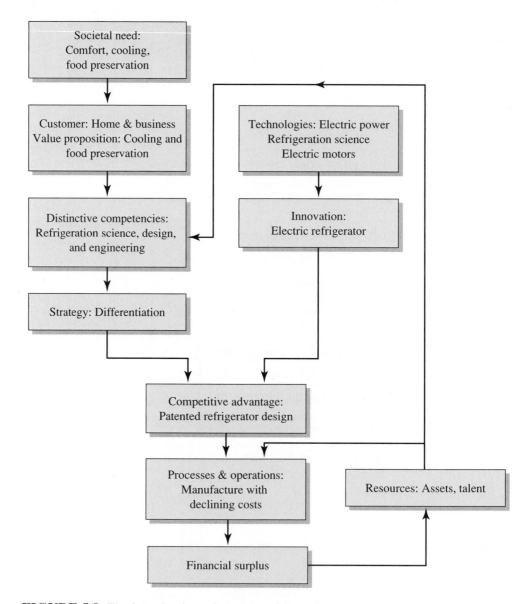

FIGURE 5.9 The introduction of electric refrigeration.

5.7 AgraQuest

Fungi and bacteria are finding their way more and more into California's groves and vineyards. Biofungicides—new products based on naturally occurring microorganisms or other plant derivatives—are bringing growers tough disease-fighting tools while making a very slight environmental impact.

Biofungicides are just one of a larger category of products known as biological pesticides, or "biopesticides." Biopesticides are pesticides derived from natural materials, including animals, plants, and bacteria. Many biofungicides are produced by fermentation, a process in which a microorganism with fungicidal properties is grown, much the same way yeasts grow in the fermentation of beer or wine.

Using a proprietary technology, each week AgraQuest researchers analyze hundreds of potential naturally occurring microbes for a novel ability to destroy or impact various undesired bacteria, fungi, insects, and nematodes, all enemies of crop production. To date, the company has screened more than 20,000 microorganisms and identified 23 that display high levels of activity against insects, nematodes, and plant pathogens. AgraQuest has selected a set of these candidates for further development.

One of the more promising discoveries that AgraQuest has licensed is a stinky fungus from Honduras that may provide farmers with a natural alternative to methyl bromide. A recent federal law requires that use of the ozone-damaging gas be eliminated, except for very limited purposes. Methyl bromide is used as a soil fumigant by growers of strawberries, tomatoes, and other vegetables that are AgraQuest's prime market.

The registration process for biopesticides and other such bioproducts through the Biopesticides and Pollution Prevention Division of the U.S. EPA's Office of Pesticide Programs tends to be shorter and more efficient than for chemical fungicides. Therefore, AgraQuest has a shorter time-to-market for a new bioproduct.

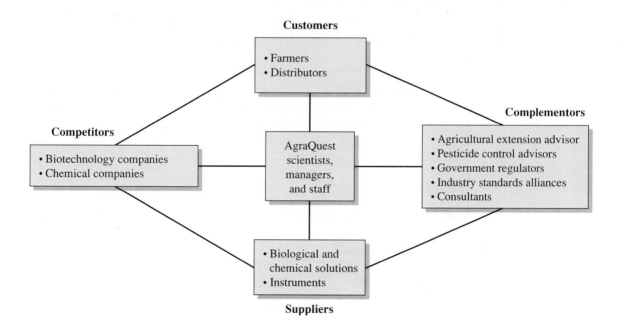

FIGURE 5.10 Value network for AgraQuest.

The biopesticide industry is only emerging and should grow over the next decade. The window of opportunity has opened, and many companies are competing for leadership. AgraQuest experiences significant delays in development and approval of products. It takes about two to three years and $6 million to get one product to market.

AgraQuest's innovation strategy is based on proprietary processes and patents. The firm holds 20 U.S. patents and 3 U.S. patent applications, along with 9 foreign patents and 95 foreign patent applications. The patents cover the microbe and its use as well as novel natural product compounds and mixtures.

AgraQuest's value network is shown in Figure 5.10. AgraQuest is dependent on its complementors to help it succeed. For example, the advice of the extension advisors and pesticide application firms is highly valued by the farmers.

5.8 Summary

Successful innovative firms strive to time their entry into markets. They balance a sense of urgency with a deliberate buildup to action. Working with partners—firms and individuals—most firms can enhance their capabilities and strengths for creativity, invention, and innovation. Almost all firms build an innovation strategy that strives to provide them with a sustainable action plan.

- A first-mover strategy can lead to significant benefits in an emerging market, but is not a guarantee of success.
- An innovation strategy creates a road map for continual commercialized invention.
- An ambitious venture can strive to design a product or service that is a disruptive application ("killer app.") that reshapes an industry.

Principle 5

An innovation strategy builds on creativity, invention, and technologies, acting within a value network, to effectively commercialize new products and services for customers.

Video Resources

Visit http://techventures.stanford.edu to view experts discussing content from this chapter.

Three Types of Innovation	Judy Estrin	JLabs
Two Weird Ideas That Work	Robert Sutton	Stanford University
Technology as Medium, Not Content	Jesse Fink, Steve Blank	MissionPoint Capital
Navigating AgraQuest's Value Chain	Pam Marrone	AgraQuest

5.9 Exercises

5.1 Name and describe the strategies of a company that was successful as a start-up being a first mover. Contrast that with a company that was successful being a fast follower.

5.2 Select an industry of interest to you and then try to find a good candidate for imitation. Describe the opportunity and tell how you will reap the benefits of imitation.

5.3 Go to a university's website and determine if a technology licensing office exists (e.g., http://otl.stanford.edu at Stanford University). Explore its website and featured technologies. Would you consider any of the feature technologies new-venture opportunities? Does the technology licensing office encourage innovation at this university? If so, how?

5.4 An inventor brings you a new design for an electric toothbrush with an oscillating head and a tilted handle that appears to meet the American Dental Association criteria. The inventor has filed a preliminary application for a patent. Also, you have tried the brush and found it easy to use. Using the factors of Table 5.4, provide a brief review of this invention. Would you recommend proceeding with commercialization?

5.5 Determine and describe the enabling technology used by Take-Two Interactive to develop its interactive software games (www.take2games.com). Describe Take-Two Interactive's value network as described in Section 4.6.

5.6 Gentex Corporation designs and manufactures automatic-dimming automotive rearview mirrors. Its safety mirrors use sensors and electronics to detect glare from trailing approaching vehicles at night and darken accordingly (www.gentex.com). Describe the invention and technology that Gentex uses. Draw a value network as described in Section 4.6 for Gentex and name its partners.

5.7 Zebra Technologies Corporation provides bar-code labeling solutions for use in automatic identification and data collection systems (www.zebra.com). Describe the technology of Zebra in terms of the three dimensions of technological inventions: importance, radicalness, and patent scope.

5.8 The E-Stamp Corporation was first to market in 1997 with the ability to sell stamps over the Internet to consumers who print the stamps on their printers (www.estamp.com). By 2001, however, E-Stamps' 31 patents and other intellectual property were purchased by Stamps.com (www.stamps.com). Study this acquisition and determine why being first to market was not a winning strategy for E-Stamp.

VENTURE CHALLENGE

1.   Describe your venture in terms of timing of entry as illustrated by Figure 5.1.

2.   Describe your creative process as outlined in Figure 5.3.

3.   Discuss your type of innovation as defined in Section 5.4.

4.   Summarize your technology and innovation strategy.

5.   Is your product or service a disruptive innovation?

PART 2

Venture Formation and Planning

The venture team seeks to build a plan that will mitigate the associated risks while using innovation to increase the chances of success. If possible, a business design will achieve economies of scale and scope as growth is experienced. The creation of a formal business plan will provide a valuable tool for investors, the venture team, and allies. This plan is a description of the opportunity, product, strategy, team, needed resources, and potential financial outcomes. The venture techniques described in this book can be used by entrepreneurs to build an independent business as well as a corporate venture emerging within an established firm. Large corporations can learn to confront the innovators' dilemma and build a viable corporate venture. Knowledge management is a powerful tool for an entrepreneur building an important, innovative new business. Furthermore, product design and prototype methods can foster the creation of outstanding products. The new venture organization carefully selects its legal form and protects its intellectual property. ■

6

Risk and Return

Our greatest glory is not in never falling but in rising every time we fall.
Confucius

What determines the success of entrepreneurial efforts and how can risks be managed?

A new venture that creates a novel solution to a problem will be subject to uncertainty of outcome. An action in an uncertain market is sure to experience a risk of delay or loss. It is the entrepreneur's task to reduce and manage all risks as much as possible.

Attractive new ventures can be designed to grow as demand for their products increases. Furthermore, it is hoped that economies of scale will be experienced so that as demand and sales grow, the cost to produce a unit of product declines. Additionally, it is desirable to have economies of scope so that costs per unit decline due to the spreading of fixed costs over a wide range of products. Many industries established on a network format exhibit network economies resulting in a reinforcing characteristic leading to the emergence of an industry standard. ■

6.1 Risk and Uncertainty

The pursuit of important opportunities and big goals by entrepreneurs requires them to assume more risks than they might take on working for a mature company or the government. Introducing a novel product into a new market has an uncertain outcome. An outcome resulting from an action is said to be **certain** in that it will definitely happen. Something certain is reliable or guaranteed. For example, it is certain that if we drop a rock, it will fall to the ground (and not float upward).

An outcome resulting from an action is said to be **uncertain** in that the outcome is not known or is likely to be variable. **Risk** is the chance or possibility of loss. This loss could be financial, physical, or reputational. When Christopher Columbus embarked on his first voyage to the New World, he risked financial, reputational, and bodily harm. Farmers are a group whose fortunes are vulnerable to unpredictable outcomes due to drought, flood, or other weather conditions.

Most, perhaps almost all, people are risk-averse or risk-avoiders. Logically, an entrepreneur seeks to avoid or reduce the risk of an action. For example, farmers purchase insurance to mitigate the effects of uncertain weather. A simple measure of a person's risk aversion is provided in the following example of a coin toss game [Bernstein, 1996].

Example: A Coin Toss Game

You have a choice of receiving $50 for certain or an opportunity to play a coin toss game in which you have a 50 percent chance of winning $100. The $50 gift is certain, if you elect it. The game's outcome is uncertain, but the expected outcome over the long run is (if you play it many times) also $50 since, for a true coin, the probability of winning, P, is 50 percent. Surely, you would play the game if the probability, P, of winning were 100 percent. Would you play if the probability were 70 percent? What is the probability, P, that you would require to play? If your P=70 percent, you are risk-averse, and if your P=40 percent, you are a risk-seeker. Of course, your willingness to play this game will also depend on your overall wealth in relation to the $50 outcome of loss and the fun of playing this simple game. This game illustrates the fact that most people, perhaps like you, are risk-averse.

The ability to successfully choose the risks worth bearing is a form of human capital based on experience and good judgment [Davis and Meyer, 2000]. We usually assume that the elevated risk of an entrepreneurial venture may provide a higher return on this human capital. Entrepreneurs often have a capability to limit the downside risks of a venture by applying their skills that are useful for mitigating the risks. One mental model may be the circus high-wire walker with a strong net below. In fact, most successful entrepreneurial firms create value by taking calculated risks and possess the core competency to mitigate or manage these risks. All of life is the management of risk, not its elimination [Brown, 2005].

TABLE 6.1 Four levels of uncertainty.

Uncertainty level	Level of risk	Example
1. A clear, single outcome	Very low	Purchase a Treasury Bill
2. A limited set of possible outcomes	Low	Set up a pushcart for selling sports items at the Olympics
3. A wide range of possible future outcomes	Medium	Launch an improved product in a new market
4. A limitless range of possible outcomes	High	Establish a firm to attempt to design and sell a revolutionary power source based on a fuel cell breakthrough

The entrepreneur is in many ways like an investment manager who chooses to pursue selected opportunities and not others. When we analyze risk, we look forward in time and try to estimate the potential outcome and the variability of that outcome. Risk is a measure of the potential variability of outcomes that will be experienced in the future. Furthermore, risk is the chance or possibility of loss. Table 6.1 lists the four possible levels of uncertainty [Courtney, 2001]. The level of return on investment we might expect should be commensurate with the level of uncertainty.

Entrepreneurs practice in the realm of opportunity in a manner similar to what financial investors do in the stock market [Sternberg et al., 1997]. They pursue opportunities with expected levels of risk and attempt to "buy low and sell high." Buying low means pursuing ideas that are not widely recognized or are out of favor. Selling high means finding a cash buyer for the successful venture, convincing the buyer that its worth will yield a significant return and it is time to harvest some of the wealth already created. As with any investment, "invest in a business you understand" is a good principle. Thus, pursuing the hope of fusion power should be left to those few who know and understand this big but risky opportunity. The entrepreneur-investor assumes the risk of the venture and should be willing to take on sizable risks with the knowledge that he or she can manage or mitigate them.

The best method for most entrepreneurs is to use a form of experimentation— trial and error. Identify possible new ventures and take a few steps toward a business and then evaluate the early feedback. If it looks good, keep going forward. Entrepreneurs will most likely take the risk of failure if the losses are constrained on the downside and the potential rewards are high on the upside [Sull, 2004]. Henry Ford said: "Failure is the opportunity to begin again, more intelligently."

Entrepreneur-investors should consider the concept of **regret,** which we define as the amount of loss a person can tolerate. People treat regret of loss differently than the potential of possible gain. How much regret people feel varies depending on their circumstances of wealth, age, and psychological well-being. Reconsider the simple coin toss game. If you play the game and lose

$50 on the first turn, it may be disturbing. Your level of regret may be $200. If you play a sequence of four turns and lose each of them, you will reach your level of regret. Thus, entrepreneurs need to evaluate their regret level of acceptable loss and limit their investment in any entrepreneurial venture. If an entrepreneur can forgo one year of income ($50,000) and willingly invest his or her savings of $60,000 in the venture, then the entrepreneur's level of regret is $110,000.

The amount of risk entrepreneurs will endure varies, but most retain some personal financial reserve so that failure will not equate to homelessness or starvation. It is easier to be a risk taker if one has some reserves to fall back on.

The risk-adjusted value of a venture, V, is

$$V = U - \lambda R \tag{6.1}$$

where U = upside, λ = risk-adjusted constant, usually greater than 1, and R = downside or regret. The larger the value of λ, the more risk-averse is the entrepreneur. We are neutral at $\lambda = 1$ but risk-averse at $\lambda = 2$ (Dembo and Freeman, 1998). If your regret is R = $110,000 and $\lambda = 2$, to proceed requires V > 0, or

$$U > \lambda R \tag{6.2}$$

Therefore, the required upside, U, is U > $220,000. Entrepreneurs can use scenarios and economic analysis to estimate the potential upside, U, of a venture.

The strategic response to uncertainty is to build a venture in stages, reserving the right to adjust your core competencies and strategies, and play again at the next stage. Thus, reconsidering the case of the entrepreneur who is risk-averse with $\lambda = 2$, he or she can choose to proceed for six months with the venture so that R = $55,000, and the minimum upside required is then only $110,000. After the six months, the entrepreneur adjusts the business strategy to improve the business performance and based on a new calculation of the upside, U, proceeds to a second stage of activity or decides to terminate the venture.

To perform the kinds of analyses appropriate to high levels of uncertainty, most firms will need to enhance their strategic capabilities. Scenario-planning techniques can be useful for determining strategy under conditions of uncertainty.

Risk reflects the degree of uncertainty and the potential loss associated with the outcomes, which may follow from an action or set of actions. Risk consists of two elements: the significance of the potential losses and the uncertainty of those losses. In most new ventures, it is the significance or size of the potential losses, *hazard,* and the *uncertainty* that are estimated by new venture entrepreneurs and their investors. We propose a measure of risk as:

$$\text{Risk} = \text{hazard} \times \text{uncertainty} \tag{6.3}$$

The hazard, H, is the size of the potential losses as perceived by the entrepreneurial team. Hazard is an entrepreneur's income forgone (opportunity cost, OC) plus the financial investment, I, which he or she will need to make. Therefore,

$$Risk = (I + OC) \times UC \tag{6.4}$$

The uncertainty, UC, is measured by the variability in anticipated outcomes, which may be described by their estimate of the probability of loss (failure). Based on these factors, the entrepreneurial team may make a selection of a new venture (or to proceed or not), as shown in Figure 6.1. High levels of hazard may not deter entrepreneurs from choosing ventures with potentially high levels of returns. The decision to proceed or not depends on the level of risk adversity of the team members and their perception of the extent of the uncertainty. Of course, the estimate of potential returns is subject to the assumptions underlying their calculation.

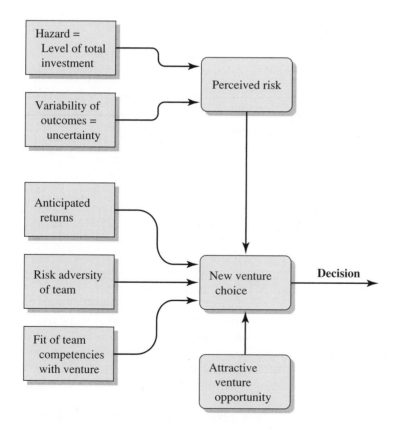

FIGURE 6.1 Risks and new venture choice.

TABLE 6.2 Sources of uncertainty.

1. Market uncertainties

   - Customer

   - Market size and growth

   - Channels

   - Competitors

2. Organization and management uncertainties

   - Capabilities

   - Financial strength

   - Talent

   - Learning skills

   - Strategies

3. Product and processes uncertainties

   - Cost

   - Technology

   - Materials

   - Suppliers

   - Design

4. Regulation and legal uncertainties

   - Government regulation

   - Federal and state laws and local ordinances

   - Standards and industry rules

5. Financial uncertainties

   - Cost and availability of capital

   - Expected return on investment

An entrepreneurial venture is launched with a degree of uncertainty due to the novelty of the product to the market, the novelty of the production processes, and the novelty to management [Shepherd et al., 2000]. Novelty to the market concerns the lack of uncertainty of the market and customer uncertainty. Novelty of production processes is dependent on the extent of knowledge of the processes by the venture team. Novelty to management concerns the venture team's lack of the necessary competencies. Regulation and legal changes are also a source of uncertainty. Sources of uncertainty are listed in Table 6.2.

The risk of failure or poor performance is significant and should not be understated. According to the U.S. Small Business Administration, about one-half of all small businesses are acquired by another firm or leave the market

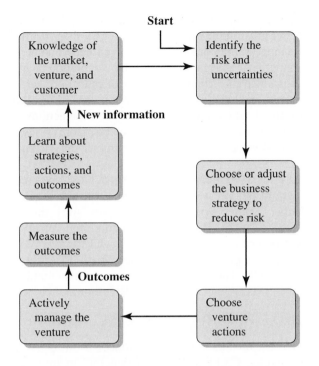

FIGURE 6.2 Managing risk and uncertainty.

within four years. Of course, being acquired at a good price may be the success one is seeking. It is fair to say that one-fourth of all business start-ups are discontinued within four years, but it is unclear at what cost or return.

Technology ventures usually have four types of risk: technology, market, financial, and team. Every effort should be made to test the assumptions about these four sources of risk. Strategic risk management involves setting up strategies that anticipate the downside of risk. The key to surviving risks is assessment and response [Slywotzky and Drzik, 2005].

Technology entrepreneurs are often less concerned with certainty than with getting into the game quickly and learning how to participate. The uncertainty and associated risks decline as the novelty in the three dimensions of market, production, and management declines. Novelty is synonymous with uncertainty, and, thus, we expect uncertainty to decline as knowledge of the market, the processes, and management competencies improves. Thus, the liability of newness declines over time. A process for managing risk and uncertainty is shown in Figure 6.2.

Whenever there is uncertainty, there is usually the possibility of reducing it by the acquisition of information. Indeed, information is essentially the negative of uncertainty. The acquisition of information and knowledge improves an organization's chances of adaptation and performance. The entrepreneurs are continuously making decisions about highly uncertain environments and the new venture's internal structures that in turn modify its performance outcomes. New venture managers may learn from past choices about how to perform better in the future. This learning can facilitate adaptation to changed environmental conditions since the strategic choices of managers help define the outcomes. An appropriate strategy for risk reduction uses new information learned from experience to adjust the business strategy and the actions taken to execute the strategy. Appropriate strategies might include adding new people to the team, creating new alliances, reducing costs, or improving customer relationships, among others.

Arthur Pitney received a patent in 1902 for a hand-operated postage meter that he hoped would be used to replace stamps for mass mailings. Pitney recognized his primary source of risk was regulatory since the U.S. Post Office controlled postage services. By 1918, after continual rejections, Pitney was joined by Walter Bowes to form Pitney Bowes and try again to get Post Office approval. With Bowes's persuasive skills, they finally received a license for the postage meter in 1920. They ultimately overcame the regulatory hazards to build up their firm. Many entrepreneurs underestimate their various risks.

Symantec: Risk and Reward

In 1981, Gordon Eubanks, a software pioneer in the personal computer industry, cofounded C&E Software to develop an integrated database management and word processor product. In 1983, John Doerr, who was a venture capitalist with Silicon Valley's Kleiner Perkins Caufield and Byers (KPCB), approached Eubanks regarding one of his investments. Symantec was an artificial intelligence software firm that was struggling to stay afloat despite having an interesting technology known as natural language recognition. Doerr suggested that Eubanks merge with Symantec and incorporate Symantec's technology into C&E's product (see Figure 6.2). Although Eubanks viewed this technology integration as a compromise, he was persuaded by the upside of the deal— a substantial percentage of ownership for C&E in the merged venture, an additional cash investment from KPCB, and a chance to lead the new company. In late 1984, Symantec Corporation was reborn as a firm that developed a natural language database manager with word processing capabilities.

Even with the additional venture capital funds, it became apparent that Symantec would not survive as a one-product company. A few months before the first version of Symantec's database manager shipped in 1985, Eubanks hired Tom Byers to search for new revenue streams and to diversify the product line. Byers realized that a market existed for software utilities that added features to the then-popular Lotus 1-2-3 spreadsheet. Symantec used a strategy similar to book publishing with these products: in return for the rights to package and sell the software,

Symantec paid developers a royalty as the product's author. Eubanks had adjusted Symantec's strategy in response to risk and uncertainty.

In 1987, Eubanks further diversified the firm. Symantec purchased Breakthrough Software for minimal cash and a $10 million note payable in cash if and when Symantec went public. Timeline, Breakthrough's project management software, quickly doubled Symantec's incoming revenue, which was important because sales of Lotus utilities were rapidly declining.

Also in the late 1980s, Eubanks quickly acted on two emerging trends in the industry. First was the proliferation of computer viruses as more and more computers were connected in networks. Second was the rapid advancements of graphical user interfaces (GUI) in personal computers such as the Macintosh. To rapidly respond to these changing conditions, Eubanks made several key decisions and acquisitions.

Ted Schlein, who worked for Byers in Symantec's publishing division, identified a market for antivirus software and pushed for Symantec to publish Symantec Antivirus for Macintosh (SAM), which became the first successful commercial antivirus product. In following years, Symantec made numerous acquisitions in this category, including Peter Norton Computing; network security eventually became Symantec's core business. In addition, Eubanks acquired two Macintosh software companies whose developers were very experienced in GUI-based software. These acquisitions helped prepare the company for the subsequent decade in which all PC software was based either on Windows or Macintosh user interfaces.

Finally, marketing and distributing Symantec's growing line of software became vital to its success. Recognizing that the software industry was rapidly maturing with large corporations becoming most important, Eubanks hired John Laing as head of global sales and Bob Dykes as CFO. Laing and Dykes, experienced professionals who had worked in large business operations in the past, created the necessary systems and processes to handle the upcoming rapid growth.

In 1989, Symantec went public with 264 employees, $40 million in sales, net income of $3 million, and 15 products. During the 1990s, Eubanks focused Symantec on network security technology. In 1999, Eubanks stepped down as CEO of Symantec, and John Thompson from IBM took the reigns. Thompson continued the tradition of focusing the entire product line around enterprise security. Symantec expanded into the market of overall data reliability by merging with Veritas in 2005, becoming one of the largest software companies in the world. In 2009, Symantec had revenues of nearly $6 billion, more than 17,500 employees, operations in 40 countries, and a market capitalization of over $12 billion. Industry leaders like Symantec continuously manage risk and return.

TABLE 6.3 Estimating risk and reward.

1. Describe the most likely scenario, the expected reward, and its estimated probability.

2. Describe the worst-case scenario, the expected loss, and its estimated probability.

3. Describe the best-case scenario, the expected reward, and the estimated probability of it occurring.

4. Determine how much the entrepreneurial team and their investors can afford to lose. Include their investment and opportunity costs.

Business can manage the problem of unpredictable customer behavior by following the ideas of portfolio management. The portfolio of customers should be diversified so as to produce the desired returns at the particular level of uncertainty the firm can tolerate. Customers are risky assets. As with stocks, the cost of acquiring them is supposed to reflect the cash-flow values they are likely to generate. The concept of risk-adjusted lifetime value of a customer has a transforming power [Dhar, 2003].

To the uninitiated, successful new ventures appear to be the right idea at the right time. However, entrepreneurs put the pieces together so that while it looks like a happenstance, it is actually the entrepreneur who makes it happen with good calculations based on sound information. As Charles Kemmon Wilson, the founder of Holiday Inn, said [Jakle et al., 1996]: "Opportunity comes often. It knocks as often as you have an ear trained to hear it, an eye trained to see it, a hand trained to grasp it, and a head trained to use it."

One way to calculate an estimate of a new venture's potential risk and reward is to answer the four questions posed in Table 6.3. In general, the entrepreneur seeks a venture where the return is expected to significantly exceed the potential losses.

6.2 Scale and Scope

In this section, we consider the strategic impacts of the scale and scope of a firm. The **scale of a firm** is the extent of the activity of a firm as described by its size. The scale of a firm's activity can be described by its revenues, units sold, or some other measure of size. **Economies of scale** are expected based on the concept that larger quantities of units sold will result in reduced per-unit costs. Economies of scale are generally achieved by distributing fixed costs such as rent, general and administrative expenses, and other overhead over a larger quantity, q, of units sold. This effect is portrayed in Figure 6.3. The cost per unit decreases, reaching a minimum at q_m. Often, the cost per unit will increase for $q > q_m$, since the complexity of coordination may increase costs per unit for a high number of units.

When significant economies of scale exist in manufacturing, distribution, service, or other functions of a business, larger firms (up to some point) have a cost advantage over smaller firms. Thus, smaller, new-entrant firms need to

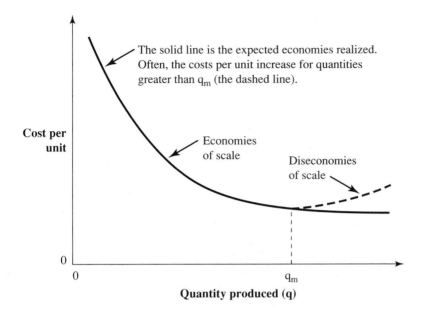

The solid line is the expected economies realized. Often, the costs per unit increase for quantities greater than q_m (the dashed line).

Cost per unit

Economies of scale

Diseconomies of scale

0

0

q_m

Quantity produced (q)

FIGURE 6.3 Economies of scale.

differentiate their product on qualities other than price. As the smaller, new-entrant firm grows in size, it can also learn to reduce its costs per unit and price competitively with larger firms.

Google's Competitive Advantage

Google's sustainable competitive advantage is tied to its innovation underlying its search engine. Google offers great results and keeps people coming back. Revenues for Google were estimated at $100 million in 2001, only three years after starting up. In 2004, Google raised $1.2 billion in its initial public offering. By 2009, its revenues were approaching $23 billion. Its primary revenues are from advertising on its pages. It gets its fees by selling rights to given keywords so an ad shows up first when those words are entered. Search engines for the Web are based on computer science algorithms and need to search without failure [Hardy, 2003]. Google now benefits from tremendous economies of scale.

Another issue related to scale is the concept of scalability. **Scalability** refers to how big a firm can grow in various dimensions to provide more service. There are several measures of scalability. They include volume or quantity sold per year, revenues, and number of customers. These dimensions are not independent, as scaling up the size of a firm in one dimension can affect the other dimensions. Easily scalable ventures are attractive, while ventures that are difficult to grow are less so.

The consequence of growth is the necessity to respond by increasing capacity. **Capacity** is the ability to act or do something. Any firm has processes, assets, inventory, cash, and other factors that must be expanded as the company grows its sales volume. A firm that can easily grow its capacity is said to be readily scalable. For example, as a firm grows, its working capital requirements will grow. **Working capital** is a firm's current assets minus its current liabilities. Sources of working capital for an emerging firm can include long- and short-term borrowing, the sales of fixed assets, new capital infusions, and net income. The ability for a new firm to grow will be influenced by its access to new capital and assets. Managing a firm for scalability is important to its success. To preempt or match competitors, a firm must attempt to foresee increases in demand and then move rapidly to be able to satisfy the predicted demand. This strategy can be risky since it involves investing in resources before the extent of the demand is verified. The total cost, TC, of the production of units is described as:

$$TC = FC + VC$$

where FC is the fixed costs that do not vary with the quantity of production. The variable costs, VC, do vary with the quantity produced where $VC = c \times q$, c being the cost/unit and the quantity being q. This relationship is shown in Figure 6.4. Table 6.4 describes the scalability and economies of scale for four types of businesses.

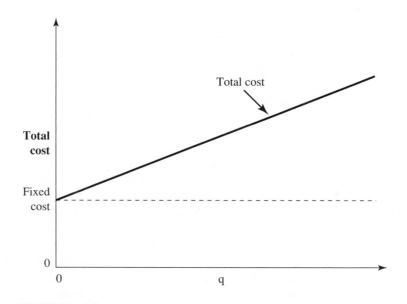

FIGURE 6.4 Total cost as fixed costs plus variable costs.

TABLE 6.4 Scalability and the effects of fixed and variable costs for four types of businesses.

Type of business	Economies of scale	Scalability	Fixed costs	Variable costs	Primary strategy	Funds required by start-up
1. Based on talent; e.g., consulting	Low	Medium	Low	High	Recruit talent	Low
2. Based on talent and knowledge assets; e.g., plastics, clean tech	Medium	Medium	Medium	Medium	Secure physical assets and talent	Medium
3. Based on physical assets, knowledge, and materials; e.g., biotechnology, semiconductors	High	Low	High	Low to medium	Secure physical assets	High
4. Based on information with few physical assets; e.g., software, movies	High	High	High	Low	Secure talent to create the software or the movie	High

The advantage of a talent-based business such as a consulting firm is that the start-up funds are low. The firm is scalable as long as new talent can be recruited for expansion, but it has few economies of scale. The advantage of a firm based on a mixture of talent and physical assets is that it can expand as long as it can secure the necessary funds. A physical asset-based business such as a steel company must secure new plants and equipment as it grows, requiring capital infusion. An information-based business must invest funds upfront to create the software or a movie. It has low variable costs and high economies of scale.

The **scope of a firm** is the range of products offered or distribution channels utilized (or both). The sharing of resources such as manufacturing facilities, distribution channels, and other factors by multiple products or business units gives rise to **economies of scope.** For example, the cost per unit of Procter & Gamble's advertising and sales activities is low because it is spread over a wide range of products. Procter & Gamble's disposable diaper and paper towel businesses demonstrate a successful realization of economies of scope. These businesses share the costs of procuring certain raw materials and developing the technology for new products and processes. In addition, a sales force sells both products to supermarket buyers, and both products are shipped by means of the same distribution system. This resource sharing has given both business units a cost advantage compared to their competitors [Hill and Jones, 2001].

> **Facebook: Seeking Economies of Scope**
>
> Facebook is a very popular social networking site with a significant market share in the United States and Europe. Seeking to take advantage of its large user base, it launched the Facebook Platform in 2007. This allowed any developer to program small Web applications for users to add to their profiles. These applications take advantage of Facebook's comprehensive social connectivity capabilities and include games, event publicizing, gift-giving, and video sharing among others. By 2009, over 37,000 applications had been created. Facebook is attracting developers seeking to take advantage of the massive economies of scope resulting from Facebook's success in the market.

The economies of scale and scope both reduce the cost per unit. For a factory, its **throughput**—the amount processed within a given time—needs to be consistently high. The introduction of the railroad to the United States and Europe reduced the travel time between markets and supply, increasing the flow of raw materials to factories. The revolution in railroad transportation and telegraph communication resulted in great increases in throughput. In 1870, the Union Pacific Railroad joined the Central Pacific at Promontory, Utah, spanning the United States [Beatty, 2001]. The United States became the world's leading industrial producer by 1929. The economies of scale and scope helped the United States to become a low-cost producer and distributor of goods.

The strategy of a new firm must incorporate plans for economies of scale and scope. Ausra is an emerging California company whose fundamental business model depends on economies of scale. Ausra uses a new technology, called the compact linear fresnel reflector, to gather sunlight and generate electricity by boiling water and spinning a turbine. Solar thermal power plants require large numbers of mirrors to concentrate adequate heat on the boiler. This need for a large mirror farm requires Ausra's power plants to be much bigger than most other solar start-ups. In 2008, Ausra constructed the first U.S. manufacturing plant for solar thermal systems to test out its new technology in the field.

6.3 Network Effects and Increasing Returns

In recent years, realization has been growing that network economies are an important element of competitive economics for new entrepreneurial firms. **Network economies** arise in industries where a network of complementary products is a determinant of demand (also called network effects). For example, the demand for telephones is dependent on the number of other telephones that can be called with a telephone. As more people get telephones, the value of a telephone increases, thus leading to increased demand for telephones. This process is called a positive feedback loop since as more people use the process,

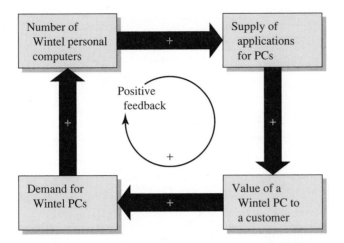

FIGURE 6.5 Increasing demand for Wintel personal computers due to positive feedback.

the value to users of the process increases and thus demand goes up, leading to more people using the process.

The positive feedback process for Windows-Intel (Wintel) personal computers is illustrated in Figure 6.5. As the number of Wintel PCs increases, the incentive for the development of software applications increases (a complementary product). With more application software available, the PC is more valuable to a user. As the value of a PC increases, the demand for Wintel PCs increases, leading to an increased number of Wintel PCs.

Networks include telephone networks, railroad networks, airline networks, fax machine networks, computer networks, ATM networks, and the Internet, among others. The overall tendency is toward "bigger is better." As Figure 6.6 indicates, over time, a winner emerges (company A) and the competitors decline. In the PC industry, Wintel captured the largest market share, with Apple holding about 10 percent of the market. In general, network effects exhibit reinforcing characteristics, as shown in Figure 6.7.

The value of a network, according to Bob Metcalfe [Shapiro and Varian, 1998], is approximately:

$$\text{value of network} = kn^2$$

where k is a constant to be determined for each industry sector and n is the number of participants in the network. Based on this model, the value of a network grows rapidly as n grows. This simple model assumes all participants are equally valuable, which is not always true. The incentive to a firm in a network economy is to secure market share, eventually taking off toward dominance. This is the underlying theory behind Amazon.com and other Internet start-ups. Of course, market share can grow rapidly while profitability may be

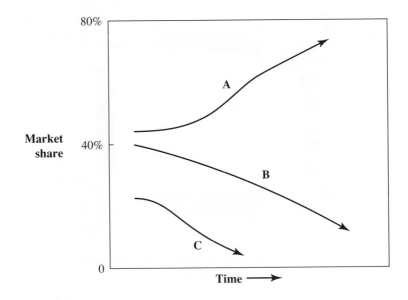

FIGURE 6.6 Emergence of a dominant firm, A.

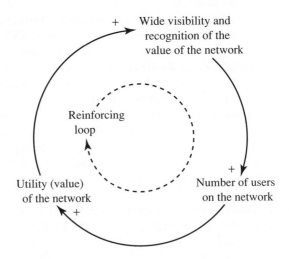

FIGURE 6.7 Reinforcing characteristic of a positive loop exhibiting network effects.

elusive. A balance of both market share and profitability can lead an entrepreneurial firm to eventual success if it has a product that has high value for its customers and strong alliances with its complementors.

Network economies work when revenues grow faster than costs. In the late 1990s Webvan tried to become the online grocer of choice in the United States.

It had to invest in warehousing, trucks, and logistic systems, which led to increased costs and inventories, which caused its costs to spiral upward faster than its revenues, eventually sinking the firm into bankruptcy.

Increasing returns mean that the marginal benefits of a good or of an activity are growing with the total quantity of the good or the activity consumed or produced [Van den Ende and Wijaberg, 2003]. Increasing returns is the tendency for a company that is ahead, firm A in Figure 6.6, to get farther ahead. The theory is that the firm that has a successful product that is increasingly becoming the standard for the industry will experience increasing returns as increasing quantities are sold. However, there is no guarantee that increasing market share (firm A in Figure 6.6) will experience profitability. Furthermore, it is not possible to predict ahead of time which firm will attain market share dominance. If a product or a company or a technology—one of many competing in a market—gets ahead by quality of offering or clever strategy, increasing returns can magnify this advantage, and the product or company or technology can go on to lock in the market. Microsoft DOS became dominant after a protracted battle with CPM and Apple for PC operating system leadership.

Many new firms may enter a new industry with the potential to eventually dominate, but only one or two survive. Products do not stand alone but depend on complementary products to make their use valuable. Examples of network industries that exhibit increasing returns are airlines and banking. As an airline expands the number of cities it serves, the value to a customer of using the airline increases. Another prominent example is eBay, which has dominant market share in online auctions. Because it was the first to connect individual buyers and sellers over the Internet in an auctionlike format, it grew very quickly. As more rare items came up for auction on eBay, more buyers were attracted to the website to bid on those items. Those extra bidders attracted still more sellers, thus leading to dominance. Note that eBay was profitable from the first year onward.

While Metcalfe's law illustrates the general idea of the value of a network, it is only an approximation to reality. The value of each node (participant) will vary. Furthermore, some links will be strong while others are weak. Customers value the number of nodes in the network but also key links in the network. A network of five nodes and eight links is shown in Figure 6.8. Note that not all nodes are connected by a link to all other nodes in this example. Consider a bank network with 100 branches. Most people do not visit many other branches except their local branch and perhaps one near their work. For many customers, the link to their account at their local branch and online or via telephone is what they value. Thus, designers of a network business must analyze their customers' needs and build their business on the best information.

Consider Southwest Airlines. It has no physical branches, nor does it use travel agents; instead it uses telephone and online links to its network. It offers strong incentives to its customers to use the online rather than the telephone

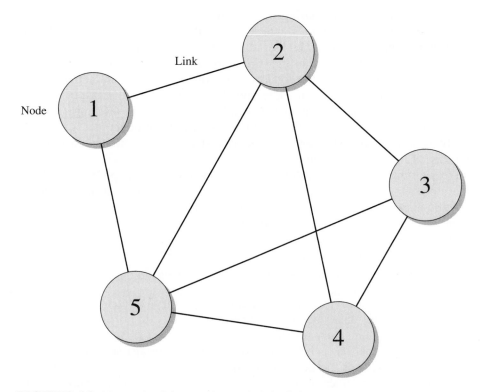

FIGURE 6.8 Network of five nodes and eight links.

links. Physical branches may be necessary for banks but not for airlines. Wells Fargo is putting minibranches in grocery stores on the theory that its customers value physical nodes (branches) as well as Internet links.

In general, knowledge-based products exhibit increasing returns. The up-front development costs are high, but the per-unit production costs thereafter are low. Knowledge-based products also exhibit network effects; that is, the more people who use these products, the more valuable they become.

Network Effects at Facebook

During the early 2000s, numerous social networking sites competed for users. Mark Zuckerberg, then a sophomore at Harvard University, launched "The Facebook" in 2004. It was intended to allow students in the same class to find one another and study together. Within one month, more than half of Harvard undergraduates had registered for the service. At first, only Harvard students were allowed access. Gradually, the site expanded, extending membership to other colleges in the Ivy League. The site was then opened to all university and high school students. In September 2006, Facebook was opened to all users.

Since then, it has become the most popular social networking site on the Internet. As of 2009, Facebook had more than 300 million users as one of the most-visited websites in the world. It claims that 85 percent of U.S. undergraduate students are registered. Facebook, Orkut, and Friendster are competing for dominance in the social networking market around the globe. Which of these three companies do you expect to reap the benefits of network effects?

6.4 Risk Versus Return

Reaching for higher returns carries higher risk. Assuming the entrepreneurs and their investors are rational beings, they will demand higher potential annual returns for higher-risk ventures. We illustrate a risk-reward model in Figure 6.9. The expected return varies as

$$ER = R_f + R$$

where ER = expected return, R_f = risk-free rate of return (T-bill), and R = risk. High-risk and high-return investments will be expected to return in excess of 30 percent annualized over a period of years, T. For a high-risk venture, T may range from three to seven years [Ross et al., 2002]. A disruptive application or radical innovation is expected to return in excess of 40 percent annualized over T years (point a in Figure 6.9).

Compact fluorescent lights (CFLs) are an energy-efficient replacement for traditional incandescent bulbs. They use up to 75 percent less power than incandescent lights and could be the replacement for most home uses. Ellis Yan, a native of China, came to the United States and recognized the opportunity presented by CFLs. He took a risk by making the bulbs when CFLs were largely ignored by large manufacturers and others in the market. CFLs are now seen as the climate-friendly solution for traditional lighting. By 2008, his company, TCP Inc. in Ohio, captured over half the U.S. market and generated revenue of almost $300 million dollars for a large return on his investment.

Often, a more moderate return and risk venture (point b in Figure 6.9) can be achieved by using an incremental technology change rather than a disruptive technology in combination with a change in the business model. A moderate change in technology plus a moderate change in business model can lead to an attractive risk-return profile [Treacy, 2004].

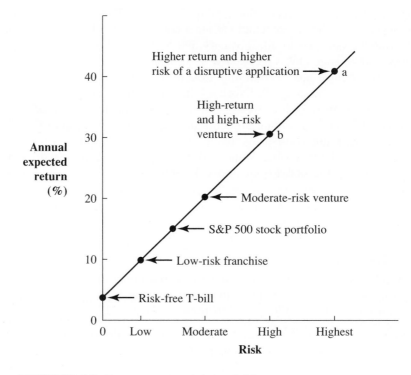

Annual expected return (%)

Higher return and higher risk of a disruptive application ⟶ ● a

High-return and high-risk venture ⟶ ● b

Moderate-risk venture

S&P 500 stock portfolio

Low-risk franchise

Risk-free T-bill

Risk

0   Low   Moderate   High   Highest

FIGURE 6.9 Return-versus-risk model for a new venture.

6.5 Managing Risk

New technologies and innovations create serious challenges for an existing firm. Change creates risks for competitors and a firm needs to manage risk. Business risk is growing in intensity and leaders of new ventures need to manage strategic risk: the possibility of loss due to hazards, operating risks, and competitive challenges. Managing risk requires anticipating threats and reversing them [Slywotzky, 2007]. Risk and reward can be decoupled if a firm learns how to manage risk. Defeating risk is based on a keen knowledge of the customer, a unique value proposition, and a winning profit model.

As industrial and technological change occurs, a well-managed enterprise is ready to quickly adapt to any challenge. Firms can manage risk by constantly asking about potential challenges to product, brand, and business model. Video-rental store Blockbuster failed to manage risk and lost to a new business model introduced by Netflix. Motorola, as another example, lost market share to Nokia after 2000 as mobile phone customers became more interested in ease of use and digital features. Ford and General Motors experienced great risk to their business design when Toyota introduced a hybrid auto called Prius. The Prius offers energy efficiency and a high-tech brand that eclipses the inefficiencies of traditional Fords and GMs.

During an economic recession, a heightened sense of risk of capital exists. However, a new venture success is sensitive to risks other than an economic downturn. The key factors for a start-up are the management quality of the founders and their earlier experience in the industry, as well as the timeliness of the new technology innovation. It is difficult to find investors during a recession, but they are available. Great entrepreneurs persevere by creating a frugal plan for attracting new customers with reasonably priced products. For many entrepreneurs with a good innovation, the time to act is during difficult economic times [Graham, 2008].

Innovations are the core element for the future success of a new venture in any economic conditions. Innovations can occur in times of capital market anxieties when resources are limited as well as during good times and financial exuberance. For example, many believe that innovation in the health-care system industry will occur in the period 2009–2012 in spite of the limited financial resources available [Mangelsdorf, 2008].

6.6 AgraQuest

From its inception, AgraQuest has experienced great uncertainty and risk. Risks arise from market uncertainty, organizational uncertainty, and process and regulatory uncertainty. AgraQuest has the potential to provide several environmentally friendly solutions to the pesticide problems of farmers. However, the market size and the willingness of farmers to use a new product are sources of significant risk. In addition, regulatory and legal controls of pesticides are large potential sources of risk.

AgraQuest stated in its original plan that it expected to receive approval for each product from the Environmental Protection Agency in 12 months. Actually, the EPA required a two-year approval process. On the other hand, from the beginning, AgraQuest has experienced few problems managing its organizational and management uncertainties. AgraQuest underestimated the market and regulatory risks but was very successful in mitigating its organizational and processes risk. Its scientific and technological capabilities and product selection and design led to few risks in the building of its technological success. Unfortunately, the market and regulatory problems resulted in a slower growth curve than planned for.

From its start in 1995 until 2002, AgraQuest was able to grow its annual revenue to only about $3.4 million. It had not yet been able to achieve economies of scale. Furthermore, only one product, Serenade, became available in 2003. Therefore, its product scope was very limited. Serenade was a foliar biofungicide that is approved for use on crops such as grapes, apples, pears, cherries, peanuts, and tomatoes. Complementary fungicides and insecticides, including Sonata and Requiem, were recently approved. (The musical names represent the movements in AgraQuest's integrated pest management symphony.) The firm is now scalable with proven, solid processes for product development and production; it is ready to grow.

AgraQuest sought to build a product line that could generate a disruptive application as chemical pesticides replaced natural pesticides. This was an attempt to develop a "killer app" in the agricultural pesticide industry.

6.7 Summary

A new venture is subject to risks due to uncertainty of outcomes in the marketplace. It is the entrepreneur's job to manage and reduce all risks. As ventures grow, they may experience economies of scale and scope, resulting in lower costs per unit produced. Furthermore, attractive ventures are those that can readily expand their capacity in response to demand and are said to be readily scalable. Many industries operate in a network of participants and exhibit network economies. As some network industries move toward an industry standard, a few firms may experience increasing returns. Significant value can be added to a product when a creative design leads to the embodiment of critical details.

Principle 6
The entrepreneur seeks to manage risks and attain economies of scale, scope, and networks while achieving scalability of the business.

Video Resources

Visit http://techventures.stanford.edu to view experts discussing content from this chapter.

Types of Risks	Jerry Kaplan	Winster
Managing the Risk/ Reward Trade-off	Bill Sahlman	Harvard Business School
Risk Is a Necessity for Exploration and Growth	Peter Diamandis	X PRIZE

6.8 Exercises

6.1   Select a well-known start-up or one that personally interests you. Use the sources of uncertainty outlined in Table 6.2 to discuss three key risk areas for that company in the immediate future. Find a recent article about the company and list which risks are discussed in that article.

6.2   An investor is asked to invest $10,000 in a new venture today. The expected return in three years is $28,000 with a probability of occurrence of 70 percent. Would you recommend this investment? Describe your reasoning.

6.3 A new entrepreneur is relatively risk-averse with a risk-adjusted constant $\lambda = 2$. Her opportunity cost is $100,000 before earning a regular salary from the venture in its second year. She also invests her savings of $50,000. Calculate the minimum annual return that will be acceptable to her in the second or third year.

6.4 Verify the empirical validity of Figure 6.9. Research the following questions using your favorite finance website or newspaper. What was the previous year's risk-free T-bill rate? What rate of return has the S&P 500 generated over the last year? Select a start-up that recently has gone public and determine its rate of return in the last year (or since the IPO).

6.5 The timing of the first revenue or customer shipment is an important milestone for any start-up. Research the average time until first revenue for Internet, biotech, and clean tech companies. If you were an investor in these three types of businesses, how would this impact your view of these businesses? How could this time risk be managed?

6.6 Amazon.com and Borders are both in the book and music-selling business. Contrast the scalability of each business.

6.7 Describe how social networking sites such as Facebook, MySpace, or Friendster have leveraged network effects to expand.

6.8 Section 5.1 discussed the first mover's versus follower's approaches. How does the importance of this decision change in a market with network effects and increasing returns?

6.9 Choose a recent technology standards battle (e.g., BluRay vs. HD-DVD, Flash vs. AJAX, WiMAX vs. 4G LTE, WiMedia vs. UWB Forum, 802.11 TGn vs. WWiSE, etc.). Succinctly outline the key differences between the two camps (technology or otherwise). Is there a start-up driving one of these two camps with an innovative technology? How was the eventual standards winner selected?

VENTURE CHALLENGE

1. Describe the major risks for your venture. How can you reduce these risks?

2. What is the potential for economies of scale and scope? Is this business scalable?

3. Describe the venture's potential for creating network effects.

CHAPTER 7

The Business Plan

The method of enterprising is to plan with audacity and execute with vigor.
Christian Bovee

How are ventures actually formed and what is the role of the business plan?

Entrepreneurs respond to attractive opportunities by forming new ventures. In this chapter, we consider the five-step process for establishing a new enterprise. One particularly noteworthy step in the process is the development of a story and a business plan. The story is a compelling synopsis of why this venture is needed at this moment of time and how it can achieve success. We then detail the task of writing a business plan, which is a significant and challenging effort for entrepreneurs. ■

7.1 Creating a New Business

Table 7.1 shows the five-step process for starting a new venture. The new venture will follow the steps to prepare a business plan that is suitable for the team as well as for the investors and business partners. This process is broad enough to apply to all types of businesses: independent or corporate, small or large, niche or broad, family or franchise, nonprofit, and one attempting a radical versus incremental innovation.

The corporate venture will also prepare a business plan suitable for review by the parent corporation to secure the necessary resources and assistance from its parent. The five-step process of Table 7.1 can be used for corporate ventures where the investors of step 5 will normally be the parent firm. We defer until Chapter 8 and Chapter 10 the discussion of corporate ventures and the appropriate legal form a venture should adopt.

A talented leader of a new venture has a vision and a plan to implement it. He or she can motivate people and manage information and resources so that the business can create a profit. The performance of a new venture is a consequence of factors that encompass the dimensions of a business as portrayed in Figure 7.1. The quality of the opportunity and its fit with the vision lead to the accumulation or creation of the distinctive competencies based on resources and capabilities. Following from a set of competencies, a business strategy is created based on novelty or innovation within an industry context. The attractiveness of the industry with respect to its business opportunities affects the profit potential and the expected return of the venture. Access to resources and the ability to attract the entrepreneurial team will depend on the attractiveness of the new venture. The industry environment will determine the amount of resources available to the venture since capital typically flows to industries in which opportunities are abundant and attractive. Furthermore, the industry-related competencies of the entrepreneurial team, which refers to the

TABLE 7.1 Five-step process for establishing a new venture.

1. Identify and screen opportunities. Create a vision and concept statement, and build an initial core entrepreneurial team. Describe the initial ideas about the value proposition and the business model.

2. Refine the concept, determine feasibility, and prepare a mission statement. Research the business idea and prepare a set of scenarios. Create a story and the outline of a business plan with an executive summary.

3. Prepare a complete business plan including a financial plan and the legal organization suitable for the venture.

4. Determine the amount of financial, physical, and human resources required. Prepare a financial model for the business and determine the necessary resources. Prepare a plan for acquiring these resources.

5. Secure the necessary resources and capabilities from investors, as well as new talent and alliances. Launch the organization.

FIGURE 7.1 Building a new venture.

level of experience and knowledge with the industry, will lead to the expectation of greater success.

The identification and acquisition of required resources and capabilities are crucial for a firm's success. For a fast-growing firm based on continuing innovation, the intellectual resources are critical to success. The securing of the necessary resources and capabilities may occur in stages, as required.

The creation of a business plan focuses on the stages of building a business, as portrayed in Figure 7.1. The initial formation of a suitable structure for the business follows the business strategy. The ability to remain competitive and innovative while operating through an appropriate structure can lead to enduring profitability.

The greatest risk in creating a business is the failure to complete all the steps of Table 7.1 and Figure 7.1. Some entrepreneurs who possess strong technical skills and capabilities unfortunately skip the formation of a business strategy that will lead to profitability. Another risk is that the business plan may have an inadequate plan for the organizational structure and the management of processes and talent. The road to success starts with a new venture story.

7.2 The New Venture Story

Stories are an integral part of the process by which founders start and build new ventures, acquire needed resources, and generate new wealth [Lounsbury and Glynn, 2001]. Storytelling is applicable to corporate new ventures as well as independent new ventures. Stories play a critical role in the process of the emergence of a new business. Stories can lead to favorable interpretations of the social benefits and wealth-creating possibilities of the venture, thus enabling resources to flow to the new enterprise. Furthermore, they can legitimize new ventures, thus helping to build acceptance of them. A good entrepreneurial story attracts financial and human resources, and builds industry acceptance. It can help amass support for a new venture. Successful entrepreneurs must shape interpretations of the nature and potential of their new venture for those who may supply needed resources.

A story is a narrative of factual or imagined events. It depicts a course of challenge, plan, actions, and outcomes in a manner similar to a plan. A good story and business plan define an opportunity, a concept, cause and effect, and an outcome all held together in a holistic way. Stories use plot lines and twists to capture the imagination and interest of the listener. They tell the ultimate goal of the venture, the ideological challenge, and the means of achieving the goal. The creation of an appealing and coherent story may be a useful form of communication for entrepreneurs in the attempt to attract interest and support for their idea and plan. To be effective, the content of the story must align with the interests and background of the listener. A well-crafted story about a new venture emphasizes the goals and merits of the venture. Telling a compelling story inspires belief in a venture's motives, character, and capacity to reach its goals [Ibarra and Lineback, 2005].

As founder, Jim Clark provided the story for both Netscape and Healtheon as mentioned earlier when recounting to Michael Lewis [2000]:

> It dawned on Clark that the food chain of capitalism was missing a link, and that, if he summoned the nerve to hoist himself up, he could be that link. And that if he didn't have the nerve to do so he would make a mockery of his entire remarkable climb. . . . His role in the valley was clear: *he was the author of the story*. He was the man with the nerve to invent the tale in which all the characters—the engineers, the [venture capitalists], the managers, the bankers—agreed to play the role he assigned to them. And if he was going to retain his privilege of telling the stories, he had to make sure that the stories had happy endings.

In constructing a legitimate identity for a new venture, the storyteller tries for a balance of alignment with existing challenges and the potential for distinctiveness. The investor may be drawn by the credibility of the story and the storyteller to seriously consider investment.

New firms should not rely solely on a pure bullet-point presentation. A list of bullet points reduces a set of issues to a few points but provides little fabric or motivation. Bulleted lists are often generic in meaning and leave challenges and relationships unspecified. Furthermore, they often leave critical assumptions unstated. A good story includes all the challenges, relationships, and assumptions in the very fabric of the narrative. A

TABLE 7.2 Three stages of the story.

1. **Set the stage:** Define the current situation, the current players, and the opportunities coherently and clearly. Life and business are in a delicate balance and the listener cares about the situation.

2. **Introduce the dramatic conflict:** An inciting incident or need throws life out of balance. Describe the challenges and opportunities, and the need for a coherent plan to proceed toward success and a new balance.

3. **Reach a resolution:** Portray a coherent plan describing how the new venture can overcome the obstacles and succeed by following the plan.

presentation can use some bullet-point slides but should emphasize the story it wants to tell.

The new venture story consists of three elements, as shown in Table 7.2 [Shaw et al., 1998]. The first step is the setting of the stage by describing the current conditions of the industry, society, and the existing relationships and opportunities. The story should be about a person or group whose challenges the listener can relate to.

Next, the storyteller introduces the dramatic conflict by describing the challenges confronting the venture, the need for a plan to overcome the challenges at this turning point, and the critical obstacles and issues. Finally, in the third step, the storyteller describes the plan to overcome the obstacles, secure the necessary resources, and move on to resolution and success [Ibarra and Lineback, 2005].

Jim Clark contracted an illness in late 1995 that gave him personal experience with the health-care industry, its bureaucracy, and resulting paperwork demands. He responded with a concept: The patient would have a password and a digital record, and the doctor would use the Internet for billing forms with no hassle for both parties. He drew a diagram of the four players in the health-care industry, as shown in Figure 7.2. He then placed his new venture in the center of the diagram as the solution, via the Internet, for the entire industry. Clark told a story of critical importance to his nation, described the dramatic challenges and opportunities, and then portrayed the solution: his new venture. With his compelling story, Clark went on to raise millions of dollars to build Healtheon, which later was sold to WebMD [Lewis, 2000].

A well-told narrative plan that shows a difficult situation and a novel solution leading to improved market conditions can be galvanizing. When listeners can locate themselves in the story, their sense of commitment and involvement is enhanced. By conveying a powerful impression of the process of succeeding, narrative plans can motivate and mobilize resources and investors.

Stories help entrepreneurs deliver direction and inspiration more powerfully than a logical argument. Facts convey information, and stories convey

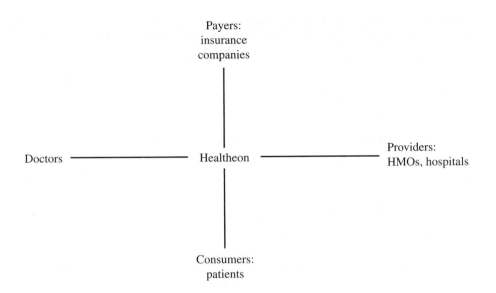

FIGURE 7.2 Healtheon diagram.

meaning. Entrepreneurs need facts to back up their plans and a story to convey the goals and meaning of the venture [Gargiulo, 2002].

Like a good book or movie, a story needs an "inciting incident," or turning point, and a good ending. With the help of colleagues and allies, the entrepreneur turns challenges and obstacles into new opportunities and an outcome desired by all.

BET: Selling a Story

In 1980, Robert Johnson created an innovative plan to produce television programs targeted to black viewers. Johnson met with several investors and told them his story of the opportunity to serve black viewers with TV programs that resonated with their values and experiences. The power of this story generated significant financial and cable channel resources. As a result, Black Entertainment Television (BET) was launched successfully in the United States (see www.bet.com).

A great story well told will pack enough power to be memorable. We can forget bullet points, but a great story is unforgettable. It displays the struggle of expectation and hope versus reality and challenge [McKee, 2003]. An example of an entrepreneur's story follows:

My father and I were very close. In 1999, he exhibited congestive heart disease that appeared to be untreatable. While in the hospital for further tests, he died in the middle of the night. These tests were inadequate and failed my father and me. I have found and licensed a patent for a new blood test, but the Food and Drug Administration has been slow to respond, and the inventor has gone on to other issues. Our early tests have shown good results for a low-cost test that can illuminate paths to a cure.

Our firm, Heartease, needs $1 million to complete our FDA certification. We have the data and proof of the test's efficacy. We need you on our team with your insights and money. Together, we can save your father or mother.

7.3 The Business Plan

Once an entrepreneurial team has selected an opportunity that is attractive and feasible, it often develops a detailed plan for describing the elements of the proposed business. A **business plan** is a document that describes the opportunity, product, context, strategy, team, required resources, financial return, and harvest of a business venture. The elements of the business plan are listed in Table 7.3. This business plan can be used for many purposes, such as attracting talented individuals and resources to the venture. Of course, there is no one right way to organize and write a business plan. Many successful companies never had a formal business plan [Bhidé, 1994]. Although plans are never perfect, they help entrepreneurs nail down key details by forcing them to formalize their ideas.

As stated by Daniel Hudson Burnham (1909):

Make no little plans. They have no magic to stir men's blood and probably themselves will not be realized. Make big plans; aim high in hope and work, remembering that a noble, logical diagram once recorded will never die, but long after we are gone will be a living thing, asserting itself with ever-growing insistency. Remember that our sons and grandsons are going to do things that would stagger us. Let your watchword be order and your beacon beauty.

TABLE 7.3 Elements of a business plan.

■ Executive summary	■ Entrepreneurial team: capabilities, commitment
■ Opportunity: quality, growth potential	
■ Vision: mission, objective, core concept	■ Financial plan: assumptions, cash flow, profit
■ Product or service: value proposition, business model	■ Required resources: financial, physical, human
■ Context: industry, timeliness, regulation	■ Uncertainties and risks
■ Strategy: entry, marketing, operations, market analysis	■ Financial return: return on investment
■ Organization: structure, culture, talent	■ Harvest: return of cash to investors and entrepreneurs

The business plan is a blueprint for your business. The clarity of this plan will be enhanced by the entrepreneurial team over a period of weeks or months. It enables the team to see clearly the plan of action and understand all the elements. Single entrepreneurs might not need a written plan if they are using their own resources, but they still will benefit from writing the plan. They can show it to advisors who can help them shape a better plan.

If the venture requires outside financing, a business plan will be required by most investors. Listing the assumptions underlying the financial plan will be helpful to all participants. Most useful plans number less than 20 pages with backup and supporting material available on request. A business plan is a reflection of its authors and should be the result of a team effort.

A business plan is an important part of the business building process. However, we recognize that many small businesses start with a modestly crafted plan and build it up over the first months after launch. A business plan is a necessity, but it may not be required in a formal and complete form in order to start. For some, the business plan may be conceived in conversations and recorded as notes by the entrepreneur. Informed action is what is needed for entrepreneurship. Eventually, however, the entrepreneur will want a formal business plan to show to investors, bankers, or potential executives. In a dynamic industry such as semiconductors or nanotechnology, flexibility is key to success, and rigid adherence to a plan may be too limiting [Gruber, 2007]. The writing of a plan may be a useful exercise for the founder team, but the team must recognize that the plan will need to be updated periodically.

After the business plan is outlined or drafted, an executive summary is written to explain the venture's story. It should be concise and about two pages in length. Some investors and potential team members will read the summary to determine if they wish to proceed further. For many investors, the executive summary will be sufficient to initiate discussions. As explained above, often entrepreneurs will write an initial narrative of the story as a first step and later revise it to become the executive summary.

Creating a business plan teaches the team about the market, customers, and each other. Working through creating a business plan will enable the entrepreneurs to estimate when cash flow will turn positive. A plan will probably turn up at least one or two big gaps, which can be corrected. Table 7.4 lists 10 common gaps that show up in a business plan. The entrepreneurs can respond to these gaps or mistakes and seek to fix them.

An example of a successful and solid business plan was that used by Juniper Networks. Pradeep Sindhu had over 10 years experience in the high performance personal computing industry at Xerox PARC and Sun Microsystems. In 1995, he left Xerox in order to develop a business idea and plan that would capitalize on his experience and skills and that would not be easily imitated. Eventually, Sindhu decided to focus on wide area networking and to

TABLE 7.4 Ten common mistakes or gaps in business plans.

- Solutions or technologies looking for a problem
- Unclear or incomplete business model and value proposition
- Incomplete competitor analysis and marketing plan
- Inadequate description of the uncertainties and risks
- Gaps in capabilities required of the team

- Inadequate description of revenue and profit drivers
- Limited or no description of the metrics of the business
- Lack of focus and a sound mission
- Too many top-down assumptions such as "we will get 1 percent market share"
- Limited confirmation of customer demand or pain

work with the packet-switching technology known as Internet Protocol (IP) to create routers. Router technology fit Sindhu's skill set, and was an attractive area because the number of Internet users was growing almost exponentially. Despite the rapid growth in Internet usage, Sindhu recognized that the hardware underlying the Internet was outdated.

Sindhu began by approaching Vinod Khosla at Kleiner, Perkins, Caufield & Byers (KPCB), a prestigious venture capital firm, and using his experience at Sun to demonstrate his knowledge in the area. Before meeting with KPCB, Sindhu had identified the top 10 software engineers in the world that had expertise in the areas he would need to create his routers. With KPCB's assistance, Sindhu was able to recruit the needed engineers and an experienced CEO, Scott Kriens, in order to complete the business plan and gain additional venture capital funding for the venture.

Despite humble beginnings, Juniper was able to take on industry giant Cisco. Sindhu's careful approach to the business plan enabled Juniper to successfully launch an innovative IP-routing platform, which challenged Cisco's existing platform both in terms of capacity and speed.

Entrepreneurs like Sindhu create and build new business ventures. They break old rules and make new ones. They exploit tools and technologies and create new markets. They are particularly adept at matching the opportunities with the competencies of the new venture team. They observe and understand customers and how their needs change over time. Their passion emerges as they find opportunity.

The business plan can be used to align the interests of all the participants of a new venture, as shown in Figure 7.3. The business plan explains how the people, the resources, and the opportunity can be linked to a deal that will hopefully benefit all stakeholders—employees, investors, suppliers, and allies [Sahlman, 1999]. See Section 2.3 for a further discussion of proper fit and alignment necessary for success.

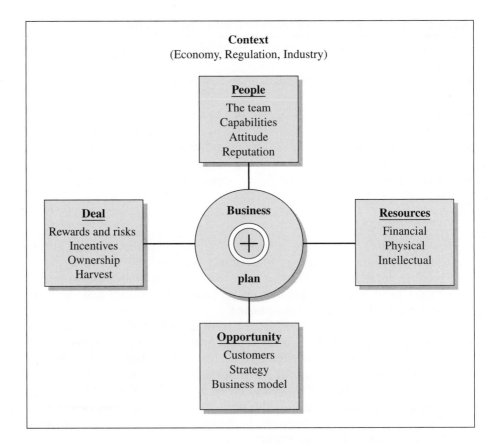

FIGURE 7.3 The business plan serves as the alignment tool for a new business venture.

7.4 An Annotated Table of Contents

Writing a complete business plan forces the entrepreneurs to crystallize key business issues in their minds. There are many ways to structure a business plan, with various references, structures, and templates available. The business plan process focuses on the venture's key success factors by posing questions that must be addressed for an idea to become a true opportunity. The remaining content of this section proposes questions for each part of a typical business plan. An example of a well-prepared plan is provided in appendix A. See appendix C and the textbook's websites for additional plans and resources.

Executive Summary

The executive summary is the most important part of the business plan. Many investors make their decision to proceed with further discussions (e.g., due diligence) based on this single section. The executive summary should encapsulate the key positioning and reasoning found in the rest of the business plan. Both the vision and mission statements should assist in succinctly communicating a compelling opportunity.

- Why is this a big problem and why are customers willing to pay for solutions?
- How does the venture plan on solving the customer problem or need?
- Why is this venture uniquely positioned to do this?
- How attractive are the economics? Why is this an exciting growth opportunity?
- Who is the team, and what key partnerships are in place?

I. Opportunity and Market Analysis

Investors like funding big problems representing large opportunities. Start strong by demonstrating a solid understanding of the customer and why this problem or pain is important to him or her. Performing customer segmentation will convince readers that the venture can grow to the size of the addressable market.

- What is the problem or need being solved by the venture?
- Who is the customer or customer segment(s)?
- How large is the total addressable market, and how much is it growing?
- How is the current market context favorable or not?

Reference: Chapters 2, 3, and 4

II. The Solution and Concept

Many product descriptions lose credibility by being too "big picture," relying on market hype, or being too product-focused in the hopes that simply explaining the technology will make it evident why the idea is valuable. It is important to balance the use of technical or industry-specific lingo with common, everyday language. In addition, detailing a value proposition and business model will ensure that the economics of the business are sound from the start.

- What is the product or service?
- Describe a "day-in-the-life" of the customer before and after adoption of the solution. What is the value proposition to a customer, and why it is compelling for the customer?
- Which customers have validated the product and are willing to pay for it?
- What is unique and defendable about the business?
- What is the business and economic model? How attractive are the financial margins?

Reference: Chapters 3, 4, 5, and 16

III. Marketing and Sales

This section of the business plan should clearly communicate an understanding of how to successfully market and sell your product to the identified customer segments. Understanding and communicating your customer development strategy are as important as your product development strategy. It should be done in sync with product development to improve the odds of success. Your business model and pricing strategy should also extend clearly into the selected sales strategies.

- What are the most appropriate marketing mediums to reach the customer segment(s)?
- What is the most appropriate type of sales channel for the product (e.g., direct vs. indirect sales)?
- Who is the customer decision maker with purchasing power, and who influences that person to buy?
- What is the expected sales cycle length?
- Are there partnerships that can be leveraged to advertise and sell?

Reference: Chapters 4 and 11

IV. Product Development and Operations

At this point, the reader should be convinced that the entrepreneurs have identified a compelling market and know how to generate revenue. This section focuses on product development and how the product will be made marketable. Any key technologies being leveraged for development should be explained clearly (e.g., diagrams are helpful). Demonstrate that continued revenue growth is planned by specifying long-term product goals. This section largely drives the amount and timing of cash required for the business, making it a critical component of the financial model.

- What is the current state of development of the product(s)?
- What resources will be required to finish and ship the product? Be specific about what types of resources will be required (e.g., engineering, tools, suppliers, materials, partners, and customer involvement).
- What are the planned development timelines and key milestones targeted?
- What are the key risks that will be mitigated at each milestone?
- What does the value chain look like for production and product delivery?
- Do any patents, trade secrets, or other defendable advantages exist?
- Are there any regulatory hurdles that must be cleared?

Reference: Chapters 5, 13, and 14

V. Team and Organization

Building a team is a critical part of beginning a new venture and communicating credibly to outside parties. Understanding how the current team fits into the broader venture vision will help investors and partners understand what roles remain to be filled and how they can potentially assist.

- What are the backgrounds and roles of the founders and early key employees?

- Describe the passions and skills of the team and why the team is committed to the opportunity.
- What key hires must be made to fill out the team?
- What head count levels are forecasted in each functional department?
- Does the company have advisors or board members that strengthen its story?

Reference: Chapters 10, 12, and 20

VI. Risks

A new venture is confronted with four major types of risk: technology/product, market/competition, management/team, and financial. Many of the opportunity-specific risks are interwoven into earlier parts of the business plan. For instance, potential competitive threats should be considered when framing both the product and go-to-market sections. In addressing company-specific risks, it is important to think clearly about how each risk factor can be managed in the coming year or two. Quantitative analysis will also aid the reader. It is critical to identify which risks need to be reduced so that the reader will be confident that the entrepreneurs understand how to build a business.

- What are the key product development risks and external dependencies?
- What is being done to mitigate product execution risks?
- Who are your main competitors, and how are you differentiated from them in the marketplace?
- Can large players easily enter the market? Are there product substitutes?
- What customer, partner, or product strategies can be used to mitigate competitive threats?

Reference: Chapters 4 and 6

VII. Financial Plan and Investment Offering

Although the financial plan is considered last, the implications of financial decisions appear throughout. If the company executes successfully across product development, marketing, sales, and other company functions, the financial results must be attractive enough to make an investment. Ensure that any financial assumptions and results are feasible by citing an enterprise that is analogous to the planned venture. Investors want to know how much funding is required and what measurable milestones will be reached. Staged financing allows both investors and entrepreneurs to better manage the risk associated with a new venture. Include a timeline that integrates company sales and product milestones, planned funding events, and cash flow position.

- What is the required funding to meet the market and product milestone goals? What amount is being asked for?
- When is the venture forecasted to be cash-flow positive?
- What is the growth opportunity of the business if successful?

- What are the forecasted initial and steady-state financial margins?
- What other companies exhibit margins and growth similar to this venture?
- What are the key financial assumptions?

Reference: Chapters 15, 16, 17, 18, and 19

Appendix. Detailed Financial Plan

A more detailed set of financial projections and assumptions are generally included in an appendix. The forecasted financials and assumptions will serve as a starting point for valuing the venture. Be sure the methodology used to arrive at the financials is transparent to the reader.

- Five-year detailed cash-flow statement, income statement, and balance sheet (monthly for first year and quarterly or yearly thereafter).
- Financial assumptions made in the construction of the financial estimates (e.g., customer penetration rates, pricing, and working capital assumptions).
- Are purchasing decisions cyclical in this industry?
- What are the largest costs of the business (e.g., engineering development, regulatory trials, manufacturing, and marketing)?
- How will product and sales costs change as volume grows?
- Has customer support and maintenance been factored in?

Reference: Chapters 16 and 17

7.5 AgraQuest

Natural pesticides potentially offer a host of benefits, from easier regulation to flexible application timing. But for all growers, the most important element is field performance. You cannot sell a product just on its environmental friendliness. It has to be efficacious, easy to use, and reliable. AgraQuest prepared a plan to develop natural pesticides as an alternative to chemicals.

AgraQuest completed its first business plan dated May 5, 1995, with the purpose of raising $3.6 million from private investors. The 40-page plan had a table of contents, as shown in Table 7.5. The 1995 executive summary is shown in Table 2.16.

AgraQuest's plan described its unique competitive advantage as follows:

Successful natural product discovery is not a simple matter of screening any group of microbes against a range of targets. It requires sophisticated knowledge of the specific microbial groups, locations, types of chemistry, and screening (isolation, fermentation, bioassay) techniques that yield the highest numbers of bio-active natural products. In addition, the type and style of people management can either enhance or hinder natural product discovery and development. For each step of the discovery process, AgraQuest has unique know-how and proprietary techniques that generate more novel, proprietary pesticidal natural products faster than others in the field.

The business plan identified the uncertainties and risks and described the keys to success, as listed in Table 7.6.

TABLE 7.5 Table of contents of AgraQuest business plan (1995).

Table of Contents
■ EXECUTIVE SUMMARY
■ THE COMPANY—Company Ownership, Products, Facilities
■ SIZE OF MARKET AND MARKET TRENDS—World Pesticide Market
■ TECHNOLOGY
■ DEVELOPMENT STRATEGY AND PROCESS
■ MANUFACTURING AND PRODUCT COSTS
■ MARKETING STRATEGY
■ AGRAQUEST'S COMPETITIVE ADVANTAGE
■ BUSINESS STRATEGY AND STRATEGIC ALLIANCES
■ MILESTONES/GOALS
■ ORGANIZATIONAL STRUCTURE
■ PERSONNEL PLAN
■ MANAGEMENT TEAM
■ SCIENTIFIC ADVISORY BOARD
■ BOARD OF DIRECTORS
■ KEYS TO SUCCESS
■ FINANCIAL PLAN AND PROJECTIONS

TABLE 7.6 Keys to success for AgraQuest from the 1995 business plan.

1. Recruitment of talented, experienced scientists who can function in a team environment.
2. Effective teamwork in the management group.
3. Good board–management team working relationship.
4. Enough money to hire a critical mass of experienced scientists, build sufficient laboratories, and purchase necessary equipment.
5. Development of strategic relationships with large agrochemical companies.
6. Proprietary protection on discoveries.
7. EPA registration as microbials or biochemicals.
8. Aggressive licensing activities to bring in product candidates from outside the company.

7.6 Summary

Building a business is described as a process that can be learned and mastered by talented and educated entrepreneurs by developing the following descriptions:

- Opportunity, vision, value proposition, and business model.
- Concept, feasibility, and story.
- Financial plan, legal form, and business plan.
- Resource acquisition plan.
- Execution and launch process.

We note that entrepreneurs have a story and business plan that helps them to codify and communicate their roadmap to achievement. Most entrepreneurs benefit from putting this plan into written form to sharpen their thinking and consistently communicate it to others.

> **Principle 7**
> Entrepreneurs can learn and master a process for building a new venture and they communicate their intentions by developing a story and writing a business plan.

Video Resources

Visit http://techventures.stanford.edu to view experts discussing content from this chapter.

Purpose of a Business Plan	Tom Byers	Stanford
Make Meaning in Your Company	Guy Kawasaki	Garage

7.7 Exercises

7.1   TerraPass sells an investment in energy projects or credits used to offset environmental impact. For example, drivers of gas-guzzler cars can exhibit a decal showing their environmental investment. Create a summary for this business (www.terrapass.com).

7.2   A new firm develops and distributes electronic games for mobile devices. These games teach children to read, recognize symbols, and perform mathematics. This new venture needs $1 million to launch a nationwide campaign for its products. Prepare a short story that will persuade a venture capitalist to support the firm.

7.3   Evaluate the business plan from appendix A using tables and figures from Section 7.3.

7.4   For whom is a business plan written and why? Using Figure 7.3, explain how a business plan serves as an alignment tool for key stakeholders in the business.

7.5   What information on the competitive landscape should be included in a business plan? What frameworks would you use?

7.6   What information on the key venture risks should be included in a business plan? Is it important for the business plan to be pessimistic or optimistic in regard to these risks?

VENTURE CHALLENGE

1.   Create a draft table of contents for your venture's business plan.

2.   Describe the process you will use to create a business plan.

3.   Write an executive summary for your opportunity.

CHAPTER 8

Types of Ventures

Even if you are on the right track, you'll get run over if you just sit there.
Will Rogers

What forms do new businesses take and what are corporate ventures?

The appropriate legal and organizational format used to establish a new venture will vary according to several factors such as context, people, legal and tax consequences, and cultural and social norms. In this chapter, we consider the various organizational and legal forms that entrepreneurs employ to achieve their objectives. New ventures can range from small business or consulting services to high-growth, high-impact enterprises. Other organizations start out operating in a niche market and grow into a broader one. Important social enterprises are often established as nonprofit ventures. Two other forms include family-owned businesses and franchises.

An important contrast to these independent ventures is the corporate new venture. It emerges within larger existing enterprises and is granted autonomy so it can fulfill its promise. Corporate new ventures are an important part of the entrepreneurial world and account for many new innovations. Often constrained by existing commitments and capabilities, corporations can fail to respond to significant new opportunities. Well-planned corporate ventures, however, help refresh and strengthen large corporations. ■

8.1 Independent Versus Corporate Ventures

Many purposes exist for establishing new ventures in different formats. Table 8.1 describes five types of new ventures. Each of these types has a set of characteristics that distinguishes it. We can describe a **small business** as a sole proprietorship, a partnership, or a corporation owned by a few people. Examples include consulting firms, convenience stores, and local bookstores. Typically, a small business has fewer than 30 employees and annual revenues less than $3 million.

A **niche business** seeks to exploit a limited opportunity or market to provide the entrepreneurs with independence and a slow-growth buildup of the business. This business might employ fewer than one hundred employees and have annual revenues of less than $10 million. On occasion, a niche business can grow over time into a large, important enterprise.

A **high-growth business** aims to build an important new business and requires a significant initial investment to start up. A **radical-innovation business** seeks to commercialize an important new innovation and build an important new business. These enterprises are the primary focus of this textbook.

A **nonprofit organization** is a corporation or a member association initiated to serve a social or charitable purpose. Thousands of new nonprofits are organized every year to serve important social needs throughout the world. A well-known nonprofit is the International Red Cross (www.ifrc.org).

Another type of new venture is started by existing corporations for the purpose of building an important new business unit as a solely owned subsidiary or a spin-off as a separate company. This activity can be called a **corporate new venture (CNV).**

TABLE 8.1 Five types of new ventures.

Type	Revenue growth	Planned-for most likely size	Description	Objective
1. Small business	Slow	Small	Sole proprietorship, family business	Provide independence and wealth to partners by serving customers
2. Niche	Slow to medium	Small to medium	Exploits limited opportunity or market	Provide steady growth and good income
3. High growth	Fast	Medium to large	Needs large initial investment; could seek disruptive innovation	Important new business
4. Nonprofit organization	Slow	Small to medium	Serves members or a social need	Serve a social need
5. Corporate new venture	Medium to fast	Large	Independent unit of an existing corporation	Build an important new business unit or separate firm

An **independent venture** is a new venture not owned or controlled by an established corporation, which includes the first four types in Table 8.1. An independent venture is typically unconstrained in its choice of a potential opportunity, yet is usually constrained by limited resources. The corporate venture is usually constrained in choice of opportunities to those consistent with the parent business. The corporate venture, however, usually has access to the significant resources of the parent firm [Shepherd and Shanley, 1998].

While independent and corporate ventures both face the same external context, their different competencies and resources cause them to develop different strategies. The independent venture has more flexibility and potentially requires fewer resources than the corporate venture. Furthermore, the independent venture has access to a wide range of advisors, while the corporate venture is advised and controlled by the parent company. Thus, the independent venture has the advantage of flexibility, adaptability, and high incentives, while the corporate venture is usually advantaged by its access to valuable capabilities and resources.

The bulk of this chapter focuses on corporate new ventures. First, we examine nonprofits and family-owned businesses.

8.2 Nonprofit and Social Ventures

The purpose of a new venture is to create wealth for society. Often, wealth is seen as financial wealth. But, many entrepreneurs seek to provide social wealth for their society. The product of a nonprofit hospital is a healthy patient. Only for the tax collector does it make a difference whether the hospital is a nonprofit or for-profit. A nonprofit organization is a corporation, member association, or charitable organization that provides a service but does not earn a profit. A nonprofit organization is permitted to generate a financial surplus but may not distribute this surplus to officers, investors, or employees. Furthermore, nonprofits have no owners. Any surplus must be used for the approved nonprofit mission of the organization. Today, nonprofit organizations are often called not-for-profits or NGOs. One out of every two Americans is estimated to work as a volunteer in the nonprofit sector. See appendix C for websites about nonprofits.

Nonprofit organizations have traditionally operated in the social sector to solve or mitigate such problems as hunger, homelessness, pollution, drug abuse, and domestic violence. They have also helped provide certain basic social goods, such as education, the arts, and health care, that society believes the marketplace may not adequately supply. Nonprofits have supplemented government activities, contributed ideas for new programs, and functioned as vehicles for private citizens to pursue their own vision of the good society.

The product of the Girl Scouts is a mature young woman who has values, skills, and respect for herself. The purpose of the Red Cross is to enable a community hit by natural disaster to regain its capacity to look after itself. In this way, the nonprofit venture forms to respond to social need.

The decision to start a nonprofit venture will be determined by the nature of the social opportunity and the creation of an innovative response that cannot or should not be performed for profit. Social functions that depend primarily on volunteers or members such as churches, museums, theaters, social clubs, industry associations, credit unions, and farmers' cooperatives are usually formed as nonprofits.

The establishment of a nonprofit organization should follow the five steps of Table 7.1. The vision for the organization is defined in terms of the creation of social value rather than economic value. Many nonprofits can be described as socially conscious service organizations. Thus, the entrepreneurial team must be committed to the social values of the new venture with its attendant risks and uncertainties.

Once the business plan is written, the required financial and human resources must be determined. How will the necessary funds be acquired and the human talent attracted? The challenge is finding donors whose special interests match those of the new nonprofit venture and who also have the expertise and commitment to provide an independent check on management judgment.

Organizations that satisfy the conditions of section 501 (c)(3) of the Internal Revenue Service Code are called charitable organizations. These organizations have religious, educational, scientific, literary, and/or charitable aims. Donations to these organizations are tax deductible for the donor and exempt from estate taxes. Noncharitable nonprofit organizations are primarily set up to serve the purposes of their members and are also tax exempt, but donations are not normally tax deductible. Nonprofits often contemplate the establishment of a related business that will generate a net surplus. Often, however, they underestimate the costs and are overly optimistic regarding revenues [Foster and Bradach, 2005].

Leading a nonprofit requires the competencies needed for most businesses mixed with a commitment to the entity's social cause. Michael Miller leads Goodwill Industries of the Portland, Oregon, area and has built it to $46 million in revenues by using his entrepreneurial skills [Kellner, 2002]. The strategies and approaches discussed throughout this book are important for nonprofits. For example, nonprofits should leverage technology to improve the efficiency of operations and the reach of their activities. They should also consider partnering with complementary organizations. The nonprofit Women in Technology partners with IBM to cohost an engineering camp for middle-school girls, drawing on IBM's resources and expertise [Austin et al., 2007].

Nonprofits also face special challenges. For example, nonprofits often find it difficult to agree on who their customer is. For the Red Cross, is it the hospital, the blood donor, or the financial donor? Who is the ultimate beneficiary of the service? Since nonprofits are not subject to the market in the same way that traditional firms are, it is critical that they constantly evaluate the effectiveness and efficiency of their activities [Bradley et al., 2003]. Managed properly, the nonprofit sector can spawn very big, high-impact ventures in the same way that the for-profit sector does. For example, the nonprofit sector delivers much of the health care for most nations.

A unique form of nonprofit corporation is a consumer cooperative, which is a business that belongs to the members who use it. The member/owners establish policy, elect directors, and often receive cash dividends. Examples of cooperatives include credit unions, housing co-ops, food co-ops, and utility co-ops. In 1938, mountain climbers Lloyd and Mary Anderson joined with 23 fellow climbers in the Pacific Northwest to found Recreational Equipment, Inc. (REI). The group formed a consumer cooperative to supply themselves with quality gear, clothing, and footwear selected for performance and durability for outdoor recreation, including hiking, climbing, camping, bicycling, and other sports. After more than six decades, REI has grown into a supplier of specialty outdoor gear currently serving more than three million active members through 90 retail stores in the United States and direct sales via the Internet (www.rei.com), telephone, and mail.

Social ventures sometimes take the form of a nonprofit. As discussed in Chapter 4, a concern for all stakeholders, including society at large, is crucial to the long-term success of any venture. Social ventures are distinguished from other ventures by their core value proposition. The value proposition for a traditional venture is designed to create financial profit by organizing to serve markets that can afford a new product or service. Social ventures aim for large-scale and transformational value that accrues to some segment of society or to society at large. They are cognizant of financial realities. But, they do not anticipate or organize to obtain significant financial profits [Martin and Osberg, 2007]. A social venture may take the form of a nonprofit or a for-profit.

Founded by Larry Brilliant, Google.org is the charitable branch of the Internet company Google. Its mission is to fight global poverty, fund new energy solutions, and protect the environment. The organization has over $75 million in investments spread throughout a number of different enterprises. It was granted three million shares during Google's IPO and Google continually contributes 1 percent of its annual profits to Google.org. Some of its main projects include a plug-in electric vehicle that gets 100 miles per gallon and a renewable energy project designed to produce electricity at a profit from wind and solar sources. A long-term goal for the organization is to develop a way to fund Google's massive power consumption from renewable energy sources. Google's data centers consume an enormous amount of electricity every year. Google.org is hoping to fund new technologies that can help reduce this cost.

David Green: Social Entrepreneur

David Green has started several small businesses in developing countries to make inexpensive medical devices such as intraocular lenses and hearing aids. These profitable businesses make low-cost devices that meet the needs of the poor without sacrificing quality [Kirkpatrick, 2003]. See www.aurolab.com.

A **social entrepreneur** is a person or team that acts to form a new venture in response to an opportunity to deliver social benefits while satisfying environmental and economic values. Social entrepreneurs focus on the social welfare of their customers or clients, while remaining cognizant of the economic and environmental costs and benefits. The goal is to harness innovation for social and public good [Jackson and Nelson, 2004].

Social entrepreneurs are, first and foremost, entrepreneurs. Thus, like all entrepreneurs, they recognize and relentlessly pursue opportunities; they act boldly without being limited by their current resources; and they engage in a process of continuous innovation, adaptation, and learning [Dees et al., 2002]. Social entrepreneurs focus on the scalability of their ventures, ultimately seeking to cause a large-scale transformation and not just a local effect [Martin and Osberg, 2007]. For this reason, they are especially cognizant of the importance of technology in facilitating their efforts.

Key Success Factors for Social Entrepreneurs

In an annual competition, Fast Company magazine and the Monitor Group pick 20 outstanding leaders who have established successful social ventures. Nominations are judged according to five key criteria: entrepreneurship ("the ability to do a lot with a little"), innovation ("a unique and dramatic idea"), social impact ("both immediate impact and broader systemic change"), aspiration ("thinking big, but with pragmatism"), and sustainability ("building an organization that adapts to change").

One of the 2005 winners was KickStart (formally ApproTEC), because of its The MoneyMaker treadle irrigation pump. It allows poor farmers in Kenya and other countries to irrigate their crops in an efficient and modern manner. Before KickStart introduced the MoneyMakers, farmers had to irrigate using buckets since motorized pumps were too expensive and were designed for larger fields. KickStart introduced its pump in order to address the smaller farmers. It is sold through for-profit companies. The MoneyMaker is an example of a for-profit product created for a social development purpose [Dahle, 2005]. See www.kickstart.com.

Social entrepreneurs hold certain advantages over entrepreneurs in other circumstances. With a mission as the guiding vision, social entrepreneurs strive to organize and deploy diverse resources. They can engage volunteers, customers, partners, and investors through a sound business plan that furthers the organization's mission. In the social sector, success in the enterprise equals significance through improved lives and healthy communities. Social entrepreneurs focus on creating social value [Dees et al., 2002].

An excellent example of social entrepreneurship is the nonprofit organization Trees, Water, & People (TWP), which has an environmental and social mission. (See www.treeswaterpeople.org.) Its mission is to reforest degraded areas and plant fast-growing trees in Central America. It decided to view the problem from the demand side and seek to reduce the demand for fuel wood. TWP teamed up with Aprovecho Research Center in Oregon to introduce a fuel-efficient stove that burns 50 to 60 percent less wood than traditional open fires by using an insulated, elbow-shaped burning chamber. The Justa stove also saves lives by removing toxic smoke through a chimney. TWP gives these wood-conserving stoves to farmers in El Salvador as an incentive to reforest their land. Fuel conservation, health improvement, and reforestation are accomplished together.

8.3 Family-Owned Businesses and Franchising

A **family-owned business** is one that includes two or more members of a family who hold control of the firm. Perhaps 80 percent of the businesses in the United States and Canada are family-owned and family-owned enterprises dominate the business landscape in most of the world. While most family-owned businesses are small to medium sized, some are very large. In fact, about 25 percent of the Fortune 1,000 businesses are family controlled. Many of the great twentieth-century companies were originally family enterprises, including IBM, Marriott, Merck, McGraw-Hill, and Wal-Mart. In many cases, family businesses are transferred to the children. For example, Fidelity Investments—FMR Corporation (www.fidelity.com)—was founded by the late Edward C. Johnson II in 1946. Ned Johnson, his son, was named president in 1972. Ned's daughter, Abby Johnson (born in 1962), was named president in 2001. A list of some family-owned or family-managed businesses is provided in Table 8.2. See appendix C for family business websites.

TABLE 8.2 Some family-owned or family-managed businesses.

Anheuser-Busch	Food and beverages
Archer Daniels Midland	Foods
Cargill	Foods
Carlson Travel	Travel
Clear Channel	Radio
Fabbrice D'Armi Beretta	Firearms
Ford Motor	Autos
Gallo	Wine
ILX Lightwave	Laser instruments
NASCAR	Auto races
Pruitt	Health care
S. C. Johnson	Cleaning products

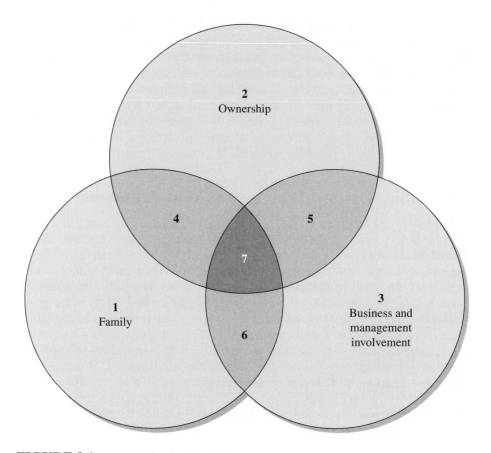

FIGURE 8.1 Model of a family business.

The family business has ownership, family, and business-management activities that overlap, as shown in Figure 8.1 [Gersick and Davis, 1997]. The family member who has ownership (area 4) will have different incentives than the owner-manager in area 5. Another issue is whether to give some ownership to a family member (area 1) when he or she is not active in the business. Should the successor CEO come from area 6 or 7—or elsewhere? These nonalignments can show up during difficult times.

Family businesses have unique advantages when everyone works well together. More trust often exists among the family members than with outside employees [Karra et al., 2006], and customers often feel that they are getting good treatment when they can work directly with a family member. The relatively small number of owners and the tendency to stay involved for long periods of time also give some family businesses the discretion, incentive, knowledge, and resources to invest deeply in the future of the

TABLE 8.3 Advantages and disadvantages of family businesses.

Advantages	Disadvantages
■ Long-term orientation	■ Nepotism
■ Independence of action	■ Family strife
■ Resilience and commitment	■ Financial strains
■ Family culture	■ Succession problems
■ Natural succession, if appropriate	■ Limited access to capital markets
	■ Concentrated risk in case of business failure

firm [Le Breton-Miller and Miller, 2006]. Family businesses, however, have a unique set of problems because family issues often carry over into the business operations.

The advantages and disadvantages of family businesses are listed in Table 8.3. Perhaps the most significant disadvantage is family strife over equitable compensation, fair treatment, and succession. The great advantage can be the continuity and commitment of a family to its business partners and employees.

A **franchise** is a legal agreement in which the owner of a business has licensed aspects of that business to an individual or a local firm, called a **franchisee**. The **franchisor** is the organization that owns and operates a firm that controls the business format and its associated trademarks and logo. Franchises focus on replicating products or services over a wide geographic area [Castrogiovanni and Justis, 1998]. They are particularly common where goods or services are produced and consumed at the same location, as with restaurants such as McDonalds and motels such as Super 8 [Carmen and Langeard, 1980]. By one estimate, franchise chains are responsible for more than 40 percent of retail sales in the United States [Combs et al., 2004].

There are three different franchise forms. A **business-format franchise** involves the provision of a complete business method, including a license for the trade name and logo, the products and methods, the form of the physical facility, the strategy, and the purchasing system [Shane, 1996]. This type of franchise is typified by Subway and Holiday Inn. A **trade-name franchise** primarily involves a brand name such as Western Auto or ACE Hardware. A **product-distribution franchise** is a license to sell specific products under a manufacturer's trademark and brand. For example, in 1898 General Motors lacked the capital to hire salespeople for its new automobiles, so it sold franchises to prospective car dealers giving them exclusive rights to certain territories.

At the heart of a franchise agreement is the desire by two parties to earn money while reducing risk. The franchisor wants to expand an existing company without spending its own funds. The franchisee wants to start

his or her own business without going it alone and risking everything on a
new idea. Thus, the franchisor provides a brand name, a business plan,
expertise, and access to equipment and supplies. In exchange, the franchisee
pays an initial fee and ongoing payments to the franchisor, and does the
work.

Like any other new business venture, the franchise business must possess
some unique competitive advantage, such as brand and a quality product. The
franchisor, franchisee, and partners strive to find an appropriate balance of
operating to suit the needs of each party. The franchisor seeks rapid growth,
while the franchisee seeks attention to quality and execution.

A franchisee is obligated by the contract to follow the prescribed methods of operation of the franchise [Bradach, 1997]. While the franchisee can
operate as an independent businessperson and share the benefits of the franchise, he or she is constrained by the format and the rules of operation.
Thus, the franchise offers significant benefits to the franchisee but con-
strains the amount of innovation that the franchisee can introduce into his
or her business [Kidwell et al., 2007]. For some franchisees, this arrangement is a perfect fit with their needs and attitudes. Others may find the controls and fees to be excessive and believe they can do better by themselves.
The advantages and disadvantages for the franchisee are summarized in
Table 8.4. The advantages and disadvantages for the franchisor are summarized
in Table 8.5.

Practitioners often recommend franchising as a method that entrepreneurs
can use to assemble resources to create large chains rapidly [Michael, 2003].
Technology companies that have used franchising to expand their business
include FASTSIGNS (www.fastsigns.com). Another franchise, The UPS Store
(formerly Mail Boxes, Etc.), is heavily dependent on technology applications.
An example of a services-oriented franchise based on intellectual property is
Dale Carnegie and Associates (www.dalecarnegie.com). See appendix C for
websites about franchising.

TABLE 8.4 Advantages and disadvantages for the franchisee.

Advantages	Disadvantages
■ Training	■ Franchise fees
■ Continuing guidance	■ Control of the format by the franchisor
■ Proven business format	■ Unfulfilled promises
■ Brand appeal	■ Rigidity of rules
■ Satisfaction and independence	■ High start-up costs
■ National advertising	■ Restrictions on purchasing of supplies
■ Site selection assistance	■ Unfair restrictions of the contract on resale
	■ Financial failure of the franchisor

TABLE 8.5 Advantages and disadvantages for the franchisor.

Advantages	Disadvantages
■ Enhanced ability to expand	■ Potential for uncooperative franchisees
■ Expanded geographic reach	■ Varying regulations from state to state
■ Use of the capital of the franchisee in expanding the business format	■ Unable to innovate quickly
■ Attracts owner-managers for the franchise chain	

8.4 Corporate New Ventures

A new venture started by an existing corporation for the purpose of initiating and building an important new business unit or organization can be called a corporate new venture. Some people refer to this process as intrapreneurship. The building of the new business enterprise depends on an entrepreneurial team leading the effort. Corporate entrepreneurship is focused on the identification and exploitation of previously unexplored opportunities that utilize the resources and competencies of an existing corporation. Corporate venturing is usually involved with the birth of new businesses and the associated revitalization of a corporation [Wolcott and Lippitz, 2007]. We differentiate a corporate venture from a project by (1) its newness to the corporation, and (2) its independence from the existing activities, organizational units, and products of the corporation. The characteristics of a corporate new venture (CNV) are summarized in Table 8.6. Corporate new ventures are distinguished from projects and product development efforts by having a limited relationship to existing business units, autonomy, innovation, and entrepreneurial leadership.

Corporate new ventures differ from independent start-ups along a variety of dimensions, as detailed in Table 8.7. The factors for success, however, are essentially the same between the two types of ventures: opportunity, vision, commitment, capabilities, resources, technology innovation, strategy, and execution. Success of corporate new ventures has been shown to be positively associated with growth and profitability of the firm [Sull and Spinosa, 2005]. A representation of this relationship and corporate ventures is shown in Figure 8.2. Mature corporations that engage in new business venturing are innovative, continuously renew themselves, and proactive.

TABLE 8.6 Characteristics of corporate new ventures.

■ Newness and novelty of the product relative to the firm's existing products	■ High potential for significant innovation
■ Independence or semiautonomy from existing corporate structure	■ Unique entrepreneurial team leadership capabilities

**TABLE 8.7 Contrasts between independent ventures
and corporate ventures.**

Dimension	Independent venture	Corporate venture
Team	Best in industry	Best available in firm
Scope	Entire company focused	Development team hands to corporate operations
Culture	Driven, team-oriented	Company's culture
"Contract"	Explicit—business plan	Increasingly explicit
Incentives	Equity (entire team)	Varies: bonus, career growth
Oversight		
Who?	Board	Upper management
When?	Monthly	Design reviews
External feedback	Customers, new investors	Customers
Financial goal	IPO or M&A value versus investment	Break-even date, return on investment
Changes to plan	Quick OK from board	Multiple levels for approval

FIGURE 8.2 Corporate new venture model.

An entrepreneur within an existing firm, like HP or Intel, starts with a description of an opportunity, the required resources to pursue it, the value that would be created, and a plan to pursue it. The entrepreneur needs a sound understanding of the technology and the customer and forms a sound strategy to move forward [Sull and Spinosa, 2005].

There are four general models of corporate entrepreneurship. In the Opportunist Model, intrepid "project champions" toil against the odds to create new businesses inside a corporation. In the Enabler Model, employees are willing to develop new concepts if they are given adequate resources. In the Producer Model, the corporation assumes organizational ownership, but limits resources, at least initially, to those provided by a specific business unit. Finally, in the Advocate Model, the corporation provides both resources and organizational ownership to encourage corporate entrepreneurship. Figure 8.3 illustrates these different approaches to corporate entrepreneurship. The appropriate model for any particular company depends on its specific objectives, such as whether the corporation is seeking change across the organization or only in a specific division [Wolcott and Lippitz, 2007].

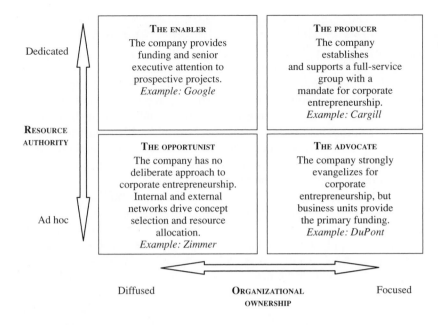

FIGURE 8.3 Four models of corporate entrepreneurship.

Source: Adapted from Wolcott and Lippitz, 2007.

Conventional wisdom says that large firms are weak at transforming opportunities into viable new businesses. The perceived reason for this weakness is the effects of bureaucracy and inflexibility that exist in large firms. Existing corporations are managed on a tightly defined strategy and within highly controlled boundaries. Corporate inertia can flow from successes that reinforce the rigidity of assumptions, processes, relationships, and values. All these factors are difficult to challenge: thus, the need for new independent organizations, such as new venture units, subsidiaries, or start-ups.

Often, the challenge for large firms is to protect the CNV from the pressures and controls of existing units within the firm and to "forget" the parent company's business model [Govindarajan and Trimble, 2005a]. Typically, a CNV needs to be established as a relatively autonomous unit that can accommodate the entrepreneurial team and its business plan. At the same time, large firms possess resources and capabilities that would be the envy of individual entrepreneurs trying to strike out on their own [Katila et al. 2008]. Unlike individuals, firms usually possess a share of the resources, capabilities, and knowledge necessary for innovation. Therefore, the challenge is how to achieve the right balance of integration. Table 8.8 outlines a number of methods that can be used to achieve this balance. Companies should develop strategy through disciplined trial and error, build on existing strengths, and work to integrate while preserving autonomy [Garvin and Levesque, 2006].

Apple: Corporate Venturing

In 1985, Apple Computer faced several challenges. Its impressive Macintosh computer was in danger of being eclipsed by the new IBM PC and its clones. While Macintosh enjoyed outstanding margins, Microsoft and Intel were quickly commoditizing personal computing. Apple was hoping that its Graphical User Interface (GUI) would distinguish the company, but Microsoft introduced Windows. Another dilemma was that the Macintosh platform was dependent on Microsoft for applications. Apple did have its own modest applications group, but it could not afford to alienate other independent applications developers.

In 1986, Apple spun off its applications group and appointed Bill Campbell to lead the new venture called Claris. He decided to attack Microsoft directly and immediately recruited talented executives from within Apple. Apple transferred its applications software with some employees to Claris. Apple agreed to provide up to $20M in working capital in return for 80 percent of the new company with the option to buy back the other 20 percent anytime. The Claris culture eschewed any connection to Apple. All employees were given Claris stock options. Salaries and benefits were reduced to those befitting a Silicon Valley start-up. Claris was going to succeed or fail on its own and there would be no safety net. Claris's strategy was to create a suite of applications that worked well together. This contrasted with Microsoft's emphasis on features rather than usability. Within three short years, Claris has grown to nearly $90 million in profitable, worldwide sales through acquisitions and internal development.

Claris's ambition became its undoing. Campbell and his team were frustrated with Apple's inability to grow sales volume. Apple was more concerned with high margins than market share, and Claris had saturated the Macintosh market. As the company prepared for an IPO in 1990, Claris revealed a controversial growth strategy to enter the Windows applications market. When Apple executives became aware of this strategy, they were unhappy. They worried that Claris would make Microsoft's Windows more attractive than its own Macintosh. To Claris's dismay, Apple decided to exercise its option to spin Claris back inside. Apple tried to retain the executive team, but within months its members had all dispersed to start new companies. Claris remains a whole-owned subsidiary of Apple today with no significant impact on the greater applications market [Komisar, 2000].

In this decade, it is interesting to contrast the Claris saga with that of Apple's iPod. The recent revitalization of Apple is dependent on entrepreneurial efforts to build the iPod into an entire line of breakout consumer products, including wireless versions, portable media devices, and home network products [Sloan, 2005]. In 2008, Apple launched the iPhone. It combines the features of an iPod with a smart phone. Apple has successfully become a collection of corporate technology ventures that often redefine and recreate their business. What did Apple learn from the Claris experience?

TABLE 8.8 Strategies for corporations to grow new businesses.

Balance trial-and-error strategy formulation with rigor and discipline.
■ Narrow the range of choices before diving deep.
■ Closely observe small groups of consumers to identify their needs.
■ Use prototypes to test assumptions about products, services, and business models.
■ Use nonfinancial milestones to measure progress.
■ Know when—and on what basis—to pull the plug on infant businesses.
Balance operational experience with invention.
■ Appoint "mature turks" as leaders of emerging businesses.
■ Win veterans over by asking them to serve on new businesses' oversight bodies.
■ Consider acquiring select capabilities instead of developing everything from scratch.
■ Force old and new businesses to share operational responsibilities.
Balance new businesses' identity with integration.
■ Assign both corporate executives and managers from divisions as sponsors of new ventures.
■ Stipulate criteria for handing new businesses over to existing businesses.
■ Mix formal oversight with informal support by creatively combining dotted- and solid-line reporting relationships.

Source: Adapted from Garvin and Levesque, 2006.

8.5 The Innovator's Dilemma

As introduced in Chapter 6, disruptive innovation can revolutionize industry structure and cause existing firms to decline or fail [Christensen and Tedlow, 2000]. Incumbent firms listen to their existing customers, who usually do not express a desire for radical innovation before it is introduced. Eventually, the new innovation improves and starts to challenge the existing methods, but it may be too late for the incumbent firm to catch up. By the time the new innovative firm has improved the innovation and captured the market, the existing firm has lost market share. The disruptive innovation gets its start outside of the mainstream of the market, and then its functionality improves over time. The existing firm ignores new, potential innovations at its own peril.

Another problem for all successful firms is **cannibalization,** which is the act of introducing products that compete with the company's existing product line. When companies decline to try to cannibalize their own products, they operate under the delusion that if they do not develop the new product, no other firm will do so. When new opportunities open up, new entrants into an industry can be more flexible because they face no trade-offs with their existing activities [Burgelman et al., 2004].

Intel potentially cannibalized its business when it released the Celeron in 1998 as an alternative to its powerful Pentium line of processors. The Celeron featured lower performance, but it quickly became popular due to its reduced cost. Intel gained access to different segments of the market with no adverse effect on the Pentium and its successors.

For many large firms, the pursuit of innovation must take a backseat to the effective exploitation of existing competencies, the satisfaction of existing customers, and the continuous improvement of existing technologies. Innovations eventually become breakthroughs when the web of people, ideas, and technologies that surrounds them grows and evolves, forcing change farther and farther into the established systems they emerged from. Existing organizations will tend to exploit an innovation that is related to their primary forms of work when the innovation enhances their existing competencies. When an innovation is unrelated, new organizations will typically emerge to capitalize on the innovation.

Established companies can offer disruptive new products of their own that capture new customers and produce new revenue growth. Instead of waiting for the threat of new products to appear, the established firm can create its own response to an anticipated threat. It is important for existing companies to invest in the development of disruptive innovations. This is a form of hedging one's bets [Bhardwaj et al., 2006]. A disruptive innovation, however, often requires new competencies, resources, and value net relationships. Thus, the best strategy for an existing firm may be to establish an autonomous corporate new venture unit or subsidiary with its own mandate, vision, people, and incentives.

The strengths and weaknesses of a corporate new venture are summarized in Table 8.9. Corporate new ventures benefit from ready access to the capital, people, suppliers, technologies, and brand of the parent firm.

On the other hand, CNVs may be limited by the budgetary and control practices of the parent. Furthermore, the parent firm may not have the technologies or people that the new venture requires. Recent studies show that the resource advantages of the existing corporation do not necessarily translate into higher performance for CNVs [Shrader and Simon, 1997]. To enable a CNV

TABLE 8.9 Strengths and weaknesses of a corporate new venture.

Strengths	Weaknesses
Ready access to capital	May be subjected to a corporate budget process
Access to capabilities of corporate employees	Multiple control and review levels
Suppliers willing to help in the design process	Limited autonomy
Emphasis on the marketing plan	Limited access to strong entrepreneurial talent
Gain from brand equity of the parent firm	Risk-reward may be less attractive than for an independent entrepreneur
Access to processes and technologies of the parent	Limited to parent firm's technologies and processes

to be most successful, it may be necessary to give the new unit more autonomy, separating it from the controls and limitations of its parent.

Capacity for creative innovation within a corporation is a function of the ability of the entrepreneurs to (1) obtain nonredundant information from their networks and (2) avoid pressures for conformity and sustained trust in developing novel innovations [Ruef, 2002]. CNVs are likely to successfully emerge from entrepreneurial teams drawing on a diverse set of functional roles in a firm. Corporate entrepreneurs can benefit from the knowledge resources of the firm but must avoid excessive conformity to the established means and norms of the firm.

The history of technology-based businesses is marked by large changes in cost effectiveness [Grove, 2003]. Traditional competition theory does not account for these transformations, which change the context of the industry as well as the competitive challenges. An example of a large transformation was the introduction of the Boeing 747 in 1996, which reduced costs dramatically.

Several companies, such as Apple and Novartis, have learned to exploit the present and explore the future. They separate their new, exploratory units from their traditional ones. Each new unit has its own processes, structures, and cultures, but remains integrated into the existing senior management structure [O'Reilly and Tushman, 2004].

Established companies, when confronted by disruptive technologies, should consider creating a completely separate organization and giving it a charter to build a new business [Christensen and Raynor, 2003]. When IBM was confronted by minicomputers, it created an autonomous business unit in Florida to develop and sell a personal computer.

The confluence of technologies could transform the health care industry by using genomics and proteomics along with computer technologies to create new health solutions. Perhaps a corporate new venture at Amgen or Johnson and Johnson will create an enterprise that will impact the health care industry.

8.6 Incentives for Corporate Venture Success

Can once-mighty giants of industry restore their health after they mature and decline in performance? Can these mature companies use corporate new ventures to transform their performance? Many studies point to significant difficulties in transforming large firms [Majumdar, 1999]. Structural factors, such as their intrinsic complexity, formality, and rigidity, are not conducive to either high performance or reorientation. Not only are larger firms and organizations structurally sluggish, but, with the passage of time, their culture becomes rigid and hard to change because of commitments to particular ways of doing things. As Dee Hock, founder of Visa, stated: "The problem is never how to get new, innovative thoughts into your mind, but how to get old ones out."

Large firms possess large collections of knowledge and intellectual capital. Furthermore, these firms have many talented staff who have entrepreneurial tendencies and the ability to exploit the intellectual capital of the firm.

Absorptive capacity is the ability of a firm to exploit external knowledge for the production of innovations. Thus, corporate new ventures can be based on both internal and external knowledge to the extent that the ability to absorb and exploit it is rewarded. A firm's successful use of innovations depends on its ability to exploit its existing base of knowledge while learning about technologies that lie outside its existing competencies [Cohen and Levinthal, 1990].

Majumdar [1999] studied the performance of large and small companies in the dynamic U.S. telecommunications industry and showed that size was not a material factor in performance. Large firms are able to effect change in a dynamic industry. With a larger variety and pool of resources available, larger firms can undergo transformation as effectively as smaller firms through a process of dynamic learning. Large firms can be transformed and rigid cultures can be made flexible by using the right methods, such as a corporate venture program.

Within every firm, there are a few individuals who find unique ways to look at problems and propose new ways to solve them using new technologies and processes [Pascale and Sternin, 2005]. The existing firm needs to identify them and help them to act as entrepreneurs. Furthermore, the new entrepreneurial unit needs to learn and create new ways by experimenting [Govindarajan and Trimble, 2005b]. The transfer of knowledge and expertise to the CNV can confer a competitive advantage to the venture.

One important incentive for corporate venturing is tying executive compensation to initiation and support of corporate new ventures. Another incentive is to encourage ownership of stock in the firm by its executives. When executives own stock in the companies they manage, they become motivated to increase the long-term value of their firm using corporate ventures [Zahra et al., 2000].

Mature firms need to exploit opportunities for novel innovations by increasing their commitment to corporate new ventures. Larger, mature firms also need to recognize, however, the barriers to CNV: familiarity, maturity, and propinquity [Ahuja and Lampert, 2001]. Familiarity exhibits itself in a firm's tendency to favor the routine or common knowledge and ways of doing things. Maturity refers to favoring fully developed knowledge rather than novelty. Finally, propinquity refers to favoring a search for solutions similar to existing solutions.

The challenge of identifying applications in the early stage is complicated not only by the limitations in technology performance, but also by the fact that attention focused on the search for a market application is attention diverted from immediate development. Many large firms bring together varying groups from time to time to address the questions of what they should do differently and what new products they should develop. People at these sessions are urged to suggest all ideas without hesitation [Drucker, 2002].

Mature firms should strive to build CNVs to experiment with novel, emerging, and pioneering innovation to create new dynamic growth. To pursue opportunities, firms need to identify and encourage corporate entrepreneurs. Corporate entrepreneurs are employees of a firm who take leadership responsibility for driving a venture in the firm. Art Fry, an entrepreneur within 3M, pushed the commercialization of Post-It notes through the firm. 3M has a guideline that its researchers can spend 15 percent of their time working on an idea without approval of management.

Corporations can use old ideas as the raw material for new applications. They can use old ideas and knowledge in new combinations or new ways. They can take an idea commonplace in one area and move it to a context where it is not common at all [Hargadon and Sutton, 2000]. This often can occur by just moving ideas from one division of the firm to another. They can also use a special internal consulting group dedicated to facilitating knowledge brokering within the firm. What distinguishes entrepreneurial firms from others are the actions they take when information is still incomplete. Entrepreneurial ability is not a function of simply gathering information but of having both the ability to make early judgments and the confidence to act on these judgments.

Because leading new CNVs always carry the risk of failure, many potential corporate entrepreneurs avoid joining them. They fear the loss of status that failure can bring. In light of these individual risks, a number of features can support corporate entrepreneurship. These include rewards, explicit management support, resources, organizational structures, risk acceptance, work design, and intrinsic motivation [Marvel et al., 2007].

Incentives for corporate entrepreneurs can be stock ownership, bonuses, or promotion within the firm if the expected performance is attained. A firm might have a goal that it will launch four new ventures each year and expect that at least one new venture will create an important new business.

A few years ago, an employee of Virgin Atlantic noticed some empty curb space at Heathrow Airport. In a matter of days, he secured the rights to the space and laid out a plan for Virgin to start a curbside check-in kiosk business unit. As a result, Virgin became the first airline at Heathrow to offer its business class passengers the advantage of getting a boarding pass without having to stand in a check-in line. As a result of his effort, the employee received a promotion [Hamel, 2001].

A key issue is the appropriate reward to a corporate entrepreneur. The corporate entrepreneur who is offered large financial gains is usually resented by his or her associates because the entrepreneur relies on corporate resources that he or she did not create as an independent entrepreneur would [Sathe, 2003]. A list of possible incentives for corporate entrepreneurs is provided in Table 8.10. Incentives include social incentives such as recognition, support, and a culture that favors individuals or teams to take the initiative to create new ideas and explore new opportunities. Employees can be provided slack time, which is time on the job for exploring as-yet-to-be-approved projects. A significant degree of autonomy and effective financial and promotion awards provide incentives for corporate entrepreneurs. Entrepreneurs are less risk-averse and seek independence of activity [Douglas and Shepherd, 2002]. These preferences can be exploited by CNVs.

TABLE 8.10 Incentives for corporate entrepreneurs.

■ Support for and recognition of employees who create and champion new ideas and opportunities	■ Significant degree of autonomy
■ A culture that favors individual or team initiative to create new ideas	■ Effective rewards such as promotion, stock ownership, or bonuses
■ Slack time for exploring not-yet-approved projects	

8.7 Building and Managing Corporate Ventures

An existing company is wise to attempt to exploit a new opportunity through some form of a new business. The types of new business arrangements that an existing corporation can use include a new independent venture, a spin-off of a new corporation, a transfer of the opportunity to the existing company's product development department, or authorization of a small project. Figure 8.4 shows the four types of business arrangements and their relationships to operational relatedness and strategic importance. Operational relatedness refers to how the new business organization couples to the existing operational resources and competencies. Strategic importance refers to the critical nature of the long-term results of this new organization to the success of the parent firm. Internal corporate new ventures are most useful for high operational relatedness and high strategic importance (quadrant 1 of Figure 8.4). It is best to view CNVs as a source of insights that can inform the strategic direction of

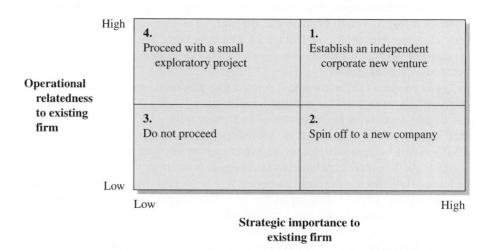

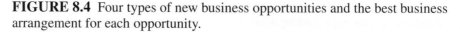

FIGURE 8.4 Four types of new business opportunities and the best business arrangement for each opportunity.

the parent company as well as provide the potential for attractive returns [Burgelman and Valikungas, 2005].

An opportunity with low strategic importance and low operational relatedness (quadrant 3) most likely calls for declining to proceed with this project or proceeding with a modest project until the strategic importance becomes clear.

An opportunity with high strategic importance and low operational relatedness may call for a spin-off to a new company (quadrant 2). A **spin-off unit** is an organization that is established within an existing company and then sent off on its own. The parent provides some resources and capabilities and sets the spin-off toward independence. Often the parent retains less than majority ownership of the spin-off. An opportunity with high relatedness and low strategic importance (quadrant 4) is a good candidate for a modest exploratory project.

Cisco Systems established a spin-off named Andiamo Systems, which makes switching gear. Cisco loaned Andiamo $42 million and committed to up to $142 million more. Cisco held 44 percent ownership of Andiamo. Cisco purchased the remaining 56 percent from the 300 employee-shareholders for $750 million of Cisco stock in 2004 [Thurm, 2002].

For the effective creation of a spin-off or new internal corporate venture, the opportunity needs a champion in the parent company [Greene et al., 1999]. The **champion** is an executive or leader in the parent company who advocates or provides support and resources as well as protection of the venture when parent company routines are breached. The champion helps, describes, and defends the venture and secures the necessary resources. The champion expresses confidence about the CNV, persists under adversity, and helps getting the right people involved [Howell et al., 2004]. The champion enables the resource transfer process, as shown in Figure 8.5 [Lord et al., 2002].

Corporate ventures are managed differently than traditional in-house corporate research and development. A corporate venture may be riskier and less subject to rigid management of internal costs than conventional corporate product development. Indeed, protecting venture investments from such controls is one reason that corporate new ventures and spin-offs are often housed outside the corporation's walls.

Furthermore, in corporate venturing, returns are part financial and part strategic, whereas with pure venture capital, investors' expected financial

FIGURE 8.5 Resource transfer process and the champion.

returns are paramount. Corporate ventures should follow the best practices of venture capital, but the dual objectives of financial and strategic returns must be balanced in ways that do not concern venture capitalists.

Large corporations can make room for radical, low-cost innovations by establishing a process for finding and funding new ideas [Wood and Hamel, 2002]. Table 8.11 describes a three-step process for finding, evaluating, and funding new entrepreneurial entities. First, expand the conversations about new opportunities widely throughout the firm. Then, establish a process for selecting and funding the best ideas. Finally, keep the budgetary control within the new venture and avoid letting traditional managers begin to control the budget of the new venture.

To extract maximum value from corporate new ventures, managers should adhere to the set of principles outlined in Table 8.12. The process of overseeing corporate new ventures is different from that used to maximize gains from existing business units.

Corporate entrepreneurs gain acceptance for new ideas by influencing organization members' perceptions about the nature of organizational interests. For an initiative to be accepted as part of the official company strategy,

TABLE 8.11 Establishing conditions for corporate new ventures.

- Increase the sources for innovation: New ideas tend to evolve and expand through conversation. The more people you can get involved, the more high-quality ideas you will generate.

- Establish a process for collecting and evaluating ideas: Establish a forum for assessing the merits of various proposals

to ensure that the most worthy ideas receive funding.

- Do not let traditional executives control the budget: Many executives are protecting their own departments and are unwilling to risk small amounts of resources on new and untested ventures.

TABLE 8.12 Extracting value from corporate venturing.

1. Protect new ventures from short-term pressures

2. Recognize that not all employees who volunteer to work with the corporate new venture are a good fit for the new venture

3. Don't expect the same results from the corporate venture that are expected from the core business

4. Manage with a portfolio mindset, not a project mindset

5. Be prepared to learn, since new markets are seldom like existing ones

6. Set milestones and manage the new venture in stages

7. Stop failing ventures early—and maximize the lessons learned

8. Continually evaluate learning-transfer mechanisms to ensure that ideas and lessons are shared

Source: McGrath et al., 2006.

belief in the idea must be linked to corporate organizational goals. Most large companies prefer to separate new venture efforts from their core business. This permits the new venture to focus on the new opportunity and readily gather and coordinate the necessary capabilities and resources [Albrinck et al., 2002].

The elements of a business as practiced by the parent firm and the corporate new venture are shown in Table 8.13. In general, the parent firm has developed assets, revenues, reward systems, and management practices that tend to support growth, fairness, and policies that lead to orderly progress. The corporate new venture needs to leverage its assets to create new revenue streams by rewarding entrepreneurial actions and flexibility. Furthermore, the CNV wants to attract the best talent of the parent company. Separating the CNV from the parent company enables the CNV to act quickly with flexibility to seize new opportunities.

Many companies use a portfolio strategy for holding ownership in several corporate new ventures as subsidiaries or spin-offs. These companies also use a new venture development process based on the development of a business plan and the analysis of what form of CNV is appropriate. This process is summarized in Table 8.14 [Albrinck et al., 2002]. At each step of the process, the parent company must evaluate the best next steps. Step 1 helps the CNV take shape as a venture champion and entrepreneurial team are identified. Step 2 includes the development of an initial concept statement and the outline of the elements of a business plan. The next step is to complete the development of a comprehensive business plan. Step 4 is focused on selecting the best organizational form for the CNV, based on the long-range objectives of the parent [Miles and Colvin, 2002]. Finally, in step 5, the corporate new venture is established with the requisite resources, talent, and capabilities transferred from the parent company.

TABLE 8.13   Elements of a business as practiced by the parent firm and the corporate new venture.

Element	Parent firm	Corporate new venture
Assets	Protect and use	Leverage
Revenues and growth	Growth of existing revenue stream	Create new revenue stream
Management	Adhere to policies and procedures	Act decisively with flexibility
Rewards	Maintain fairness and equity	Reward entrepreneurship and performance
Talent—people and knowledge	Retain talent and knowledge	Attract the top talent and transfer the best knowledge from the parent company to the CNV

TABLE 8.14 Five-step process for establishing a corporate new venture.

1. Identify and screen opportunities. Create a vision. Designate a venture champion and an entrepreneurial team.

2. Refine the concept and determine feasibility. Prepare the concept and vision statement. Draft a brief business plan summary or outline for review and to gather support.

3. Prepare a complete business plan. Identify the person to lead the new venture.

4. Determine the best form of the corporate new venture: internal new venture unit, spin-off, subsidiary, or internal project.

5. Establish the corporate new venture with talent, resources, and capabilities transferred from the parent company.

The selection of the appropriate form (step 4) should try to fit the needs and strategy of the parent company. For example, 3M usually incorporates the CNV within an existing or new division. Conversely, Barnes and Noble, when considering the establishment of an online unit, decided to spin off a new company into the stock market.

The Virgin Group, under the leadership of Richard Branson, has created 200 new businesses in many industries such as media, airlines, and music. Business ideas come from anywhere in Virgin Group, and Branson remains accessible to employees who have proposals. Branson also hosts gatherings for employees where they can give him their ideas. One employee proposed a bridal planning service including wedding apparel, catering, air travel, and hotel reservations. She became the CEO of Virgin Bride. Virgin Group works to start independent new ventures (see www.virgin.com).

Landmark Communications launched the Weather Channel as an internal new venture in 1981 [Batten, 2002]. With Landmark's strong corporate support and commitment, the Weather Channel became a top weather information source. With the help of Landmark's talent, knowledge, resources, and capabilities, the new venture took off through several deals with cable operators. By 1996, the Weather Channel was also available online. The Weather Channel succeeded because of the investment of significant resources of Landmark. The Weather Channel was started in the face of widespread skepticism but prevailed because of the assets and capabilities of Landmark.

Existing firms have the capability to organize a market, turning an idea into something that can be economically produced, marketed, and distributed to the customer. Entrepreneurs are able to explore new technologies quickly and effectively, and make the creative leap from technological possibility to something that meets consumer needs. Effective firms that meet the challenge of change possess people who are capable of both tasks.

Many new innovations are introduced by pioneer firms, and a learning phase is started in the market. Existing corporations can recognize these new innovations and quickly join in the innovation-commercialization phase, exploiting their capability to produce, market, and support new products.

Guidant: A Successful Spin-off of Eli Lilly

In the early 1990s, the pharmaceutical giant Eli Lilly built a series of internal ventures focused on medical devices. By 1994, Lilly had created four internal corporate venture divisions in the medical devices area concerned with cardiac and vascular issues. By September 1994, Lilly incorporated these units into a new company, Guidant, and consummated an initial public offering of its common stock. By September 1995, Lilly disposed of its ownership, and Guidant was a separate company. Guidant is an excellent example of growing from a group of internal corporate ventures to became a leading company. By 2006, Guidant had annual revenues approaching $4 billion and was purchased by Boston Scientific for $27 billion.

The life cycle of a market, such as the telephone industry, can be portrayed by Figure 8.6. The first stage is the introduction of a disruptive technology such as Bell's telephone. Then the key application, step 2, is identified and exploited. A dominant design emerges for the product and the market starts to grow (step 3). Process innovation, step 4, occurs as the product usage grows. When the marketplace matures, experimental innovation occurs in step 5. Later in maturity, the customer relationship processes are improved. In decline, business model innovation is used (step 7). Finally, structural innovation capitalizes on disruption to industry relationships. These eight types of innovation are

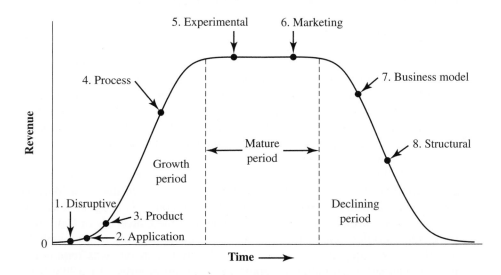

FIGURE 8.6 Eight types of innovation and the life cycle of a market.
Source: Moore, 2004.

TABLE 8.15 Eight types of innovation for periods of the market life cycle.

Type of innovation	Period	Description
1. Disruptive	Very early	Technological discontinuity
2. Application	Early	Technology application creates new market—Killer app
3. Product	Start of growth	Improved performance, dominant design
4. Process	Later growth	More efficient and/or effective processes
5. Experiential	Mature	Improved customer experience
6. Marketing	Mature	Improved marketing relationships
7. Business Model	Declining	Reframes the value proposition or value chain
8. Structural	Declining	Reponds to structural changes in the industry

summarized in Table 8.15 [Moore, 2004]. In order to renew an existing firm, the leaders must choose the appropriate innovation depending upon the life cycle of the industry.

Many critics depict incumbent firms as going into decline in the face of radical technological innovation. This tendency is not universal, however, nor should it be inevitable. Corporations can respond effectively to new technological innovations when they are prepared and organized to do so. Firms that have a history of navigating turbulence and creating loosely coupled, standalone divisions, and possess a critical complementary asset have a good chance of managing the challenge of a radical innovation [Hill and Rothaermel, 2003].

8.8 AgraQuest

Pamela Marrone was the entrepreneurial leader of two corporate new ventures before incorporating an independent start-up, AgraQuest. After she received her doctorate in 1983, she was recruited to start a new corporate venture by Monsanto of St. Louis. Her unit was a separate research and development project within an existing division of the agricultural business organization of Monsanto. She reported to a division head and enjoyed moderate independence for the direction of her new venture unit.

In 1990, Marrone was recruited by Novo Nordisk, a Danish company, to start a wholly owned subsidiary called Entotech. Novo Nordisk and Marrone chose Davis, California, as the location for this new company because of the significant research work being done in entomology at the University of California, Davis. She recruited scientists and staff and built the firm up to 50 people by 1995. Marrone enjoyed day-to-day control of Entotech but was required to travel to Denmark each month to report on plans and progress. By 1995, Novo Nordisk decided that Entotech was outside its core business segment and sold the new venture corporation to Abbott Labs.

Rather than move with the unit to Abbott Labs, Marrone decided to launch her own firm, incorporating it as AgraQuest. Her earlier experiences from 1983 to 1995 served her well in moving on to found a new company. She sought independence as well as the chance to create the contribution that a natural pesticide firm could offer to agriculture around the world. Often, as in Marrone's case, the entrepreneur finds the controls and limits of an existing company are burdens uncompensated by the availability of the resources of the larger parent firm. Many entrepreneurs who seek their own career path find that it requires a difficult but important decision to establish their independence and make their own way in the world of commerce by starting a new independent venture.

8.9 Summary

There are five types of new ventures: small business, niche, high-growth, non-profit, and corporate. Important contributions have been made by small and niche businesses, especially when they later grow and extend their mission globally. High-growth ventures including radical innovation start-ups are very important to creating growth and jobs as well as providing an important service or product that makes a difference. A special form of new venture, called a nonprofit, enables an organization to meet an important social purpose. Other special forms include family-owned businesses and franchises.

Finally, corporate new ventures make important contributions of novelty and creativity that provide new vigor for existing large enterprises. For many firms, the pursuit of innovation and creation of a new venture independent of the existing structures may renew the vigor of the firm. A corporate new venture needs the right amount of slack, independence, and resources to create a novel business.

Principle 8
An important, vigorous new business venture can emerge from a large firm when afforded the appropriate balance of independence, resources, and people to respond to the opportunity.

Video Resources

Visit http://techventures.stanford.edu to view experts discussing content from this chapter.

Sustainability for Nonprofit Organizations	Kavita Ramdas	Global Fund for Women
Classes of Innovations in the Product Leadership Zone	Geoffrey Moore	MDV

8.10 Exercises

8.1 Research the number of new ventures created in the last year. Try to segment the data you collect into the venture types outlined in Table 8.1. Compare the number of new ventures of each type. What growth rates have different venture types exhibited? Can this be explained by broader economic trends?

8.2 A partnership between a social entrepreneurship course at Stanford University and the nonprofit Light Up the World Foundation (www.lutw.org) worked to bring safe, affordable lighting to people in Mexico, China, and India. Students in engineering and business worked to design a lamp appropriate to the needs of villagers. Develop a brief plan for a social entrepreneurship project with an international nonprofit for your school.

8.3 In 2000, three graduate students at Harvard University launched a nonprofit called New Leaders (NLNS). This venture recruits, trains, places, and supports principals in U.S. urban school districts (www.nlns.org). The three founders met while enrolled in a class at Harvard on entrepreneurship in the social sector. Determine the mission of NLNS and describe the accomplishments of the enterprise.

8.4 Zimmer Holdings (www.zimmer.com) was incorporated in January 2001 as a wholly owned subsidiary and CNV of Bristol-Meyers Squibb Company. Zimmer designs and markets orthopedic and surgical products. The subsidiary was created from its parent in August 2001, with shareholders receiving 1 share of Zimmer for each 10 of Bristol-Meyers they owned. Study the origins of Zimmer and determine if the spin-off was the right action for Bristol-Meyers.

8.5 The traditional newspaper industry is in a declining phase and much has been written on how newspapers should reinvent themselves. How are various newspaper organizations addressing this challenge? What is the best next step in the newspaper industry? See Figure 8.6 for potential options.

8.6 Many new clean tech ventures have relied on funding and partnership from established corporations. Select a recently funded clean tech venture with corporate venture involvement. Did the funding impact the structure of the new venture? What does the new venture expect to gain from the backing by the larger corporation? What does the larger corporation expect to gain from being involved in the new venture?

8.7 WiMAX is an emerging wireless broadband technology that promises to bring high-speed Internet to both developed and developing countries. Intel Capital has made significant marketing and corporate venture investments in this technology. List three of the start-up companies Intel has invested in within the WiMAX space. How much money has Intel Capital invested in WiMAX more broadly? Why is Intel Capital placing these bets and how does Intel stand to gain from the success of WiMAX?

8.8 Describe the investment philosophy of a corporate venture capital firm such as Intel Capital, BlueRun Ventures (Nokia), Google, Dow Corporate Venture Capital, or General Electric (GE) Equity. How are they synergistic with their parent companies' strategic direction?

VENTURE CHALLENGE

1. Using Table 8.1, describe the specific type of new venture selected by your team.

2. Assuming your venture was developed as a corporate venture, describe the advantages and disadvantages of this approach using Table 8.9.

CHAPTER 9

Knowledge, Learning, and Design

Knowledge and human power are synonymous, since the ignorance of the cause frustrates the effect.
Francis Bacon

How can a new organization access and use knowledge in order to build its new venture?

Knowledge is power. Knowledge assets and intellectual capital are potential sources of wealth. The creation and management of knowledge can lead to new, novel applications and products. Sharing knowledge throughout a firm can enhance the firm's processes and core competences, thus making the firm more innovative and competitive. Most technology ventures are based on knowledge and intellectual property that must be enhanced and managed. A learning organization is skilled at creating and sharing new knowledge and uses this knowledge to do a better job.

Product design and development, which is concerned with the concrete details that embody a new product or service, can add significant value to what is offered to the customer. Prototypes are models of a product or service and can help a new technology venture to learn about the right form of the product for the customer. Scenarios are used to create a mental model of a possible sequence of future events or outcomes. Knowledge acquired, shared, and used is a powerful tool for the entrepreneur to build a new venture organization. ■

9.1 The Knowledge of an Organization

Assets are potential sources of future benefit that a firm controls or can access. Knowledge is an asset that is a potential source of wealth, as described in Chapter 1. The creation and management of knowledge leads to new, novel applications and products that can result in wealth creation. **Knowledge** is the awareness and possession of information, facts, ideas, truths, and principles in an area of expertise. Intellectual capital is the sum of the knowledge assets of a firm. These knowledge assets include the knowledge of its people, the effectiveness of its management processes, the efficacy of its customer and supplier relations, and the technical knowledge that is shared among its people. It can be thought of as best practices, new ideas, synergies, insights, and breakthrough processes. Thus, the firm's intellectual capital (IC) is the sum of its human capital (HC), organizational capital (OC), and social capital (SC), as described in Chapter 1.

From the generation of new ideas through the launch of a new product, the creation and exploitation of knowledge are core themes of the new product development process. In fact, the entire new product development process can be viewed as a process of embodying new knowledge in a product [Rothaermel and Deeds, 2004].

Knowledge is one of the few assets of a firm that grows when shared among its people. A new venture is wise to strive to acquire, store, manage, and share its knowledge throughout its organization. Intellectual capital is knowledge that has been formalized, captured, and leveraged to produce an output that has great value. In many ways, we can view a product as embodied knowledge. For example, an ATM machine embodies all the knowledge necessary for completing most banking transactions.

The knowledge creating and sharing activities of a firm can be represented by Figure 9.1 [Leonard-Barton, 1995]. The value of commercial knowledge is in its use, not its possession. The value of knowledge compounds when it is shared. Using current knowledge for cooperative problem-solving is the first of the four knowledge activities of a firm. The second is the implementation of new processes and tools within the firm. The third activity is experimenting and learning in order to build the knowledge base. The fourth activity is acquiring knowledge from outside the firm.

As a result of creating and sharing knowledge, a firm can enhance its people's skills and capabilities, as well as the knowledge embedded in its processes and managerial systems. Knowledge finds value in practice and use. The strategic approach of a new venture is linked to a set of intellectual assets and capabilities. Thus, if a firm has an opportunity that requires certain knowledge and it is not yet available, we can state that the firm has a knowledge gap. Acquiring the knowledge to fill the gap will be critical to the future success of this firm.

The knowledge of a firm encompasses (1) cognitive knowledge, (2) skills, (3) system understanding, (4) creativity, and (5) intuition. The first three forms of knowledge can be codified and stored. The last two forms of knowledge are types of trained intellect that people possess but are difficult to codify.

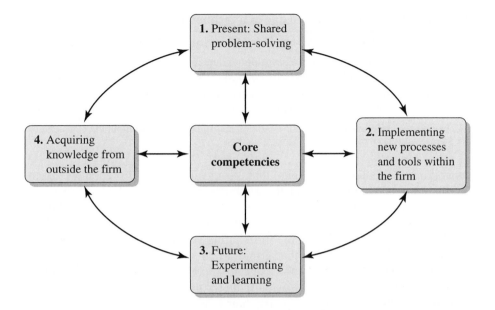

FIGURE 9.1 Knowledge creating and sharing activities of a firm.

Knowledge can be *prepositional,* which deals with beliefs about natural phenomena, such as scientific discoveries and practical insights into the properties of materials, waves, and nature. *Prescriptive* knowledge is all about techniques— the manipulation of processes and formulas, such as how to write a piece of software. The growing interplay between these two forms of knowledge transformed the world economy during the twentieth century [Mokyr, 2003].

Knowledge can be seen as a source of innovation and change leading to action. Also, it provides a firm with the potential for novel action and the creation of new ventures. With increased flow of new information, firms need to develop the means to convert that information into insight [Ferguson et al., 2005]. Knowledge creates real wealth for a new venture through multiple applications, which can have breadth across an organization. The knowledge represented by patented inventions, software, marketing programs, and skillful employees makes up 70 to 90 percent of the assets held by corporations like Microsoft, Amgen, and Intel.

9.2 Managing Knowledge Assets

The growth of a new venture rests, in part, on the increasing value of a knowledge base within the emerging firm. **Knowledge management** is the practice of collecting, organizing, and disseminating the intellectual knowledge of a firm for the purpose of enhancing its competitive advantages. The four steps for managing knowledge in a new venture are given in Table 9.1. The first step is to identify and evaluate the role of knowledge in the firm. How is knowledge created, stored, and shared? The second step is to identify the expertise,

TABLE 9.1 Managing knowledge in a technology venture.

1. **Role:**	Identify and evaluate the role of knowledge in the firm.
2. **Value:**	Identify the expertise, capabilities, and intellectual capital that create value in the form of products and services.
3. **Plan:**	Create a plan for investing in the firm's intellectual capital and exploiting its value while protecting it from leakage to competitors.
4. **Improve:**	Improve the knowledge creation and sharing process within the new venture.

capabilities, and intellectual capital that create value for a firm. Then, we examine the uniqueness and value of these intellectual assets.

The third step is the creation of an investment and exploitation plan for maintaining and harvesting the value of the knowledge assets. Finally, the firm improves the process of creating and sharing its knowledge. Though knowledge is one of the few assets that grows when shared, the new venture needs to carefully determine what knowledge to share and what knowledge should be protected or kept secret. This is particularly true for technology ventures for which intellectual property is usually their key asset.

Most professionals are unable to keep up with all they need to know. One method of knowledge access is to embed knowledge into the technologies used by the professionals. For example, when designing a product, the databases required can be linked directly to the design tools [Davenport and Glaser, 2002].

Part of an emerging firm's knowledge base is information about competitors. This knowledge is useful in responding to competitive changes and challenges. **Competitive intelligence** is the process of legally gathering data about competitors. Competitor intelligence may include securing data about competitors' products, services, channels of distribution, pricing policies, and other facts. Legal means of acquiring competitor intelligence include gathering company reports, news releases, and industry reports, and visiting competitor websites and trade show booths.

Knowledge is worthy of attention because it tells firms how to do things and how they might do them better [Davenport and Prusak, 1998]. The key skill for an emerging start-up is the ability to turn knowledge into products and services. Knowledge turns into action as it is embedded in the products, routines, processes, and practices of the new technology venture. Knowledge embedded in a company's activities can be a sustainable competitive advantage since imitating it is difficult for competitors.

9.3 Learning Organizations

New ventures grow powerful from learning and adapting to new challenges and opportunities. A **learning organization** is skilled at creating, acquiring, and sharing new knowledge and at adapting its activities and behavior to reflect new knowledge and insights. A technology venture creates and acquires knowledge and shares this knowledge among its people. As a result of this new knowledge, the organization adapts its actions and behavior. Learning organizations are skilled at five activities: systematic problem-solving, experimentation with new approaches, learning from their own experience and past history, learning from the experiences and best practices of others, and transferring knowledge quickly and efficiently throughout the organization. The learning organization is active, imaginative, and participative. It attempts to shape its future rather than react to forces. A learning firm adapts itself to its learning and increases opportunities, initiates change, and instills in employees the desire to be innovative. A learning organization confronts the unknown with new hypotheses, tests them, and creates new knowledge. Thus, the learning organization creates innovation and new knowledge that is used by the technology venture to develop new products and services.

Information that does not enable an action of some kind is not knowledge. Knowledge comes from the ability to act on information as it is presented. It truly is power, giving an organization the ability to continuously better itself. The power of knowledge depends on the company's ability to provide a supportive environment: a culture that rewards the sharing of knowledge across various barriers. The company that develops the right set of incentives for its employees to work collaboratively and share their knowledge will be successful in its knowledge management effort. Knowledge management has several benefits: it fosters innovation by encouraging a free flow of ideas, enhances employee retention rates, enables companies to have tangible competitive advantages, and helps cut costs.

The decisions of an entrepreneurial firm are the result of the firm's ability to process knowledge and learning [Minniti and Bygrave, 2001]. Knowledge acquired through learning-by-doing takes place when entrepreneurs choose among alternative actions whose payoffs are uncertain. Over time, entrepreneurs repeat only those choices that appear most promising and discard the ones that resulted in failure. Thus, entrepreneurship is based on a process of learning that allows entrepreneurs to learn from successes as well as failures. Jack Welch [2002] described the learning process thus: "In the end I believe we created the greatest people factory in the world, a learning enterprise, with a boundary-less culture."

Siemens, a global organization, uses the ShareNet network to enable 19,000 technical specialists in 190 countries to help each other in solving problems. In one case, a project manager in South America was trying to find out how dangerous it was to lay cables in the Amazon rain forest in order to determine the type of insurance his project needed. He posed the question on ShareNet and within hours a project manager in Senegal who had encountered a similar situation responded. Getting the right, *actionable* information before the cables went underground saved the company several million dollars in insurance costs [Tiwana and Bush, 2005].

Managers and entrepreneurs often have distorted pictures of their businesses and their environments. Busy among the trees, they can lose sight of

TABLE 9.2 Entrepreneurial learning process.

Step	Question	Outcome or action required
1. Identify the problem or opportunity	What do we want to change?	Desired specific result
2. Analyze the problem or opportunity	What is the key cause of the problem?	Key cause identified
3. Generate potential solutions	How can we make a positive change?	List of possible solutions
4. Select a solution and create a plan	What is the best way to do it?	Establish a criteria, select the best solution, and set a plan to accomplish it
5. Implement the selected plan	How do we implement the plan effectively?	Monitor the implementation
6. Evaluate the outcome and learn from the results	How well did the outcome match our desired result?	Verify that the problem is solved. Why did it work?

the forest. They can review the impacts of their actions, however, and then modify their approach accordingly. The greatest asset that entrepreneurs can bring to knowledge and learning is their willingness to seek and make wise use of feedback [Mezias and Starbuck, 2003].

The entrepreneur's learning process is based on the six steps outlined in Table 9.2 [Garvin, 1993]. At each stage in the development of the new business, the entrepreneur encounters a set of challenges or problems that require resolution. A firm can use the method shown in Figure 9.2 to resolve issues and learn from its successes and failures.

Organizational learning looks at an organization as a thinking system. Organizations rely on feedback to adjust to a changing world. Thus, organizations engage in complex processes such as anticipating, perceiving, envisioning, and problem-solving in order to learn. This approach is very important for new technology ventures to adopt and improve.

Process improvement projects can produce two types of learning. *Conceptual learning* is the process of acquiring a better understanding of cause-and-effect relationships by using statistics and scientific methods to develop a theory. *Operational learning* is the process of implementing a theory and observing positive results. Conceptual learning yields know-why—the team understands why a problem happens. Operational learning yields know-how—the team has implemented a theory and knows how to apply it and make it work. It is useful to design projects that are more likely to deliver both conceptual and operational learning [Lapre and Van Wassenhove, 2002].

For learning to be widely used in a firm, knowledge must spread widely and quickly throughout the organization. Ideas carry maximum impact when they are

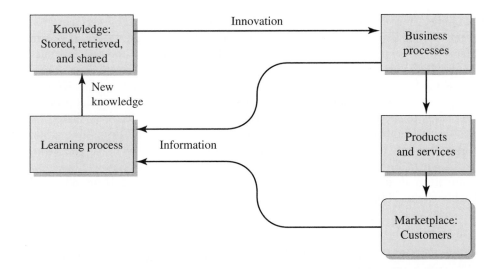

FIGURE 9.2   Knowledge and learning within a technology firm.

shared broadly rather than retained by a few people. A variety of mechanisms enable this process, including written, oral, and visual reports, knowledge bases, personnel rotation programs, education, training programs, and formal and informal networks. The organization needs to foster an environment that is conducive to learning. The new venture must strive to set aside time for reflection and sharing. Furthermore, boundaries that inhibit the flow of knowledge must be reduced so that learning is shared. A new venture can profit from efforts to eliminate barriers that impede learning and to place learning high on the organizational agenda.

Genmab was established in 1999 to develop products based on human antibodies for the treatment of a number of life-threatening and debilitating diseases. The company was founded by Lisa Drakeman, who spotted the work of a Dutch scientist while working at another company. She proposed starting a company, but American venture capitalists were unwilling to invest. Genmab was thus founded in Copenhagen, with research facilities in the Netherlands. The company had its initial public offering in 2000. In 2008, its market value reached $2.5 billion as one of the world's top 20 biotechnology firms. The venture created powerful intellectual property (knowledge) via its learning processes. It then converted that value into financial assets through an IPO.

A learning organization, properly managed, can enable a firm to meet the challenge of change by constantly reshaping competitive advantage even as the marketplace rapidly shifts. The learning organization is able to improvise or adapt to balance the structure that is vital to meet budgets and schedules with flexibility that ensures the creation of innovative products and services that meet the needs of changing markets [Brown and Eisenhardt, 1998]. The firm that is most responsive to change and capable of learning is the one that succeeds [Galor and Moav, 2002].

Knowledge is stored in documents, databases, and people's minds. Knowledge created in a learning process as depicted in Figure 9.2 is a social process that leads to increasing knowledge [McElroy, 2003]. Knowledge is shared by people and embedded within the business processes of the firm. Innovation flows through the business processes, products, and services of the firm, as shown in Figure 9.2. As the firm learns and creates new knowledge, new innovation is created and new opportunities are identified [Lumpkin and Lichtenstein, 2005].

Genentech: Learning from Prior Experience

In the early 1970s, the biotechnology industry was just beginning to emerge when Cetus was formed by Ronald Cape (a Ph.D. biologist with an MBA), Donald Glaser (Nobel laureate in physics), Peter Farley (a physician with an MBA), Calvin Ward (a scientist), and Moshe Alafi (a venture capitalist). Being one of the first firms in the industry and having the backing of a Nobel laureate, Cetus was able to attract a star-studded advisory board. Unfortunately, neither Cetus's employees nor its advisors knew what a biotechnology firm should look like or what it should do. Therefore, Cetus took money and formed partnerships with whoever was willing, and attempted to be all things to all people. The end result was that Cetus worked on projects that spanned from health sciences to agriculture to finding better processes for making industrial alcohol. In the early 1980s, Cetus recognized that it needed to tighten its focus. It channeled 70 percent of its R&D spending into the health care field, it brought in a professional manager to run the firm, and it was more forthright with analysts and the media. Unfortunately, by then, Cetus had lost many of its supporters and investors.

Frustrated with Cetus's lack of direction and armed with their experience, several of the board members left Cetus to form their own biotechnology companies. They were convinced—and investors appeared to agree—that a more focused strategy was a better way to go. One individual who was familiar with Cetus's business strategy was Robert Swanson, a young venture capitalist with Silicon Valley's Kleiner Perkins. In 1976, just five years after Cetus was founded, Swanson approached Herbert Boyer, then a professor at University of California at San Francisco (UCSF), about starting a new biotechnology company based on work that

Boyer had completed with Stanford University professor Stanley Cohen. What Boyer had scheduled as a 20-minute polite conversation turned into a 3-hour meeting as Swanson won Boyer over with his enthusiasm. By the time Swanson and Boyer left the bar, they had agreed to form Genentech (for "genetic engineering technology").

Swanson left Kleiner Perkins and delved right into learning the science and becoming a hands-on, deeply involved CEO. Boyer also became deeply involved and took a leave of absence from UCSF. Swanson and Boyer worked hard to build a creative firm that was in many ways the antithesis of traditional pharmaceutical firms. To lure postdoctoral candidates away from academia, they offered their employees stock options and structured the R&D portion of their firm to resemble an academic lab: scientists worked flexible hours, dressed casually, and were allowed to publish their research.

In 1980, Genentech became the first biotechnology firm to go public, pricing its million shares at $35 each. Within half an hour of trading, the stock hit $89 per share and closed the day at $70. Genentech's public offering broke many previous IPO records. Today Genentech continues to be a well-managed learning organization.

Sources: Lax, 1985; Swanson, 2001; Teitelman, 1989; Robbins-Roth, 2000.

9.4 Product Design and Development

One of the early tasks of a new venture is the design and development of the new product. The entrepreneurial team wants to develop a new product or service that can establish a leadership position. One of the strengths of a new venture is that the leadership of the venture plays a central role in all stages of the development effort. Furthermore, the small new firm is able to integrate the specialized capabilities necessary for the development of a successful product [Burgelman, 2002].

In recent years, product complexity has dramatically increased. As products acquire more functions, the difficulty of forecasting product requirements rises exponentially. Furthermore, the rate of change in most markets is also increasing, thereby reducing the effectiveness of traditional approaches to forecasting future product requirements. As a result, entrepreneurs need to redefine the problem from one of improving forecasting to one of eliminating the need for accurate long-term forecasts. Thus, many product designers attempt to retain flexibility of the product characteristics as the development proceeds. A design and development project can be said to be flexible to the extent that the cost of any change is low. Then, project leaders can make product design choices that allow the product to easily accommodate change [Thompke and Reinertsen, 1998]. Uncertainty is an inevitable aspect of all design and development projects, and most entrepreneurs have difficulty controlling it. The challenge is to find the right balance between planning and learning. Planning provides discipline, and learning provides flexibility and adaptation. Openness to learning is necessary for most new ventures that are finding their way into the market [DeMeyer et al., 2002].

Design of a product leads to the arrangement of concrete details that embodies a new product idea or concept. The design process is the organization and management of people, concepts, and information utilized in the development of the form and function of a product. The role of design is, in part, to mediate between the novel concept and the established institutional needs. For example, Thomas Edison designed and described the electric light in terms of the established institutions and culture. As a result, he succeeded in developing an electric lighting system that gained rapid acceptance as an alternate to the gas lamp. A new product needs to be advanced, yet it should not deprive the user of the familiar features necessary for understanding and using the product. As new products are designed, the challenge ultimately lies in finding familiar cues that locate and describe new ideas without binding users too closely to the old ways of doing things. Entrepreneurs must find the balance between novelty and familiarity, between impact and acceptance [Hargadon and Douglas, 2001].

Palm Pilot: Synchronization Breakthrough

A prototype model of a Palm Pilot was built by Donna Dubinsky and Jeff Hawkins and shown to potential buyers in 1995. A key issue was addressed when they showed the docking cradle and explained that their device could connect to a personal computer (PC) with the touch of a button. Nobody had done that before this breakthrough. It seems basic now, but no one had made the logical leap that this organizer was a PC accessory, not a stand-alone PC. This allowed Palm to establish the market for personal digital assistants (PDA)..

Good, effective products or services are the outcome of a methodology based on solid, proven design principles [Brown, 2008]. Innovation is powered by a thorough understanding of how people want products made, packaged, marketed, sold, and supported. The overall development process is shown in Figure 9.3 [Thompke and Von Hippel, 2002]. The overall development process can include design of the product and its architecture, its physical design, and testing. The iPod and the BMW auto are examples of the outcome of a creative, artistic process of design. Part of the user experience is the look and feel of a product. A good product is attractive to see and easy to use and understand. Furthermore, customers want a product that does a few things really well. Fortunately, customers can participate fruitfully in the product design process when the innovations are incremental

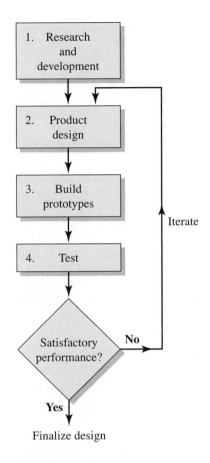

FIGURE 9.3 Overall
development process.

[Nambisan, 2002]. Good designers think about the qualities of a product as
well as its soft benefits such as warmth, status, and community.

Design includes aesthetics as well as basic needs. A beautiful glass must
be functional as well as attractive. However, design also includes compromises
and limits. Even the Maglite flashlight is flawed by the spot in the middle of
the beam.

Successful product design and development requires commitment, vision,
improvisation, information exchange, and collaboration, as listed in Table 9.3
[Lynn and Reilly, 2002]. These five practices may be easy to achieve in a start-
up where collaboration is the order of the day. The product team, which may
be all of the employees of a start-up, needs to clearly understand the vision for
the product and work together effectively.

TABLE 9.3 Five practices of good product development.

■ Commitment of senior management to the design process	■ Improvisation and iteration to develop a prototype
■ Clear and stable vision and goals for the product	■ Open sharing of information
	■ Collaboration of everyone on the team

Source: Lynn and Reilly, 2002.

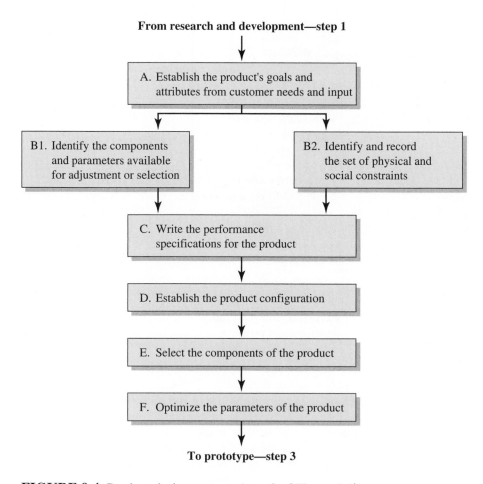

From research and development—step 1

A. Establish the product's goals and attributes from customer needs and input

B1. Identify the components and parameters available for adjustment or selection

B2. Identify and record the set of physical and social constraints

C. Write the performance specifications for the product

D. Establish the product configuration

E. Select the components of the product

F. Optimize the parameters of the product

To prototype—step 3

FIGURE 9.4 Product design process (step 2 of Figure 9.3).

The product design process [step 2 in Figure 9.3] is shown in Figure 9.4. The first step is to establish the goals and attributes of the product expressed as the required performance and robustness of the product (step A). When possible, the potential customer should be included in the design process. The voice of the customer can communicate the insights needed for the best products [Lojacono and Zacoai, 2004]. Potential customers can suggest ideas for new products and can be involved throughout the product development process to provide continuous feedback [Ogawa and Piller, 2006]. Firms may also find that *observing* potential customers, rather than simply surveying or interviewing them, can yield important information about their product needs.

In step B of Figure 9.4, the components and parameters available for adjustment are identified, and specifications for the product are agreed upon. Specifications are the precise description of what the product has to do. In addition, the set of physical and social constraints should be determined. Next, the product configuration is established, and the components of the product are recorded. Finally, the parameters of the product are optimized to achieve the best performance and robustness at a reasonable cost [Ullman, 2003]. A **robust product** is one that is relatively insensitive to aging, deterioration, component variations, and environmental conditions. Preparing a robust design implies minimizing variation in performance and quality. All designs involve trade-offs between performance, cost, physical factors, and other constraints [Petroski, 2003]. The success or failure of any design is ultimately determined in the marketplace.

Usability is a measure of the quality of a user's experience when interacting with a product. Usability is a combination of the five factors listed in Table 9.4. Examples of a common product with poor usability are most DVR and DVD players. New products should pass the five-minute test, which requires that the product is simple enough to use after quickly reading the instructions and then trying it for a few minutes. Information technology products with excellent usability are the iPhone, Skype, Twitter, Gmail, and Wikipedia.

Many system designs use a combination of modules within a specified architecture. A **module** is an independent interchangeable unit that can be combined with others to form a larger system. In modular designs, changing one component has little influence on the performance of others or on the system as a whole. An example is the iPod, which Apple's engineers first developed from a wide range of standard, interchangeable parts and modules. Design methods using independent modules make product design more predictable. Of course, the predictability inherent in modular design increases the chances that competitors can develop similar products.

TABLE 9.4 Five factors of usability.

1.	**Ease of learning:**	How long does it take to learn the product's operation?
2.	**Efficiency of use:**	Once experienced, how fast can the user complete the necessary steps?
3.	**Memorability:**	Can the user remember how to use the product?
4.	**Error frequency and severity:**	How often do users make errors, and how serious are these errors?
5.	**Satisfaction:**	Does the user like operating the product?

Realistically, most products consist of modules that possess some dependency between them. For example, an auto is a product that consists of wheels, engine, body, and controls that are relatively interdependent. Products made up of modules with intermediate levels of interdependence are harder for competitors to duplicate and may also provide better performance than a design based on purely independent modules [Fleming and Sorenson, 2001].

Designers strive to create new products different enough to attract interest but close enough to current products to be feasible to make a market. Many new designs flow from changing the components, attributes, or integration scheme to create a new product [Goldenberg et al., 2003]. The designer asks what can be rearranged, removed, or replicated in new ways.

Over time a dominant design in a product class wins the allegiance of the marketplace. A **dominant design** is a single architecture that establishes dominance in a product class. An example is the IBM-compatible personal computer, which is a dominant design because it is viewed as superior in the marketplace. Eventually, a dominant design becomes embedded in linkages to other systems. The VHS system became the dominant design in its competition with the Betamax system.

A **product platform** is a set of modules and interfaces that forms a common architecture from which a stream of derivative products can be efficiently developed and produced. For example, Google's Android and Apple's iPhone seek to be the leading platform for smartphone applications. Firms target new platforms to meet the needs of a core group of customers but design them for ready modification into derivative products through the addition, substitution, or removal of features. Well-designed platforms also provide a smooth migration path between generations so neither the customer nor the distribution channel is disrupted. A good example of a platform is Hewlett-Packard's electronics and software used for its wide range of printers.

9.5 Product Prototypes

Whenever possible, new business ventures should create a prototype of their product. A **prototype** is a physical model of the product or service. It is a model that has the essential features of the proposed product but remains open to modification. It can be used to identify and test requirements for the product. Prototypes are incomplete models that can be used by the new venture team to elicit comments from designers, users, and others to learn more about the product. Prototypes can be pictures, sketches, mock-ups, or diagrams that can be collaboratively studied. New ventures can use prototypes to redefine their business models and strategies. Prototypes can be used to create a dialogue between people that leads to innovation [Schrage, 2000]. Testing a prototype on a small group has been a common approach for many new products.

The computer software industry uses prototypes called beta versions of software to elicit response from lead customers. Microsoft sent out tens of thousands of copies of its Vista operating system to beta testers. Prototypes can be physical, digital, pictorial, or some combination of media. Innovative prototypes lead to innovative conversations, which potentially lead to better products.

Ford: The Power of Prototypes

Henry Ford planned to build a horseless carriage. However, no one could be persuaded to invest in it. A key turning point came when Ford built a prototype car for the Grosse Pointe automobile races. Ford entered the races, drove the car himself, and won decisively. He repeated the feat the following year in 1902. The victory attracted investors, and the Ford Motor Company was up and running.

In the creation of a movie or play, many innovators use sketches, storyboards, and videos to describe the product. The designers of a movie or play want to see how it works and engage in a collaborative redesign. The iterative procedure for prototype development is shown in Figure 9.5. Two or three iterations of the process may be sufficient to arrive at a satisfactory prototype.

New technologies such as computer simulations can make the creation of a prototype fast and cheap. **Rapid prototyping** is the fast development of a useful prototype that can be used for collaborative review and modification. An initial prototype can be rough since it enables the team to view the product and improve it. The ability to see and manipulate high-quality computer images helps create innovative designs. BMW uses computers to help engineers visualize automobile design and the results of crash tests [Thompke, 2001]. Personal fabrication systems, clusters of tools and software that function as complete job shops, are available [Gershenfeld, 2005].

Product-development firm IDEO believes that prototypes should be "rough, ready, right." While working with Gyrus ENT to develop a better sinus-surgery tool, IDEO employees demonstrated the value of having a prototype to show customers. During one discussion, 10 surgeons struggled to explain the discomfort involved in using the existing tool. An IDEO manager picked up a film canister, a white-board marker, and a clothespin and taped them together as a prototype. The physical prototype helped to move the conversation along, allowing the surgeons to hold and adjust it. The rough prototype was not so pretty or final that nobody wanted to break it or modify it. Creating a rough prototype allowed customers to engage in the development of the product, and to enthusiastically adopt it in surgery. Eventually, IDEO

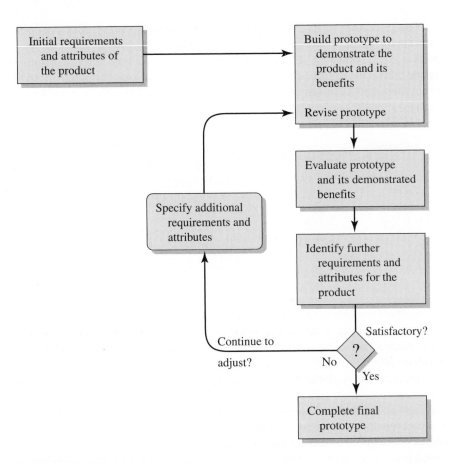

FIGURE 9.5 Prototype development process.

and Gyrus produced a tool that is used in over 300,000 procedures in the United States every year.

It is often best to carry multiple product concepts into the prototyping phase and to select the best of those designs later in the process [Dahan and Srinivasan, 2000]. Keeping multiple product concept options open and freezing the concept late in the development process affords the flexibility to respond to market and technology shifts. It is possible to create static and dynamic virtual prototypes that are displayed at a website for review and testing by suppliers, customers, and designers. Virtual prototypes cost considerably less to build and test than their physical counterparts, so design teams using Internet-based product research can afford to explore a much larger number of concepts. Furthermore, Internet-based prototypes can help to reduce the uncertainty in a new product introduction by allowing more ideas to be tested in parallel.

SaltAire Sinus Relief provides a nose wash to relieve the symptoms of sinusitis and allergies (see www.saltairesinuswash.com). A bottle and pump are used to spray a supersalty solution into the sinus chamber and relieve the symptoms [Ridgway, 2003]. Two New York physicians founded the firm in 1997 and created a series of prototypes for people to try. Based on that knowledge, they showed a revised product to other physicians and launched the product in 2000. The patented dispenser bottle the firm developed won an award for innovative design.

For the innovator, a prototype is a mechanism for teaching the market about the technology and for learning from the market how valuable that technology is in that application arena. Uses for robots in situations too dangerous for people have long been imagined. Many robotics companies tried and failed to create robots that could successfully enter and explore disaster zones and other dangerous environments. For example iRobot first demonstrated a prototype of the Urbie robot in 1997. This was the first commercially available robot that was able to climb stairs. This prototype showed the market that iRobot's products had overcome many of the fundamental limitations of other contemporary robots. By 2009, iRobot's revenue was over $300 million, and its products were available in over 7,000 retail outlets. Robots like the Urbie were used to explore the rubble of the World Trade Center, and have been used by the military to explore situations that would be extremely dangerous for troops. Over three million units of the Roomba, the consumer version of the Urbie, have been sold. The Roomba uses similar technology as the Urbie to sweep and mop floors.

Many firms have developed their products by entering potential markets with early versions of the products, learning from the tests, and probing again. These firms ran a series of market experiments, which introduced prototypes into a variety of market segments. The initial product design was not the culmination of the development process but rather the first step, and the first step in the development process was in and of itself less important than the learning and the subsequent, better-informed steps that followed. Software products lend themselves to rapid prototyping and early tests with potential customers.

9.6 Scenarios

Any new venture can benefit from creating a set of scenarios to address complex, uncertain challenges as it develops its strategy. A **scenario** is an imagined sequence of possible events or outcomes, sometimes called a mental model. A few realistic scenarios based on the industrial context and a few associated possible sequences of events help a planner to plan for the future. Each scenario tells a story of how the various elements might interact under a variety of assumptions. It paints vivid narratives of the future. The goal of scenario planning is not to forecast what is going to happen but to encourage an openness of mind, a flexibility of response, and a habit of questioning conventional wisdom. As Stephen Covey and A. R. Merrill [1996] stated: "The best way to predict your future is to create it."

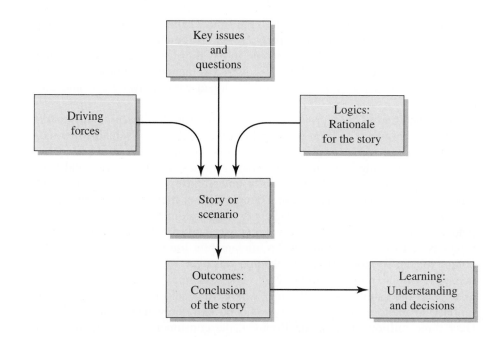

FIGURE 9.6 Elements of a scenario.

Scenarios lead to learning in a two-step process: constructing a scenario and using the content of the scenario to learn [Fahey and Randall, 1998]. The key elements of a scenario are shown in Figure 9.6. A scenario attempts to answer key questions and is based on a statement of the driving forces and the rationale for the story. The outcomes or results of the story lead to understanding and a useful decision. A scenario is an internally consistent picture of what the future might bring. It is not a forecast, but rather one possible outcome. Creating four or five scenarios will help portray the range of potential outcomes to core questions facing any organization.

Entrepreneurs will often weigh whether a new technology will be radical or nonlinear and have a profound impact on the marketplace. A scenario can help define the impact and time frame for a new technology.

An example of the outline of a scenario for the growth of electric auto sales is shown in Figure 9.7. The structure of the story for electric vehicles can be used to build several possible scenarios that can be used to learn about the opportunities in this market.

Scenarios can sometimes become a mirage. By 2001, the futurists—George Gilder and others—had created a scenario for the future of telecommunications that was overblown and ill-timed. This rosy, nirvana-like scenario missed the regulatory issues and the concept of excess capacity. Scenarios often can be too rosy [Malik, 2003].

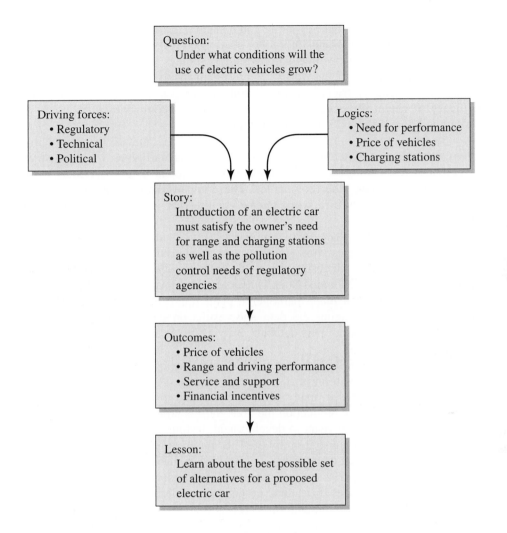

FIGURE 9.7 Elements of scenarios for electric cars.

9.7 AgraQuest

With two other former executives, Duane Ewing and Bruce Holm, and three former Entotech scientists, Pamela Marrone launched AgraQuest in 1995 in a small Davis, California, lab, furnished with $35,000 in used furniture and equipment. By 2003, the company had grown to 50 employees with 19 scientists.

Hunting for new natural products for pharmaceutical purposes has a long history. For example, penicillin and streptomycin are natural-product antibiotics from microorganisms. Little concerted effort has been made to hunt for new natural-product pesticides. Only approximately 200 insecticidal/miticidal, 30 herbicidal, and fewer than 10 nematicidal natural products are known, compared to the tens of thousands of known pharmaceutical natural products. Some natural products have become commercial pesticides. AgraQuest has tapped into the vast diversity of the world's microorganisms to discover and sell novel and environmentally friendly natural products for pest management with the efficacy of chemical pesticides. It is one of the first companies to invest in a continued and sustained effort to find them.

By focusing on unique groups of microorganisms, unique automated in-vivo screening, and technically difficult and unusual types of chemistry, AgraQuest has obtained a greater number of pesticidal natural products with fewer resources than other companies. The company knows how to organize and manage scientists in a work team structure that increases the output of novel molecules.

AgraQuest, like other small companies, can readily share and manage knowledge created by its work. Furthermore, it can quickly add new products to its product line by licensing them. In 2002, AgraQuest licensed *M. albus* from Gary Strobel of Montana State University. This product, when fully developed, will help control many pathogenic plant fungi. AgraQuest uses a computer system for managing the data and results on the screening of 20,000 microbes for 10 diseases. This database is available to all 19 scientists and the executives of the firm.

AgraQuest created its first prototype product, Laginex, for killing mosquito larvae, in 1998. It drew a lot of interest from mosquito control districts, and the product was fully tested in many sites. It was found to have a short shelf life, however, and alternative production methods for this product remain under study.

By 2002, after three years in development, testing, and government review, Serenade was granted registration by the U.S. Environmental Protection Agency (EPA) and its California counterpart, and became available for commercial sales a month later. Serenade is an environmentally friendly fungicide that can be used to ward off diseases that attack grapes, vegetables, nuts, and hops.

To satisfy the EPA's regulations and requirements, AgraQuest must accurately maintain test results and reports from toxic laboratories. All this information is provided to the EPA for review.

9.8 Summary

Knowledge is power, and the creation and management of knowledge can lead to novel applications, markets, and products. Sharing and managing knowledge wisely and efficiently with a technology venture can help build competitive and innovative skills. A new entrepreneurial firm seeks to build a sound knowledge management system that supports a learning organization.

Prototypes are models of a product or service and can help a new venture learn the right form and function of a product by showing it to customers and letting them observe it or try it. Furthermore, scenarios can be used to examine possible future outcomes based on specific actions.

> **Principle 9**
> Knowledge acquired, shared, and used is a powerful tool for the entre-
> preneur to build a learning organization that can design innovative
> products and grow effectively.

Video Resources

Visit http://techventures.stanford.edu to view experts discussing content from
this chapter.

Design Is Risk-Taking	David Kelley	IDEO
Ideas Come from Everywhere	Marissa Mayer	Google
Product Development Process: Observation	David Kelley	IDEO

9.9 Exercises

9.1   Examples of larger learning organizations are Microsoft, Hewlett-
Packard, Medtronic, Pfizer, and Starbucks. Choose a company and
describe its learning process. How would learning in an
entrepreneurial venture be the same? How would it be different?

9.2   What role should the "end customer" have in the product design and
development process? Do customers always know what they want?

9.3   The magazine *Fast Company* produces an annual list called "The Fast
Company 50, the World's Most Innovative Companies." Choose one of
the younger companies on the list and use Figure 9.1 as a guide in
describing the core competencies of the firm and how knowledge is
created and shared within the firm to spur product or service innovation.

9.4   Capstone Turbine is a developer, assembler, and supplier of
microturbine technology. Its primary customers are in the on-site
power production and hybrid-electric car markets
(www.capstoneturbine.com). Using the format of Figure 9.6, describe
a scenario for the growth of Capstone over the next five years.

9.5   A new firm plans to design and sell fuel cells for vehicle use. The
firm has received a $1 million grant from the U.S. Department of
Energy and is free to exploit the intellectual property developed
during the research and development grant. Prepare a knowledge
management plan that will enable the firm to file for patents.

9.6   The advantages of the Web as a distribution platform are many.
Describe some of the impacts the Web has had for the delivery of
services and content on product prototyping and product design and
development.

9.7   A number of software development methodologies exist to encourage rapid design and implementation (e.g., agile software development, extreme programming, etc.). Select two of these methodologies and compare and contrast the specific product design and development processes each is attempting to address and improve.

9.8   An *IEEE Spectrum* magazine article in September 2005 (www.spectrum.ieee.org/sep05/1685) featured some of the most expensive software failures ever. Examine the "Software Hall of Shame" and select your favorite. Describe why failure of knowledge management and the lack of behaving as a learning organization led to this result.

VENTURE CHALLENGE

1.   Describe the means of managing knowledge and learning that will be used in your venture.

2.   Discuss the robustness and usability of your product.

3.   Discuss your plans for developing a prototype of your product.

CHAPTER **10**

Legal Formation and Intellectual Property

When one door closes, another door opens; but we often look so long and so regretfully upon the closed door that we do not see the ones which open for us.
Alexander Graham Bell

What early decisions should an entrepreneur make about legal and intellectual property issues?

When entrepreneurs establish a new business, they must make some critical decisions about the detailed elements of the firm. The first steps for establishing a new corporation are illustrated in Figure 10.1. The choice of a legal form, name, logo, and other formal elements is critical to a successful future. The right name and logo can be the key to building a significant brand. Consider Sony, Intel, and IBM as examples of firms that built a notable brand.

The legal form of the venture should match the objectives of the entrepreneurs, customers, and investors of the enterprise. Furthermore, there should be a plan to build and protect the intellectual property of the new venture. The proper array of trade secrets, patents, trademarks, and copyrights can add up to a set of very valuable proprietary assets. For many new firms built on innovation and technology, intellectual property can provide a competitive advantage in the marketplace. ■

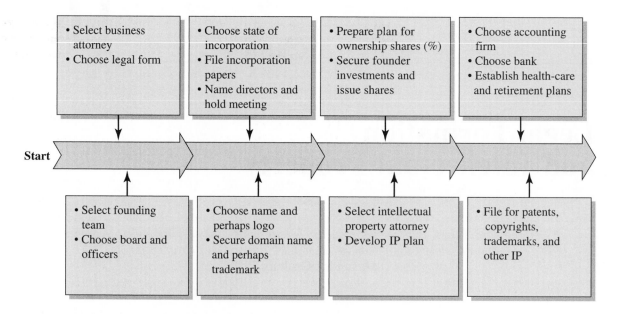

FIGURE 10.1 First steps of establishing a new corporation.

10.1 Legal Form of the Firm

When establishing a new technology venture, the entrepreneur needs to choose the legal form of the organization. The entrepreneur should choose a legal form that will facilitate the business, tax, and capital-raising objectives of the new company. The choice of a legal form depends, in part, on how the firm and its owners want to handle federal taxes. For tax purposes, we will address two types of forms: **regular taxable corporations** and **flow-through entities.** The elements of taxable corporations and flow-through entities are summarized

TABLE 10.1 Legal form of the firm.

Type	Taxation
1. Regular taxable corporation: C-corporation	Taxation of the corporate profits as well as taxation of any corporate distributions to owners

2. Flow-through entities
    ■ Sole proprietorship
    ■ Partnership
    ■ S-corporation
    ■ Limited liability company (LLC)
	All profits or losses flow through to the owners and are not separately taxed to the firm

in Table 10.1. A **corporation** is a legal entity separate from its owners. A flow-through entity, sometimes called a pass-through entity, is one that passes all losses or gains through to the owners of the firm. The profits of a regular corporation are taxable to the corporation, and any distributions are taxable to the owners. This results in double taxation of any distributions such as dividends. Most technology ventures choose the taxable corporation form since they expect to seek venture capital and corporate investors who are restricted in the types of entities in which they may invest.

There are four main types of flow-through entities to choose from. They are the sole proprietorship with one owner, which is the simplest form; the partnership; the S-corporation, which is taxed much like a partnership and is named for the U.S. Internal Revenue Code subchapter that covers it; and the limited liability company (LLC).

A **sole proprietorship** is a business that is owned, and usually operated, by one person. This is a simple form of doing business but exposes the owner to unlimited liability for all debts of the business. A **partnership** is a voluntary association of two or more persons who act as co-owners of a business. Each partner who participates in management is liable for the acts of the business. Liability for all acts of a business is all-encompassing, and this factor encourages most entrepreneurs to set up a LLC or corporation.

Many new flow-through firms are formed as LLCs rather than partnerships or S-corporations. Because they limit liability, it is often wise to use them for sole proprietorships. The Internal Revenue Service (IRS) in the United States treats single-owner LLCs automatically as sole proprietorships and multiple-owner companies as partnerships, unless they elect treatment for tax purposes as corporations.

Most entrepreneurs will, with their attorney, consider the regular corporate form or the LLC. Both of these forms offer limited liability for the owners of the entity. The personal liability of a regular corporation or an LLC is limited to the amount of capital contributed to the entity by that person.

Many firms that start as a small business should consider a LLC or a subchapter S-corporation form. These forms allow the initial business losses to flow through to the owners, and these losses typically can be used to offset income from other sources. As the firm grows, it may be wise to consider converting the LLC or S-corporation to a regular corporation since a regular corporation has several potential advantages. It can be sold or merged into another corporation with a tax-free exchange of stock. Other factors that should be considered with an attorney are the number of owners and investors, as well as the need to raise capital, the long-term goals of the business, and issues related to accounting and health-care/retirement plans.

A corporation or LLC can be created under state laws and usually requires some legal steps, including registering the name of the firm and the owners. The corporation is seen as an artificial person created by law to act as a business with limited liability. The owners receive shares of the corporation called **stock.** The process of forming a corporation is called incorporation. Usually, a company incorporates in its home state, but many incorporate in another state for reasons of law or ease of doing business. A corporation will file its articles of incorporation, adopt bylaws, issue shares, and establish a board of directors and officers as required by law. The limited liability feature of a corporation arises from the fact the corporation is itself a legal "person," separate from its owners. If a corporation fails and proper formalities have been followed, creditors have a claim only on the corporation's assets, not on the owners' personal assets.

An **S-corporation** is a corporation that is taxed as a flow-through entity. To qualify, a firm must meet certain requirements regarding its owners and types of stock. The S-status is established by filing with the IRS and may later be converted to a regular C-corporation. Some entrepreneurs prefer this election to the LLC. The key elements of the five types of legal form of a new business are summarized in Table 10.2.

The Hewlett-Packard Company was first established as a partnership in 1937, with William Hewlett and David Packard as equal partners. Their first sale was eight oscillators to Disney Studios in 1939 [Packard, 1995]. In 1947, Hewlett-Packard was incorporated to provide for continuity of life for the firm as well as limited liability for the owners.

In general, many businesses start out as sole proprietorships or partnerships but soon migrate to LLCs or a corporate form. With unlimited liability as a risk of the proprietor or partnership form, it may be unwise to continue in that form beyond the initial period necessary for completing a business plan. Most investors will only be willing to invest in a corporation or LLC since they wish to avoid any liability beyond the amount of their investment.

If the intention of the new business is to raise a significant amount of funds to start the venture and eventually to build it to a significant size, it is wise to start from the beginning as a regular C-corporation. A **C-corporation**

TABLE 10.2 Key elements of the five types of legal form for a new business in the United States.

Factors	Sole proprietorship	General partnership	Regular C-corporation	S-corporation	LLC
Owners' personal liability	Unlimited	Unlimited	Limited	Limited	Limited
Taxation	Proprietor's personal tax forms	Partners' personal tax forms	Profits taxed at corporation and owners pay tax on distributions	Profits or losses flow through to owners	Profits or losses flow through to owners
Continuity of business	Terminated by proprietor	Dissolved by partners	Perpetual	Perpetual	Varies
Cost of formation	Very low	Low	Moderate	Moderate	Moderate
Ability to raise capital	Low	Moderate	High	Moderate	Moderate

provides limited liability, unlimited life, the ability to accept investments from venture capitalists and other corporations, and greater flexibility when sold.

The majority of U.S. venture-backed companies are incorporated in a handful of states such as California, New York, or Delaware. In addition to the large number of companies resident in California and New York, relative to other states, California, New York, and Delaware have corporate laws that are well developed, stable, and transparent, all characteristics that ultimately reduce the risk to investors. Moreover, because venture capitalists and their counsel are familiar with conducting financings under such laws, the speed and efficiency with which a transaction moves will likely be improved.

The **limited liability company** offers an ideal form of ownership for small companies. It offers limited liability to the owners along with the tax advantages of a sole proprietorship or partnership. An LLC is particularly attractive to a family business that receives investments of a family's funds since it offers continuity of life, limited liability to participants, and advantages in handling tax issues.

The LLC's articles of organization establish the company's name, its duration, and the names and addresses of organizers. The operating agreement is similar to the articles of incorporation bylaws of a corporation in that it outlines the rights of the owners and the way the LLC will operate. The owners of an LLC are called *members,* and their ownership interests are known as *interests.* These terms are equivalent to *stockholders* and *stock.* Unlike the S-corporation, there is no limitation on the number of members or their status, and the LLC may have foreign investors. Like the S-corporation, the LLC will typically need to convert to a regular corporation to accept any venture capital or issue stock in a public market.

The regular C-corporation form will be often chosen by firms that intend to seek investment from numerous professional investors and other corporations. The C-corporation allows for various classes of stock, such as common stock and preferred stock. In particular, investors in a C-corporation can purchase *convertible preferred stock,* which is the most common form of venture capital investment.

The legal steps required to form a corporation in the United States are straightforward. For example, if a team selects California in which to incorporate, it files articles of incorporation with the secretary of state and then pays a fee. The certificate of incorporation states the name of the corporation; the broad business purpose; the authorized capital, including the total number of shares and the class of shares; the name and address of the registrant; and provisions for reimbursing certain damages and expenses that directors, employees, and officers may incur on behalf of the corporation [Bagley and Dauchy, 2007].

10.2 Company Name

The name of the new company is important. It should be memorable, related to the product or service, and attractive. It is also helpful if it can be used as the company's website domain address. The right name can evoke a sense of the company's character, bestow distinction, and make a powerful impression. Ideally, the name tells the prospective customer what the product's major benefit is. Many firms are named after their founders, for example, Dell and Wrigley. Others use creative names, such as Kodak or Exxon. Some firms use a locational name, such as Silicon Valley Bank or Allegheny Technology.

The right name can deliver a subtle message about the firm's unique features. Tinker Toys evokes a spirit of play. If possible, a name will serve as a marketing tool and will be easy to remember, spell, and say. Jeff Bezos chose the name Amazon.com because it conveyed the idea of a huge entity and did not limit him to one product [Leibovich, 2002].

Good examples of names of Web companies that immediately convey a purpose are LinkedIn and Facebook. LinkedIn is a social networking site targeting professionals. Facebook, originally designed for college students, was inspired by college yearbooks.

It is best to test the proposed name of your firm on others since it is important to avoid negative connotations. Founder Scott Cook of Intuit, the makers of Quicken, tried and rejected the name Instinct for his new company because it sounded like "it stinks" [Taylor and Schroeder, 2003]. Once a suitable name is chosen, a name search will be required to ensure that no one else has already claimed the name. Then, the name is registered with the appropriate state office. If national and international operation is envisioned, it may be wise to register the name as a trademark with the U.S. Trademark Registry Office.

In late 1984, Leonard Bozack and Sandra Lerner initially financed their own new firm with funds from their credit cards. They named the company cisco, as in the end of the name of the city of San Francisco. Lerner designed the logo in the form of the Golden Gate Bridge. Eventually, they capitalized the name to become Cisco [Bunnell, 2000].

A new venture should make sure that the name does not translate into something embarrassing or negative in a foreign language, and that the name carries no other undesirable connotations. Another factor is its pronounceability. A software company supporting Linux chose Red Hat as a memorable name and created a red hat logo. Other examples of good, memorable, robust names include Wal-Mart, Intel, General Electric, and Microsoft. On the articles of incorporation, the name typically must include the word "Incorporated," "Corporation," or "Company." Moreover, the name usually cannot contain certain words like "insurance" or "bank" without meeting additional criteria [Bagley and Dauchy, 2007].

Once the new venture chooses its name, it should reserve a domain name for its website and e-mail address. New ventures should check into the availability of a domain name early on, because doing so often eliminates certain choices for company names and trademarks. The best situation is when you can use the same legal name and domain name. Good examples of memorable corporate names that are also used as domain names are Google and Yahoo.

10.3 Intellectual Property

Within intellectual assets is a subset of ideas, called intellectual property, that can be legally protected [Davis and Harrison, 2001]. Property is defined as something valuable that is owned, such as land or jewelry. Furthermore, we can distinguish real property (or physical property) from intellectual property. **Intellectual property (IP)** is valuable intangible property owned by persons or companies. As discussed in Chapter 9, knowledge is the awareness and possession of information, facts, ideas, truths, methods, and principles in an area of expertise. This knowledge is a valuable asset of the firm and is called intellectual property. A comparison of the qualities of physical and intellectual property is provided in Table 10.3.

Since knowledge and innovation are keys to competitive success, the management of intellectual property is important to most firms. For many firms, intellectual assets are the wellsprings of wealth and competitive advantage. The protection of intellectual property can lead to the possession of valuable assets; for example, the secret formula for Coca-Cola.

The protection and enforcement of legal ownership of intellectual property is more difficult than for physical property. How can a firm tell that another firm has used or taken its intellectual property? Unauthorized copying or illegal use of intellectual property can be difficult to discern and prove. The owner of a textbook or a CD purchased at a store has the right to share it with another person but is precluded by law from copying it for sale to another person.

As intellectual property is difficult to defend, it may be useful to develop a strategy to deter misappropriations [Anand and Galetovic, 2004]. Suitable strategies include overwhelming competitors by continually out innovating them and/or licensing the IP to create cooperation with competitors.

The purpose of intellectual property law is to balance two competing interests: the public and the private. The public interest is served by the creation and distribution of inventions, music, literature, and other forms of intellectual expression. The private interest is served by rewarding people for creating and publicly disclosing these works through the establishment of a time-limited monopoly granting exclusive control to the creator.

During the course of working for a firm, an employee has an idea for a new product that is outside of the scope of the business of the firm. Who owns

TABLE 10.3 Comparison of physical and intellectual property.

Factor	Physical property	Intellectual property
Multiuse	Use by one firm precludes simultaneous use by another	Use by one firm does not prevent unauthorized use by another
Physical depreciation	Depreciates, wears out	Does not wear out
Protection and enforcement from encroachment	Generally can enforce and protect ownership	May be difficult or expensive to enforce and protect ownership

this intellectual property, the employer or the employee? Does it matter whether the new idea was conceived on the weekend? Entrepreneurs should avoid any potential complication or dispute of ownership of intellectual property by assiduously following the legal and moral laws of property. They should also reread all employment agreements that may state restrictions on their ownership of intellectual property. For example, most universities require that graduate students and faculty sign a form assigning intellectual property rights to the university for any invention made using university resources or made in the course of university-sponsored research.

Clearly, if a firm plans to apply for a patent or use some technical advance of a proprietary nature, the firm and its employees should not at first reveal the details to prospective investors (or anyone else). If these investors become serious about investing in the new business, then the firm may need to reveal more information about the proprietary asset.

Many technology start-ups take four to seven years to reach the market and become profitable. Typically, these new ventures are founded on a significant array of intellectual property such as patents. When the entrepreneur's personal knowledge is perceived to be a critical portion of the intellectual property, then this person will be expected to remain with the firm for several years. Arrangements such as employment agreements and stock vesting terms will help ensure this active involvement [Lowe, 2001].

Protecting and managing intellectual property is important. Analysts estimate the intellectual property market to be $100 billion annually. At IBM, patents and licenses represent 15 percent of revenues. A useful reference for an entrepreneur is *Patent, Copyright, and Trademark,* by Stephen Elias and R. Stim [2003]. Professional legal assistance is also advisable. Technology law firms often advise new ventures and defer payment of cash fees until the company receives its first investment. These law firms often help the venture get under way for three to six months in return for a small percentage of equity (e.g., 1.0 percent) [Henderson et al., 2006]. If the venture succeeds like an Electronic Arts or Google, the attorneys will reap great financial benefits.

Intellectual property may be protected in a variety of different ways. The most common types of protection are trade secrets, patents, trademarks, and copyrights.

10.4 Trade Secrets

A **trade secret** is a confidential intellectual asset that is maintained as a secret by the owner and provides the owner with a competitive business advantage. A trade secret may include knowledge, methods, ideas, formulas, or the like. The period of life for a trade secret is potentially indefinite. The formula for Coca-Cola has been a trade secret for over a century. Trade secret protection may be lost, however, upon any unauthorized disclosure, such as theft, violation of confidentiality, independent recreation, or reverse engineering. The potential protection offered by secrecy depends on the attributes of the intellectual property and the circumstances of its use. Secrecy is valuable for formulas, algorithms, and know-how that can be implemented by a firm without its being known by other than a few people. If the knowledge must be widely shared throughout the firm, it will be difficult to protect it from those who would copy or imitate it.

Many production processes can be protected behind the walls of the firm. For example, methods for manufacturing integrated circuits are widely available, but the best production process for making them is quite complicated. Several semiconductor firms keep their competitive advantage by maintaining the secrecy of their methods as well as their processes. The risk always exists that an employee will learn the secrets of the methods and the process and decide to start a competitor firm.

A firm will have to balance the need to protect secrets with the necessity to widely share information among employees. Employees must be informed that they are dealing with secrets that are the property of the firm and they are expected to protect these secrets. For many firms, common knowledge among employees of the methods and procedures is necessary for success. For them, it is the execution of the total business process that provides the competitive advantage.

10.5 Patents

Abraham Lincoln called the introduction of the U.S. system of patent laws "one of the three most important events in the world's history" [Schwartz, 2002]. A **patent** grants inventors the right to exclude others from making, using, or selling their invention for a limited period of time. In the United States, this is generally 20 years from the date of filing once the patent issues. A patent for an invention is the grant of a property right by the country in which the application is filed. Patents may be granted to new and useful machines, manufactured products, and industrial processes, and to improvements of existing ones. Patents also may be granted for new chemical compounds, foods, and medicinal products, as well as for the processes for producing them. See appendix C for patent websites.

Utility patents are issued for the protection of new, useful, nonobvious, and adequately specified processes, machines, and articles of manufacture. Examples include the patent for the safety razor and the rolling bag that is widely used by air travelers.

Design patents are issued for new original, ornamental, and nonobvious designs for articles of manufacture. For example, the new design of a computer case could be submitted for a patent. **Plant patents** are issued for certain new varieties of plants that have been asexually reproduced.

A **business method patent** is a type of a utility patent and involves the creation and ownership of a process or method. While U.S. Courts had more broadly allowed business method patents in the past, under current law the extent to which these patents may be allowed has been curtailed.

An invention must be considered novel and useful to be considered for patentability. It must also represent a relatively significant advance in the state of the art and cannot merely be an obvious change from what is already known. Such requirements are meant to reduce the number of inventions that modify existing products in minimal ways. Patents are often granted for improvements of previously patented articles or processes if the requirements of patentability are otherwise met. However, the granting for an improvement does not by itself give the holder any rights to the underlying patent. In general, patents tend to work well in industries where the core technology is biological or chemical [Shane, 2005].

A patent is recognized as a type of property with the attributes of personal property. It may be sold or assigned to others or mortgaged, or it may pass to the heirs of a deceased inventor. Because the patent gives the owner the right to exclude others from making, using, or selling the invention, the owner may authorize others to do any of these things by a license and receive royalties or other compensation for the privilege. If anyone makes use of a patented invention without authorization, the infringement can be brought to court in a suit filed by the patent holder, who may request monetary damages as well as a court injunction to prevent further infringement [Elias and Stim, 2003]. A patent only protects what is stated in its claims. An inventor tries to make multiple claims, but must often write narrow claims to avoid conflict with a previous patent.

King C. Gillette desired to invent a product that would be used and then discarded so that the customer would keep returning for more. While sharpening a permanent, straight-edge razor, Gillette had the idea of substituting a thin, double-edged steel blade placed between two plates and held in place by a handle. Though the invention was received with skepticism because the blades could not be sharpened, the manufactured product was a success from the beginning. Gillette filed for a patent in 1902 and started sales in 1903. By the end of 1904, Gillette's company had produced 90,000 razors and 12.4 million blades.

The patent process requires an application that includes a clear, concise description of the invention. It also defines the boundaries of the exclusive rights that the inventor claims. Furthermore, patents are territorial so inventors must apply for patents in each country where they wish to protect their inventions.

Once a patent has been issued, the owner may start a patent protection program that includes issuing notices and labeling products or services covered by the patent, monitoring uses of the patent, and pursuing known or suspected infringers of the patent. The patent provides the owner the right to exclude others from using the patent without compensation. However, the owner is responsible for enforcing that right by sending notices to infringers and possibly resorting to litigation, if necessary. Sometimes imitators are able to design new products or methods that circumvent the existing patent.

The value of patents can be high. Medtronic purchased patent rights for the inventions of Dr. Gray Michelson for $1.35 billion in 2005. Dr. Michelson is a back surgeon in Los Angeles who held approximately 220 patents [Burton, 2005]. The patents awarded to Stanley Cohen and Herbert Boyer in 1980 covering gene-splicing techniques, a basic part of biotechnology, earned more than $250 million for their owners, Stanford University and the University of California. Revenues in the United States from the licensing of patent rights have grown from $15 billion in 1990 to more than $110 billion in 2000.

Patents have proved to be very effective for inventions in the pharmaceutical and medical instruments industries. A laptop computer may include up to 500 patented inventions held by many firms. On the other hand, a pharmaceutical drug will normally be covered by a single patent. Thus, drug companies typically enjoy strong intellectual property positions.

In many industries, firms eager to capture gains from their innovations are filing patent applications at an unprecedented rate. Several legal changes and court decisions in the 1980s provided more protection for patents. The 1985 case in which Polaroid won more than $900 million in damages from Kodak for instant-camera patent infringement provided strengthened precedence for patent infringement litigation. More recently, Research In Motion, makers of the Blackberry portable email device, paid $612.5 million to NTP in 2006 to settle a patent infringement lawsuit. Court decisions have also expanded patentable subject matter to include genetically modified organisms, software, and in certain cases business methods.

Growth in the number of patents issued in the United States in recent years is shown in Table 10.4. In 2007, the U.S. Patent and Trademark Office issued nearly 183,000 patents. Companies are increasingly building their innovation strategy around patents and intellectual property. IBM, for example, was granted more than 3,000 patents in 2007.

Patenting an innovation can be expensive since a single patent application can cost $20,000. In addition, the cost of infringement litigation can be very high. Proving patent infringement requires documentation and analysis of the infringing product or process. Fledgling entrepreneurs cannot expect to have the funds required to file for patent protection for all intellectual property and to litigate all possible infringements. Therefore, they must determine when it makes sense to pursue a patent. First, entrepreneurs should evaluate the core technologies that are fundamental to the business's success. If a technology is outside of this core, it probably should not be patented. Second, the entrepreneur should consider whether a competitor could easily invent an alternative to the patented technology. If so, then the technology probably should not be patented. Entrepreneurs should

TABLE 10.4 U.S. patents issued.

Year	1980	1990	2000	2007
Number issued (thousands)	66.2	99.1	176.0	182.9

Source: U.S. Patent and Trademark Office, 2007.

TABLE 10.5 Developing a patent strategy.

1. Identify the goals of a patent portfolio.

2. Identify the intellectual assets and gather supporting documents.

3. Identify those assets most suitable for patent applications.

4. Draft invention disclosures and patent applications.

5. Develop a plan for licensing, enforcing, and enhancing patents.

Source: Fenwick & West LLP, R. P. Patel.

also consider whether other, less costly, forms of protection might be effective, such as trade secrets.

At the same time, it is important to recognize that patents can be useful when bargaining with other companies and can be a positive symbol of innovative capacity when raising money. Therefore, it is wise for entrepreneurs to consider the role of patents beyond the simple protection of a particular invention. In developing and maintaining its patent portfolio, a new venture should follow the steps outlined in Table 10.5.

10.6 Trademarks

A **trademark** is any distinctive word, name, symbol, slogan, shape, sound, or logo that identifies the source of a product or service. The holder of a trademark gains rights as the trademark is used. In the United States, registering a trademark gives the holder advantages in enforcement. A registered trademark is renewable indefinitely as long as commercial use is proven. A new venture should consider trademarking its company name, symbol, or logo. Commonly known trademarks include Kodak, Apple, Google, the NBC logo, and Yahoo.

The trademark owner has the right to bring legal action to halt any infringing use for damages and recovery of profits. Trademark rights are often among the most valuable assets of an emerging new venture in today's competitive marketplace. The goodwill and consumer recognition that trademarks represent have great economic value and are therefore usually worth the effort and expense to properly register and protect them.

A good trademark is an integral part of a firm's brand. To possess good value, a trademark should readily be associated with and exclusive to the firm. Excellent examples of a powerful trademark are the Apple logo and the Intel Inside logo. There were 210,000 trademarks issued in the United States in 2008, and there are an estimated total of 1.5 million registered trademarks in the United States. Companies may also file an "intent to use" trademark application. However, the company must actually be using the trademark before it can be registered.

A firm may lose the exclusive right to a trademark if it loses its unique character and becomes a generic name. Aspirin, thermos, and cellophane are examples of names that have become generic. Coca-Cola and Xerox have successfully protected their trademarks.

10.7 Copyrights

A **copyright** is a right of an author to prevent others from printing, copying, or publishing any of his or her original works. For copyrights created today, the life of the copyright is for the life of the author plus 70 years after the author's death. Because copyright protection automatically attaches upon creation of a work and the process of registering a work with the U.S. Copyright Office requires only the completion of a simple form, the process of obtaining copyright protection demands very few resources.

A copyright extends protection to authors, composers, and artists, and it relates to the expression rather than its subject matter. This is important, because a copyright only prevents duplicating or using the original material. This does not prevent use of the subject matter. Therefore, software programs, books, and music are protected from copying, but the ideas in these forms may be used by others.

The protection provided by copyright is somewhat limited. In the software field, for example, courts have narrowed the scope of copyright protection. Copyright is most effective against wholesale copying of all or a significant portion of a program. It has limited protection for functional aspects of software products, such as the underlying algorithms, data structures, and protocols of multimedia technology. Copyright may also protect fanciful aspects of the graphical user interface of a program.

10.8 Licensing

Licensing is a contractual method of exploiting intellectual property by transferring rights to other firms without a transfer of ownership. A **license** is a grant to another firm to make use of the rights of the intellectual property. This license is defined in a contract and usually requires the licensee to pay a royalty or fee to the licensor.

Many firms have a large number of unexploited or underexploited patents that a licensee may be able to exploit. IBM, for example, widely grants licenses, and its royalty income amounts to more than $1 billion each year. IBM holds more patents than any other U.S. company and licenses its software patents widely. But, while most new firms realize that intellectual property can be among their most valuable and flexible assets, they remain unaware of the earning potential of their patent holdings.

Licensing can form the core of a business model. For example, licensing is widely used to provide software to users. Microsoft derives most of its revenues from license fees for its Office suite and Windows operating system. Dolby Laboratories Inc. gets much of its revenue from licensing its products to electronics makers. It succeeds this way partly because of its well-known technology and the reasonable price it charges other firms for use of the technology.

A new venture can derive valuable income streams by licensing its intellectual property to other firms for noncompetitive, complementary uses. The benefits to the licensor include spreading the risk, achieving expanded market penetration, earning license income, and testing new products and markets. Disadvantages of licensing may include risk of infringement and nonperformance of the licensee.

A new venture also can save time and resources by licensing another firm's technology to use in its products. The terms of the license with the third-party technology owner establish what rights a start-up has to use, distribute, modify, and sublicense the licensed technology. Licensing terms often are structured in recognition of a start-up's high potential but lack of capital. For example, a firm may waive or minimize the up-front license fee charged to a start-up. But, in return, it may demand a percentage of sales revenue once the start-up releases a product.

AMR's Patent Stream

Albany Molecular Research was founded in 1991 by chemist Thomas D'Ambra as a pharmaceutical research firm. The firm develops and patents new methods and drug compounds (see www.amriglobal.com). AMR receives royalties on patented technologies that it has licensed. It had revenues of almost $200 million in 2007.

Most start-ups founded on the basis of university-developed technologies will need a license from the university. Even if a student or professor is both the inventor and the entrepreneur who brings the technology to market, most universities own the intellectual property since they provided the lab space, salaries, and other resources to conduct the research. Thus, the inventor must obtain a license. University technology transfer offices can provide valuable marketing and intellectual property assistance. Inventor-entrepreneurs who wish to avoid licensing their own inventions from the university must be careful to work on these inventions without using university resources.

10.9 AgraQuest

Upon its formation, AgraQuest's entrepreneurial team looked for a name that conveyed that it was searching for natural solutions in the agricultural industry. It first proposed the name Agrisearch, but that was already taken. Its next choice was AgraQuest, and that was available. The team registered the name as a trademark internationally. Then the AgraQuest team selected a logo depicting a hummingbird searching in a flower for nectar, as shown in Figure 10.2.

AgraQuest incorporated in Delaware, since it anticipated venture capital investments and an initial public offering. The company possesses many trade secrets, primarily in the manufacturing of its products. It also has a proprietary method for identifying, screening, and developing the microbes for its products.

FIGURE 10.2 AgraQuest logo of a hummingbird searching for nectar in a flower.

As of 2008, AgraQuest held 27 patents. Each patent covers a microbe and its use. In addition, AgraQuest has licensed *M. albus* and a collection of microbes for testing.

The intellectual property of AgraQuest is managed by CEO and the director of research. The firm engages a large law firm for business and intellectual property legal matters.

10.10 Summary

The legal form of the venture should match the needs of the entrepreneurs, customers, and investors. For most high-growth ventures, the regular C-corporation will be most appropriate. However, for many new organizations, other legal forms may be suitable. Ten mistakes with legal matters are provided in Table 10.6.

TABLE 10.6 Ten mistakes with legal matters.

- Failing to secure legal assistance
- Delaying the handling of legal issues
- Delaying the intellectual property management process
- Issuing founder shares without vesting provisions
- Failing to incorporate early
- Disclosing intellectual property without a nondisclosure agreement
- Starting a business while employed by a potential competitor
- Overpromising and exaggerating claims in the business plan
- Failing to register the name of the firm early in the start-up process
- Failing to develop confidentiality, nondisclosure, and noncompete agreements

Source: Bagley and Dauchey, 2007.

The plan to acquire, build, and protect the intellectual property of the new venture should be clear to all the participants. The proper array of trade secrets, patents, trademarks, and copyrights can come together as a strong set of valuable proprietary assets. For high-growth, technology-based companies, intellectual property can be used to build a competitive advantage. An important step after deciding to launch a business is to choose a name and logo for the business that conveys the right image and represents the business well.

> **Principle 10**
> The name, logo, and intellectual property of a new venture can provide a proprietary advantage leading to success in the marketplace.

Video Resources

Visit http://techventures.stanford.edu to view experts discussing content from this chapter.

The Sarbanes-Oxley Act	Karen Richardson	E.piphany
The Role of Lawyers in the Start-up Ecosystem	Gordon Davidson	Fenwick & West

10.11 Exercises

10.1 Three friends have decided to form a firm to design and manufacture nanotechnology devices for medical applications. Michael Rogers has worked for Hewlett-Packard for 12 years and on his own has designed and submitted a patent claim for a nanotechnology manufacturing technology. Steve Allegro, a graduate student, has a software program he has developed for the design of nanotechnology medical devices. Alicia Simmons, CFO of Alletech Software Inc., is a skilled and experienced manager. Shall they incorporate immediately? What is the problem, if any, of using Rogers's patent ideas? Should they incorporate in their home state of Alabama? Simmons has knowledge of several manufacturing trade secrets of Alletech. Can she use these secret methods at her new firm?

10.2 Continuing exercise 10.1, if the three founders expect to be able to fully bootstrap fund the new venture, is there a preference for what legal form of the firm makes most sense? What if the three founders expect to require angel or venture capital investment in the future? How would medical device legal risks influence this decision?

10.3 The three founders of the new firm described in exercise 10.1 are looking for a name for their firm. One idea is Advanced Nanoscience

& Technology. Another is Nanoscience Applications. What do you think of these names? Can you suggest a better name?

10.4 Apple Inc. and Apple Corps have had trademark disputes over the use of the name Apple associated with the music business. Apple Inc. has a thriving iPod and online music store. Apple Corps is a multiarmed multimedia company formed by the Beatles in 1968, consisting of the following subsidiaries: Apple Records, Apple Electronics, Apple Films, Apple Publishing, and Apple Retail. Briefly describe the arguments for both sides. Why is it important for a company to challenge and protect its trademarks?

10.5 Describe the advantages and disadvantages of a company name representing more than just a company name (e.g., Xerox, Kleenex, and Google).

10.6 Mayo Clinic has filed an application for a broad method patent that gives it control over a new generation of treatments for chronic sinus inflammation (sinusitis). The patent, in effect, blocks others from selling an antifungal agent to treat the condition without Mayo's approval. Mayo will soon try to license this patent to a pharmaceutical company. Are patents helpful in the process of developing a cure for diseases?

10.7 Headwaters Inc. develops and licenses technologies for turning coal and other fossil fuels into higher-value products. Is this an appropriate name for the firm? Revenue is generated through the licensing of the firm's patented chemical processes (www.hdwtrs.com). Examine the patents of Headwaters and its intellectual property protection. How would you describe the growth potential of this firm?

10.8 In March 2005, *Research in Motion (RIM) vs. Network Technology Partners (NTP)* concluded a prolonged patent fight. Describe RIM's service offering and what NTP patents they were violating. How did RIM address the patent challenge? What is NTP's business model?

VENTURE CHALLENGE

1. Describe your venture's legal form and provide its company name.

2. Go to the State of Delaware's website (http://corp.delaware.gov/howtoform.shtml) and download the "How to Incoporate" packet. Determine what type of incorporation best fits your venture and fill out the appropriate form(s). Alternatively, locate similar forms for another state or nation.

3. Create a company logo.

4. What are the key elements of the intellectual property of your firm?

Detailed Functional Planning for the Venture

The creation of a plan for marketing and selling a product is based on clearly describing the target customer and how the product will be priced, communicated, delivered, and supported. An organizational plan that supports a collaborative, performance-based culture and a sound compensation scheme must be created to attract good talent. The acquisition of resources and capabilities and facilities will be planned for and initiated in order to build momentum for the venture. The management of operations, processes, and manufacturing will be described in an operations plan. A plan for outsourcing some activities and acquiring necessary assets and technologies will facilitate the early growth of the firm. Finally, the venture team will describe the potential for acquisitions, if any, and the plan for operating internationally in order to further stimulate growth. ■

CHAPTER 11

The Marketing and Sales Plan

Successful salesmanship is 90 percent preparation and 10 percent presentation.
Bertrand R. Canfield

What is the best way to attract, serve, and retain customers?

Marketing and sales are critical to the success of a new firm since the firm normally starts without any customers. A new business must create a marketing and sales plan, which describes its target customers for its product offering. A sound marketing plan is built on solid information obtained through market research. The new firm creates a product position and a mix of price, product, promotion, and distribution channels that will attract and satisfy the customer. Gaining recognition and acceptance in a target market requires the following steps in sequence:

- Describe the product offering
- Describe the target customer
- State the marketing objectives
- Gather information through market research
- Create a marketing plan
- Create a sales plan
- Build a marketing and sales staff

251

11.1 Marketing

Marketing is a set of activities with the objective of securing, serving, and retaining customers for the firm's product offerings. Marketing is getting the right message to the right customer segment via the appropriate media and methods. It is the task of the marketing function to help create the product and the terms of its offering as well as communicate its value to the customers. Ideally, marketing "merges the minds" of customers and product developers, facilitating the identification of unspoken but important needs [Lassiter, 2002].

The purpose of the **marketing plan** is to describe the steps required to achieve the marketing objective. The marketing plan is a written document serving as a section of a new venture's business plan and contains action steps for the marketing program for the products. Peter Drucker [2002] has said, "Because its purpose is to create a customer, the business has two basic functions: marketing and innovation. Marketing and innovation produce results; all the rest are costs."

In Chapter 3, we described the creation of a value proposition and business model for the identified customer. In Chapters 4 and 5, we described the elements of an overall business strategy and market analysis. Given these business elements, we need to develop a marketing strategy and build a marketing plan. The six elements of a marketing plan are shown in Table 11.1.

The first element of the marketing plan is a clear statement of objectives. The second element is the identification of one or more customer target segments. The goal of target segments is to carefully select the appropriate customers and to focus the marketing activities on those segments. The third element is the description of the product and the terms and conditions of its formal offering. Given our knowledge of the product and its offering, we need to determine what the response of the customer might be and how we can develop a strategy to attract and retain the customer. Next, we describe the marketing mix consisting of price, product, promotion, and place (channels). Finally, we describe plans for relating to our customer in the sales and service activities.

The marketing plan will be implemented through a marketing program. The plan will describe how we will take the product to market and attract, serve, and retain satisfied customers. The marketing and sales activity is portrayed in Figure 11.1. The new venture communicates information about its product and how it sells and services the product for the customer. When the

TABLE 11.1 Six elements of the marketing plan.

- Marketing objectives
- Target customer segments
- Product offering description
- Market research and strategy
- Marketing mix
- Customer relationship management

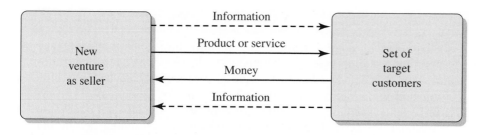

FIGURE 11.1 Marketing and sales activity of the new venture.

customers purchase the product, they provide some useful information about the purchase and the use of the product for the seller.

The marketing and sales plan will flow from the opportunity and the business model, as shown in Figure 11.2.

11.2 Marketing Objectives and Customer Target Segments

The **marketing objectives statement** is a clear description of the key objectives of the marketing program. Objectives may include sales goals, market share, profitability, regional plans, and customer acquisition goals. Objectives should be quantified and given for a time period, such as "the firm will sell 1,000 units in the initial sales phase in Texas and Oklahoma in the first year of activity."

A clear understanding of who the customers are and why they will buy is critical. Selected markets or groups of customers are often called customer target segments. A **market segment** consists of a group with similar needs or wants who reference each other and may include geographical location, purchasing power, and buying attitudes. **Market segmentation** divides markets into segments that have different buying needs, wants, and habits. Different segments will require different marketing strategies. For example, a business based on Internet sales should know that different age groups, which are one type of segment, vary dramatically in their propensity to shop and in the amount they spend [Abate, 2008]. Often a new venture identifies one target segment for its initial marketing effort, carefully describing the customer in that segment in terms of geographic, demographic, psychographic, and other variables [Winer, 2000]. Geographic variables include city, region, and type, such as "urban." Demographic variables include age, gender, income, education, religion, and social class. Psychographic variables include lifestyle and personality variables that influence a customer's wants and needs. Good segmentations identify the groups most worth pursuing and they tell companies what products to sell to these groups. Good segmentations also change over time, along with customers [Yankelovich and Meer, 2006].

FIGURE 11.2
Building a marketing plan and a sales plan.

InVision: The Right Product at the Right Time

InVision Technologies designs and manufactures electronic baggage screening systems (see www.invision-tech.com). InVision Technologies was founded in 1990 to provide airport security devices by using computed tomography to detect explosives. Its target segment, U.S. airport baggage screening, was slow growing before the terrorist attacks on September 11, 2001. It sold 250 machines in the preceding decade but 750 machines in the two years following the attacks. InVision also planned to enter the international market. InVision's success and expertise led to its acquisition by GE in 2004 for $900 million.

The target segment for Research In Motion's Blackberry can be described as adults who need constant wireless connectivity while on the go. The target customer for the Blackberry is a professional who wants a pocket-size device that is effective at delivering enterprise applications including e-mail. With a clear description of the target customers, a new venture can devise a plan to attract and retain them. Marketing to a segment enables the new firm to narrow the marketing strategy and put all its effort into acquiring new customers in the target market. Often, new firms reach for too many market segments in their early efforts, thus dissipating their resources before they can build up a customer base. Table 11.2 identifies four critical questions to ask about target markets.

Redefining Flexcar's Customer Segments

Car sharing is aimed at people who want to use a car for a short time but do not need to own it. This scheme is particularly attractive in urban centers. In general, people are moving toward access rather than ownership of autos in dense urban areas with high auto costs. Neil Peterson started Flexcar in 1999 in Seattle and then expanded to Los Angeles, San Francisco, Washington, D.C., and San Diego.

As the venture grew, the average cost of owning or leasing a new car became $625 a month. The average member in a car-sharing program spent less than $100 a month on car expenses. Flexcar initially identified its target segment as individuals. It soon discovered, however, that the biggest growth came not from individuals but from small and midsize companies that did not want to maintain their own fleets of vehicles [Stringer, 2003]. In 2007, Flexcar merged with Zipcar (see www.zipcar.com).

It is wise to figure out who will be your best customer and then pursue that segment. A **best customer** is one who values your brand, buys it regularly whether your product is on sale or not, tells his or her friends about your product, and will not readily switch to a competitor. Entrepreneurs identify their customer segment and position their product to serve them very well [Ettenberg, 2002].

TABLE 11.2 Four crucial questions to ask about target markets.

- Is there a target market segment where the company can enter the market and provide clear customer benefits at a price the customer is willing to pay?

- Do customers perceive that these benefits are superior to other solutions/options?

- How large is the target market segment and how fast is it growing?

- Will entry into the target market segment serve as a springboard for entry into other segments?

Source: Mullins, 2006.

11.3 Product and Offering Description

The product's features and primary attributes are typically described early in a business plan. If possible, a product positioning map should be developed. All products can be differentiated to some extent by communicating the most highly valuable benefit to the buyer. **Positioning** is the act of designing the product offering and image to occupy a distinctive place in the target customer's mind [Ries and Trout, 2001].

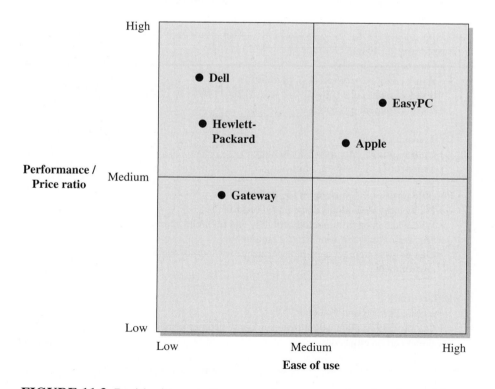

FIGURE 11.3 Positioning map for personal computers showing the position of a new product, EasyPC.

Positioning of a product enables the firm to differentiate it in the mind of the prospect. Volvo connotes safety, and FedEx owns "overnight." A product positioning map shows the product characteristics in relation to its competitors. Figure 11.3 shows a hypothetical new personal computer called EasyPC that is positioned as having high ease of use and a high performance-to-price ratio. It would then be the task of the marketing effort to clearly communicate that position to the target customer.

Positioning a product focuses on a few key attributes of the value proposition. A positioning statement, as shown in Figure 11.4, helps to define the positioning of the product. Once we have a product position, we seek to build a powerful product offering [Moore, 2002]. A **product offering** communicates the key values of the product and describes the benefits to the customer. The **unique selling proposition** is a statement of the key customer benefit of a product that differentiates it from its competition. The unique selling proposition for the EasyPC could be:

> EasyPC delivers high performance at a reasonable price and is easy for anyone to use.

The unique selling proposition of FedEx is:

> We deliver your packages overnight—guaranteed.

- **Positioning statement**
  - **For** (target customer)
  - **Who** (statement of need or opportunity),
  - (Product name) **is a** (product category)
  - **That** (statement of benefit).   **(a)**

- **Differentiation**
  - **Unlike** (primary competitive alternative),
  - **Our product** (statement of primary differentiation).

- **Possible positioning statement for Tesla Motors**
  - For wealthy individuals and car aficionados
  - Who want an environmentally friendly and high-end sports car,
  - The Tesla Roadster is an electric automobile
  - That delivers unprecedented performance without damaging the environment.   **(b)**

- **Differentiation**
  - Unlike Ferraris and Porsches,
  - Our product has fantastic mileage, unparalleled performance, and no direct carbon emissions.

FIGURE 11.4 **(a)** Positioning statement format, and **(b)** for Tesla Motors

11.4 Market Research and Customer Development

Market research is the process of gathering the information that serves as the basis for a sound marketing plan. Once a target market is selected, the entrepreneur needs some information about customers' preferences and behavior as well as competitors' products. The objective of market research is to learn how to attract and retain customers for a product. Market research can provide critical information to the new venture team. Without complete information, a new venture may launch a product only to ultimately determine that the customer does not value the product. A key question is: Does the target segment want the perceived value that our positioning is trying to deliver more than other segments? If so, how can we reach this segment efficiently?

The market research effort can consist of four steps, as shown in Table 11.3. Using these steps, entrepreneurs can develop an understanding of their customers, including their preferences and behavior. The first step is to determine the needed information and research objectives. Then, the new venture team develops a plan for gathering information from the targeted customer segment. It is helpful to use printed sources such as trade data, magazines, and trade journals. Corporate reports and news about the industry will be available. A search on the Internet will lead to many valuable sources. See appendix C for marketing research sources.

Primary data are collected for your specific research objective. Secondary data sources that were collected for another research purpose are already available. Primary data sources are very valuable, and entrepreneurs should avoid relying solely on secondary sources. Talking to the actual customer and other channel participants is very important. A popular form of research uses the **focus group,** which is a small group of people from the target market. These people are brought together in a room to have a discussion about the product. This discussion can be led by someone from the venture team or a professional moderator. Other methods of collecting data include surveys and observation of customers. The movie business uses free previews to test viewers' reactions. The studio then uses that information to revise or take out scenes or characters or change the ending. Focus groups have limits since no potential customer ever asked for ATMs, traveler's checks, or personal computers.

TABLE 11.3 Market research process.

1. Define the product and its unique selling proposition. Identify the customer segment. Develop a set of questions that will provide the necessary data on customer preferences and behavior.

2. Collect the data using surveys, published sources, focus groups, interviews, and other means.

3. Analyze and interpret the data to determine if the product meets the needs or wants of the customers and determine whether they will pay the selected price.

4. Draw conclusions about the customers and their needs, preferences, and behavior.

Many technology companies use the "a day in your life" format that takes customers through their day before and after the new product launch to expose the benefits of the new product.

Technology ventures are well served if they can identify and work with some key customers who are also innovators. These lead users are knowledgeable people who are willing to donate their time to work cooperatively with the new venture [Von Hippel, 2005].

Customer development is the process of the discovery, validation, and creation of customers leading to company building [Blank, 2006]. This model for the development of the sales and marketing plan runs in parallel with the product development process. It uses several iterations of each step to arrive at a successful plan and promotes continuous evaluation of who the customer is as shown in Figure 11.5.

An important use of market research is to estimate the **market potential** for maximum sales under expected conditions. Then the new venture team can estimate a realistic **sales forecast.** Market potential for a new product is often overestimated. For example, we might estimate that the potential market for the EasyPC is 10 million units per year, based on the sales of Dell and HP (see Figure 11.3). Then, an optimistic estimate of actual sales in the first year might be 1 million units. Clearly, that forecast is subject to many unstated assumptions—which may be in error. A sales forecast should be a realistic estimate of the amount of sales to be achieved under a set of assumed conditions within a specified period of time. Many sales forecasts are unrealistic. A sales forecast for a new venture needs to be conservatively developed within a statement of assumed conditions.

Being first, being best, and being correct may not matter as much as providing what the customer really wants right now. Finding out what the customer really wants is a very important and difficult task. Often people will not or cannot verbalize their true motivations and attitudes. In creating a marketing plan, the attitudes and preferences may not be clearly reported. One widely used method is **conjoint analysis,** which provides a quantitative measure of the relative importance of one attribute as opposed to another. In conjoint analysis, the respondent is asked to make trade-off adjustments and decisions. This method requires an investment of time and money in the research process but may be worth it to avoid misreading the customers' preferences [Aaker et al., 2001].

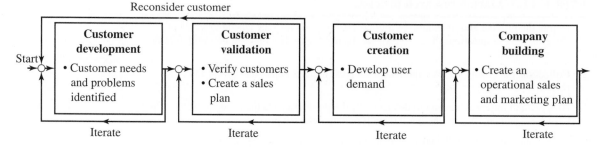

FIGURE 11.5 The customer development process.

11.5 Brand Equity

The new venture should have a competitive advantage such as low cost, quality, customer relationship, or performance advantage. Many new technology ventures differentiate themselves from competitors by doing a better job of convincing their customers that they have a better product characteristic, such as performance, reliability, or quality. A brand is a combination of name, sign, or symbol that identifies the goods sold by a firm. A brand accurately identifies the seller to the buyer. A **brand** is something that resides in the minds of consumers. Well-known brands include Intel, HP, and Dell. Brand equity is the brand assets linked to a brand's name and symbol that add value to a product. **Brand equity** is the perceived worthiness of the brand in the mind of the customer and may be portrayed as the sum of four dimensions, as shown in Table 11.4 [Aaker and Joachimsthaler, 2000]. Brand awareness or familiarity is the first step in building a brand. The perceived quality of the product and respect for the product will help build brand equity. The quality of the product and its perceived vitality will build an image of the brand. A brand association is how the customer relates to the brand through personal and emotional associations. This dimension is present in the emotional relationship that Harley-Davidson owners have with the motorcycle brand. In other words, brand loyalty responds to promises kept by the seller. Technology firms with significant brand equity include eBay, GE, Genentech, Intel, Microsoft, and Nokia.

A brand's promise of value is the core element of differentiation. This promise of value is tied to the customer, and loyalty will follow from good customer experiences. Many customers are willing to pay more for some badge of identification—Apple's rainbow-colored logo, for example—that makes them feel they are part of a community. A strong corporate brand lets customers know what they can expect of the whole range of products that a company produces. The most successful corporate brands are universal and facilitate differences of interpretation that appeal to different groups. This is particularly true of corporate brands whose symbolism is strong enough to allow people across cultures to share symbols even when they don't share the same meaning.

The brand of a company is customer centered and focuses on the product or service offered. By contrast, the reputation of a company focuses on the credibility and respect among a broad set of constituencies such as suppliers, regulators, employees, the media, and the local communities. Brand equity depends on the delivery of a good product to customers. Reputation depends

TABLE 11.4 Four dimensions of brand equity.

- Brand awareness or familiarity
- Perceived quality and vitality of the product
- Brand associations: connects the customer to the brand
- Brand loyalty: a bond or tie to the product

on the goodwill of the communities and stakeholders it interacts with [Etten-son and Knowles, 2008]. Both brand equity and reputation are important to a venture's success.

Some brands, such as Nike, Harley-Davidson, and BMW, become icons [Holt, 2003]. A brand becomes an icon when it offers a compelling story that can help people resolve tensions in their lives. One of the most potent stories is the depiction of a group of rebels. For example, Nike appeals to rebel youth who want to stand out as different from the crowd.

One approach to building a brand is to identify the differentiating benefit that is important to the target customers and describe the attributes that imply this benefit. Intel identifies superior quality as its benefit. Successive marketing campaigns have informed consumers that Intel integrated circuits have reliable high performance and are leading-edge products promising superior quality and performance.

11.6 Marketing Mix

The four elements of the marketing mix are shown in Table 11.5. The **product** is the item or service that serves the needs of the customer. The marketing plan describes the key methods of differentiating the product. Coca-Cola, for example, differentiates regular Coke by using a distinctive trademarked bottle with ribbing. Some auto companies use their warranty to distinguish their product. Nordstrom distinguishes its products by quality and its liberal return policy. Kodak's EasyShare digital camera is distinguished by its ease of use. Intel's Pentium chip is distinguished by its high-speed performance.

Pricing policies can be used to distinguish a firm's offering. Warren Buffet said it clearly, "Price is what you pay, value is what you get." For example, Amazon.com offers 30 percent off most books' list prices and free shipping for orders over $25. Price is a flexible element, and various discounts, coupons, and payment periods can be tested in test markets. The price can be initially

TABLE 11.5 Four elements of the marketing mix.

Product	Price	Promotion	Place
Product variety	List price	Public relations	Channels
Quality	Discounts	Advertising	Locations
Design	Credit terms	Sales force	Inventory
Features	Payment period	Direct messages	Fulfillment
Brand name			
Packaging			
Warranties			
Returns policy			

set by estimating demand, costs, competitors' prices, and a pricing method to select the price. Effective pricing requires gathering and integrating information about the firm's strategic goals and cost structure, the customer preferences and needs, and the competition's pricing and strategic intent [Nagle and Hogan, 2006]. The pricing method or strategy can seek market share, premium pricing, or maximum profit. The cost to make the product is a floor under the price, and an estimate of the total value to the customer sets a ceiling on the price (see Figure 3.5). After studying competitors' prices, the new venture can test a price on a set of test customers.

Consider the setting of a price for a textbook where the total market demand is 10,000 books per year. Competitors have established a retail price in the range $60 to $80, and the demand per year for a new textbook may be described by

$$D = 10,000 - kP \qquad (11.1)$$

where D = demand in units, k = estimated sensitivity constant, and P = price in dollars. The fixed cost to produce the new book is $30,000, and the variable cost is $10 per unit. The book is differentiated from its competition by quality and clarity. What price, P, would you select within the established range of $60 to $80? To maximize market penetration, one would select the lowest price, P = $60, since this will result in the largest demand. If market research shows that the market is price-sensitive and k = 90, then when the price is set at $60, the demand is for 4,600 units. Then the gross profit = (revenues − cost of goods) is

$$GP = R - (VC \times D) = (D \times P) - (VC \times D) = (P - VC)D \quad (11.2)$$

where R = revenues and VC = variable cost. When P = $60 and k = 90, the gross profit is $276,000. As shown in Table 11.6, if you raise the price to $70, the gross profit declines. The calculation of the best price to obtain the maximum gross profit depends on the estimated sensitivity constant. If we change our assumptions so that k = 80, then we obtain the gross profit for the book as shown in Table 11.6. Note that $70 would be the price to maximize profit when k = 80. Note that k is an estimate obtained through experience and research and will change over time.

TABLE 11.6 Gross profit for selected values of the sensitivity constant, k, and the price of the new book (gross profit in thousands).

		Price		
		$60	**$70**	**$80**
Sensitivity constant, k	**90**	$276	$259	$166
	80	$230	$234	$222

In many industries, customers demand low prices, and the competitors have little pricing power. Pricing power accrues to companies without wide competition, such as universities that raise tuition, hospitals that increase fees, or virtual monopolies such as cable-television operators. Most mature companies operate in a world of flat or falling prices due to an excess number of providers. A new venture can pick its pricing strategy from the three shown in Figure 11.6. Many new ventures use value pricing since demand will be sensitive to price and the new firms possess little brand equity. Demand-oriented approaches look at the demand for the product at various price levels and try to estimate a price that will provide a good market share and profitability for the long term. Many technology ventures with a new breakthrough product will use a premium pricing strategy.

New technology companies usually offer new, value-oriented products. A new product or service is, by its unknown nature, difficult to price. Many new products are characterized by quality and performance uncertainty. To attract customers to a new product, it may be useful to offer a warranty—a contract or guarantee of a specified performance. Another possibility is to offer quality-contingent pricing that specifies a price rebate for poor performance [Bhargava, 2003].

Since customers often use price as a signal of quality when they are unfamiliar with a product, companies should be careful not to underprice offerings [Marn et al., 2004].

Using a traditional model for growth, firms can take advantage of the demand for new goods and services by creating and marketing products that

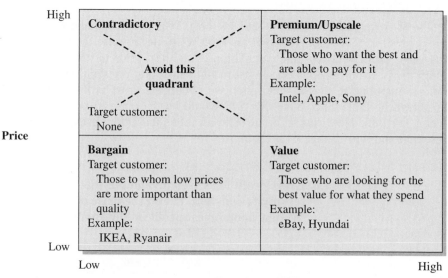

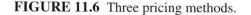

FIGURE 11.6 Three pricing methods.

satisfy a demonstrated need in the marketplace. As their customer bases grow and the products become more and more popular, profits begin to emerge. The profits are then reinvested in projects that will provide new sources of revenue and income. A portion of the profits is retained to build brand value, which can be created through a variety of techniques, not the least of which is aggressive pricing, savvy promotion, and advertising.

Promotion includes public relations, advertising, and sales methods. Selecting the message for advertising and the media for transmitting the message is a complex activity. *Advertising* is the art of delivering a sales proposition and positioning the product uniquely in the customer's mind [Roman, 2003]. The initial product message is used to attract customers to the new venture. Advertising can use print, radio, television, or the Internet. Charles Revson, co-founder of Revlon, once said: "In our factory, we make lipstick. In our advertising, we sell hope." Many products sell hope. All the purveyors of weight-control products sell hope. Matchmakers and dating services also sell hope. By contrast, Microsoft and Intel sell reliable performance. Advertising can enhance brand name recognition, create value, and enhance return for a new technology venture [Ho et al., 2005].

A list of marketing media is given in Table 11.7. Sending direct messages via mail or telemarketing can be a useful method. Public relations normally takes the form of an article in the print media or an interview on radio or television that delivers the product message. Many firms find the use of a sales force necessary to carry their message to the customer. Social networks have become important as Facebook and MySpace together attracted more users than the population of the United States [Hayes and Malone, 2008]. Word-of-mouth (buzz) promotion is particularly important for movies, toys, recreational activities, and restaurants. The buzz around the Harry Potter books and movies was large. Other products such as pharmaceuticals also can generate a lot of buzz [Dye, 2000]. Products that merit a buzz campaign have some unique, attractive attribute, such as the BMW's Mini Cooper or a new anticancer drug. Furthermore, they should be highly visible. The latest fashion in clothes or accessories often runs on buzz with teenage girls.

Word-of-mouth marketing is often called **viral marketing.** The concept is based on an age-old phenomenon: people will tell others about things that interest them. The Internet is an important avenue for finding passionate tastemakers who will carry a message forward. They have their own networks, primarily reached through e-mail lists, blogs, and social networking sites. As consumers increasingly use digital video recorders to skip commercials, listen to podcasts and downloaded music instead of radio, and use e-mail filters, word-of-mouth promotion will become more important.

The Tesla Roadster is a high-performance sports automobile with a retail price of over $100,000. It is a fully electric vehicle that can travel over 200 miles per charge. Tesla generated buzz around its product by getting celebrities and technology pioneers excited about it. At one point, the waiting list was over 1,000 people including Google's cofounder and Governor Arnold Schwarzenegger

**TABLE 11.7
Marketing media.**

Radio and podcasts

Newspapers

Magazines

Television and video

e-mail

Telemarketing

Catalogs

Infomercials

Websites

Social networks

Blogs and wikis

Presentations
   and speeches

of California. Journalists spread photos of celebrities driving the vehicle throughout traditional and popular Internet media outlets. Despite the $5,000 deposit required to be placed on a waiting list, this "buzz" has helped Tesla Motors experience tremendous interest in this car.

Place means selecting the channels for distribution of your product and, when appropriate, the physical location of your stores. Channels of distribution are necessary to bring your product to the end user. A publisher sells a book through multiple channels such as bookstores, direct sales to the end user, and Internet bookstores, such as BarnesandNoble.com. Each industry has a some sort of distribution system. Differential advantages can accrue to sellers who creatively use different channels.

In the personal computer industry, Dell Computer sells direct to the end user via phone or the Internet, while Hewlett-Packard sells primarily via retail stores and value-added resellers. When several parallel channels are used, channel conflict can occur due to the divergence of goals between channels and domain (territory or customer) disagreements.

Many technology ventures will sell their product to other manufacturers who will incorporate the product as a module or component within the final product. For example, Intel provides microprocessors that Dell incorporates within its PC.

The use of the Internet as a distribution channel will cause a shift in the relationships between consumers, retailers, distributors, manufacturers, and service providers. It presents many companies with the option of reducing or eliminating the role of intermediaries and lets those providers transact directly with their customers. Before launching an e-commerce effort and bypassing its traditional distribution channels, however, a business should analyze which products are appropriate for electronic distribution. Those most appropriate are digital, such as information products.

"Intel Inside" Campaign

In the late 1980s, Intel decided to redirect some of its advertising efforts away from computer manufacturers to actual computer buyers. The consumer's choice of a personal computer was based almost exclusively on the brand image of the manufacturer, such as HP, Dell, and IBM. Consumers did not think about the components inside the computer. By shifting its advertising focus to the consumers, Intel created brand awareness for itself and its products, and built brand preference for the microprocessor inside the PC.

The first step was to create a new advertisement using the slogan "Intel: The Computer Inside." Second, Intel chose a logo to place on a computer—a swirl with "Intel Inside." Then it chose a name for its new microprocessor—Pentium. As a result, Intel became a leader in the PC boom of the 1990s. Intel was successful at branding a component.

Many firms are using the Internet for selling and experimentation. Procter & Gamble (see www.pg.com) is using its corporate website to invite online customers to sample and give feedback on new prototype products [Gaffney, 2001]. This approach permits P&G to test new products and their marketing mix. P&G conducts at least 40 percent of its tests online. In August 2000, when P&G was ready to launch Crest Whitestrips, a home tooth-bleaching kit, it tested its proposed price of $44. P&G ran TV and magazine ads to attract people to the test. It also sent e-mails to people who had signed up to sample new products. In the first eight months, it sold 144,000 whitening kits online.

Microsoft, Yahoo, and Google have emerged as key online advertising platforms [Vogelstein, 2005]. Internet advertising is displacing newspaper and magazines. Google and others place an advertisement in front of users when they are looking to buy or research an item online.

An emerging firm has to decide how and where to spend its marketing dollars. It may have several product categories and numerous regions on which to spend its limited marketing budget. An emerging firm should collect information on each regional or international market and allocate its resources based on the regions and products that offer the best opportunity for profit [Corstjens and Merrihue, 2003].

11.7 Customer Relationship Management

The quality of the relationship that a firm has with its customers directly influences the intrinsic value of the firm. **Customer relationship management** (CRM) is a set of conversations with the customer. These conversations consist of (1) economic exchanges, (2) the product offering that is the subject of the exchange, (3) the space in which the exchange takes place, and (4) the context of the exchange [McKenzie, 2001]. For the customer relationship to be fruitful, the attraction of the customer, the conversion or sale of the customer, and the customer retention process must all be managed well. These relationships are managed through conversations in real time—that is, without delays. The firm and the customer usually engage in a series of brief conversations that help build a relationship. The conversations take place between the customer and the firm in a relationship space, as shown in Figure 11.7. The first part of a conversation is the economic exchange based on a product offering that is communicated to the customer. The space in which the exchange takes place could be physical, such as a grocery store or furniture showroom, or a website displaying goods (e.g., Amazon.com). The context of the exchange includes all that is known about the customer and the situation with the customer.

A necessary step to a CRM system is the construction of a customer database. This is relatively easy for banks and retail firms since they have a high frequency of direct customer interaction. It is more difficult for firms that do not interact directly with the end customer, such as semiconductor and auto manufacturers [Winer, 2001]. The customer database can be used for CRM

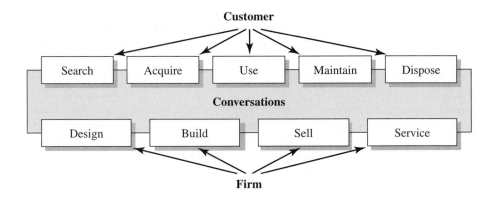

FIGURE 11.7 Customer-firm relationship as a conversation.

activities such as customer service, loyalty programs, rewards programs, community building, and customization.

CRM, when properly constructed, allows firms to gather customer data quickly, identify valuable customers, and increase loyalty by providing excellent service. Through the CRM process, customers become a new source of competencies engaged in building the firm's products and services as they provide ideas via the conversational process [Prahalad and Ramaswamy, 2000]. Unfortunately, too many companies distance themselves from customers by using phone loops that trap and frustrate customers seeking aid.

CRM is best operated when the customer and the CRM employee are fully engaged in conversation. A process that fully engages the customer and the employee can deliver more effective outcomes for the firm and the customer [Fleming et al., 2005].

Progressive Corporation uses CRM to relate to its customers 24 hours a day. It sells auto insurance both directly and through agents. Progressive's information systems allow customers to manage their accounts online, including paying their bills electronically and adding a vehicle or driver. It has a highly functional website, telephone call centers, and claims service available 24 hours a day, seven days a week. Progressive's claims agents travel quickly to the scene of an accident. The agents are equipped with notebook computers that communicate wirelessly to the corporate network, which lets them key in information on site.

The CRM and the total marketing effort are depicted in Figure 11.8. CRM helps improve marketing research, customer retention, and the marketing mix.

Customers who say they are satisfied are not necessarily repeat customers because satisfaction is a measure of what people say, whereas loyalty is a measure of what they actually do. Customer surveys measure opinions but are unreliable predictors of future behavior. **Loyalty** is not a matter of opinion [Klein, 2003]. Loyalty is a measure of a customer's commitment to a product or a company's product line. Loyalty measurements are more difficult to obtain than satisfaction measures. Good satisfaction measurement can help identify what's

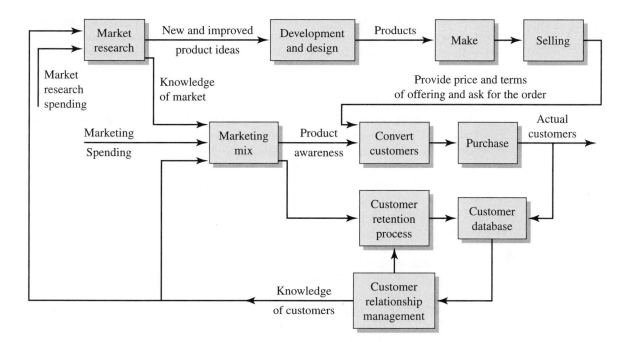

FIGURE 11.8 CRM and the total marketing effort.

broken in your business today. Good loyalty measurement is a forward-looking tool that firms can use to devise strategies to hold on to customers they want to keep.

> **FrontRange: Attracting Customers**
>
> CRM tools can be used to collect and organize the activities involving a firm's customers. For example, FrontRange (www.frontrange.com) enables a firm to track current or potential customers and provide service, sales, and support management. Another leading CRM provider is salesforce.com.

Customization, sometimes called one-to-one marketing, is a process that enables a product to be customized (changed) to a single customer's specifications. A firm uses a CRM system to elicit the information from each customer specific to his or her needs and preferences. Customization allows the company and the customer to learn together about the customer's needs. Dell Computer popularized the concept with its build-to-order website. Other companies such as Levi Strauss and Nike have developed processes and systems for creating customized products according to customers' tastes. For a good example of customization, see Dell Computer at www.dell.com. Customization is easy to do with digital goods such as music files, but other manufacturers can also tailor products to provide customization [Winer, 2001].

11.8 Diffusion of Technology and Innovations

Most entrepreneurial ventures have some novelty or innovation embedded in their product. Customers will adopt one innovation earlier than others based on their perception of its advantages and its risks. The **diffusion of innovations** describes the process of how innovations spread through a population of potential adopters. An innovation can be a product, a process, or an idea that is perceived as new by those who might adopt it. Innovations present the potential adopters with a new alternative for solving their problem, but they also present more uncertainty about whether that alternative is better or worse than the old way of doing things. The primary objectives of diffusion theory are to understand and predict the rate of diffusion of an innovation and the pattern of that diffusion. Innovations do not always spread quickly. The best ideas are not always quickly adopted. The British Navy first learned in 1601 that scurvy, a disease that killed more sailors than warfare, accidents, and all other causes of death, could be avoided. The solution was simple (incorporating citrus fruits in a sailor's diet), and the benefits were clear (scurvy onboard was eradicated), yet the British Navy did not adopt this innovation until 1795, almost 200 years later [Rogers, 2003].

The diffusion of innovations depends on a potential adopter's perception of five characteristics of an innovation, as listed in Table 11.8. The adopter's perceptions of these characteristics strongly influence his or her decision to adopt or not. Consider the introduction of black-and-white television in 1947. By 1950, 10 percent of all households had adopted this innovation, and by 1960, 90 percent of households had a TV. This rapid adoption of TV was due to its relative advantage compared to radio, its high compatibility within the home,

TABLE 11.8 Five characteristics of an innovation.

- **Relative advantage:** the perceived superiority of an innovation over the current product or solution it would replace. This advantage can take the form of economic benefits to the adopter or better performance.

- **Compatibility:** the perceived fit of an innovation with a potential adopter's existing values, know-how, experiences, and practices.

- **Complexity:** the extent to which an innovation is perceived to be difficult to understand or use. The higher the degree of perceived complexity, the slower the rate of adoption.

- **Trialability:** the extent to which a potential adopter can experience or experiment with the innovation before adopting it. The greater the trialability, the higher the rate of adoption.

- **Observability:** the extent to which the adoption and benefits of an innovation are visible to others within the population of potential adopters. The greater the visibility, the higher the rate of adoption by those who follow.

its relatively low complexity, its easy trialability, and its ready observability in TV store windows and friends' homes. On the other hand, consider the slow adoption of the personal computer in the home. PCs were introduced into the home market by 1982, but by 2007, only two-thirds of households had a PC. The high complexity of a PC discourages many consumers from adopting it in the home. Also, the perceived advantage is not clear to many would-be users.

As the PC example highlights, to understand a customer's likelihood of adoption, it is necessary to compare his or her current "pain" to the perceived pain of the solution presented by your company's product or service. For example, without a PC and word processor, a potential customer feels pain when typing documents that cannot be easily changed and when tracking expenses by hand rather than by using a spreadsheet. But, adoption of a PC also imposes its own perceived pain. This perceived pain is not only price, but also time and effort required to read instruction manuals, research products, wait in line, install software, and so on. Customers will adopt only when their perceived benefit exceeds this perceived pain [Coburn, 2006].

In a rapidly changing technology area, customers may wait to adopt if they think that a better technology is just around the corner. For example, many people did not purchase Sony's MiniDisc recorders because they suspected that CD recorders would be available soon. If customers wait long enough, they can "leapfrog" entire technologies, as with communities in Africa and Asia that skipped landlines and adopted mobile phones [*Economist*, 2006].

The adoption of any innovation usually follows an S curve, as shown in Figure 11.9. When the adoption follows the S curve, then the distribution curve of adopters follows a normal distribution, as shown in Figure 11.10, where Sd = standard deviation. The five categories of adopters are shown in Figure 11.10 and described in Table 11.9 [Rogers, 1995].

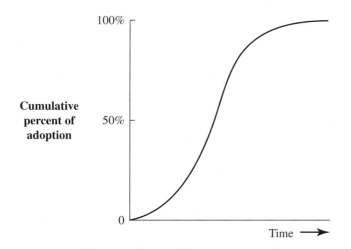

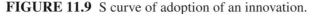

FIGURE 11.9 S curve of adoption of an innovation.

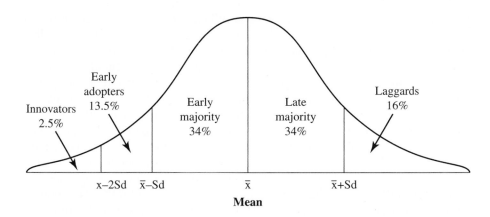

FIGURE 11.10 Innovation adoption categories when Sd = standard deviation.

TABLE 11.9 Five categories of adopters of an innovation.

■ **Innovators** want to be on the leading edge of business and are eager to try new innovations. They have an ability to work with complex and often underdeveloped ideas as well as substantial financial resources to help them absorb the uncertainties and potential losses from innovations.

■ **Early adopters** are more integrated with potential adopters than innovators and often have the greatest degree of opinion leadership, providing other potential adopters with information and advice about an innovation. They are visionaries.

■ The **early majority** adopts just ahead of the average of the population. It typically undertakes deliberate and, at times, lengthy decision-making. Because of its size and connectedness with the rest of the potential adopters, it links the early adopters with the bulk of the population, and their adoption signals the phase of rapid diffusion through the population. They are pragmatic.

■ The **late majority** is described as adopting innovations because of economic necessity and pressure from peers. While it makes up as large a portion of the overall population as the early majority, it tends to have fewer resources and be more conservative, requiring more evidence of the value of an innovation before adopting it.

■ **Laggards** are the last to adopt a new innovation. They tend to be relatively isolated from the rest of the adopters and focus on past experiences and traditions. They are the most skeptical when it comes to risking their limited resources on an innovation.

11.9 Crossing the Chasm

The transition from the early adopters to the early majority is difficult since it requires attracting pragmatists, as shown in Figure 11.11. This large gap between visionaries and pragmatists is called a **chasm** [Moore, 2000]. The early adopters or visionaries are independent, motivated by opportunities, and quickly appreciate the nature of benefits of an innovation. However, the early majority or pragmatists are analytical, conformist, and demand proven results

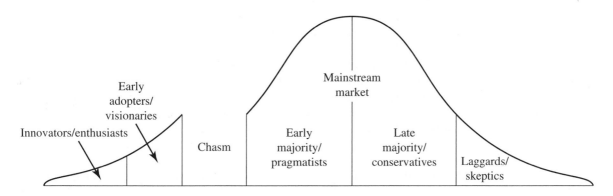

FIGURE 11.11 Chasm model.

from an innovation. Once a product crosses the chasm, others will purchase it since they will readily observe and try it [Rohlfs, 2001]. Crossing the chasm is a challenging task for a new business. An example of an innovation first available in the 1960s is the videophone. Another example is three-dimensional movies that first launched in 1953. The inconveniences involved in 3-D movies have outweighed the benefits, and the adoption of 3-D stalled in the 1990s [Huntington, 2003]. These innovations have not yet made it across the chasm.

In 1513, Machiavelli wrote in *The Prince* about change and disruption:

> There is nothing more difficult to carry out, nor more doubtful of success, nor more dangerous to handle than to initiate a new order of things. For the reformer has enemies in all who profit by the old order, and only luke warm defenders in all those who profit by the new order. This luke warmness arises partly from the incredulity of mankind, who do not truly believe in anything new until they have actual experience of it.

Hybrid cars were introduced in U.S. markets in 2001. They were rapidly accepted by enthusiasts and visionaries. Currently hybrid model cars have been increasingly popular and have crossed the chasm into the mainstream of auto sales.

Digital photography stalled at the chasm for over a decade (1985–1995). Factors that held it were the lack of personal computers and software that could easily manipulate digital images, as well as the absence of inexpensive printers for printing photos [Ryans et al., 2000].

Many high-technology products are weak on both compatibility and complexity. They are difficult to know how to operate and understand. Thus, they require the would-be user to learn how to use them and how they work. The firm offering these products needs to educate the user on both these matters. In some cases, this education effort can be expensive and time-consuming.

All innovators are not educators; however, anyone who adopts an innovation has to be a learner. Even the most transparent and intuitive designs present learning challenges. A bicycle looks easy to ride. But for most adults, learning to ride a bicycle is difficult because they may have to fall down before they learn how. Teaching an adult to ride a bicycle is often an exercise in frustration. Too many innovators rely not on sound training sessions and documentation to support their users' learning, but rather on Web pages that list frequently asked questions. Encouraging users to read the directions just doesn't work. Innovators often try to transfer the costs of teaching and learning to their customers. As a result, adoption of an innovation is delayed or halted.

Innovators with good ideas have no choice but to be educators, through either the medium of their innovations or teaching. They should recognize that if they really want to overcome customer resistance, they need to balance ease of use with ease of learning [Schrage, 2002].

The chasm model and the diffusion characteristics of a product help to explain the **diffusion period** required to move from 10 percent to 90 percent of the potential adopters. Table 11.10 shows the diffusion period for selected innovations.

Early on, some suggested that it could take a decade or more for the DVD format to be adopted. By 2003, purchases of DVD players and discs had doubled over the previous year. DVD sales rapidly overcame two potential hurdles: (1) the higher purchase price of DVD discs compared to VCR tapes and (2) the inability of the DVD player to record. Currently, video-on-demand delivered over broadband cable or satellite TV poses a challenge to continued growth in the DVD and Blu-ray market.

Many innovators face a chicken-and-egg problem since to use a new device, the user needs a widely available infrastructure. However, it doesn't pay others to build an infrastructure for just a few users. This is true for the world of wireless Web devices. Another example is the push for fuel-cell vehicles. Their use requires hydrogen fuel stations, but who will build them until

TABLE 11.10 Diffusion period for selected innovations.

Innovation	Diffusion period (Years)
Electric power	40
Telephone	70
AM radio	20
Automobile	60
Black-and-white television	12
Color television	16
Videocassette recorder	17
Artery stents (for relief of heart congestion)	8

Note: Diffusion period is the time required to diffuse an innovation from 10 percent to 90 percent of the potential adopters.

many fuel-cell vehicles will use them? It can be difficult to get some customers to adapt innovations because markets are taking on the characteristics of networks. The interconnections among today's companies are so plentiful that often a new product's adoption by one player depends on its systematic adoption by other players [Chakravorti, 2004].

Some products spread like an epidemic once they reach critical mass. The emergence of fashion trends or best-selling books can be described in terms of a virus outbreak. These epidemics have three characteristics: contagiousness; little causes have big effects; and a big change happens, not gradually, but at one moment [Gladwell, 2000]. A **tipping point** is the moment of critical mass or threshold that results in a jump in adoption. This type of jump happens in networks where, at some point, enough people have the product that its value jumps significantly and the product takes off. The low-priced fax was introduced in 1984, and by late 1987, about two million fax machines were in use—enough to make it worthwhile to get one. As a result, 1988 was the tipping point.

An epidemic is spread by a message that is meaningful and emotional to the receiver and motivates the buyer who finds the message sticks in his or her mind and passes it on to others. The persuasive message is communicated by a trusted agent, and it moves the buyer to act. A tipping point can occur as a result of viral marketing. A contagious message that is memorable, motivating, and delivered by a trusted agent can take a product across the chasm by helping a product reach its tipping point.

Twitter: Reaching for the Tipping Point

Tipping points are especially evident in social networking applications. Twitter, which is a microblogging website and mobile application, experienced explosive growth in 2009. Its user base grew to over 18 million US adults who accessed Twitter on any platform, which was a 200% increase over 2008. Awareness of the site was largely spread by trusted agents with a contagious message. For example, many *New York Times* columnists began to publish on Twitter, and some of their readers were enticed to join as well.

The entrepreneur is the agent who creates the vision for the new product and builds a marketing plan that will help potential adopters to understand and value the innovation and respond to the message communicated by the firm. This message must be persuasive, believable, and understandable. New ventures create the resources and necessary strategy to overcome the potential adopters' lack of knowledge and understanding of the product. They are more likely to receive funds from investors and succeed in bringing their product across the chasm and to the remaining adoption categories.

Marketing can be described as taking the actions necessary to create, grow, and maintain your firm's place in your chosen market. To cross the chasm, the marketing strategy has to attract and retain pragmatists. They care about quality, service, ease of use, reliability, and the infrastructure of complementary products (often called the whole product).

To get across the chasm, new ventures must determine the characteristics of the pragmatist customers and build a marketing plan for them. For a small start-up, this may mean focusing on a particular market segment or acting locally and then expanding as sales grow. Crossing a chasm means assembling the whole product, through alliances with partners, to satisfy the pragmatic buyer in a specific target market [Moore, 2002].

Although crossing the chasm is often associated with disruptive technologies for industrial markets, it may also apply to consumer markets with "disruptive" business models. A new venture must cross a monetization chasm instead of an adoption one by focusing on viral marketing and volume operations.

11.10 Personal Selling and the Sales Force

All businesses involve **selling,** which is the transfer of products from one person or entity to another through an exchange mechanism. It includes identifying customer needs and matching the product or solution to those needs. For many technology ventures, this process is called sales and business development. Most firms employ a sales force to make the contacts with the purchaser. New ventures should develop a selling strategy and a plan of action. Then they locate target customers and recruit, train, motivate, compensate, and organize a field sales force. They also have to manage the interactions between customers and salespeople. This dialogue is influenced by the buyer's needs and the salesperson's skills. The results of successful salesperson-customer interactions are orders, profits, and repeat customers. The salesperson implements the marketing strategy. In a small start-up, the salespeople may have other roles, such as product development or market planning.

Selling a technology product is difficult since the product is less tangible than a house or a suit. Buying a technology product takes longer and the salesperson must inspire a buyer to take action. The technology sales person must fully understand the product offering and be able to communicate its benefits clearly.

Especially in industrial markets where the customers are other businesses, the buyers may actually be multiple decision-makers. The ultimate user of the technology product or service is certainly one of them. However, others could include those who make the recommendation of which solution to buy such as the information technology staff and those who actually negotiate the contract such as the purchasing agent. This can complicate and delay the sale. This length of time from the first contact until a sales transaction is completed is called the **sales cycle.** It can be as short as one day (e.g., purchasing ad listings on Google or auction listings on eBay) or many months (e.g., evaluating and choosing sophisticated MRI equipment for a hospital's radiology lab). Technology ventures must estimate the length of their sales cycle as they develop their business model and financial plan.

IBM's success from 1955 to 1990 was due to a very knowledgeable, well-trained, and highly motivated sales force. IBM's salespeople had real experience with computers as well as understanding of their clients' needs. Thomas Watson, Sr., former CEO of IBM, noted that great technological innovation combined with a powerful sales force was unbeatable.

For a new technology venture with an innovative product, the salesperson must fully understand the product and the idea of creating a solution for the customer. It is the responsibility of everyone in a new venture to (1) identify and create a purchaser, (2) offer a creative solution, and (3) make a profitable sale. In many start-ups, the staff is comfortable with steps 1 and 2 but shies away from step 3. Without actually making the sale, the start-up is destined to fail [Bosworth, 1995]. The goal is to determine the purchaser's needs, or latent pain, create a solution to meet that need, and then sell it to the purchaser.

The solution sales process is shown in Figure 11.12. The salesperson identifies the target market and makes contact with a potential purchaser. Then the salesperson determines the customer's problem and needs. Based on these needs, the solution to the customer's problem is created and presented. The benefits of the solution must be clearly communicated. Then the salesperson asks for the order and, with a positive response from the customer, confirms the order. You may have experienced this process when shopping for new clothes. The salesperson makes a contact and determines your needs. Then the salesperson shows you one or more solutions (options), and you try them on for size and appearance. The salesperson aligns his or her comments on each solution in a discussion with you. When you both see a solution, the salesperson asks for the order. If you agree, the salesperson writes the order at the cash register. This process is the same, although more complex, for a purchaser seeking a new computer system for a government agency or an electronics firm. Salespeople sell themselves, show they care, and provide proof of product, consistency of message, authority, and scarcity. The sales process rests, in part, on the skills of persuasion, as later described in Section 13.2.

New ventures often use their own people to manage the sales process but engage others, called sales representatives, under contract, to actually sell the product. The advantages and disadvantages of using company salespeople versus independent representatives are shown in Table 11.11. The choice of the right balance of company salespeople and independent representatives is a critical issue for a new business.

It is also important to grow the sales force at the right pace. Often, companies beef up their sales force capacity too early, when new products are not quite ready. A better approach is to start with a small group of salespeople who learn as much as they can about customers' responses to your product. Then, use these responses to refine the product and the sales/marketing strategy, and expand the sales force only as sales themselves accelerate [Leslie and Holloway, 2006]. A new venture needs to focus on its customer rather than product features. The sales force needs to find out what the customer needs, which will

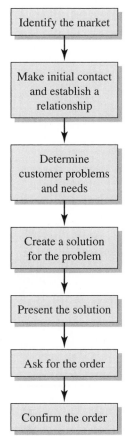

FIGURE 11.12
Solution selling process.

TABLE 11.11 Selling via company salespeople versus independent representatives.

	Advantages	Disadvantages
Company salespeople	■ Know your product well ■ Relatively easy to manage ■ Provide feedback from customers ■ Paid salary plus commission	■ High fixed cost ■ Low geographical dispersion ■ Time and costs to hire and train ■ Travel costs
Independent representatives	■ Paid on commissions ■ Lower hiring and training costs ■ Geographical dispersion ■ Have established relationships with customers ■ Low fixed costs	■ Sell for several firms, making it difficult to get their attention ■ Difficult to manage ■ Low feedback from customers ■ May have limited understanding of your complex product

be a combination of products, services, and the product elements. They should provide a solution to the customer needs [Charan, 2007].

The emerging technology business may initially use a focused, direct sales force to create demand and penetrate to the primary target segment. Then, as growth accelerates, a transition to other segments and sales channels may be appropriate. It is important to clearly identify the primary target segment and key customers [Waaser et al., 2004].

New businesses encounter sales resistance due to competition and lack of knowledge of their product and its quality. One method to overcome this is to utilize trial periods, warranties, and service contracts. Many new ventures do an excellent job of building a good product and developing a solid marketing plan but then fail to make the forecasted sales.

We will cover the issues of international marketing and sales in Chapter 15.

11.11 AgraQuest

AgraQuest prepared a sketchy marketing plan in 1998 as it looked forward to launching its first product. Based on a review of this plan, prepared by Pamela Marrone and the vice president for product development, it moved to recruit a vice president of marketing and sales. The marketing plan described the market segment (customers) as farmers in the United States and Chile. AgraQuest planned to use distributors as the channel to distribute the natural biopesticide Serenade. The positioning map for Serenade is shown in Figure 11.13. The effectiveness of Serenade and chemicals was the same for most fungi; Serenade, however, had higher resilience (retention of effectiveness over time).

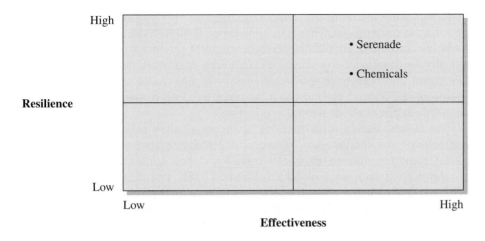

FIGURE 11.13 Positioning map for the biopesticide Serenade versus chemical pesticides and fungicides.

AgraQuest did some market research and realized that many farmers distrusted new bioproducts because of excessive or unfulfilled claims by other natural-product firms. It used field trials to demonstrate the product's effectiveness and overcome the distrust. It also carried out a pricing study and decided to price its product at the same level as those of its chemical company competitors.

In 1998, AgraQuest hired its first vice president for marketing, who launched a large, traditional advertising campaign that consumed $500,000 over two years. The positioning statement was:

The Best Biopesticide on Earth!

The advertising and sales campaign was viewed by farmers as arrogant, considering the product was unproven, and failed to attract buyers. The first marketing VP left, and a second was engaged in 2000. He then proceeded to redo an advertising campaign to build image, brand, and acceptance, which also failed. Both VPs had worked for large chemical pesticide companies that used advertising campaigns to launch new products. AgraQuest, however, was a small start-up that needed a marketing plan that would help it attract farmers to try its product.

AgraQuest's natural pesticides need to cross the chasm to the pragmatic farmer who is slow to adopt new tools and products. This slow, deliberate adoption process cannot be accelerated by advertising. It needs a series of deliberate trials and proof of its key advantage: resilience of the effectiveness of the product. AgraQuest installed a third vice president of marketing in 2003.

AgraQuest has a huge chasm to cross to convince mainstream customers that they should use Serenade. Success has come slowly by doing extra field trials and demonstrating the efficiency and effectiveness to farmers in their own fields. Farmers will not use something if they have not seen it work. Breaking into the pesticide market with a natural product is very difficult. The industry is dominated by chemical pesticides that work very well. Farmers like them, and they have a 50-year history of use. A lot of companies have overpromised natural products that have not lived up to the pitch. So farmers are very skeptical about natural products.

AgraQuest uses a sales force of seven on a salary and bonus compensation scheme. Its markets are, in order of priority: grapes (United States and Chile), tomatoes, lettuce, bananas (Costa Rica), and apples.

AgraQuest had sales of $2.5 million in 2002 and $13 million in 2008. Serenade has a unique advantage because only natural products can be used in the weeks just prior to harvest. If it rains during that time, Serenade, unlike chemicals, can be used to prevent fungi. Harvest seasons in 2005 to 2008 were dry, however, so this competitive advantage is not constant over time or likely to apply very often.

11.12 Summary

Any new firm needs to build a marketing plan that describes how it will attract, serve, and retain the customers targeted for its products. Since a new firm normally starts without established customers, it must carefully identify the target market that will value its product. Market research can provide the information about the customers, appropriate distribution channels, and communication methods for attracting the customer.

The new firm creates a product positioning statement and selects a mix of price, product, promotion, and channels to attract and satisfy the customer. Most new firms are challenged to cross a chasm in the diffusion process that enables their product to attract the pragmatic and skeptical potential customer. The marketing process consists of describing or implementing the following elements:

- Product offering
- Target customer
- Marketing objectives
- Market research
- Marketing plan
- Sales plan
- Marketing and sales staff

Principle 11

A sound marketing and sales plan enables a new firm to identify the target customer, set its marketing objectives, and implement the steps necessary to sell the product and build solid customer relationships.

Video Resources

Visit http://techventures.stanford.edu to view experts discussing content from this chapter.

Passion and the Customer	Vic Verma	Savi
Marketing a Start-up	Donna Novitsky	Big Tent

11.13 Exercises

11.1 There continues to be a disparity between the advertising dollars spent on reaching TV viewers and Internet users versus the amount of time that is spent interacting with each media. A large degree of Google's success is attributable to taking advantage of this large gap. Research to determine (a) TV advertising dollars and Internet advertising dollars spent, and (b) the amount of time spent watching TV versus using the Internet. How has Google taken advantage of this disparity? What other major societal trends are forecasted that will continue to shift advertising dollars to new mediums, and why?

11.2 With the explosion of mobile handsets worldwide, many marketing and advertising firms are looking at how to take advantage of the ubiquity of a communications device carried everywhere. Describe why marketers view the mobile handset as such a valuable marketing platform. What types of mobile advertising challenges do you foresee arising?

11.3 What is viral marketing? Provide an example of a start-up using viral marketing to promote and sell its product or service. Why does it work (or not work)?

11.4 Facebook and MySpace are rapidly growing social networking sites based in the United States but with global reach. Prepare a positioning map for these two firms.

11.5 BusinessWeek with Interbrand conducts an annual worldwide brand survey ranking the top 100 global brands. Examine the most recent survey and choose a new entrant to the list. Describe that company's marketing objectives and customer target segments.

11.6 Powerful brands are built on innovativeness and advertising. Examine the brand value for Genentech, Merck, and Apple, and describe the reasons for their brand power.

11.7 Electronic Arts (EA) Sports has gained large brand value by securing an exclusive five-year license from the NFL in the United States. Determine if EA Sports has reaped the rewards from acquiring the rights to use the valuable NFL brand.

11.8 HDTV is an emerging consumer electronic technology. Discuss the marketing challenges associated with HDTV (e.g., building an ecosystem of content, players, and content distribution). How do you see this influenced by (a) competing with DVD (the last technology generation) and (b) emerging substitutes like online digital content, PCs, and mobile devices? Use the categories of Table 11.9 to describe where the market response to HDTV has been to date and how long adoption has taken for each group.

11.9   Identify a high-tech firm that uses an indirect sales channel model. What is the model used? Why was the indirect sales model chosen for its particular products or services?

11.10 What is the best way to reward salespeople: salary, commission, or a mix of the two? How do the rewards motivate different selling behaviors? What is the best method for a new emerging medical technology business such as Enzo Biochem (www.enzo.com), or a clean-tech business such as Ausra (www.ausra.com)?

VENTURE CHALLENGE

1.   For your venture, describe the customer and the target segment you have identified.

2.   Develop a positioning statement using the template in Figure 11.4.

3.   Using Table 11.3 and Figure 11.5, describe the market research and customer development plans.

4.   Briefly describe the marketing mix for your product.

5.   How will your venture sell its product and develop customer relationships?

6.   Research the industry for your venture and determine the length of its sales cycle.

CHAPTER 12

The New Enterprise Organization

Two people working as a team will produce more than three working as individuals.
Charles P. McCormick

How can entrepreneurs best organize and reward the people who will lead their venture to success?

After recognizing an opportunity and deciding it is attractive, usually one or two founders assemble a new venture team to build a plan and an organization to execute it. This initial team creates or designs an organizational arrangement to respond to the opportunity. The leaders of the venture are identified early in the organization's development. These leaders are able to inspire and motivate others to join the new venture and work on tasks of the venture. They build a team that is collaborative and possesses diverse competencies. As an organization grows, managers will be needed to carry out the tasks that keep the organization running well. A leader is a team's emotional guide and exhibits solid emotional intelligence.

As the organization grows, the firm works to build an organizational culture and trust among team members. Leaders and teams strive to build social relationships and networks to foster collaboration. One of the methods of creating an ownership culture is to facilitate ownership for all people in the firm through stock options or restricted stock. A new firm also builds a set of advisors and a board of directors to help it move forward. ■

281

12.1 The New Venture Team

The first step toward forming a new venture is often taken by one or two individuals who recognize a good opportunity and then develop a business concept and vision to exploit it. After a short period, it becomes clear that a team is required in order to have all the necessary capabilities in the leadership group. We define a **team** as a few people with complementary capabilities who are committed to a common objective, goals, and approach for which they hold themselves mutually accountable. In a true team, as opposed to a group, members are fully integrated and feel responsible for their collective output. Think of the contrast, for example, between a basketball team that has difficulty functioning without all players and a group of tennis players who act independently toward a common goal.

The **new venture team** is a small group of individuals who possess expertise, management, and leadership skills in the requisite areas. Thus, the team will incorporate people with skills and knowledge in finance, marketing, product development, production, and human resource management. Since they can draw on this broad expertise, new team-founded ventures tend to achieve better performance than individually founded ventures [Beckman et al, 2007].

Typically, a team of two to six people is required to develop a business plan, secure the financing, and launch the firm into the marketplace. It is important to ensure that each team member contributes fully and serves a critical role. Jeff Bezos, the founder of Amazon.com, argues that team size should follow the "two pizza rule": just two pizzas should feed the entire team. Otherwise, the team is probably too big.

The capabilities of the one or two lead entrepreneurs are critical to the new venture since others are willing to join the team based on the integrity, experience, and commitment of these lead entrepreneurs. Often we call the lead entrepreneurs the **founders.** In other cases, all the members of the initial leadership team are called the founders. The founders display all the characteristics of capable entrepreneurs: passion, commitment, and vision. Entrepreneurs understand the long-term implications of the information-based, knowledge-driven, and service-intensive economy. They know what new ventures require: speed, flexibility, and continuous self-renewal. They recognize that skilled and motivated people are central to the operations of any company that wishes to flourish in the new age.

Furthermore, these new ventures must exhibit adaptability and readiness to change as the context of the business evolves. The organizational arrangement of the firm must evolve as the market and the customer change. Newly formed firms are challenged by this necessity to change since they usually have limited resource and capability bases. Strong teams ensure that the firm is able to constantly reorganize in terms of strategies, structures, systems, and resources. The new venture team must have the skills to balance the needs for change, efficiencies, alignment of effort, and timeliness. The team must include one or more persons who can gain access to external sources of funds. One of the advantages of a new organization is that people can use active thinking rather than precedent as a basis for action [Pfeffer and Sutton, 2000].

Most of the members of the new venture team are known to the lead entrepreneurs and usually include family, friends, and business associates. The members of a team should be selected for their demonstrated skills and abilities. They also can be selected for their ability to gain access to information, knowledge, financial capital, and other resources required by the new venture.

The leadership team at Apple Computer in 1978 consisted of three diverse persons, all with specific skills and personalities. Steve Jobs was the charismatic leader who could motivate the employees and talk directly to computer lovers. Mike Markkula was the business and marketing leader. Stephen Wozniak was the engineering leader and creator of the company's computers. This balanced and powerful team helped create Apple Computer's place in business history.

12.2 Organizational Design

Organizational design is the design of an organization in terms of its leadership and management arrangements; selection, training, and compensation of its talent (people); shared values and culture; and structure and style. The nine elements of an organization are listed in Table 12.1. Note that the last four elements in the list can be considered as the elements of an organizational design.

The **talent** consists of the people, often called employees, of an organization. The leadership team and the firm's managers are responsible for communicating and leading the firm in the appropriate direction. The shared values and corporate culture are the guiding concepts and meanings that the members of an organization share. The structure of a firm is its formal arrangement of functions and activities. Style is the manner of working together—for example, collegially or team-oriented.

TABLE 12.1 Nine elements of an organization.

1. Mission and vision
2. Goals and objectives
3. Strategy
4. Capabilities and resources
5. Processes and procedures
6. Talent
7. Leadership team and management
8. Shared values and culture
9. Structure and style

A good organizational design leads toward the reduction of bureaucratic costs so that a low-cost advantage can be achieved. Furthermore, a good design can maximize a firm's value creation capabilities, leading to differentiation advantages and good profitability. A good design requires a clear definition of the customers and the value offered to them.

What is the best way to organize a group of people so as to maximize productivity and innovation? Most successful innovative organizations include many small units having free communication with each other, significant independence in pursuing their own opportunities, and freedom from central micromanagement. Innovation grows most rapidly under conditions of an intermediate degree of fragmentation. Excessive unity and excessive fragmentation are both ultimately harmful. The best organization design is one of teams or units that compete and generate different ideas but maintain relatively free, open communication with each other [Diamond, 2000].

Competency-based strategies depend on talented people operating in a loose-tight structure. Thus, hierarchical structures need to be replaced by networks, bureaucratic systems changed into flexible processes, and control-based management roles evolved into relationships [Bartlett and Ghoshal, 2002]. Flexible organizations that effectively adapt to change are often called **organic organizations.** As summarized in Table 12.2, the strategic resources necessary for sustainable competitive advantage are human, organizational, and intellectual capital.

New ventures serious about obtaining profits through people will expend the effort needed to ensure that they recruit the right people in the first place. The organization needs to be clear about what are the most critical skills and attributes needed in its applicant pool. The notion of trying to find "good employees" is not very helpful. New ventures need to be as specific as possible

TABLE 12.2 Objective, tools, and resources of organizational design.

Objective	Sustainable competitive advantage
	Continuous renewal of the talent
Tools and methods	Vision
	Values
	Flexibility
	Innovation
	Entrepreneurship
Strategic resources	Human capital
	Organizational capital
	Intellectual capital

about the precise attributes they are seeking. Technology start-ups tend to seek people who have strong technology expertise but also are flexible, and willing and able to assume a number of key roles in a new venture. The talent in a new venture needs to know what to do and be capable of doing it. The idea is to hire a "great athlete who has already run the race before." The qualities of a good new member of a new venture include flexibility, experience, technical knowledge, and self-motivated creativity. Robert Noyce and Gordon E. Moore spawned Intel, William H. Gates and Paul Allen built Microsoft, and Robin Li and Eric Xu started Baidu in China. All these leaders possessed the requisite characteristics.

Organizational performance is the result of individual actions and behavior. Successful firms have people who take the right actions in concert with others. Thus, the form of the new enterprise is often a network characterized by relationships within the firm. In general, the firm starts off as a single team, and as it grows, it evolves into a series of cross-functional teams.

Southwest Airlines has been productive due to its effective use of its major assets—its aircraft and people. Southwest uses **relational coordination** (RC), which describes how its people act as well as how they see themselves in relationship to one another [Gittell, 2003]. RC requires frequent, timely problem-solving carried out through shared goals, shared knowledge, and mutual respect. Three conditions that increase the need for RC are reciprocal interdependence, uncertainty, and time constraints—all common to new business ventures.

One model of an organizational design is shown in Figure 12.1. The three activities of an organization are operations, innovation, and customer relationship management (CRM). These activities all support the key objective of the organization to create and maintain a sustainable competitive advantage. The integration of innovation, operations, and CRM can lead to a strong competitive advantage. The Internet and an intranet can help provide low interaction costs between the three activities. Most new ventures have an advantage because their newness permits them to easily integrate the three activities shown in Figure 12.1, thus rapidly gaining a competitive edge.

Most, if not all, new enterprises design a flat organization that facilitates speed of action. They avoid layers and bureaucratic structures, and keep communication flowing and a bias toward action [Joyce et al., 2003].

A new venture normally starts out with a team or a **collaborative structure** that primarily consists of teams with few underlying functional departments. In a collaborative structure, the operating unit is the team, which may consist of 5 to 10 members. The best collaborative structures are self-organizing and adaptive. A **self-organizing organization** consists of teams of individuals that benefit from the diversity of the individuals and the robustness of their network of interactions. This collaborative effort coupled with the self-organizing behavior of the network can lead to benefits that exceed the sum of the parts— often called synergy.

Perlegen's Experienced Leadership Team

Perlegen Sciences is a biotechnology firm that was spun off from Affymetrix in 2000. Venture capitalists invested $100 million in Perlegen. Perlegen's mission is to apply biochip technology to scan human genomes and illuminate DNA variations that render people genetically different [Stipp, 2003] (www.perlegen.com). Perlegen's leading advantages are its founders and leadership team, which includes Stephen Fodor, who coinvented the Affymetrix biochip process, and legendary serial entrepreneur Alejandro Zaffaroni. Zaffaroni provided a wealth of knowledge and industry experience. He had built on his early success from commercially developing the birth control pill by founding over 30 biotech start-ups. These included Alza, which was the first maker of drug-delivery patches such as Nicoderm, and now Perlegen.

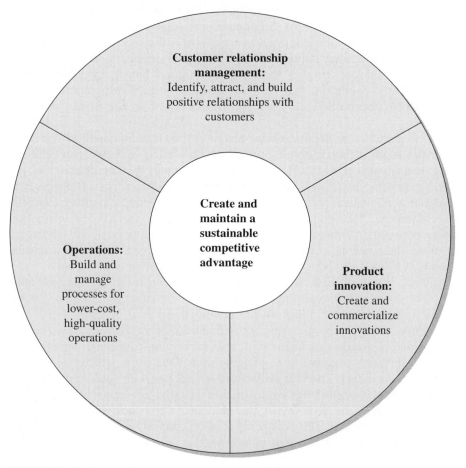

FIGURE 12.1 Model of an innovative organization.

12.3 Leadership

Leadership is the process of influencing and motivating people to work together to achieve a common goal by helping them secure the knowledge, power, tools, and processes to do so. Leadership is critical to an entrepreneurial venture and is normally provided by one or two leaders of the new venture. The leader of a new venture can be thought of as a leader of a jazz band that is known for its ability to play familiar and new music while creating and improving new variations and collaborative music. Mobilizing an organization to adapt its behaviors in order to thrive in changing business environments is critical. Responses to challenges reside in the collective intelligence of employees at all levels who need to use one another as resources, often across boundaries, and together learn their way to those solutions.

A good leader is hopeful about the venture's goals and can readily describe the vision of the venture. They communicate a clear vision and the value of the venture and are convinced they are the one to make it happen. Good leaders make good judgement calls about people, strategy, and challenge [Tichy and Bennis, 2007]. Judgement is the nucleus of leadership and is based on a practiced process of naming and framing the issues, exercising good judgement, and listening to the team. Most leaders are skilled at framing the issue and creating the case for change, strategy, and actions needed. Then the team proceeds to the execution phase and makes the required steps for success.

The leader of the new venture responds to routine work and challenging work issues in different ways, as shown in Table 12.3 [Heifetz and Laurie, 2001]. The most important capability of the entrepreneur-leader is the ability to cultivate and make use of the competencies of the talented team members [Davidsson, 2002]. Responding to challenges and adapting the effort of the

TABLE 12.3 Leadership of routine and adaptive work.

A leader's role in:	Routine issues	Challenging and adaptive work
Direction	Define problems and possible solutions	Define the challenge and the issues
Team and individual responsibilities	Clarify and define roles and responsibilities	Define and discuss the necessity to adapt roles and responsibilities to changing needs
Conflict	Restore order and reduce conflict	Accept useful conflict and use it to define new approaches and strategies
Norms and values	Reinforce norms and values	Reshape norms and values
Teaching and coaching	Provide training and skill learning for existing employees	Teach and coach new people

talent is the role of a leader. Challenging problems confronting an organization require the members of the organization to take responsibility for solving the problem. Thus, the leader helps the team members confront the challenges and learn new ways to solve the problems.

Leaders build companies through a blend of personal humility and professional will [Collins, 2001]. They are ambitious, but primarily for the organization, not themselves. Leaders have the drive to build great companies through the efforts of the team members. They seek sustained results and facilitate new approaches to challenging situations while maintaining clear goals and methods for routine work. Leadership is the ability to acquire new organizational methods and capabilities as the situation changes. The leader stimulates discussion so that people contribute and understand the issues, ultimately leading to a shared strategy for sustained advantage.

There are four styles of leadership, as shown in Figure 12.2 [Northouse, 2001]. A leader's behaviors are both directive (task) and supportive (relationship). *Directive behaviors* assist group members in goal accomplishment through giving directions, establishing goals and methods of evaluation, setting time lines, defining roles, and showing how the goals are to be achieved. Supportive behaviors involve two-way communication and responses that show social and emotional support to others.

The *supporting* style is used when a leader does not focus exclusively on goals but uses supportive behaviors that bring out the employees' skills around

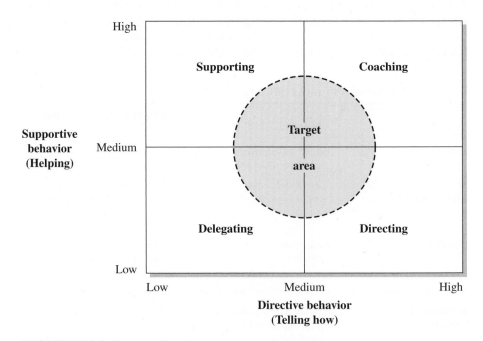

FIGURE 12.2 Four leadership styles.

the task to be accomplished. The *directing* leader gives instructions about what and how goals are to be achieved by the subordinates and then supervises them carefully. The *coaching* style calls for a leader to focus on goal achievement and give encouragement to subordinates. The *delegating* style occurs when the leader is less directive and facilitates employee confidence. Most effective leaders adopt all four styles of leadership depending on the needs of the situation and the team members while operating within the central target area. The leader of a new technology venture will most likely use a directing-supporting style in the early period of the venture. Later, in the growth period of the firm, the leader will probably use a mix of all four leadership styles.

Leaders exhibit seven traits, as recorded in Table 12.4. Leaders are seen as authentic, decisive, focused, caring, coaching, communicative, and improvement-centered [Collins and Lazier, 1992]. Leaders articulate a clear, compelling vision for the venture and stimulate the team to achieve high performance. They avoid letting talk substitute for action. They also strive to develop a sustainable competitive advantage through building new competencies and products in a timely way. Leaders have a bias for simple concepts that can be clearly understood and acted on. Informed action is their goal. The entrepreneurial leader uses a collaborative style while setting high standards and driving toward achievement.

Teaching is at the heart of leading. In fact, it is through teaching that leaders lead others. Teaching is how ideas and values are transmitted in an organization. Leading is helping others to see a situation as it really is and to understand what responses need to be taken so that they will act in ways that will move the organization toward where it needs to go. Organizational performance is the result of individual actions and behaviors. In successful companies, people do the "right things." Those companies have effective leaders who create conditions under which their people have the information, authority, and incentives to make the right decisions. When leadership is effective, behavior at all levels of the organization is both aligned and adaptable; thus, the organization performs to its potential.

Leaders have a guiding vision and passion that allow them to communicate a sense of hope to followers. The key to this communication is integrity and credibility. Leaders with both have a heightened sense of self-awareness and a

TABLE 12.4 Seven traits of leaders.

■ Authenticity: consistent actions and words	■ People skills: offer helpful feedback and good coaching to all team members
■ Decisiveness: willing to act on limited, imperfect information	■ Communication: stimulate conversation and communicate vision
■ Focus: create a priority list and stick to it	■ Continuous improvement: keep learning and energy flowing in the firm, retain optimism
■ Care: build relationships and social capital	

Source: Adapted from Collins and Lazier, 1992.

strong understanding of what they believe in and value. The five functions of leadership are challenging the status quo, inspiring a shared vision, enabling people to act, modeling through personal example, and motivating people to act.

Leaders display an inner strength and a constant set of values that everyone knows and can rely on. They avoid self-aggrandizement, inspire others, and exhibit a combination of modesty and extraordinary competence.

What makes a great leader? One study reveals a key attribute: the capability to handle adversity and to learn from such experiences [Bennis and Thomas, 2002a]. Bennis and Thomas call these shaping experiences *crucibles*. These experiences make leaders stronger and more confident. Leaders are formed by a combination of their individual personalities and the events of the era in which they spend their formative years, which are then transformed in a crucible of experience and challenge. They organize the meaning of these experiences into capacities to respond to future challenges. These great leaders also evidence a capability to adapt as well as resilience, and are not necessarily charismatic [Khurana, 2002]. They are likely to be capable people with solid experiences that helped shape and build their leadership skills.

12.4 Teams

Much of the success of new ventures can be explained by the social interaction (collaboration) within entrepreneurial teams [Lechler, 2001]. A team is a small number of people with complementary skills who are committed to a common set of goals and tasks for which they hold themselves mutually accountable. Team members are given the power, freedom, and responsibility to control their work. Together, the team develops direction, momentum, and commitment by working to shape a meaningful purpose, such as designing a new product.

Teams offer multiple advantages. First, teams substitute peer-based control for hierarchical control of work. Instead of management devoting time and energy to controlling the workforce directly, workers control themselves, thereby removing layers of hierarchy. Moreover, teams permit employees to pool their ideas to come up with better and more creative solutions to problems.

In knowledge-intensive, dynamic industries, entrepreneurial teams outperform single entrepreneurs, since the new venture requires more capabilities than one individual is likely to have [Beckman et al., 2007]. It is the combination of complementary capabilities that leads to success. Thus, the advantage of a team results from bringing together people with diverse characteristics, skills, knowledge, and capabilities, each of whom is an expert in his or her own particular area [Fischer and Boynton, 2005]. While some similarities between team members are important to facilitate communication and fast execution, these differences allow for greater breadth and creativity [Beckman, 2006].

Effective teams also depend on shared understandings and good team processes. Team members should support the team effort and engage in "buy-in" without respect to the entrepreneurial task [Shepherd and Krueger, 2002]. Each member should feel comfortable making suggestions, trying things that

TABLE 12.5 Characteristics of an effective team.

- All members share leadership and ownership of the team's tasks.
- Communication is continuous among members in an informal atmosphere.
- Tasks and purposes are well understood.
- People listen to each other and are comfortable with disagreements within the team.
- Most decisions are reached by consensus.
- Feedback on performance is frequent.
- The division of tasks and work effort is clear.
- A collaborative effort is the norm.
- Members set their own, shared interim deadlines for project stages.
- Team members rely on each other and hold each other accountable.
- Teams learn and share their learning.

might not work, pointing out potential problems, and admitting mistakes. The team should create a frequent and rich flow of ideas and meet regularly to trade these ideas. The use of trained facilitators may help them with important creative tasks [Thompson, 2003]. The innovation process consists of the appropriate use of questioning, trust, and openness of the team. The team consistently seeks the best potential product that will create significant value [Estrin, 2009]. Table 12.5 lists several characteristics of an effective team.

Migrating Geese

Geese heading south for the winter fly in a **V** formation. As each bird flaps its wings, it creates uplift for the bird immediately following it. By flying in a **V** formation, the whole flock can fly at least 71 percent farther than if each bird flew on its own. People who share a common direction can get where they are going more quickly and easily if they cooperate.

Whenever a goose falls out of formation, it feels the resistance of trying to go it alone and quickly gets back into formation to take advantage of flying with the flock. Teams will work with others who are going the same way. When the lead goose gets tired, it rotates back in the formation, and another goose flies on the point. It pays to take turns doing hard jobs for the team. Perhaps the geese honking from behind are even the cheering squad to encourage those up front to keep up their speed.

Source: Muna and Mansour, 2005.

12.5 Management

Management is a set of processes such as planning, budgeting, organizing, staffing, and controlling that keep an organization running well. Managers are concerned with the allocation of resources and may be particularly focused on routine tasks. The management of a new venture firm will work to accomplish all the tasks required to keep the company operating. Management of a new firm is important and should not be undervalued compared to entrepreneurial leadership—both are valuable. The goal of the manager is to make a business carry out activities efficiently and on time. Managers work hard on focused goals in order to implement the strategy of the firm. They make deliberate choices about resources. One reason that purposeful managers are so effective is that they are adept at husbanding resources. Aware of the value of time, they manage it carefully.

Managers use their personal contacts when they need information or help. These informal networks create part of the social capital of an organization. People who link and connect people through a business network can be valuable managers who cross boundaries, help build subnetworks, and make the organization work [Cross and Prusak, 2002].

Managers are good at pattern recognition—making generalizations out of inadequate facts. Good managers are also prepared, frugal, and honest. Getting employees to stick to important strategic initiatives—and to give those initiatives their undivided attention over time—is crucial to competing successfully today. Great organizations have outstanding strategies and capabilities as well as the mechanisms necessary for executing the firm's strategies. The follow-through with operations and processes is part of a firm's competitive advantage. Managers focus on performance, feedback, and decision-making. They are also synthesizers who bring resources together in a timely manner. Given the rate of change of the competitive environment, they can make a great contribution by managing resources on a rolling time scale of 12 to 18 months.

Managers set expectations by describing the desired outcome, not the path. Talented people will determine the best path once they know the desired result. Most good managers motivate their people by offering them the opportunity for achievement, recognition for achievement, the work itself, responsibility, and growth or advancement [Harzberg, 2003]. Managers need to tell employees the span of their job, the support they can expect, and what they will be accountable for. Then, as they achieve good results, they should receive the credit due to them [Simons, 2005].

Good managers turn the team members' talent into good performance. They engage people in decisions that directly affect them, explain why decisions are made the way they are, and clarify what will be expected of them after changes are made. They also provide feedback to team members regarding their performance. Organizations profit when employees ask for feedback and deal well with criticism. Openness to feedback is critical for all staff in a new venture facing many transitions. As people begin to ask how they are doing relative to management's priorities, their work becomes better aligned with organizational goals [Jackman and Strober, 2003].

All successful managers excel in the making, honoring, and remaking of commitments. A commitment is any action taken in the present that binds an organization to a future course of action. Managerial commitments take many forms, from capital investments to personnel decisions to public statements, but each exerts enduring influence on a company. A commitment may impede a response to changing conditions. Managers can learn to recognize when commitments have become roadblocks to needed changes. They can then replace those roadblocks with new, rejuvenating commitments. The ability to make and rewrite commitments is an important managerial skill [Sull and Spinosa, 2005].

Management of a new venture is complex. Managers balance five perspectives: reflection, analysis, contextual dynamics, relationships, and change [Gosling and Mintzberg, 2003]. Reflection can lead to understanding. Analysis can result in better organizational arrangements and performance. Understanding of context (or environment) can lead to regional and global activities. Collaboration with other firms and sound industry relationships are important. Also, a bias toward action and beneficial change is another valuable perspective for a manager. The effective manager weaves together all five perspectives into a fabric for growth and performance.

Managers can make better business decisions if they ask for evidence of the efficacy of a proposed action [Pfeffer and Sutton, 2006]. The first step is to demand evidence (facts) that proves that a proposed action or process will work. Look for gaps in logic, inference, and applicability to your situation. Managers should consider running trial programs, pilot studies, and small experiments to help provide facts and insights that can lead to better management.

12.6 Emotional Intelligence

The emotional task of the leader is primal; that is, it is both the original and the most important act of the leader [Goleman et al., 2002]. The leader is a venture's emotional guide. **Emotional intelligence** (EI) is a bundle of four psychological capabilities that leaders exhibit: self-awareness, self-management, social awareness, and relationship management, as described in Table 12.6. Self-awareness refers to the ability to understand one's moods, emotions, and motivations, as well as their effect on others. Self-management is the ability to control emotions, as well as exhibit optimism and adaptability. Social awareness is the empathetic sensing of other people's emotions and awareness of the social currents within an organization. Relationship management includes inspiration, influence, conflict management, and teamwork.

According to Goleman and colleagues, leaders and managers who possess these capabilities have a high degree of EI and tend to be more effective. Their self-awareness and self-management help to elicit the trust and confidence of their colleagues. Strong social awareness and relationship management skills can help to earn the loyalty of their colleagues. Empathetic

TABLE 12.6 Four elements of emotional intelligence.

■ Self-awareness of one's emotions, emotional strengths and weaknesses, and self-confidence; the ability to read one's own emotions	■ Social awareness of empathy and organizational currents, and recognition of the needs of followers and clients
■ Self-management of honesty, flexibility, initiative, optimism, and emotional self-control; the ability to control one's emotions and act with honesty and integrity	■ Relationship management achieved by communicating through inspiration, influence, catalyst actions, conflict management, and collaboration

and socially adept persons tend to be skilled at managing disputes between people, better able to find common ground and purpose among diverse constituencies, and more likely to move people in a desired direction than leaders who lack these qualities.

People with high emotional intelligence tend to (1) behave authentically, (2) think optimistically, (3) express emotions effectively, and (4) respond flexibly in their relationship styles.

Leaders and managers who are in touch with their colleagues are said to be in resonance, which is the reinforcement of emotion. People are in resonance when they are "in synch" or on "the same wavelength." Resonant leaders use their EI skills to spread their enthusiasm and resolve conflicts [Goleman et al., 2001]. Good teams work to establish group resonance by building emotional and social awareness and management.

12.7 Organizational Culture

Organizational culture is the bundle of values, norms, and rituals that are shared by people in an organization and govern the way they interact with each other and with other stakeholders. An organization's culture can have a powerful influence on how people in an organization think and act. **Organizational values** are beliefs and ideas about what goals should be pursued and what behavior standards should be used to achieve these goals. Values include entrepreneurship, creativity, honesty, and openness.

Organizational norms are guidelines and expectations that impose appropriate kinds of behavior for members of the organization. Norms (informal rules) include how employees treat each other, flexibility of work hours, dress codes, and use of various means of communication such as e-mail. **Organizational rituals** are rites, ceremonies, and observances that serve to bind together members of the organization. Examples of rituals are weekly gatherings, picnics, awards dinners, and promotion recognition.

In innovative firms, the values and beliefs favor collaboration, creativity, and risk-taking [Jassawalla and Sashittal, 2002]. These firms employ stories and rituals to reinforce these values and beliefs.

Intel: Staying Entrepreneurial

Intel, which has become the semiconductor standard for personal computers, uses three principles to keep its entrepreneurial spirit alive. First, leaders must be willing to solve complex problems. There is no substitute for deep domain knowledge in technology industries, and at Intel, new products are continually being released. Second, the most effective managers are not afraid to change the rules. The "Intel Inside" branding campaign was a highly unusual marketing idea for a company with little previous direct contact with consumers. Last, Intel rarely tries to convince anyone to take a job or assignment. Managers must display a raw enthusiasm to try something new and a clear passion of their own [Barrett, 2003].

Culture is expressed in community. Communities are built on shared values, interests, and patterns of social interaction. The fit of the company culture in the business environment is critical to the company's competitive advantage. Most entrepreneurial start-ups have the founders and employees bound together by strong values and norms that include long hours working closely together. The sense of solidarity of a new venture is very high. The goals are uniformly shared—survival and early success being foremost. Start-ups are often founded by friends or former colleagues and exhibit high sociability within the organization. Employees possess a high sense of organizational identity and membership. In the early days, employees of Apple thought of themselves as "Apple people." Entrepreneurial firms often sponsor social events that take on ritual significance.

Hewlett-Packard Company fostered a culture known as the "HP Way" [Packard, 1995]. This culture was captured in a statement of objectives, values, and norms regarding fairness and justice. Employees were promised opportunity for security, job satisfaction, and sharing in profit. For example, during the 1981 recession, rather than lay people off, Hewlett-Packard introduced a 10 percent cut in pay and hours across every rank.

Cypress Semiconductor: Enjoying Work

T. J. Rodgers, founder of Cypress Semiconductor, explained his view of a favorable workplace of a technology venture [Malone, 2002]:

> The goal of starting the company was that I wanted to make a comfortable living. But the primary goal was to control my own environment. I wasn't happy working for other people. I wasn't happy working on projects I didn't find interesting, working with people I didn't enjoy spending time with. So it was an environmental control system I was trying to create, where I'd enjoy going to work every day, and enjoy the people I was working with, and enjoy the projects I was working on, and make a decent living.

TABLE 12.7 Seven principles of trust.

1. Trust is not blind. It is unwise to trust people whom you do not know well, whom you have not observed in action over time, and who are not committed to the same goals.

2. Trust needs boundaries. It is wise to trust people in some areas of life but not necessarily in all.

3. Trust requires constant learning. Every individual of a team must be capable of self-renewal and learning.

4. Trust is tough. When trust proves to be misplaced because people do not live up to expectations or cannot be relied on to do what is needed, then those people must go, be reassigned, or have their boundaries severely curtailed.

5. Trust needs bonding. Teams of people need to build their own bonds.

6. Trust needs touch. Personal contact is necessary, and teams need to meet in person to renew their trust and bonds.

7. Trust has to be earned. Organizations that expect their people to trust them must continually demonstrate that they are trustworthy.

Source: Adapted from Handy, 1999.

Entrepreneurial firms usually demonstrate high sociability and solidarity in their early years, built around the leadership of the founders—such as Hewlett and Packard. As the founders leave and other challenges come to bear, it may be difficult to retain the communal culture of a start-up. Yahoo built a fast-growing firm based on a set of founders who worked furiously at their jobs. Their communal culture helped them build the firm, but their insularity may have made them see the marketplace through a Yahoo lens in 2000 [Mangalindan and Hoang, 2001]. Google has succeeded due in part to a powerful culture that is maintained at all of its locations worldwide. This helps develop a consistency in user experience that customers value.

Perhaps the strongest element of an organizational culture is trust [Covey, 2006]. **Trust** is a firm belief in the reliability or truth of a person or an organization. It is critical that we can trust those with whom we work. Handy provides seven principles of trust, as listed in Table 12.7. Teams are formed with many people we already trust, but new members will need to earn our trust. By *trust,* organizations mean confidence in a person's work, competence, and commitment to the organization's goals and tasks. Every team member must be capable of self-renewal, learning, and adaptability. When trust is broken, the person needs to be reassigned, leave the organization, or have his or her boundaries constrained. Teams need to build bonds among their members to enable trust. People need to meet in person to restore the group bonds. Finally, organizations, like people, need to continually demonstrate that they are trustworthy.

Good communication is part of any sound and trusting community. Knowing when to speak up and when to keep silent is an important skill for any worker. A new enterprise needs openness and creativity to grow and become fruitful. Carefully choosing the right issues to raise and avoiding ill-chosen conflict can be important. When emotions are high and the new enterprise is challenged with

TABLE 12.8 Four principles of a performance-based culture.

1. Inspire everyone to do their best.

2. Reward achievement with praise and pay-for-performance, and keep raising the performance goals.

3. Create a work environment that is challenging, rewarding, and fun.

4. Establish, communicate, and stick to clear values.

Source: Joyce et al., 2003.

obstacles, it may be wise to keep silent. Nevertheless, managers of any enterprise must welcome ideas and comments. Organizations of all sizes and arrangements must strive to keep the ideas flowing. Breaking the silence can bring an outpouring of fresh ideas from all levels of an organization—ideas that might just raise the organization's performance to a whole new level [Perlow and Williams, 2003].

In a study of successful companies, the building of a performance-based culture was central [Joyce et al., 2003]. The study showed that winning companies built a culture on the four principles described in Table 12.8. Examples of successful firms that have a performance-based culture include Intel, Cisco Systems, and General Electric. Just about any high-potential firm needs to focus on performance, define it, and build a culture that reinforces it.

Leaders can develop a design for a workplace that can be fulfilling for all involved. With growing trust and dignity as well as wide participation on decision-making, all employees can experience ownership and pleasure in their work [Bakke, 2005].

Cisco Community Statement

Cisco's culture requires that all employees, at every level of the organization, are committed to responsible business practices. Additionally, our business strategy incorporates our dedication to corporate citizenship, which includes our commitment to improving the global community in which we operate, empowering our workforce, and building trust in our company as a whole.

Cisco was founded on, and still thrives today in, a culture based on the principles of open communication, empowerment, trust, and integrity. These values remain at the forefront of our business decisions. We express these values through ethical workplace practices; philanthropic, community, and social initiatives; and the quality of our people.

Source: Cisco Systems Annual Report, 2003.

12.8 Social Capital

Social capital consists of the accumulation of active connections among people in a network [Coleman, 1990]. Social capital was also considered in Chapters 1 and 4. Social capital refers to the resources available in and through personal and organizational networks [Baker, 2000]. These resources include information, concepts, trust, financial capital, collaboration, social structure, and emotional support. These resources reside in networks of relationships. Social capital depends on whom you know. Social capital, like financial capital, can lead to increased productivity, if used wisely. A firm can build up or deplete its social capital by its actions. Relationship networks are often as important as technology, land, capital, or other assets of a venture.

Social capital tends to be self-reinforcing and cumulative. Successful collaboration in one endeavor builds connections and trust—social assets that facilitate future collaboration in other, unrelated tasks. Firms that build up social capital demonstrate a commitment to retaining people and promoting from within. They also enable far-flung teams to meet in person periodically. They give people a common sense of purpose and keep their promises to people. Employees need to hear the same messages that an organization sends out to vendors and customers. Alternatively, social capital may be depleted by declining trust among a firm's people and the effects of people working offsite or on their own [Prusak and Cohen, 2001].

Social capital can be described as consisting of three dimensions: (1) structural, (2) relational, and (3) cognitive. The structural dimension concerns the overall pattern of relationships found in organizations. The relational dimension of social capital concerns the nature of the connections between individuals in an organization. The cognitive dimension concerns the extent to which employees within a social network share a common perspective or understanding [Bolino et al., 2002]. Social capital is valuable because it facilitates coordination, reduces transaction costs, and enables the flow of information between and among individuals. In other words, it improves the coordinated effort and organization.

Better knowledge sharing can lead to increased trust and better decisions. Teamwork can lead to inventiveness, creative collaboration, and a good spirit. Trust is the fuel of a social capital engine that in turn engenders more trust. When people in an organization say their firm is "political," they often mean that trust is low throughout the organization. An organization with strong capital is a community with shared values and good trust. Moreover, social capital extends to relationships outside of the company, which should also be characterized by shared values and trust.

Managing to develop and utilize social capital can lead to enhanced corporate performance [Lee and Kim, 2005]. Providing the appropriate networks that enable creativity can improve quality and productivity. Engaging the creative energies in a collaborative activity of the technologists and designers with customers can lead to improved results [Florida and Goodnight, 2005]. Social capital can provide greater access to other resources [Kalnins and Chung, 2006]. Or, it can substitute where other types of capital are lacking [Packalen, 2007]. For most technology ventures, the social network should be dense and redundant, and link external and internal sources to needs [Cross et al., 2005]. Firms should continually adapt their social capital to changing resource needs [Maurer and Ebers, 2006].

IBM's Guiding Principles

Louis Gerstner joined IBM in January 1993 to bring the firm back to its roots and success. In his first months, Gerstner created a set of principles for the firm that included [Gerstner, 2002]:

- At our core, we are a technology company with an overriding commitment to quality.
- We operate as an entrepreneurial organization with a minimum of bureaucracy.
- We think and act with a sense of urgency.
- Outstanding, dedicated people make it all happen, particularly when they work together as a team.

Gerstner also described the IBM culture this way: "In the end, an organization is nothing more than the collective capacity of its people to create value." By 2002, Gerstner left IBM as a powerful technology company focused on its entrepreneurial principles.

12.9 Attracting and Retaining Talent

All firms know that attracting and retaining the best people is key to their future success. However, open competition for other companies' people is now an accepted fact. Leaders know that in entrepreneurial markets, fast-moving firms are competing for the best people. New ventures pursuing important opportunities can attract talented people. By a person's talent, we mean that person's recurring patterns of thought, feeling, or behavior that can be productively applied and play a significant role in performance [Buckingham, 2005]. Typically, these people are found in the social networks of the founding entrepreneurial team members. New employees can expect to have a direct stake in the new enterprise and participate in an open, trustworthy team that will build the new business.

Attracting and retaining key employees depends on compensation, work design, training, and networks. Compensation systems include wages, incentives, and ownership options. Successful new enterprises look for people who can thrive in environments in which people trust each other and are willing to debate assumptions, share information, and express feelings. Although winning is important to them, the goal is to win as an enterprise, rather than as individuals. The goal is to get the talent to act as owners of the firm by engaging people in the setting of goals and objectives and then organizing to achieve them. Talented employees can be attracted to new ventures that offer a plan for corporate social responsibility comprising legitimate activities and vision. They want to work at companies that exhibit good corporate citizenship and have a plan for products that are green and improve the environment [Bhattacharya, 2008].

Talented people in new ventures work hard and fast, trying to get to market with the hope of big payoffs. They remain focused on the critical factors of success and resist any tendency to drift to peripheral goals. They

TABLE 12.9   Six principles for retaining loyal people and partners.

1. Preach what you practice. Communicate the vision, goals, and values of the organization. Practice what you preach is also required.

2. Partners must win also. Enable your vendors and partners to participate in a win-win venture.

3. Be selective in hiring. Select people with values consistent with the firm's. Membership on the team is selective.

4. Use teams of talented people. Use small teams for most tasks and give them the power to decide. Provide simple rules for decision-making so teams can act.

5. Provide high rewards for the right results. Reward long-term values and profitability. Provide solid compensation, benefits, and ownership.

6. Listen hard, talk straight. Use honest, two-way communication and build trust. Tell people how they are doing and where they stand.

Source: Adapted from Reichheld, 2001.

are motivated by significant responsibility, participation, and the possibility of big financial gains. Their internal commitment is, in fact, aligned with the realities of their incentives, and the potential results can be motivating.

Keeping faithful, loyal employees and partners in a new venture can be based on six principles, as shown in Table 12.9 [Reichheld, 2001]. A simple reliance on financial rewards is insufficient. Good communication, trust, and treatment are essential to retaining talent and partners. High standards of decency, consideration, and integrity are necessary for all members of the new venture. Through loyalty to ideals, the firm becomes worthy of loyalty from its people and partners.

Many, if not all, technology start-ups use an organizational design and culture that involve challenging work, peer group control, and selection based on specific task abilities. Few imperatives are more vital to the success of young technology companies than retaining key technical personnel, whose knowledge often represents the firm's most valuable asset [Baron and Hannan, 2002]. Therefore, the leaders of the enterprise need to hold to their commitments to their employees and retain their trust.

Perhaps the most important qualities of a new hire are proficiency in the skills required and demonstrated capability to acquire the attributes needed for future situations. Focusing on the raw ability to learn may be most critical for technology ventures [McCall, 1998]. Most new ventures benefit from attracting new people who possess stretchwork capabilities [Bechky and O'Mahony, 2006]. Stretchwork is the capability to readily bridge from proven competencies to new ones demanded by the needs of the new venture. These people possess strong competencies and quickly learn new ones.

To the extent possible, as a company grows, finding the best people should be accomplished by recruiting through existing team members and the firm's supply chain. To be sure about the hire, new ventures often use a trial period in a consulting role or temporary relationship on a project basis. The right team members are critical to a new firm's early success. Venture capitalist John Doerr of Kleiner Perkins says that when he looks at the business plan, "I always turn to the biographies of the team first. For me, it's team, team, team. Others might say, people, people, people—but I'm interested in the team as a whole" [Fast Company, 1997].

New organizations will need to attract the right people for the executive positions. People who lead complex organizations are difficult to identify, attract, and retain. Before considering any candidate, firms need to clearly define what they need in terms of crucial competencies and experience [Fernandez-Araoz, 2005].

The growth and survival of an emerging business depends on star players and good supporting talent [DeLong and Vijayaraghavan, 2003]. Star employees can make important contributions to performance. Yet a firm's future also depends on the capable, steady performers. These steady performers bring stability and depth to an organization's resilience. Furthermore, they are more likely to be loyal to the organization.

12.10 Ownership and Stock Options

New ventures are able to offer stock ownership to their people. They also need to provide reasonable compensation and benefits. People in a start-up may have to sacrifice financial compensation for an initial period while building the firm toward financial break-even. Thus, it is important to offer reasonable benefits and ownership opportunities. Offering health benefits may be necessary to attract people to the new venture. Most technology ventures offer health benefits to their employees before reaching break-even.

Ownership interest in the new firm will be of great interest to most new employees. Broad-based ownership, when done right, can lead to higher productivity, lower workforce turnover, better recruits, and improved outcomes [Rosen et al., 2005]. **Stock options** are offered in a plan under which employees can purchase, at a later date, shares of the company at a fixed price (strike price). Stock options take on value once the market price of a company's stock exceeds the exercise (strike) price. Stock options should vest with the recipient over a period of years. For example, a new hire may receive an option for 10,000 shares exercisable at $1 per share vesting over four years. This is equivalent to 2,500 shares each year. Stock options give employees the right to buy the company's stock in the future at a preset price, thus motivating them to work to increase productivity and innovation—and eventually, the market value of the firm.

The purpose of employee stock options is to create a noncash substitute for part of the wage compensation the firm must provide to attract and retain employees. A new, entrepreneurial firm may not be able to provide the cash compensation needed to attract outstanding workers. Instead, it can offer stock options. From the beginning, Starbucks decided to grant stock options to every employee in proportion to their level of base pay: it called these options "bean stock." Microsoft created thousands of millionaires over its first 20 years due to the issuance of wide stock options.

> **Stock Options Basics**
>
> ■ Options give employees the right to buy a certain number of their company's shares at a fixed price (strike price) for a certain period of time.
>
> ■ The strike price is usually the market price of the stock on the date the options are granted.
>
> ■ Options usually begin vesting after one year and fully vest after four years. If an employee leaves the company before his or her options fully vest, the remaining options are canceled.
>
> ■ Once an option is vested, the employee can then exercise it, that is, purchase from the company the allotted number of shares at the strike price, and then either hold the stock or sell it if the company is public.
>
> ■ The difference between the strike price and the market price of the shares at the time the option is exercised is the employee's gain in the value of the shares.
>
> ■ When an employee exercises an option, the company must issue new shares of stock.

An alternative to stock options is **restricted stock,** which is stock issued in an employee's name and reserved for his or her purchase at a specified price after a period of time—say, one, two, or three years. Some call this type of stock "reserved stock." New firms need to increase the level of share ownership through such means as offering restricted shares and requiring that employees hold their shares for certain periods of time. For example, an employee could have 10,000 shares reserved to be purchased at $2 per share as they vest in two years. If the price of the stock appreciates to $10, the gain in the stock's value is $80,000. Restricted stock still has value if the stock price falls, while options expire as worthless if the stock does not appreciate. In 2003, Microsoft switched from offering stock options to restricted stock.

12.11 Board of Directors

An incorporated firm or LLC has a board of directors. **A board of directors** is a group composed of key officers of a corporation and outside members responsible for the general oversight of the affairs of the entity. This board normally consists of the founders of the firm and one or more investor-owners. A new start-up might find a board of three owners is adequate. As other investors are added, one or more additional owners may be added to the board. The board is the overseer of the corporation with responsibility to select and approve the appointment of the chief executive officer (CEO) and other officers of the corporation. Directors should possess significant knowledge and competencies in the industry of the company. The board of directors is a legally constituted group whose responsibility is to represent the stockholders. A board of five

TABLE 12.10 Five goals for an effective board process.

1. Engage in constructive conflict—especially with the CEO.

2. Avoid destructive conflict.

3. Work together as a team.

4. Work at the appropriate level of strategic involvement—avoid micromanagement.

5. Address decisions comprehensively.

Source: Finkelstein and Mooney, 2003.

might consist of two insider executives, two representatives of investors, and one independent director. This board must approve the bylaws, officers, and annual report to the shareholders, as well as any financial offerings to investors or banking activities. The members of this board have fiduciary responsibility, meaning that they are under a legal duty to act in the best interests of the corporation and its stockholders.

The **board of advisors**, if any, is constituted to provide the firm with advice and contacts. The members have extensive skills and knowledge and provide good advice. The board of advisors is nonfiduciary and does not engage in the legal or official actions of the corporation. Thus, the advisors are free from liability as long as they refrain from any legal or official role.

Good boards provide critical strategic guidance, legitimacy, and connections, while protecting their assets [Huse, 2000]. Thus, the board of directors should consist of people who are knowledgeable, interested, and shareholder oriented. Directors must have the skills to make the appropriate decisions and must be available for the necessary official meetings. In selecting directors, a premium should be placed on a wide range of expertise and backgrounds but, above all, on people who will seek to expose the downsides as well as the upsides of every major decision. Directors should work for better governance of the firm. They should know the strategy of the firm and keep everyone focused on the firm's innovation strategy. Savvy start-ups look for directors who are fluent in one or more of the following: audit and finance, strategy, marketing, and sales. The board of advisors should consist of persons of good reputations who can give help and advice from time to time.

Good boards are those that function and work well. They are distinguished by a climate of trust, respect, and candor. Members feel free to challenge each other's assumptions. They should feel a responsibility to contribute significantly to the board's performance. In addition, good boards assess their own performance, both collectively and individually [Sonnenfeld, 2002]. Table 12.10 lists five goals for an effective board process [Finkelstein, 2003].

Compensation of directors will normally be in the form of stock. The members of the board should, after a few years, have a reasonably substantial stake in the firm. This can be achieved with the use of stock options or restricted stock that vests over a period of time.

> **CI's Board of Directors**
>
> Conservation International (CI) is a nonprofit corporation whose mission is to conserve the earth's living natural heritage and biodiversity. CI applies innovations in science and economics to protect the world's natural capital. CI's board of directors is impressive (www.conservation.org). The chair of the board is Gordon Moore, cofounder of Intel. Others on the board include actor Harrison Ford, Queen Noor of Jordan, Orin Smith, former CEO of Starbucks, and Rob Walton, chairman of Wal-Mart.

12.12 AgraQuest

The initial leader of AgraQuest was Pamela Marrone, who left Novo Nordisk's Entotech to found AgraQuest as a biotechnology start-up in 1995. She recruited three other experienced biotech leaders to join her team, which then drafted a business plan. They spent from January 1995 to March 1997 rewriting the plan and attempting to raise several million dollars in venture capital. During that 18-month period, all four members of the lead team deferred compensation and agreed to accept stock options as an alternative. One of the original team members could not financially accommodate the 18-month deferment and left for another job. Eventually, they were funded by a venture capital firm in mid-1997. At that time, three scientists from Entotech joined AgraQuest and remain there today.

AgraQuest was a firm with a strong science and technology bias and culture. Most of the staff were attracted by the science and the potential for biopesticides. The organization structure was nonhierarchical, and most employees enjoyed an opportunity to provide their own views and input on decisions. By 2001, the firm grew to 70 people, with most acting in a scientific or technical support role.

Marrone primarily used a directing leadership style, with a secondary style of coaching (see Figure 12.2). From the beginning, stock options have been awarded to all employees based on their salary level and when, in terms of risk, they joined the firm.

From 1997 to 2006, AgraQuest's board of directors consisted of Pam Marrone and representatives of the investors (venture capitalists). In 2004, Mike Mille became CEO while Marrone remained as a board member and president. Marrone left in 2006 to found another company.

A scientific advisory board, consisting of professors at prestigious universities, has served AgraQuest since 1996. In addition, a strong network of other scientists and technologists has provided the venture with access to help and information in the agricultural biotechnology industry.

AgraQuest's leaders have always attempted to create a scientific organization with a collaborate culture and compensation scheme based on equity participation using stock options. As of 2009, the firm was achieving profitability after more than a decade of diligent efforts and tenacity.

12.13 Summary

Early in the development of a firm, a leadership team is created to build a business plan and organizational plan. The organizational plan is structured to align the culture of the firm with its goals and values. The firm's leaders strive to motivate and inspire team members and foster a collaborative, innovative culture. As the firm grows, managers join to build the structure and carry out the detailed tasks of the new firm. As the firm develops, the leaders strive to exhibit emotional intelligence. The firm also puts together a compensation scheme that emphasizes "buy-in" or ownership—normally achieved by awarding stock options or restricted stock. Finally, a firm creates a board of directors and a board of advisors to help monitor and enhance the growing firm.

Principle 12

Effective leaders coupled with a good organizational plan, a collaborative performance-based culture, and a sound compensation scheme can help align every participant with the goals and objectives of the new firm.

Video Resources

Visit http://techventures.stanford.edu to view experts discussing content from this chapter.

Strength of a Team	Vinod Khosla	KPCB
Founding Teams	Ashwin Navin and Ping Li	BitTorrent
Forming the Founding Team	Kim Smith	New Schools Venture Fund
Team Composition	Kathleen M. Eisenhardt	Stanford

12.14 Exercises

12.1 Examine the beginnings of a technology venture that currently has lots of "buzz". Who were the company founders? What background, capabilities, and qualities did each bring to his or her new role? Who was hired next and why?

12.2 Matthew Smith knows his talented people are the reason for the success of ElectroMag. Smith has noticed that the costs of medical and dental benefits are escalating, and he needs to control them. With 800 employees, the firm's profitability is threatened. He has three options: (1) eliminate health benefits, (2) try to find a cheaper plan that covers fewer medical and dental procedures, or (3) withhold a fixed amount of each person's salary to use to fund the benefits. Which option would you recommend he choose?

12.3   Genentech (www.gene.com) has a unique culture known for rigorous science, guarding of industry secrets, and rigid rules. Its key principle is: good scientists make for good science make for good products make for a good company. Describe Genentech's culture in terms of norms and rituals.

12.4   Take-Two Interactive (www.take2games.com) develops entertainment software games. Revenues grew from $19 million in 1997 to $1.5 billion in 2008. Describe the organizational design for Take-Two, which has about 2,100 employees. Use Figure 12.1 to help determine its organizational model.

12.5   Getty Images, founded in 1995, offers still and moving images distributed via its website (www.gettyimages.com). Examine its 2008 annual report and determine how the firm motivates its employees. Describe its employee stock plan and its organizational design using Table 12.2.

12.6   Red Hat is the leading distributor of Linux software and services. The firm had over 2,500 employees and revenues of $650 million in fiscal year 2009. Using the concepts of Section 12.7, describe the firm's organizational culture.

12.7   24/7 Customer provides services to U.S. firms that wish to outsource call centers and CRM activities to India. Using the information provided at the firm's website, describe the founding team of this firm. Also, describe its organizational design (www.247customer.com).

12.8   Examine Google's corporate philosophy "Ten things Google has found to be true" (www.google.com/corporate/tenthings.html). What does Google's "do no evil" phrase mean? How are these broad truth statements translated into action within the Google organization and culture?

VENTURE CHALLENGE

1.   Describe the team and the organizational arrangement for your venture.

2.   Discuss your plans to build a board of directors and a board of advisors. Name some candidates for these boards.

CHAPTER 13

Acquiring and Organizing Resources

To get profit without risk, experience without danger, and reward without work,
is as impossible as it is to live without being born.
A. P. Gouthev

How can entrepreneurs efficiently acquire and organize the resources needed to launch their venture?

To tap required resources, entrepreneurs need to build credibility and legitimacy in the marketplace of resources and talent. Influence and persuasion can help entrepreneurs build their case for securing scarce resources for their venture.

Both choosing a physical location and operating as a virtual organization are viable options for a firm today. We examine the benefits of joining a cluster of interconnected enterprises operating within a geographic region. Using the Internet and related technologies, new enterprises can build a powerful virtual organization. All firms need to create a plan for outsourcing functions while maintaining critical functions within the firm. As firms strive to be innovative and competitive, they seek to control costs by outsourcing functions to those who can do them better and cheaper. However, these firms are challenged to retain the cohesion and coordination required to effectively manage these supplier partners. Seeking financial resources will be discussed in Chapter 18. ■

13.1 Acquiring Resources and Capabilities

Another definition of entrepreneurship is the pursuit of opportunity without regard to resources currently controlled [Stevenson et al., 1999]. This view stresses the idea that the entrepreneur can locate and access resources when they are needed. For example, when an entrepreneurial team needs legal counsel, it can engage a lawyer. When it needs a circuit designer, it can hire one. In fact, resources are usually scarce, and the attraction of talented employees or financial investors is not easy or guaranteed. A firm's competitive advantage flows from the combination of resources and capabilities executing a unique strategy. If these resources and capabilities are scarce, then the new venture needs to compete to secure them.

The founders of a new venture attempt to acquire resources and capabilities by contacting key organizations and people and asking them to support their venture. For example, they ask their bank, suppliers, and sources of financial capital to take some risk and support the new venture, which will be further discussed in Chapter 18.

This resource-seeking activity can be represented by the cycle shown in Figure 13.1 [Birley, 2002]. The founders are asking all the participants in the credibility cycle to believe in their opportunity, vision, and story, and invest in their venture. To move forward, entrepreneurs need to persuade someone in the cycle to believe in them. If the entrepreneurs get some talented people to commit to the venture, this will help convince the suppliers. If the entrepreneurs get some customers to tentatively commit to purchasing the product, the sources of financial capital (bankers and investors) will become more interested. The entrepreneurs travel around the cycle, slowly building their credibility. In other words, the entrepreneurs demonstrate the legitimacy or truthfulness of the new venture to the members of the credibility cycle [Zimmerman and Zeitz, 2002]. Legitimacy or credibility is evidence of a social judgment of desirability and

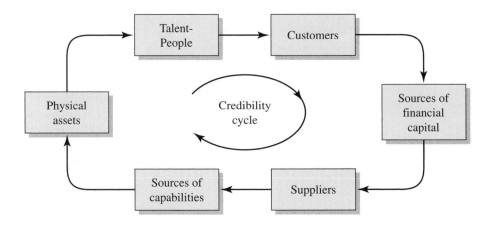

FIGURE 13.1 Credibility cycle.

TABLE 13.1 Sources of legitimacy.

- Regulatory: legal actions, accreditation, credentials
- Social: fair treatment, endorsements, networks, image
- Industry: attractive, respected industry, known and understood business model
- Talent: known, respected people
- Location: within an industry cluster, favorable location, visible
- Intellectual property: trade secrets, patents, copyrights

enables a new venture to access resources. The holders of scarce resources provide resources to new ventures only if they believe that the ventures are efficient, worthy, and needed, and the teams are competent. The greater the level of a new venture's legitimacy, the more resources it can access.

A new venture can build its legitimacy by tapping the sources of legitimacy listed in Table 13.1. A new venture can join industry associations, secure endorsements, and get commitments from talented, respected individuals. Patents, copyrights, and trade secrets also help to build legitimacy. New ventures should focus on actions that have the greatest payoff for legitimacy. A certain amount of legitimacy is required to make the credibility cycle build up the investment of resources. A successful new venture needs to acquire, build, and use legitimacy to secure the necessary resources to commence operations and grow successfully. Creating, building, and retaining a firm's credibility or legitimacy are critical tasks for the leadership team of any business. They determine the key influencers in your industry and reach out to them about the opportunity. For example, it has been shown that attaining the support of credible and reputed venture capitalists results in increased credibility of the venture [Hsu, 2004]. The leadership team of a new venture recognizes and creates new economic or social opportunities and makes decisions on the location, form, and use of resources [Wennekers et al., 2002].

To the extent possible, smart leaders of ventures manage the new venture like it is a public company [Carl, 2007]. They manage it with the idea that it will grow to become an important enterprise eventually. They hire a highly qualified law firm and accountant. They look for the best employees and talent available including compelling board members and advisors. A new venture should strive to be one of the best from the beginning.

Entrepreneurs who exhibit higher social competencies and emotional intelligence greatly improve their ability to access resources. Entrepreneurs possessing a high level of social capital (e.g., a favorable reputation and an extensive social network) gain access to persons important for their success. Once such access is attained, their social competence influences the outcomes they experience [Baron and Markman, 2003].

In order to attract resources, entrepreneurs craft and use a story about their venture that they tell to potential partners and visitors [Downing, 2005]. In addition, they can use advisors and guides to help them in building their credibility and story [Chrisman et al., 2004].

Gina Bianchini's Quest for Resources at Ning
Gina Bianchini founded Ning in 2005 to create an online platform for users to build their own social networking sites. Bianchini was a former investment banker at Goldman Sachs with knowledge of the online advertising industry. After receiving her MBA from Stanford, she teamed up with Marc Andreessen (the founder of Netscape) and designed Ning to take advantage of viral networks. Viral applications are defined as software where every new user in a network brings in more than one additional user. This causes exponential growth and can generate huge amounts of Internet traffic. Ning was founded just as viral networks became popular with angels and venture capitalists. As CEO, Bianchini was able to take advantage of their enthusiasm and has successfully raised over $100 million through 2009. Bianchini proved highly effective in attracting critical resources like Andreessen and venture capital investment.

13.2 Influence and Persuasion

Influence and persuasion play a role in the entrepreneur's acquisition of resources. They are part of the process of selling, acquiring resources, and structuring deals regarding acquisitions and investments. Every entrepreneurial team ideally needs a person who is a master of persuasion [Cialdini, 1993]. Persuasion skills exert greater influence over other's behavior than formal power relationships do.

Persuasion is governed by basic principles that can be taught, learned, and applied [Cialdini, 2001]. The six principles of persuasion are provided in Table 13.2. The principle of *liking* states that people like to please or work with those who sincerely like them also. One can uncover shared interests and bonds and offer sincere praise and compliments.

The second principle of *reciprocity* states that one can elicit the desired behavior from others by displaying it first. Offering help or information to others first can encourage them to reciprocate. The principle of *social proof* states that people look for and respond to a display of endorsements from those they trust.

The fourth principle of *consistency* states that people stick to their verified commitments—those they make as voluntary, public statements. People who make verbal public or written commitments are likely to stay with them. The principle of *authority* states that people highly regard experts. Therefore, it is useful to show and display your firm's expertise and competencies.

Finally, the principle of *scarcity* states that people want scarce or unique products. Therefore, it is necessary to explain and demonstrate the unique benefits and the chance to gain exclusive advantages.

TABLE 13.2 Principles of persuasion.

1. Liking

- People like those who like them.
- Uncover shared bonds and offer sincere praise and compliments.

2. Reciprocity

- People respond in kind to others.
- Give to others what you want to receive.

3. Social proof

- People respond to a display of endorsements by people they trust.
- Use testimonials and endorsements from trusted leaders.

4. Consistency

- People adhere to their verified commitments.
- Ask for voluntary, public commitments.

5. Authority

- People highly regard experts.
- Show and state your expertise.

6. Scarcity

- People want scarce products.
- Describe unique benefits.

Source: Cialdini, 2008.

Celtel: A Better Way

As mobile telephony grew in prominence during the last several decades, the giant telecommunications companies invested in infrastructure worldwide. In this spurt of development, Africa was largely ignored due to its widespread instability, corruption, and poverty. Companies that attempted to do business in Africa faced the unpleasant reality that bribes were often a prerequisite to conducting business. In 1998, Mo Ibrahim, a consultant with experience in African telecom companies, saw this as an opportunity.

Mo founded Celtel, which is a mobile phone service provider, and took an especially effective approach to doing business. Rather than targeting individual African markets, he eliminated inefficiencies by widening the scale of operations to a pancontinental level. This large scale also increased the respectability and political clout of Celtel, allowing him to operate without paying bribes. In this way, Mo was able to utilize the positive forces of influence to achieve his goals rather than resorting to immoral pecuniary incentives.

13.3 Location and Cluster Dynamics

Choosing a location can have long-lasting effects on a new venture. Entrepreneurs need to choose their location with their customers, future employees, suppliers, partners, and competitors in mind. While the importance of location to start-ups in the retail and restaurant businesses is obvious, location is important to all companies. Criteria for location selection are listed in Table 13.3. Knowledge-based enterprises, especially, need to locate in an area where skilled employees and complementors are readily available. The cost of doing business will be an important factor in location selection, too. Furthermore, housing that is affordable to all employees is important. A company's home should be a place that current and future employees will like. Thus, it should have good schools, a feeling of safety, and good transportation links.

Location advantages are based on the flow of knowledge, relationships, and access to institutions. Commonly known ideas and technologies that can be accessed from anywhere will be widely available to all competitors. Thus, they cannot serve as a competitive advantage. Local advantages, such as access to knowledge and research at a university, therefore can become a unique advantage [Audretsch et al., 2005]. Local companies that supply required components and technologies could also be very advantageous. Moreover, it is important for companies to be physically near some key constituents. These constituents include competitors and other companies in related fields, as well as the venture capital firms that back them.

Companies should attempt to proactively link into strong local institutions such as trade associations, universities, and professional societies. Entrepreneurs should take advantage of opportunities in their local region. Long-term competitive advantage relies on being able to avoid imitation by competitors. Location-based advantages in innovation may prove more sustainable than implementing corporate best practices [Porter, 2001].

Firms in industrial regions benefit from localization of cost economies derived from specialist suppliers, a specialist labor pool, and the spread of local knowledge [Best, 2001]. Since new ventures benefit from new knowledge, competent suppliers, and available talent, they should consider locating in a region that offers easy access to all three factors. Entrepreneurial activity differs

TABLE 13.3 Criteria for location selection.

■ Availability of potential employees and consultants	■ Costs of doing business
■ Availability of complementary firms	■ Availability of suitable facilities
■ Road and airplane transportation	■ Proximity to markets
■ Quality of life—education, culture, recreation	■ Availability of support services
	■ Affordable housing

TABLE 13.4 Selected centers of entrepreneurial activity in technology.

Western United States:
Las Vegas, San Diego, San Francisco and Silicon Valley, Seattle

Eastern and Southern United States:
Austin, Boston, New York, Raleigh-Durham, Washington, D.C.

Asia:
Bangalore, Beijing, Shanghai, Singapore, Taipei

Europe/Middle East:
London, Munich, Switzerland, Tel Aviv

significantly across regions of a country. Examples of attractive locations for new ventures are provided in Table 13.4.

A **cluster** is a geographic concentration of interconnected companies in a particular field. Clusters can include companies, suppliers, trade associations, financial institutions, and universities active in a field or industry. A good example is the Hollywood cluster of firms and infrastructure coming together for the creation of movies. If a new venture wants to enter the movie industry, it is probably wise to consider locating in Los Angeles.

An excellent example of an emerging cluster is Israel's high technology cluster, often called "Silicon Wadi." In 2007, venture capitalists invested $1.76 billion in Israel, much of which went to companies based in this cluster. Although this is only a fraction of that spent in Silicon Valley, the cluster is growing rapidly. It provides an environment conducive to technology venture formation: elite universities, venture capital firms, research centers run by big corporations, and a location attracting talented engineers.

A start-up can gain regional advantages by joining a cluster of companies that have complementary or competitive capabilities and resources. A new company within a cluster is more likely to find the employees and infrastructure that it needs [Iansiti and Levien, 2004]. Firms located in clusters have better product innovation performance, sales growth rates, and survival rates [Gilbert et al., 2008]. In general, good clusters provide access to ideas, role models, informal forums, and sources of talent [Venkataraman, 2001].

Clusters promote both competition and cooperation. Firms form alliances, recruit each other's talent, and compete, all at the same time. A cluster also provides new firms with a critical mass of talent, knowledge, and suppliers for easy entry into an industry. A cluster of independent and informally linked companies and institutions represents a robust organizational form that offers advantages in efficiency, effectiveness, and flexibility.

Furthermore, clusters are conducive to new business formation for a variety of reasons. First, barriers to entry are lower than elsewhere. Second, individuals working within a cluster can more easily perceive gaps in products or services around which they can build businesses. Finally, the formation of new ventures creates market opportunities for others. For example, a new venture may introduce a new product that creates the possibility of complementary products offered by other local enterprises.

The availability of firms providing complementary products may be critical to a new venture's success. In a tourism cluster, such as New York City, the quality of a visitor's experience depends not only on the appeal of the primary attraction but also on the quality and efficiency of complementary businesses such as hotels, restaurants, shopping malls, and transportation systems. Because members of the cluster are mutually dependent, good performance by one can improve the success of the others.

The area south of San Francisco called Silicon Valley is a cluster for electronics, medical devices, green tech, and Internet companies. Connectedness and mobility of talent and ideas are a way of life in Silicon Valley. The support structure includes entrepreneurs, venture capitalists, attorneys, consultants, board members, universities, and research centers. Technology firms in Silicon Valley prosper in a dynamic environment of novelty and innovation. This network environment is the outcome of collaborations between individual entrepreneurs, firms, and institutions focused on the pursuit of innovation and its commercialization. Silicon Valley has an openness to change and is supportive of the creative and the different [Florida, 2002]. It displays all of the cluster characteristics that support innovation, as listed in Table 13.5.

The entrepreneurial attitude prevalent in Silicon Valley is exemplified by the views of T. J. Rogers, founder of Cypress Semiconductor, who said [Malone, 2002]:

> What makes us special and different here in Silicon Valley is that we're truly capitalists. We invest. There is no safety net. You can go out of business. You can crash into the wall. There are companies, you can count them on both hands every day, that go out of business, and that's life.

Dynamic, growing industrial regions are constantly upgrading their capabilities and resources, and commercializing innovations. A cluster's boundaries may be defined by the totality of the industry participants and may reach across political boundaries. Regional clusters can be a virtuous circle leading to better opportunities, more venture capital, increasingly educated talent, and more

TABLE 13.5 Characteristics for innovation in a cluster setting.

■ High-quality human resources	■ Acceptance of globalization
■ Research in local universities	■ Successful role models
■ Availability of investment capital	■ Moderate regulation and taxes
■ Representative customers	■ Suppliers and complementors
■ Rule of law	■ Competitors
■ Sufficient infrastructure	■ Consultants, attorneys, and accountants

success. The cluster of independent activities can engender dynamic flows of cost reductions and competitive advantage.

13.4 Facility Planning

Once the entrepreneur chooses the city or region of location, the next task is to find a suitable facility. A building must fit the needs of the organization and allow for expansion as the firm grows. For most new ventures, leasing space in an existing facility is the most economic choice. For some more established firms, owning rather than renting can pay off. However, most start-ups need to use their financial resources for innovation and marketing.

The next challenge is choosing a proper layout within the facility. A **layout** is the arrangement of the facility to provide a productive workplace. This can be accomplished by aligning the form of the space with its use or function.

Today's facilities have replaced the private office and laboratory with public spaces and open-plan areas without walls. Since innovation is, in part, a social or collaborative activity, the work spaces are laid out to host team activity. New ventures are best started in an open facility with few walls and doors in order to promote collaboration. Studies have shown that communication between people declines at a rate inversely proportional to the distance between them. For example, we are five times more likely to communicate with someone who sits six feet away as we are with someone who sits 60 feet away [Allen, 1984]. Thus, a firm needs to avoid separate facilities for as long as possible. The center of the facility might be a public area with a coffee bar and meeting tables. The goal is to design the facility for flexibility and collaboration.

13.5 Telecommuting and Teleconferencing

The idea that community no longer means location has become accepted by most people. By 2002, 55 million Americans maintained an office in their homes. **Telework** refers to all kinds of remote work from home, satellite offices, and on the road. Technology has made it practical for people to work at places other than a central office. It has also made it possible to work asynchronously—not at the same time. Asynchronous access through e-mail, voice mail, video conferencing, and wikis enables groups to communicate with one another at any time.

It is important to reinforce the fact that social capital is based on face-to-face communication. Trust requires personal, face-to-face experiences. Videoconferencing and audioconferencing are useful technologies for virtual meetings. Virtual meetings work best, however, for the discussion of focused topics such as budgets, schedules, and facts. Presence is necessary for soft functions such as negotiations, planning, and restructuring of plans [Hinds et al., 2002]. People's social lives take place in a physical world, and the virtual world may be best reserved for information and content transmission.

13.6 The Internet

The **Internet** is a worldwide network of computers linking businesses, organizations, and individuals. **E-commerce** involves digitally enabled commercial transactions between and among organizations and individuals. The Internet enables worldwide communication. Computer networks link people, organizations, and knowledge, and primarily support social networks [Wellman, 2001]. Networks of workers communicate, in part, using technological means such as the telephone, fax, and Internet. Many workers participate in many networks where teams collaborate on a given task.

The World Wide Web is a popular service on the Internet and provides access to information via a system of addresses, standards, and protocols. In many ways, browsers for the Web are the "killer apps" of the Internet, serving as a valuable communication and information channel. We use *Web* and *Internet* interchangeably to describe the Web application on the Internet. The Web is a communications infrastructure and information storage system. Because transaction costs can be much lower on the Internet than in traditional channels, companies shift some or all of their business and supplier functions onto the Web.

E-commerce grew using Internet technology starting in 1995. Between 1998 and 2000, venture capitalists invested $120 billion in about 12,500 Internet start-ups, often called "dot.coms." Successful dot.coms include Amazon.com, eBay, and Yahoo. Jeff Bezos founded Amazon in 1995 as an online bookstore, raising several million dollars for the venture. Amazon offers convenience, low prices, and a wide selection of books and other items. In May 1997, Amazon raised $50 million by selling its shares to the public for the first time. By 2000, it held a clear position as the world's largest online retailer. Another use of the Internet is the distribution of information in the form of a magazine or a newsletter. The *New York Times* has successfully offered an online version in addition to its print edition.

The Internet also facilitates three important functions: personalization, customization, and versioning [Luenberger, 2006]. **Personalization** is the provision of content specific to a user's preferences and interests. It uses software programs to find patterns in customer choices and to extrapolate from them. For example, Amazon provides personalized book and music recommendations. **Customization** is providing a product customized to a user's preferences. Navigenics offers health-risk profiles customized to the genetics of their users. **Versioning** is the creation of multiple versions of a product and selling these modified versions to different market segments at different prices. The *New York Times* offers a free online version of today's newspaper but charges per article for archived material.

While personalization and customization are good ideas, there are some concerns. Many users report that they do not notice any differences in the site due to personalization. Often attempts to customize are cumbersome and imperfect. Customization is a more powerful approach if executed well since the customer is actively involved in the selection process. Customization is a powerful tool when customers want their preferences converted directly into a specific form of product.

A central activity of the Internet is the search process that enables users to find information on a seemingly infinite set of items, ideas, terms, and issues. A buyer of airline tickets can use Expedia, Travelocity, and Orbitz to search for a bargain. Travel services appear to be an ideal service/product for the Internet since travel is an information-intensive product requiring significant consumer research. Table 13.6 lists 15 exemplary websites.

Not all customers want to do business online; most prefer having a choice of various ways. The **hybrid model,** sometimes called "bricks and clicks" or "clicks and mortar," utilizes the best of the Internet as well as other channels. A hybrid model can extend a company's reach to new market segments as well as its global reach. Alliances are set up to combine the functions served by each company, such as the partnership between Drugstore.com and Rite Aid.

The advantages of e-commerce are low transaction costs, ubiquity, wide reach, and massive information. Since product and price information are readily available on the Web, the pricing power of many industries has diminished. Furthermore, many early e-commerce ventures underpriced their products in order to secure customers but never showed a profit and eventually failed. A firm must have a competitive advantage to sustain itself, and very low prices may not permit profitability. Because of the wide reach of the Internet, many

TABLE 13.6 Exemplary websites.

Website	Address (www.)	Primary offering
Alibaba	alibaba.com	E-commerce
Amazon	amazon.com	Books, videos, CDs
BBC	bbc.co.uk	News
Dell Computer	dell.com	Computers
eBay	ebay.com	Auctions
Facebook	facebook.com	Social networking
Flickr	flickr.com	Pictures
Google	google.com	Search
Kiva	kiva.org	Microfinance
National Geographic	nationalgeographic.com	Magazine
New York Times	nytimes.com	News
TED	ted.org	Videos
Yahoo	yahoo.com	Search and information
Yelp	yelp.com	Reviews

competitors can imitate successful offerings, thus eroding the competitive advantage of any one firm.

Internet ventures have the potential to offer extensive selection of their wares. For niche markets, Internet providers can offer an advantage over physical stores. Amazon and the iTunes Music Store exploit this advantage [Economist, 2005c]. Apple's iTunes has become the largest music retailer in the world.

Jeff Bezos realized that Amazon had a unique competency that could provide value to other, especially brick-and-mortar, companies. Amazon had a huge online infrastructure and significant experience with direct-to-customer shipping. With these advantages in Internet retailing, it began to offer its services to other companies. A full spectrum of retail services were rolled out for individuals, small and medium businesses, and even large corporations. Amazon Commerce is Amazon's partnership program with major traditional retailers such as Target and Sears. Through this program, Amazon creates a branded website, handles online sales, and provides customer support and order fulfillment. This arrangement yields additional revenue for Amazon and gives traditional retailers access to a critical distribution channel—the Internet.

All start-up companies should create a website. An early Internet presence can help build credibility and provide the look of an established firm. Most customers will visit a website to learn about the firm and its products. The website should clearly describe what the business does, explain the products and services, and provide complete contact information. Many new firms will also use their computer network to link with their suppliers and customers.

The use of the Internet is widespread. It is estimated that almost 70 percent of the North American population used the Internet in 2006. Similar growth of the use of Internet is occurring worldwide. The Internet is ubiquitous, cheap, and standardized, and it accommodates data, voice, video, and e-mail. It readily allows individuals to search, collaborate, coordinate, and transact. Thus, a new business venture can use the Internet to synchronize its activities through different channels, different stages of the value experience, and across different offerings [Sawhney and Zabin, 2001]. Perhaps the most revolutionary aspect of the Internet is that it gives virtually everybody access to the same information. It is this transparency that has caused the shift in power from sellers to buyers.

13.7 Vertical Integration and Outsourcing

New ventures normally have limited financial resources and are unable to internally provide all the functions required for operation of all activities. One way to identify resources and activities that have the potential for creating competitive advantages for a firm is to consider the value chain. The **value chain** of a firm is a sequence of business activities for transforming inputs into outputs that customers value, as depicted in Figure 13.2. The issue for the new venture is to decide which of the activities on the value chain will be accomplished by the firm and which activities will be provided by other firms (outsourced). A new or emerging firm will necessarily focus on a few of the activities on the value chain and outsource the others. **Vertical integration** is the extent to which a firm owns or controls all the value chain activities of a business.

In this chapter, we discuss the value chain and the decision to outsource or retain an activity within the firm. In Chapter 14, we return to the value chain and discuss how to manage and operate it to attain a competitive advantage.

The decision by a new venture to choose to carry out a value chain activity can be based on four questions, shown in Table 13.7 [Barney, 2002]. The decision to focus on a few value chain activities can be aided by an analysis of the four issues: (1) value, (2) rarity, (3) ability to be imitated, and (4) mission and organization. In many cases, it may be necessary to extend the activities of a firm to control a specific activity on the chain, especially when the firm affirmatively answers all four questions of Table 13.7.

Many companies are operating in a hypercompetitive global market where there is overcapacity in most industries. In such an environment, they are being called upon to achieve profitability by relentless cost cutting. This often entails heavy **outsourcing** to lower-cost labor and moving business abroad.

If a firm decides to outsource an activity, it plans on gaining a cost advantage or access to the supplier's superior competency or economy of scale. Access to a supplier's superior competency and cost advantage may be favorable. For example, a new venture cannot offer its employees a superior, low-cost, internally operated cafeteria and food service. When it is cheaper and easier to conduct an activity internally, then new ventures may consider taking it on. The best reason, however, to carry out an activity on the value chain is that it is strategically critical to a firm's success. Typically, product design and marketing cannot be outsourced since they are critical to the success of most new technology ventures.

A new venture in the personal computer business may choose to control product design and marketing while outsourcing the other functions after answering the four questions of Table 13.7. On the other hand, a packaged food

Concept

↓

Technology development

↓

Product design

↓

Logistics and manufacturing

↓

Marketing

↓

Sales and distribution

↓

Service

↓

Customer

FIGURE 13.2
Value chain from concept to customer.

TABLE 13.7 Questions for selecting value chain activities that will be carried out by the new venture firm.

1.	**Value**	Is the activity a primary source of product value for the firm?
2.	**Rarity**	Does the activity include a resource or capability controlled by the firm and rarely available to competing firms?
3.	**Ability to be imitated**	Do competing firms have a cost disadvantage when imitating the scarce resource or capability held by the new venture firm?
4.	**Organizational mission**	Is the activity critical to the mission of the firm, and is the firm organized to exploit this valuable, rare, and costly-to-imitate resource or capability?

business would probably attempt to control all product design, manufacturing, and marketing functions while relying on other firms to provide the technology development, distribution, and service activities [Aaker, 2001].

As companies outsource more functions, the scope for competitive differentiation narrows. Almost all routine activities are of low value, not rare, easily imitated, and not central to the mission of a firm. Thus, most new ventures outsource routine services such as payroll, accounting, and other administrative services. The transaction costs of managing outsourced services have recently declined, thanks to cheaper communication and the standardization of Web-based tools. Managing a relationship with a strategically important outsourcing agent is far more complex than coping with an ordinary supplier. In many cases, a partnership arrangement with a critical outsourcing agent is required. Reasons for failure of outsourcing include poor contracts, poor control of the function, and not planning for a termination strategy [Barthelemy, 2003]. Failures can be avoided or managed if accountability for managing the outsourced functions is clearly assigned to one or two persons in the new venture.

Transaction costs with suppliers and customers can be the most important types of costs. Transaction activities are time-consuming and prone to errors. Thus, companies use technology to automate their purchasing and contracting transactions with their suppliers and providers. Discount broker Charles Schwab & Company entered the securities brokerage market by offering a transaction cost advantage for both its customers and suppliers. Another firm that offers lower transaction costs to its customers and suppliers is FedEx [Spulker, 2004].

Although a new venture should consider outsourcing many of its activities, outsourcing can lead to problems. If an activity is outsourced and the supplier fails to deliver the required result (or activity) on time and with the required performance, the new venture can experience great difficulties. The conditions that favor the internalization of the activity within the new venture are summarized in Table 13.8. If the demand forecast for an activity or

TABLE 13.8 Conditions that favor an activity operating internally within the new venture.

Factor	Internal activity favored due to:
Costs	Total cost to produce internally is lower
Demand forecast	High uncertainty
Number of suppliers	Few powerful suppliers
Proprietary, nonpatented technology	Need to keep secret
Value-added function	Source of the firm's future sustainable competitive advantage

component is highly uncertain, it may be better to do the task internally. If there are only a few, powerful suppliers of a service or component, the danger is that they might use their power and not meet the required needs at the agreed-upon cost and time. If a firm's technology is valuable and proprietary, it may decide to keep it internal for reasons of secrecy.

Sometimes firms are led to outsource those value-added functions in which most of their profits will be made in the future. In the struggle for a sustainable competitive advantage, it is important to retain these value-added functions that will be critical to the firm's advantage in the future [Brown, 2005].

Industries with modular products and standards established for interfaces and interconnections support the use of modular architecture. As a result, vertical integration is less necessary. The personal computer industry is an industry where outsourcing of many activities takes place. A **virtual organization** manages a set of partners and suppliers linked by the Internet, fax, and telephone to provide a source or product. In this case, the value provided by the company is primarily the networking of the participating partners and outsourcing agents. This value, however, is often not rare and can be readily imitated.

Henry Ford built a vertically integrated facility at River Rouge near Detroit in 1917. The self-reliant plant made steel for auto bodies as well as all parts from engines to windshields. From Ford's own forests came wood for the paneling. For Ford, integration meant control of all the activities. A new firm cannot afford this type of vertical integration. It is too expensive and too risky. It is risky because with vertical integration comes control and commitment to a large investment in one way of doing things. Loss of flexibility is risky in a continuously changing economy. Today, Ford Motor Company is responsible only for the design, assembly, and marketing of its vehicles. All the modules and parts are provided by an array of suppliers and partners. In the near future, automobile companies may do only the core tasks of designing, engineering, and marketing vehicles. Everything else, including final assembly, may be done by the parts suppliers.

Salesforce.com's Outsourcing Model

Salesforce.com uses the "software-as-a-service" model as an application service provider (ASP). The firm offers software for "rent" that is delivered to customers online through a Web browser [Clark, 2003b]. Salesforce.com rents software online that companies use to manage their salespeople and lets other software developers rent out their software using its computers. This is a novel idea. One obstacle to this service is that most corporations do not want sensitive information stored on any computer other than their own. Should a start-up firm use this outsourcing service?

It seemed like one of those great business ideas: Do your grocery shopping online. Webvan was founded in 1999 to offer an extensive line of grocery and nongrocery items for selection online with delivery to the customer's door. Following several rounds of venture capital investments and an initial public offering, same-day delivery was offered in several cities. To provide this service, Webvan found it necessary to build distribution centers. With margins as low as 2 percent and expensive packaging and delivery functions, it struggled to make a profit. Webvan hoped to minimize costs by setting up a string of futuristic, $35 million warehouses with motorized carousels and robotic product-pulling machines. This was to help offset the enormous cost of its delivery fleets. After burning its way through more than $1.2 billion, Webvan closed its doors in July 2001. Webvan failed because it had a flawed business model that required a large investment in expensive distribution and service activities, as shown on the value chain of Figure 13.2.

Tesco of Britain offers an online grocery service that is profitable. Tesco used a new channel, the Internet, to reach its existing customers as well as new ones. Tesco provided online ordering combined with customers picking up their selected and boxed groceries at a Tesco store or paying a delivery charge. Tesco used existing stores, while Webvan built new warehouses and extensive delivery services. This illustrates why efficient operations along the value chain are critical to profitability.

13.8 Innovation and Virtual Organizations

A virtual organization manages a set of partners and suppliers in order to provide a product. Innovation and creativity can flourish with teams forming to build new ventures. These firms form, merge, and change form as required by the demands of a fast-changing industry and exhibit a high diversity of organizational forms and high rates of turnover. Ideas flow with people and recombine for new opportunities. The Hollywood movie industry demonstrates these flexible characteristics. Hollywood firms form and disband nearly continuously, causing a continual reshuffling of the human participants. The vitality of the organizational community appears to reflect its flexibility.

Movies are basically project-based enterprises that rent all their resources. Filmmakers develop their core competency in the identification and recruitment of talented people and the management of a complex project, a movie, through the steps of a value chain. Temporary organizations, such as film projects, capitalize on the specialized skills of their members while controlling the costs of coordination. Coordination of a film project is based on continuous, public, redundant communication. Additionally, as crew members carry out their roles on a project, they are strongly socialized via a culture of direct feedback, excessive gratitude, and role-directed humor, which further reinforces and makes clear role expectation [Bechky, 2002]. An excellent example of a theater company operating as a coordinated project is shown in the film *Shakespeare in Love*.

Amazon.com now operates the online stores and fulfillment activities of the online operations of Target, Toys-R-Us, Circuit City, and Borders. Amazon has become an outsourcing agent of Web services, logistics, and customer service for brick-and-mortar giants. Global Sports (www.globalsports.com) is an outsourcing company that operates e-commerce businesses for 25 sports retailers, including Sports Authority and the Athlete's Foot. Global Sports owns the merchandise and manages inventory, fulfillment, and customer service. Each retailer, in turn, receives a single-digit percentage cut of any revenues that come from its site. To consumers, it just looks like they're buying from their local retailer. Global Sports relieves the retailers of the burden of building and maintaining costly e-commerce infrastructure. Moreover, because Global Sports bears all the costs, it allows a client's e-commerce operations to be profitable from day one. Global Sports, in turn, makes money because of the scale of its business. Moreover, Global Sports spends nothing on advertising and marketing. The retailers take care of that activity.

Companies using outsourcing and networks can pull together resources to address specific projects and objectives without having to build permanent organizations. Virtual organizations use computers and networks to build an integrated system. Software applications can be "rented" from an application service provider (ASP), for example.

24/7 Customer: An Example of Outsourcing

As a U.S. startup, 24/7 Customer received $22 million in its first round of venture capital from Sequoia Capital in 2003. The firm initially catered to U.S. corporations that wanted to outsource backoffice operations to India. It now provides customer service and technical support via the Internet from India and other countries. 24/7 Customer is now profitable with over 8,000 employees. It attracts solid, middle-class college graduates to its operations [Vogelstein, 2003]. (See www.247customer.com.)

New innovative firms can access new ideas from an "idea" marketplace through licensing, alliances, and renting (subscriptions). Flexible, dynamic firms know that the best ideas are not always within their own boundaries. Importing new ideas is a good way to multiply the building blocks of innovation [Rigby and Zook, 2002]. Furthermore, more companies are willing to outsource their ideas and technologies to serve this market.

The creation of virtual organizations brings several challenges with it. Virtual firms encounter difficulty in building trust, coordination, and cohesion among the partner firms and outsource suppliers. New ventures need to invest time and resources in the task of keeping the coordination, trust, and synergy active in a virtual firm [Kirkman et al., 2002].

13.9 Acquiring Technology and Knowledge

For many firms, the effective use and management of the outsourcing function can be a competitive advantage. An open-architecture value chain can be a powerful business model [Moore, 2000]. The prudent use of other firms that provide significant contributions to a given need can be productive. For example, an electronic system might be built with an Intel microprocessor, Micron memory, and EMC storage, assembled by Solectron, and distributed by Ingram.

The new venture needs to identify which tasks are *core* and which are *context*. A task is core when its outcome directly affects the firm's competitive advantage. Everything else is context [Gottfredson et al., 2005]. The core/context ratio is a direct measure of effectiveness at generating shareholder value.

The asset base that a firm seeks to leverage through entrepreneurship has shifted over the past decade. The key assets are no longer plants and physical assets but instead are technology, science, and knowledge [Hill et al., 2002]. Entrepreneurs strive to see where new products have become feasible due to the availability of new technologies. With the advent of the new technologies, such as genomics, entrepreneurs see the opportunity for important new medical drugs, for example.

New enterprises also need to import leverage and recombine knowledge bases. Imported knowledge bases can include licensed technologies, purchased technologies, and knowledgeable employees.

A new technology venture will have developed or acquired a basic new technology. This new technology will, it hopes, be the basis for developing and contributing to countless products for many industries. For example, a new venture may possess powerful competencies and knowledge in the science and technology of superconductors. This new technology venture would look for applications in the electric power industry and the electronics industry.

The success of the embodiment of any new basic technology depends on seven characteristics, as shown in Table 13.9 [Burgelman, 2002]. These categories or characteristics are useful because they apply in all industries. For example, a powerful new superconducting technology might provide superconductivity of metal at a low initial cost to semiconductor manufacturers and be easy to use in integrated circuits with very low operating costs, highly reliable operation and serviceability, and high compatibility with normal circuits. In this case, we would have a very powerful new technology for the electronics industry.

TABLE 13.9 Seven key characteristics of a new basic technology.

■ Functional performance: an evaluation of the performance of the basic function	■ Reliability: service needs and useful lifetime
■ Acquisition cost: initial total cost	■ Serviceability: time and cost to restore a failed device to service
■ Ease of use: use factors	
■ Operating cost: cost per unit of service provided	■ Compatibility: fit with other devices within the system

> ### Cisco's Acquisition Methodology
>
> One of the most active purchasers of technology is Cisco Systems. Between 1993 and 2009, Cisco acquired more than 140 high-tech companies. Many of these acquisitions were made in 1999 and 2000. During this period, entrepreneurs would talk of founding companies in hopes that they would be acquired by Cisco. Although the number of acquisitions that Cisco made dropped dramatically, purchasing nascent technology remained an important part of its growth strategy.
>
> To facilitate the often arduous task of integrating the acquired firm, Cisco developed and used a documented template that focused on integrating both the people and the technology. Cisco thought carefully about who it would acquire, often taking months to make the decision. The firm did not believe in hostile takeovers and usually acquired geographically proximate companies with "market congruent" visions. Cisco preferred companies that were old enough to have a first product but still young enough that they were not entrenched in their ways or enmeshed with a broad base of customers. Second, Cisco had a no-layoff policy and made a point of keeping the acquired firm's staff. In the late 1990s, Cisco's turnover rate for acquired employees was well below 5 percent, and senior executives were often folded into Cisco's senior ranks.
>
> On the technology side, the R & D and product organizations were integrated with Cisco's other products and immediately labeled with the Cisco brand. In addition, any nonstandard technology was eliminated from the acquired firm, and its employees were given immediate access to Cisco's own infrastructure and core applications. The result was that most of Cisco acquisitions were fully integrated within 60 to 100 days.

13.10 AgraQuest

AgraQuest was formed in January 1995, but it did not attract venture capital investors until March 1996. AgraQuest raised $50,000 from its three founders and $420,000 from friends and family. It used those funds to work to build credibility and access to formal venture capital.

AgraQuest chose to locate in Davis since it was the home of University of California, Davis, a world leader in agricultural biotechnology. The three founders and the three scientists that came over to AgraQuest from Entotech moved into a small (750 square feet) laboratory that had been left by a biotech firm that required more lab space. AgraQuest purchased used laboratory equipment for $30,000. Eventually, AgraQuest occupied the adjoining offices as they became vacant. After three years, it moved to a new facility of 13,000 square feet built for it.

Within nine months of its founding, AgraQuest developed its first product, Serenade, and tested it at a northern California vineyard of the E. & J. Gallo Winery. In this test, it proved to control bunch rot and powdery mildew. With the prototype product and proof of concept, venture investors paid more attention to the new venture, which had successfully traversed the credibility cycle of Figure 13.1.

AgraQuest, once funded, had little trouble recruiting talented scientists. In the first four years, it outsourced the design of the production process as well as the manufacturing itself. This proved to be expensive and complex, so AgraQuest hired two production technologists in 1999. Then, in 2000, it purchased a production plant in Mexico and staffed it with a manager and workers. It was convinced that control of the design of the production process and the operation of production were critical to its success. By late 2000, only payroll, accounting, and legal services were outsourced.

13.11 Summary

Successful entrepreneurs are good at locating and acquiring the resources they need to start and build their firm. They need capital, people, and intellectual and physical assets to launch and grow their business. They do this by building credibility and legitimacy with the sources of these scarce resources. Typically, they are good at telling persuasive stories about their vision and its potential. They use their skills of persuasion to acquire the required resources in a timely way.

Entrepreneurs also create a plan for outsourcing some functions while retaining critical functions, such as product design and marketing. They use the Internet to help communicate and manage their relationships with their partners and suppliers in a virtually integrated firm.

> **Principle 13**
> Effective new ventures use their persuasion skills, credibility, and location advantages to secure the required resources for their firm in order to build a well-coordinated mix of outsourced and internal functions.

Video Resources

Visit http://techventures.stanford.edu to view experts discussing content from this chapter.

Outsourcing	Jeff Hawkins	Palm
The Silicon Valley Ecosystem	Beth Seidenberg	KPCB

13.12 Exercises

13.1 In 2005, Google released Google Earth and the Google Map APIs. Google brought compelling mapping and visualization functions within reach of all Web developers. In 2006, Amazon released a grid storage Web service called S3 that stands for Simple Storage Service. Amazon positions the service as a highly scalable, reliable, and low-latency data storage infrastructure at very low costs. In 2007, Facebook opened up its social network to third-party widgets via Facebook developer APIs. What business models are these three companies pursuing with their products? How are these types of Web services impacting the resource acquisition strategies of new ventures? Have these strategies been successful for Google, Amazon, and Facebook?

13.2 Determine who is or was one of the most persuasive people you have known. Using Table 13.2, describe how this person exercised his or her sources of legitimacy.

13.3 Four of the most popular U.S. locations for technology firms are Boston, the San Francisco Bay area, Austin, and Seattle. Using Table 13.3, determine the most attractive location for an orthopedic medical devices start-up. Do the same for a clean tech start-up.

13.4 Identify a local start-up company (or select one from another part of the world). Does this company operate in a specific industrial cluster? What local resources or local advantages does the company leverage? Why is the company located where it is?

13.5 Research In Motion develops and manufactures wireless handheld devices and provides the Blackberry service. The firm is located in Waterloo, Ontario, Canada. Study and describe the location advantages and disadvantages of the firm. Is it located in a cluster setting?

13.6 Identify a firm that operates on a hybrid model of using the Internet as well as physical facilities and stores. Describe the advantages and disadvantages of the hybrid model for the firm.

13.7 There is a strong trend in many information technology fields of outsourcing to another country and continent. What are the primary motivations for this movement? How and why are start-ups participating in this trend? What are some of the risks for a new enterprise considering outsourcing?

VENTURE CHALLENGE

1. How do you plan to attract talent and resources?

2. What location have you selected? Describe your planned use of Internet technologies and commerce.

3. Describe what functions your venture will outsource.

CHAPTER 14

Management of Operations

Real intelligence is a creative use of knowledge, not merely an accumulation of facts.
D. Kenneth Winebrenner

How do new firms build a set of operational processes that serve to create, make, and provide the product to the customer?

Most businesses build a chain of activities that add value at each section of the chain. Each element of the value chain has a capability that provides value added to the product. A new venture manages its value chain to provide the ultimate product to the customer. The firm also moves, stores, and tracks parts and materials to its value-adding partners and strives to ensure timely, efficient production of the service or product. Information flow along the value chain enables the coordination of the distributed tasks. An effective enterprise manages for operational excellence by trying to develop and communicate measurements of efficiency and timeliness.

Another way to describe a set of interrelated tasks is as a network of activities. A value web (or network) can use an Internet-based system to communicate with all the network participants. With a common schedule and associated tasks, the venture can manage the value web to achieve on-time production. ∎

14.1 The Value Chain

As discussed in Chapter 13, the purpose of a firm is to provide products that customers value. A **value chain** is a series of activities for transforming inputs into outputs that customers value. Each value chain activity adds value, as shown in Figure 14.1. Information flows back from the customer and the sales and service activities so that the value chain can maximize value for the customer. More than merely things with features, products are increasingly viewed as things with features bundled with services. Products and services are grounded in activities and relationships in a value-creating system. Furthermore, each element of the value chain has certain capabilities that can be improved over time. Capability development along the chain and the design of the chain can lead to a powerful core competency for a new venture. Furthermore, customers participate in this value-creating process by communicating their preferences and priorities. Understanding the customer enables the producer to better match the customer's needs, as described in Table 14.1.

A highly integrated company provides most, if not all, of the functions along a value chain. This approach is most suitable when proprietary interdependent activities occur at each stage of the chain. As many industries mature, the functions along the chain become independent so that modular subproducts are available at each stage. At that point in time, the value chain breaks up, and a number of independent firms participate in the activity chain. For example, today's automakers are adopting modular architectures for their mainstream models. Rather than putting together individual components from diverse suppliers, they are procuring subsystems from fewer suppliers. The architecture within each subsystem (braking, steering, chassis) is becoming progressively more interdependent as these suppliers strive to meet the auto assembler's performance and cost demands [Christensen et al., 2001].

Every industry has its own rate of evolution that erodes its competitive advantage. In a fast-changing industry, a firm must have the ability to readily redesign its value chain to find new sources of competitive advantage [Fine, 2002]. In designing or redesigning a value chain, each stage of the value chain can be assigned an economic value-added measure (EVA), which accounts for knowledge assets and strategic assets. Strategic assets are those in which the firm has relative competitive advantage. Strategic assets may include logistics, manufacturing, and distribution assets. Knowledge assets will exist primarily in the research design, marketing, and service functions. A new firm should

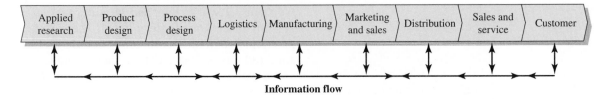

FIGURE 14.1 Value chain and information flow.

TABLE 14.1 Understanding the customer.

■ Preferences	■ Buyer behavior
■ Purchase criteria	■ Functional needs
■ Decision-making process	

retain the functions with high EVA and outsource functions with low EVA. If the industry is changing rapidly, however, a firm may decide to retain a key function to strategically respond to the change internally. Often a sound approach is to retain the high EVA activities and key strategic assets while outsourcing the low EVA activities.

Vertical integration along the value chain provides firms with an opportunity to choose the value-added stages in which it will compete. Intel is a manufacturer of both integrated circuits and circuit boards, but it also assembles personal computers under an original equipment manufacturer (OEM) agreement with PC companies.

Zara, a European clothing retail firm with 1,000 stores, has retained its manufacturing capability rather than outsourcing it so it can respond quickly to changing fashion demands. Other firms may be able to make the clothes more cheaply, but the strategic asset retained by Zara is the ability to quickly deliver new fashions to its stores. Information flows (see Figure 14.1) from the store floors back to the designers, who redesign the products to fit the customers' changing ideas and tastes. In fashion, nothing is as important as time to market [Helft, 2002]. Zara has new designs arriving in its stores every two weeks. Many new designs arrive at the stores within a few days.

Logistics is the organization of moving, storing, and tracking parts, materials, and equipment. Logistics can be the basis of competitive advantage since a firm with fast, accurate logistics will be first to respond to customers. Logistics systems usually are based on electronic networks, such as a supply chain intranet. Companies look for unique ways to service customers quickly through improved tracking, transporting, handling, and delivery in an effort to create unique competencies. Nokia, for example, uses a logistics management system to obtain all the parts for its mobile phones, build them, and get the phones out to the market where and when they are wanted.

Logistics might sound like a simple business of moving things around, but it is growing more complex as customers demand timely, customized services. New technology and greater use of the Internet opened up new ways of sharing information. Companies are also trying to build only after receiving orders from customers (known as built-to-order, or BTO), rather than estimating what will be in demand and supplying it from accumulated stocks. The BTO concept tries to avoid producing any item without a firm purchase order. Dell Computer is a leading example of a company based on BTO.

The information flow along the value chain can be facilitated by the Internet [Hammer, 2001]. Using this information and working closely with partners along the chain to design and manage the activities, efficiencies can be improved. Who captures the profits from these efficiencies? Often companies such as Wal-Mart and Home Depot have assumed the distribution, sales, and service functions, and captured significant value created by their suppliers. Furthermore, by using the Internet, companies like Amazon and eBay have become electronic sales and distribution channels. At the same time, companies like Nokia are producing a seamless manufacturing, distribution, sales, and service offering. Managing the value chain is a challenging task for all new business ventures, which can ill afford to assume many of the value chain activities.

Wal-Mart has a profit margin less than 4 percent, and many supermarket chains have a profit margin less than 2 percent. Clearly, every penny saved is a penny earned. The bar code became a mainstream success after Wal-Mart adopted it in 1980. Now Wal-Mart will require that its suppliers use radio-frequency-identification (RFID) on their supply pallets. RFID relies on a computer chip to hold and convey information. Wal-Mart manages its supply chain as if it were an orchestra.

Intermediaries in a value chain make sense only for exchanges in which the parties to the transaction can save more money by hiring the intermediary than it costs. It can cost 20 percent of revenues to "hire" a retailer to sell a product. Is it cheaper to sell direct? Can the Internet reach the same customers effectively?

Value chain speed is important to any new venture. With a long lead time from design to the customer, a new firm may wind up with inventory it cannot sell. Since many products have short life cycles, the chain must be able to move fast. Zara can design and manufacture a new clothing fashion in one week, if necessary. On the other hand, a movie studio can take a year to produce a new film and introduce it into the market. One of the best examples of value chain management is the Dell Computer system, which enables customers to specify their desired product and pay for it before Dell starts assembling the computer.

A123Systems: A Green-Tech Participant in Value Chain

A123System makes high-power lithium ion batteries based upon nanotechnology developed originally at MIT. As a leading clean technology provider, it seeks to provide a compelling combination of more power, longer life, and better safety than alternative suppliers. Applications for the batteries range from electric vehicles to power tools. Founded in 2001 with an initial public offering in 2009, the company now participates in many successful value chains, including General Electric, General Motors, Cessna aircraft, and ThinkGlobal cars (www.a123systems.com).

While goods and services flow largely down the chain, information flows in both directions. For example, information about what is demanded at successive stages passes up the chain, while information about supply conditions such as availability, pricing, time-to-manufacture, and so on passes down. Because this is an information-intensive process, the Internet holds the potential to significantly increase the amount of value created.

A new venture's lasting core competency is its ability to continuously assess industrial and technological dynamics, build value chains that exploit current opportunities, and select the high value-added activities for operation by the firm itself.

14.2 Processes and Operations Management

An **operation** is a series of actions, and **operations management** is the supervising, monitoring, and coordinating of the activities of a firm carried out along the value chain. Operations management deals with processes that produce goods and services. A **process** is any activity or set of activities that takes one or more inputs, transforms and adds value to them, and provides one or more outputs [Krajewski and Ritzman, 2002]. Processes are the series of operations, methods, actions, tasks, or functions leading to the creation of an end product or service. Processes are used to transfer value to customers in the form of products. A product could involve delivering a tangible good to customers, it could involve performing a service for customers, or it could be, and usually is, a combination of the two [Melnyk and Swink, 2002]. At a factory, a process transforms materials into products. At an insurance company, a process transforms client information into an insurance agreement. A network of processes helps create the value provided at each stage of the value chain. Business processes can add unique value to products and services.

The business processes of an organization should be aligned with its strategies and employee competencies, as shown in Figure 14.2. Alignment requires a continual rebalancing of strategy, business processes, and the competencies of the people to satisfy and retain the customers in order to keep the business theory as the clear driving force of the business, as shown in Figure 14.2. The business theory should be expressed as a clear statement of vision and purpose. Recall the vision statement of eBay: "We help people trade practically anything on Earth through an online system."

Ergonomics is making a physical task easier and less stressful to accomplish. Products should be designed ergonomically, and tools used in factories need to be ergonomic. For example, Herman Miller's Aeron chair has been acclaimed as an ergonomically pleasing product. It provides form-fitting support and maximum comfort due to a suspension system that allows for proper ventilation.

Processes bring value to customers and stakeholders. Business success comes, in large part, from a company's process performance. Therefore, a firm should strive for superior process design. An example of a simple business process is shown in Figure 14.3. Part of this process can be automated.

> **Intel's Operations Strategy**
>
> Gordon Moore and Robert Noyce founded Intel in 1968. Their first act was to recruit a director of operations. They offered the job to Andrew S. Grove, who became Intel's third employee. Though Grove had no manufacturing experience, they recognized his innate intelligence and drive. Grove was responsible for getting products designed on schedule and built within budget. The scope of this position extended into nearly every functional area at Intel, from marketing to sales to engineering. His influence, presence, and attitudes pervaded the company, and within three years of Intel's formation, it became clear that the majority of daily decision-making was passing to Grove. Operations has always played a critical role in Intel's success.

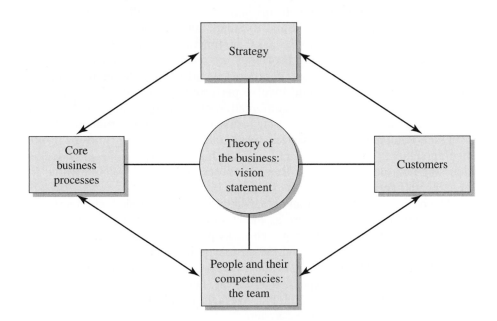

FIGURE 14.2 Business alignment.

Functions in operations management include design of processes, quality control, capacity, and operations infrastructure. The new venture needs to plan for its operations or production function, which will be led by one member of the entrepreneurial team. Both service and product firms need to design and control operational processes to achieve efficiencies, throughput, capacity availability, inventories, capital expenditures, and productivity. Unique operational management competencies can be part of the competitive advantage of a firm [Vonderembse and White, 2004].

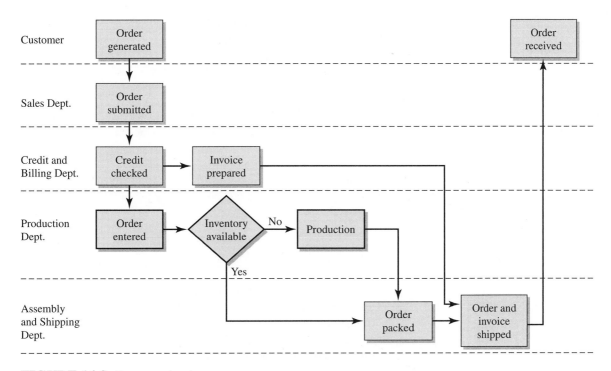

FIGURE 14.3 Common business process.

The best return on investment is often found in companies that combine operational excellence—consistently outstanding performance for customers that is brought to the bottom line—with sustained rapid growth. Operational excellence is a necessity today. Investors mercilessly punish companies that fail to meet these expectations [Lucier and Dyer, 2003]. Operational excellence can lead to lower costs, as shown in Figure 14.4. With economies of scale, unit costs will drop, and thus, prices can decline. As prices drop, the product is more attractive and sales increase. As financial resources increase, investments in marketing and operational processes will lead to better economies of scale. This can be a very powerful self-reinforcing loop.

A firm can implement four competitive capabilities: low cost, high quality, speed, and flexibility. **Quality** is a measure of a product that usually includes performance and reliability. Performance is the degree to which a product meets or exceeds certain operating characteristics. **Reliability** is a measure of how long a product performs before it fails. **On-time speed** measures the pace of lead time, on-time delivery, and product development. **Flexibility** is a measure of a firm's ability to react to a customer's needs quickly. All of these goals need to be achieved while operating with the additional goal of worker safety.

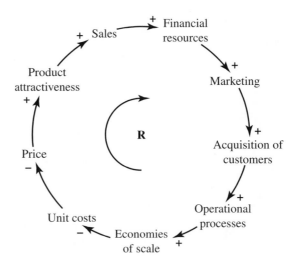

FIGURE 14.4 Self-reinforcing growth through acquisition of customers. A self-reinforcing loop, R, is assumed to work as long as the economies of scale are actually realized through effective operational processes.

Many firms have adopted a six-sigma quality goal, with the aim of getting rid of defects in a process. Six sigma is a statistical term that measures how much a process deviates from the ideal. Six-sigma quality equals just 3.4 flaws per million. The six-sigma method attempts to build low-defect products with low costs [El-Haik and Roy, 2005]. Six-sigma quality is the result of a well-defined and structured process that is highly repeatable—a process with well-defined tasks and milestones.

Consumers often react favorably or unfavorably to the experience they have with a product or service, such as the packaging, clarity of an operating manual, or ease of use. Many customers view consumption as an experience rather than a singular purchase event [LaSalle and Britton, 2003]. Most products or services include *objective value,* such as performance, and *subjective value,* such as experience. A candle purchased for light might cost only $1, while a second candle with shape, form, and scent may provide a richer experience (and sell for $5 to $10). Thus, for many products, the value of the product involves the experience of it and the associated purchase and fulfillment process. The designer needs to identify, design, and fulfill a customer experience that will register with the customer as positive.

Supply-chain management is focused on the synchronization of a firm's processes and those of its suppliers to match the flow of materials, sources, and information that meets customer demand. Today, the goal is to minimize the stock of goods or inventory required to support variability of customer demand. As products become more susceptible to changing demands, the risk grows that a given product line will have disappointing sales. But if a manufacturer decides to go lean on inventories, it risks running out of stock, losing sales, and endangering relationships with its customers.

Operations systems that are designed to create efficient processes by using a total systems perspective are called **lean systems.** Flexible or lean systems aim to reduce setup times and increase the utilization of key processes. Flexible systems can quickly respond to changes in demand, supply, or processes with little cost or time penalty. They often use a **just-in-time** (JIT) approach that focuses on reducing unnecessary inventory and removing non-value-added activities. This system uses a pull method in which the customer activates production. For example, an order for an auto chassis at a carmaker activates the manufacturing processes [Liker, 2004].

The Taguchi method is used to design and improve production systems. The Taguchi method is a technique for designing experiments that converge on a near-optimal solution for a robust system. The method uses the term *noise* to describe uncontrolled variations and states that a quality product should be robust to noise factors. The design of a firm's production system is critical to its overall success [Ulrich and Eppinger, 2004].

Companies strive for quality service at low cost. Google, the Web search firm, handled 170 million page views a day in 2003 with a hardware plant of 12,000 servers (computers) that cost about $2,000 apiece. When a server fails, Google pulls it and replaces it immediately. Google has no fix-it department; it just pulls and inserts. With this approach, Google saves funds and keeps its system up 99.9 percent of the time [Karlgaard, 2003].

Entrepreneurial firms seek to develop unique capabilities by fostering an interactive dynamic between their capabilities and market opportunities. The market provides them with signals of opportunities, and they respond with new products. Adjustment of business models, production capabilities, and skill formation enable the firm to respond to opportunity. In value networks, firms focus on their core competencies and use others for complementary capabilities.

The goal of operations **throughput efficiency** (TE) may be measured by the formula

$$TE = \frac{VA}{VA + NVA} \qquad (14.1)$$

where VA = value-adding time and NVA = non-value-adding time. Examples of NVA are waiting in a queue or system downtime. The goal is to reduce NVA.

Gentex, which makes rearview mirrors and other auto equipment, uses technology to control costs and increase throughput (see www.gentex.com). Gentex continually adds automation and monitoring systems in its factories and increased its throughput 30 percent from 2001 to 2003 [Green, 2003].

> **Improving Operations in the Operating Room**
>
> Intuitive Surgical, Inc., makes surgical robots used in prostate and heart bypass surgeries. When they are given the opportunity to test drive the da Vinci robot, surgeons quickly become enamored with the level of control they achieve, using only a small incision. Although Intuitive Surgical's robots have been enthusiastically embraced, the company continues to find ways to make the surgical experience better for doctors and patients. The small incisions improve overall efficiency in the healthcare system, since they lead to more rapid recovery times for patients. Intuitive is looking at ways to provide ultrasound and other diagnostic images on the same screen that the surgeon uses to control the robot. This would improve efficiency by letting the surgeon focus on only one image, and improve the real-time surgical experience by making sure that all relevant information is only a click away (www.intuitivesurgical.com).

14.3 The Value Web

A series of business activities can be thought of as a business process, as shown in Figure 14.5. Another way to describe a process is as a set of interrelated tasks accomplished in a network of activities. Instead of a value chain—a linear series of processes—the value-creating process can be organized as a **value web.** Webs are grids with no center but allow open communication and movement of items and ideas. In a web, each participant focuses on a limited set of core competencies [Tapscott et al., 2000]. A value web is usually based on an Internet infrastructure to manage operations dispersed in many firms. The value web consists of the extended enterprise within a network of interrelated stakeholders that create, sustain, and enhance its value-creating capacity. The long-term success of a firm is determined by its ability to establish and maintain relationships within its entire network of stakeholders. It is relationships rather than transactions that are the ultimate sources of organizational wealth [Post, 2002]. The value web of a typical firm is shown in Figure 14.6.

The value web organized and operated by Amazon includes participants such as Ingram, Target, and Toys-R-Us. Amazon takes responsibility for choosing and offering the product selections, setting prices, and ensuring fulfillment. Cisco Systems leads a value web that provides its routers and computers to its customers. Cisco designs and markets the product, while others do most of the manufacturing, fulfillment, and on-site customer service. The Cisco Systems value web is shown in Figure 14.7. Recall from earlier chapters that CRM means customer relationship management. Cisco defines the goals and coordinates the integration of the value web providers. Many new ventures will use the Internet to coordinate their value web effectively.

Consider the operations management strategy of IKEA, the Swedish furniture company that has 296 stores in 36 countries. Its business strategy is to make and sell inexpensive, solid, well-designed furniture through large stores.

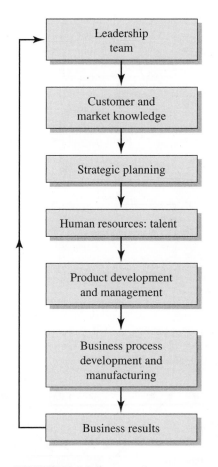

FIGURE 14.5 A business process is a series of activities.

Its business process starts with the identification of a needed product and the specification of a low target price for this item. Next, IKEA determines what materials will be used and what manufacturer will do the assembly work. IKEA buys from about 1,800 suppliers in 55 countries. The next step is to design the item and select the parts. After manufacture, the item is shipped disassembled in a flat cardboard box to one of IKEA's 18 distribution centers and ultimately to one of its stores. IKEA sells its unassembled furniture without salespeople. The customer selects the item, gets the correct box from its rack, and brings it to the checkout counter. IKEA implements a low-cost, quality strategy through a far-flung value network that provides design, parts, and manufacturing in a coordinated manner.

In the past as companies grew, they added assets. As companies grow today, they tend to add relationships and enhance their value web. Orchestrating a value

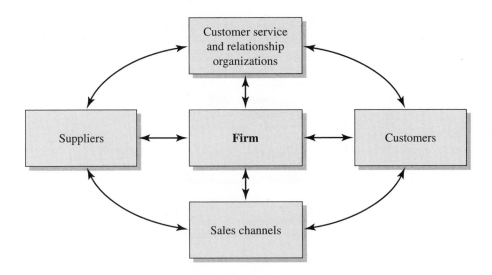

FIGURE 14.6 Value web for a firm.

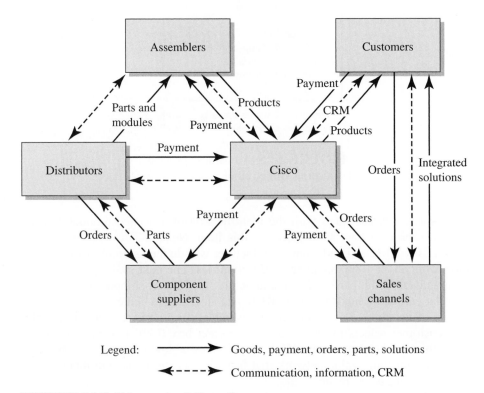

FIGURE 14.7 Value web of Cisco Systems. (Adapted from Slywotzky and Morrison, 2000.)

web is a powerful process for growth. Of course, it is possible to develop too complex a web of partners and lose control. Effective management of the value web enterprise requires a new conception of the firm as a network, rather than a hierarchy. The key to effective implementation is recognition of value web management as a core competence.

In a fast-paced, competitive world, competitive advantage can result from the effective concurrent design of products, processes, and capabilities. Designing the product, how it is produced, and a supply chain that works harmoniously is critical to a firm's success. The firm that controls the interdependent links in the value web captures the most profit [Lawrence et al., 2005]. The coordinated product, process, and supply chain system is depicted in Figure 14.8 [Fine et al., 2002].

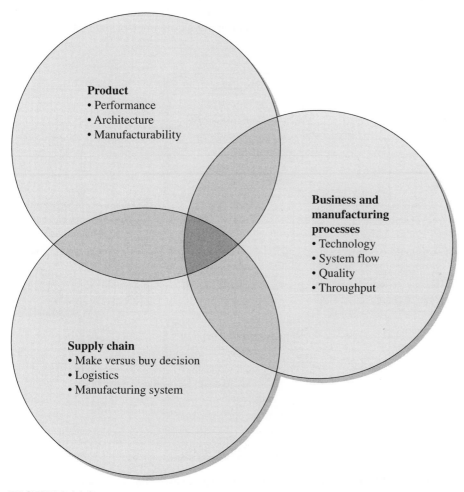

FIGURE 14.8 Coordinated system of product, processes, and supply chain.

14.4 Strategic Control and the Balanced Scorecard

Strategic control is the process used by firms to monitor their activities, evaluate the efficiency and performance of these activities, and take corrective action to improve performance, if necessary. The goal is to keep the firm's operations on track with the performance goals of efficiency, quality, and responsiveness to customers.

To evaluate the effectiveness of their strategies, some companies are developing **balanced scorecards,** a set of measurements unique to a company that includes both financial and operational metrics. This gives managers a quick yet comprehensive picture of the company's total performance. The balanced scorecard is a strategy formulation device as well as a report of performance. A successful balanced scorecard measures the tangible objectives that are consistent with meeting an organization's goals. The business operations area indicates how the operations and processes should work to add value to customers. The customer area indicates how the company's customer-oriented strategy and

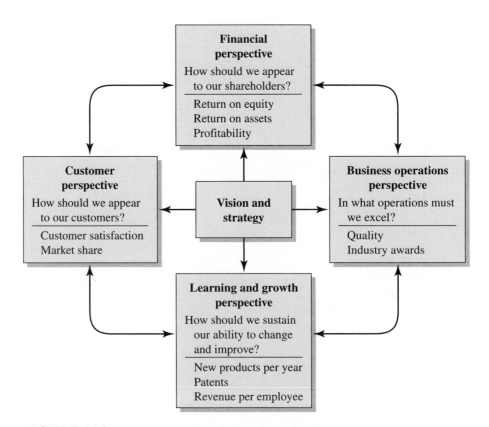

FIGURE 14.9 Balanced scorecard. Each perspective has a question and a set of measures. [Adapted from Kaplan and Norton, 2004.]

operations add financial value. The financial area measures the company's success in adding value to shareholders. The learning and growth area indicates how the infrastructure for innovation and long-term growth should contribute to strategic goals. A balanced scorecard is shown in Figure 14.9 [Kaplan and Norton, 2004].

To build an effective scorecard, a firm needs to determine the fundamental drivers of performance and measure them. Finding the right measures such as reliability, quality, or customer satisfaction is challenging.

General Electric's Digital Cockpits

General Electric Vice Chairman Gary Rogers created the idea for a digital dashboard—the continuously updated online display of a company's vital statistics. GE's "digital cockpits" now give 300 managers instant access to the company's essential data on desktop PCs and Blackberry PDAs [Tedeschi, 2003].

Jack Welch of General Electric created the idea of a boundaryless company, which eliminated the walls between suppliers, customers, and units of GE. GE's slogan was "Finding a Better Way Every Day," which addressed improvement of business processes. Then it added the idea of measuring performance of the processes. As a result, operating margins went from 1.2 percent in 1994 to 13.8 percent in 2000 [Welch, 2002]. Welch believed that the system of operations was the key to understanding, learning, and improving results.

14.5 Scheduling and Operations

New firms should develop diagrams and flowcharts that show how their operations work. Diagrams help communicate the process system to all concerned.

An operational plan outlines a number of actions that will be taken in the future. To consolidate the timing of events, the firm should prepare a schedule, in chart form, of all of the important milestones that the firm expects to reach in the near and intermediate term. A *Gantt chart* is a way to depict the sequence of tasks and the time required for each. Gantt charts, by using shaded bars on a grid, compare what was done with what was planned over time, as shown in Figure 14.10. Timelines are a visual means of comparing the actual and planned progress of a project or activity. Timelines allow participants to envision the ending of an otherwise open-ended plan [Yakura, 2002].

Effective management of operational processes requires schedules and coordination. Since Gantt charts are a means of contrasting the actual and planned progress of a plan over time, they depict both scheduling and coordination of separate tasks.

Enterprises of any size can profit from using Gantt charts to depict schedules and milestones. For example, completion of task B could represent the

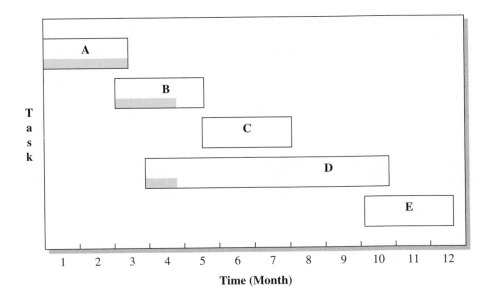

FIGURE 14.10 Gantt chart for five tasks. The actual progress is indicated by the shaded bars.

completion of a prototype, and task C could represent the testing of the pro- totype (see Figure 14.10). It is important to set milestones and then strive to reach them on time.

Another, perhaps more graphic, form of representing the activities, out- comes, and schedule is to use a milestone picture, as shown in Figure 14.11. This depicts a road map to achieve a scheduled outcome.

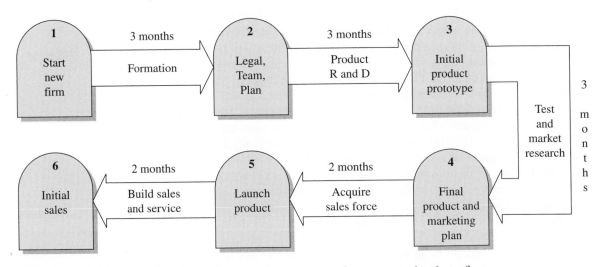

FIGURE 14.11 Example of a road map with milestones for a new technology firm.

Developing effective natural pest management products

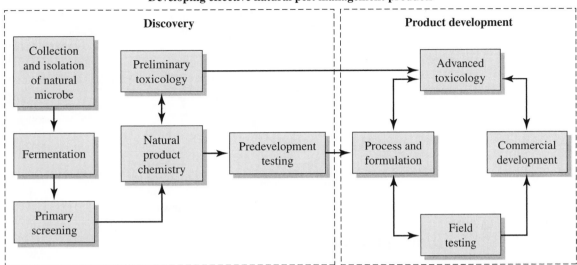

FIGURE 14.12 Product development process for AgraQuest.

14.6 AgraQuest

AgraQuest has two key operations: (1) product development and (2) product manufacturing. The product development process is shown in Figure 14.12. The two steps to develop a new product are discovery of the new natural microbe and development of a product based on that microbe. This process leads to microbial products to serve as pesticides or fungicides.

The design of the manufacturing processes takes place at the Davis, California, facility and is led by John Lin, director of process development.

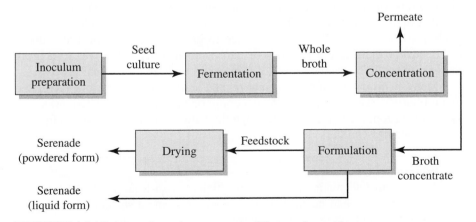

FIGURE 14.13 Manufacturing process of Serenade products.

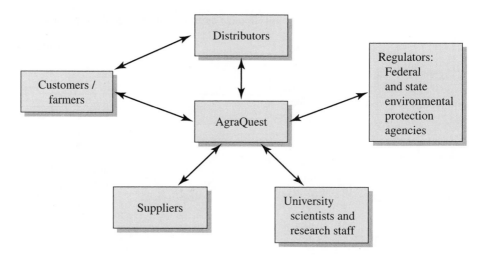

FIGURE 14.14 Value web for AgraQuest.

AgraQuest operates its own plant in Mexico, where it produces and ships its products. AgraQuest purchased the plant in December 2000 for $7 million and had 16 employees at the plant in 2006. The 208,000-square-foot plant sits on 35 acres near Tlaxcala, Mexico. The manufacturing process is shown in Figure 14.13.

AgraQuest has a system of business relationships based on the value web shown in Figure 14.14.

14.7 Summary

A new venture needs to design a set of operational processes that will enable it to build, store, and ship the products provided to the customer. New businesses build a supply chain of partners that add value at each stage of the assembly or manufacture of the product. Service companies use business processes to put together their service outputs. The new venture manages its value chain to effectively provide the final product or service to its customer. The firm also needs to effectively manage the logistics of parts and materials. It strives to achieve the best possible coordination of its partners as well as its internal processes.

Many firms establish a set of interrelated activities as a network facilitated by an Internet-based value web. With a common schedule, associated tasks, and synchronization, the venture can manage the value web to maintain an efficient, on-time business process.

Principle 14

The design and management of an efficient, real-time set of production, logistical, and business processes can become a sustainable competitive advantage for a new enterprise.

Video Resources

Visit http://techventures.stanford.edu to view experts discussing content from this chapter.

Operational Excellence Zone	Geoffrey Moore	MDV
The Scaling of Vision	Sheryl Sandberg	Facebook

14.8 Exercises

14.1 Select a start-up of your choice. Using the format of Figure 14.6, prepare a value web diagram for this company.

14.2 Flextronics manufactures products for many electronic companies (www.flextronics.com). It is a global company, headquartered in Singapore, and has over 120,000 employees. The company's core competence is lean manufacturing. Where does Flextronix fit in the value web for Microsoft, Ericsson, and Dell?

14.3 China is a world-class manufacturer and has advantages of size, scale and cost. Identify a Chinese assembler of electronic products for European companies. What advantages does China have versus manufacturing in Europe?

14.4 Prepare a road map diagram for the development and launch of a new model of a hybrid or all-electric automobile to compete with traditional automobiles. What processes are the same? Different?

14.5 Wal-Mart has invested heavily in RFID technology to improve the efficiencies of its warehouse operations and inventory control. Outline Wal-Mart's value web. How has Wal-Mart's adoption of RFID influenced its supply partners? What new opportunities have been created with RFID? What have been the results for Wal-Mart from this push?

14.6 The use of a new technology can bring new life to a mature industry such as the plastics industry. Logistics, supply-chain, and scheduling software enable large productivity increases in several mature manufacturing industries. Examples of such software firms are i2, Moldflow, Quad, and Keane. Select one of these firms and describe an actual application for operations productivity improvement.

14.7 About 5 to 10 percent of pharmaceuticals produced do not meet specifications and have to be reworked or discarded. Quality testing is done by hand, and the batch process method is widely used. A new venture has been launched to design new processes for drug makers. What new methods and approaches should it develop to sell to drug makers?

14.8 Describe and contrast the operational challenges faced by the following start-ups: (a) consumer Web services start-up, (b) iPhone application company, and (c) electronic device company.

14.9 Describe and contrast the operational challenges faced by the following start-ups: (a) new drug discovery company, (b) medical device company, and (c) biofuel company.

VENTURE CHALLENGE

1. Describe the key business processes used by your venture.

2. Draw a road map with milestones for your venture as illustrated in Figure 14.11.

3. Draft a value web for your venture as in Figures 14.5–14.7.

CHAPTER **15**

Acquisitions and Global Expansion

Opportunity is rare, and a wise person will never let it go by
Bayard Taylor

How can entrepreneurs best manage expansion via acquisitions and entry into new geographic markets?

Entrepreneurs often create a new business by acquiring an existing firm and then improving it. The acquirers attempt to create growth and new value for the firm. Another strategy is for entrepreneurs to start and build their own firm and then expand the company by acquiring other firms. A series of successful acquisitions can help build a firm into a powerful leader in an industry. The integration of the newly acquired firm within the existing firm is a large challenge, however, especially when the cultures of the two firms differ significantly.

Most new firms develop, at the appropriate time, a plan for building an international strategy for growth. The forces for globalization are powerful, and new business ventures need to plan for them. ∎

15.1 Acquisitions and the Quest for Synergy

An entrepreneur can enter a new business by acquiring an existing firm. An **acquisition** is when one firm purchases another. Usually the acquired company gives up its independence, and the surviving firm assumes all assets and liabilities. Purchasing an established business has the advantages of ongoing businesses: location, employees, equipment, and products. Obvious potential disadvantages include poor location, depleted assets, obsolete inventory, depreciated brand, and lack of profitability. Buying a business can be less risky than starting new because the business's operating history provides entrepreneurs with valuable data on the chances of its success. Finding and evaluating an acquisition candidate, however, can be time-consuming. After finding a good acquisition, the acquirer needs to arrange financing and negotiate the terms of the deal.

Often acquisitions end up eroding the value of the acquired company due to difficulties with the transition to new ownership and overestimation of the value of the acquired firm. When a transition to the acquirer is attempted, difficulties may occur in working with or changing the established culture of the acquired firm. The three main steps for acquiring a company are (1) target identification and screening, (2) bidding strategy, and (3) integration or transition to the acquirer.

Most, if not all, acquisitions are justified on the basis of an expected **synergy,** which is the increased effectiveness and achievement produced as a result of the combined action of the united firms. Suppose that you identify a firm that you determine is worth V and that a bid for its acquisition is accepted at a price of V. Then you estimate that after adding the value created by you and the entrepreneurial team, the value of the newly revitalized firm will be V_N. We then expect a synergy (Syn) as defined as:

$$Syn = V_N - V$$

The synergy is the expected value added by the acquiring party. The source of the synergy may be revenue enhancements and cost reductions due to capabilities and resources introduced into the firm by the new entrepreneur team. When new entrepreneurs acquire an existing business, the synergy is the added value of the entrepreneurial team that often replaces the management team of the acquired firm. The new team strives to add value to the acquired team's product that will be rarely available to competitors and is difficult to imitate.

Acquirers try to find firms with valuable and scarce product innovations that can be enhanced by the capabilities of the acquirer's management team. Nevertheless, it is a good rule to avoid bidding contests and to close the deal in a timely way.

eBay and PayPal

eBay and PayPal announced their combination in July 2002, promising multiple synergies. The press release added:

> In a move that will help millions of Internet users buy and sell online, eBay, Inc., the world's online marketplace, today announced

that it has agreed to acquire PayPal, Inc., the global payments platform. A natural extension of eBay's trading platform, the acquisition supports the company's mission to create an efficient global online marketplace. Payment is a vital function in trading on eBay, and integrating PayPal's functionality into the eBay platform will fundamentally strengthen the user experience and allow buyers and sellers to trade with greater ease, speed, and safety.

Source: eBay Press Release, July 8, 2002.

We will consider three common methods of valuation of a firm used by acquirers: (1) book value, (2) price-to-sales ratio, and (3) price-to-earnings ratio. The **book value** is the net worth (equity) of the firm, which is the total assets minus intangible assets (patents, goodwill) and liabilities. The price-to-sales and price-to-earnings ratios are obtained for comparative firms in a specific industry.

Consider a firm that designs and makes orthopedic devices for injured and disabled people. An accounting consultant determines that the net worth of the firm is $800,000. Annual revenues have remained at $1.2 million over the past two years. The firm has several patented products, but it has not fully exploited its marketing opportunities. Therefore, the net worth or book value sets a base value of $800,000 for the firm. With no growth in revenues, the accountant suggests a purchase value of one-half of sales, or $600,000. Earnings have held steady at $100,000 per year for the past several years. Assuming a comparable price-to-earnings ratio of nine for a zero-growth firm, the valuation could be nine times earnings, or $900,000. Assuming these three valuation methods ($800,000, $600,000, and $900,000), the buyer chooses a target price—say, $700,000—and tries to determine a suitable deal structure. One starting arrangement could be $200,000 in cash with the remaining $500,000 as a loan from the seller of the firm set at the prime rate for four years. Ultimately, the valuation and the final deal are a result of negotiation between buyer and seller.

Earl Bakken and Palmer Hermundslie founded Medtronic as a medical equipment repair company in Minnesota in 1949. The company sustained itself early on by selling and servicing medical equipment and building many of its own custom devices as well. In 1957, Medtronic developed the first wearable external cardiac pacemaker. Three years later, Medtronic purchased the exclusive rights to the first implantable pacemaker. The company grew to be a leading manufacturer of heart pacing technology worldwide. Sales outside the United States were strong, but competition was fierce. In response, Medtronic began establishing international facilities, and it gained direct control of its international operations in 1968 by purchasing the firm that had been its sales agent in Canada. Medtronic then began to acquire the firms that had been its major distributors in the United States, thus building a direct sales force to market products around the world.

In the 1980s, Medtronic purchased Johnson & Johnson's Cardiovascular Division. In addition, Medtronic acquired nearly a dozen other medical technology companies, enabling it to enter new markets. Acquisitions included a manufacturer of coronary angioplasty catheters and guiding catheters, a producer of centrifugal blood pumps, and a Dutch pacemaker manufacturer. By 1990, through a combination of internal developments and strategic acquisitions, Medtronic had successfully made the transition from a company with a limited product line to an international, diversified, medical technology corporation. By continuing to acquire market leaders with strategic mergers in the 1990s and 2000s, Medtronic has maintained its leadership with the medical technology industry.

Before entering into an acquisition, a new venture should consider the technology and customer uncertainty. If the uncertainty is high, it may be appropriate to consider an alliance, which will normally cost less and limit the firm's financial exposure. If the alliance starts showing results, then a move to acquisition may be appropriate [Dyer et al., 2004].

15.2 Acquisitions as a Growth Strategy

Acquisitions and mergers can serve as a growth strategy in fragmented industries. A **merger** refers to the fusing together of two companies. An acquisition is when one company buys another. The difference between a merger and an acquisition is the degree of control by one of the two firms; a merger may result in 50-50 control. Mergers involve a much higher degree of cooperation and integration between the partners than do acquisitions. Most of the time, mergers occur between relatively equal-sized organizations, while one organization tends to be larger and more established in acquisitions. Many mergers suffer from insufficient integration of the functions and activities of the two firms. An example of poor integration of two merged firms is AOL Time Warner.

In fragmented industries, numerous small companies are differentiated as specialists and compete for market share. Powerful forces are driving industries to consolidate into oligopolies. An **oligopoly** is an industry characterized by just a few sellers. The incentives to consolidate are significant in the technology, media, and telecommunication industries, where fixed costs are large and the cost of serving each additional customer is small.

An oligopoly, a market in which a few sellers offer similar products, is not always undesirable. It can produce efficiencies that allow firms to offer consumers better products at lower prices and lead to industrywide standards that create stability for consumers. But an oligopoly can allow some businesses to make big profits at the expense of consumers and economic progress. It can destroy the vital competition that prevents firms from pushing prices well above costs. Many industries also face large fixed costs. A typical semiconductor fabrication plant now costs between $2 billion and $3 billion, compared with $1 billion five years ago. A maker of basic memory chips must sell far more integrated circuits (chips) to justify an investment of that size. This is why makers of memory chips are eager to merge.

Industries tend to become more efficient as they undergo consolidation [Sheth and Sisodia, 2002]. In a fragmented market, the consolidator firm within the industry can realize the synergy of the economics of scale. A new venture in a mature industry ripe for consolidation can offer good opportunity. The new entrant can concentrate the resources and use them effectively in one niche and then acquire small competitor firms as resources to do so become available [Santos and Eisenhardt, 2009, 2005]. An example of a fragmented industry is the Internet service provider (ISP) sector. Every town has many independent ISPs as well as large competitors, such as Earthlink and AT&T. As these large firms consolidate their strength, the small competitors are fading.

A merger can stimulate growth if the new conjoined firm has a sound business plan for the near-term future that includes a few key measures of profitability. The merged firms should work to redeploy unproductive assets and focus on optimizing their joint activities. Acquisitions are one form of corporate entrepreneurship that can be particularly useful as an established company tries to innovate and infuse the organization with more entrepreneurial behavior or new product lines.

Most studies show that about two-thirds of mergers do not pay off with any synergy or gains. To realize the full value of a merger, the merged organizations must be appropriately integrated. A **horizontal merger** is a merger between firms that make and sell similar products in a similar market. The merger between Exxon and Mobil is an example of a horizontal merger. A **vertical merger** is the merger of two firms at different places on the value chain. The union of AOL and Time Warner is an example of a merger that has features of both horizontal and vertical mergers. By merging, AOL enhanced its potential delivery of Internet content, since it could then offer customers some of Time Warner's variety of entertaining and informative products. In the same way, Time Warner found a partner that could deliver its content to a large existing audience. The idea was that AOL Time Warner would control both the content—music and movies—and the distribution of that content through cable television as well as the Internet. It proved difficult, however, to convince AOL users to buy Time Warner content.

In 1976, Jim McCann got an urge to own his own business. He bought a flower shop in Manhattan for $10,000 and brought in a day-to-day manager. He then opened 12 new stores over the next decade. He was doing well and left his regular job to build up his floral business. He purchased the troubled 1-800-Flowers in 1987 for $2 million and assumed $7 million in debt. He then moved the firm to New York and merged it with his flower store chain. McCann then made a good set of acquisitions, enabling the expansion of 1-800-Flowers. It moved beyond flowers to include gifts in the 1990s. Flowers in 1999 acquired Great Foods, a specialty food unit. In 2001, Flowers acquired Children's Group, a maker of toys and dolls. Flowers has expanded its candy business by offering more gift-box products, higher-price candy brands, and express delivery of items. Nonfloral items such as baked goods, sweets, and jewelry are offered on its website (www.800Flowers.com).

Five different types of mergers and acquisitions are listed in Table 15.1 [Bower, 2001]. The first type aims to reduce overcapacity in a relatively mature industry. The acquirer tries to close inefficient plants and reduce costs while retaining the acquired firm's technologies and customers so that economies of scale can be realized. The Hewlett-Packard and Compaq merger in 2002 was an example of this type.

Many deals are based on acquiring the customers of the acquired company and reducing overcapacity. The goal is to identify the best customers and retain them while trying to attract new customers from the acquired firm's customer list [Selden and Colvin, 2003].

A geographic extension or roll-up occurs when a successful company rolls up (buys up) local or regional firms into a nationwide powerhouse. Roll-ups are designed to achieve geographic reach and economies of scale and scope. The third category, market extension, is designed to extend a company's product line or its entry into unserved markets. To extend its product line, eBay purchased PayPal to facilitate transactions with its customers.

A technology acquisition aims to quickly acquire new technologies and capabilities by purchasing a small firm. As discussed in Section 13.9, Cisco Systems built its performance during the 1990s on a long string of acquisitions of small firms. The final type of merger is based on a perception of a future

TABLE 15.1 Five different types of mergers and acquisitions.

Type:	Overcapacity reduction	Geographic extension (roll-up)	Product or market extension	Technology acquisition	Industry convergence
Objective	To reduce excess capacity and increase efficiencies	To extend a company's reach geographically and build economies of scale and scope	To extend the product line or reach a new market segment	To quickly add new technologies and capabilities	To establish a position during the convergence of an industry or sector
Examples	Daimler-Benz and Chrysler	Bank of America and Nations Bank	Tyco and Raychem	Cisco Systems acquired 130 companies between 1993 and 2008	Sirius and XM Radio
	Hewlett-Packard and Compaq	Waste Management and numerous local firms	eBay and PayPal	Medtronic and numerous medical device companies	Disney and ABC-TV and radio
Issues	What to eliminate in the merged company and how to get it done in a timely way	How to merge two firms with different cultures	Merging two cultures and distribution channels	Overvaluation of the technology acquisition and loss of leaders from the acquired firm	The convergence may not materialize or may be of low value

convergence of industries. Disney purchased ABC-TV and Radio, envisioning a convergence of content and media channels. Often firms use an acquisition to restore a sense of vitality to their businesses and, they hope, unleash a subsequent surge in performance [Vermeulen, 2005].

Boston Scientific and Scimed: A Successful Integration

The successful integration of two firms is an important goal of any merger or acquisition. For example, Boston Scientific Corporation of Massachusetts was a well-respected pioneer in the less-invasive medical device industry after its IPO in 1992. Its aggressive acquisition strategy soon led to a merger with Scimed Life Systems of Minneapolis, which specialized in angioplasty products for catheter-based treatment of cardiovascular disease. Scimed also had built a top-notch distribution system in Europe and Japan. After several challenging years working to fully integrate the two firms, Boston Scientific became an industry leader, because of the Scimed products, distribution system, and team.

The Hewlett-Packard and Compaq merger was actually based on both overcapacity reduction and technology acquisition. In this case, the surviving firm, Hewlett-Packard, acquired Compaq's technologies. The goal of the merger was to improve HP's competitiveness since the firm had largely missed out on the personal computer and Internet transitions [Anders, 2003].

Rules for integrating an acquired firm into the acquiring firm are provided in Table 15.2. The key step is to appoint an integration manager who works full-time for a period on integrating the two firms. The integration effort starts with a strategy and an integration plan. Then the goal should be to achieve a majority of the integration by a short period after the close of the deal—about six weeks. An important step is to build an integration team working with the integration manager. The team helps build the social connections for the merger and get early results. The role of the integration manager is to inject speed into the process, create a new structure, make social connections, and build success, as summarized in Table 15.3 [Aiello and Watkins, 2000]. The number of overall mergers and acquisitions varies depending on market conditions. Worldwide activity for technology firms is shown in Table 18.9 in Chapter 18.

The leaders of entrepreneurial firms often continue to play a vital role after an acquisition by a larger firm closes. The leaders of the buying firm are often too busy with their own business to provide effective direction to the acquired employees. Moreover, they initially may not understand the acquired business well enough to make good decisions. This creates a need for the acquired managers to continue to lead their companies, even after the deal closes. Acquired leaders can add value by focusing their employees on specific goals and timelines, and by helping to resolve problems that arise as employees are assigned to their new positions and supervisors [Graebner, 2004].

TABLE 15.2 Rules for the acquiring firm and the acquired firm.

Rules for the acquiring firm

- Use your highly valued stock as payment.
- Identify the key people of the acquired firm and get their agreement to stay.
- Decide who to keep and build relationships fast.
- Contain the tendency to act with hubris.
- Integrate the culture and the operations of the two firms.
- Appoint an integration manager or team to lead the acquisition process.

Rules for the acquired firm

- Demand cash, not stock, as payment from the acquiring firm.
- Key people will agree to stay for a short period.
- Avoid signing a noncompete agreement or keep its duration short.
- Explain the benefits that accrue to the employees and managers of the acquired firm.
- Tell people who will and will not have a job.
- Restructure with respect for people.

Acquisitions and mergers are as likely to destroy value as to create it. They only add value if they make strategic sense, if their fair value is paid based on realistic expectations, and if management stays focused on executing the plan. AOL and Time Warner failed to properly mesh the organizations and destroyed value [Klein and Einstein, 2003].

TABLE 15.3 Four roles of the integration manager.

1. Inject speed into the process.
   - Push for decisions and progress.
   - Set the pace.
2. Create a new structure.
   - Create joint teams.
   - Lead an integration team.
   - Provide the new structure framework.
3. Make social connections.
   - Interpret the cultures of both companies.
   - Actively be present in both companies.
   - Bring people together.
4. Build success.
   - Identify and communicate synergies.
   - Show short-term benefits.
   - Demonstrate corporate efficiency gains.

15.3 Global Business

A common motivation for mergers and acquisitions is to enter new markets. This strategy is often adopted by firms seeking to start doing business in a new country. The current globalization phenomenon dates from the fall of the Berlin Wall in 1989, which ended the principle division that ruled the world since 1945, when the Iron Curtain descended. If the metaphor for the Cold War period was the Berlin Wall, the metaphor for globalization is the Web. Globalization is the triumph of free-market capitalism worldwide. The previous era of globalization was built around falling transportation costs. Today's era of globalization is built around falling telecommunications costs [Friedman, 1999]. **Globalization** involves the integration of markets, nation-states, and technologies, enabling people and companies to reach around the world to offer and sell their products in any country in the world. With the Internet and globalization, a company can sell anytime, anywhere. Globalization is characterized by speed, modernization, movement, and the removal of distance.

A new venture should consider a globalization strategy, even if it is only to decide that its initial strategy is to remain local. A new design-automation firm may choose to only serve the United States in its initial years but later consider expanding internationally. We use a classification system for the strategies for globalization, as shown in Figure 15.1.

A **local** or **regional** strategy focuses all a firm's efforts locally since that is its pathway to a competitive advantage. Early-stage companies often select a regional market that they know well and attempt to be successful there first. This enables them to understand their customers and why they buy the product. When the firm launches its initial product, it wants to be close to the customer and learn about missing features. There is an obvious advantage to having your first customers close, not far away [Bhide, 2008]. Thus, a new technology firm may start locally and fashion its marketing and sales methods. Then it can go global.

Another reason to remain local initially is limited availability of resources. Restaurants, retail stores, and other local opportunities are best started locally. For example, Starbucks started in Seattle and moved to other regions of the United States only after perfecting its operational capabilities locally.

The **multidomestic** strategy calls for a presence in more than one nation as resources permit. In this case, the firm offers a separate product and marketing strategy suitable for each nation. This strategy is not cost-efficient but does enable a firm to have independent subsidiaries in many nations. Examples of companies using a multidomestic strategy are Nokia and Sony-Ericsson.

To exploit cost economies while creating differentiated products, a **transnational** strategy can be used. This strategy rests on a flow of product

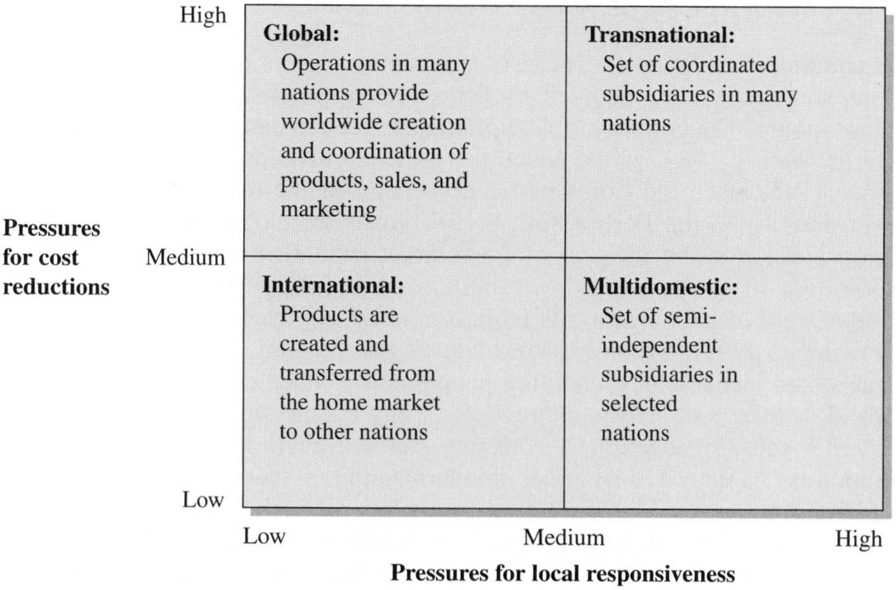

High

Global: Operations in many nations provide worldwide creation and coordination of products, sales, and marketing		**Transnational:** Set of coordinated subsidiaries in many nations

Medium

International: Products are created and transferred from the home market to other nations		**Multidomestic:** Set of semi-independent subsidiaries in selected nations

Low

Pressures for cost reductions (vertical axis)

Low   Medium   High

Pressures for local responsiveness

FIGURE 15.1 Strategies for globalization.

offerings created in any one of the countries of operation and transferred between countries. Examples of companies using a transnational strategy are ABB and Caterpillar.

An **international** strategy tries to create value by transferring products and capabilities from the home market to other nations using export or licensing arrangements. One benefit of international activities can be the exposure to new business environments where the firm can learn about different methods, products, and innovations. Examples are Microsoft and IBM. Microsoft has regularly tried to bring the same business model to other countries.

Fargo Electronics and China

Fargo Electronics is a leading maker of instant ID-card systems that is competing for sales throughout the world (see www.fargo.com). As many as 800 million identity cards with a microchip (smart cards) may be used in China alone. The smart card carries all sorts of data about an individual.

The IDs essentially are microcomputers containing a 32-bit processor and 32 kilobytes of memory that are encrypted to provide access to buildings, computers, e-mail, and other functions. They also can store information about the cardholders, such as their medical history. Fargo, based in Eden Prairie, Minnesota, markets its product in 80 countries using an international strategy, as described in Figure 15.1.

TABLE 15.4 Advantages and disadvantages of the four global strategies.

	Advantages	Disadvantages
■ Multidomestic	Ability to customize products for local markets	Failure to reduce costs and to appropriate learning from other nations
■ Transnational	Ability to reduce costs and learn from other nations	Difficult to implement due to many independent organizations
	Ability to be locally responsive	
■ International	Transfer of unique products and competencies to other nations	Low-local responsiveness
		Failure to reduce costs
■ Global	Ability to reduce costs and gain worldwide learning	Lack of local responsiveness
		Difficult to coordinate

A **global** strategy emphasizes worldwide creation of new products, sales, and marketing. The company uses facilities and organizations in several nations to create products for worldwide sales. Examples are General Motors, Intel, and Hewlett-Packard. Headquartered in Silicon Valley, Intel's international business has grown to represent 70 percent of its total revenues. Craig Barrett, Intel's former CEO, anticipates the largest growth areas in the coming years will be India, China, and Russia. These markets represent almost half the world's population and have just recently become available to U.S. technology companies [Barrett, 2003]. The advantages and disadvantages of each of the globalization strategies are listed in Table 15.4 [Hill and Jones, 2001].

Suzlon Energy's Global Strategy

Founded in India by Tusli Tanzi in 1995, Suzlon Energy is now a leading wind turbine manufacturer with an increasingly global strategy. The company has operations on five continents with sales, R&D, and manufacturing employees in locations like India, China, Germany, and Belgium. It offers complete wind power solutions including consulting, manufacturing, installation, and maintenance to its customers. It has rapidly grown into one of the largest multinational clean energy companies with sales approaching $3 billion in 2008. Its leadership challenge is now formidable because the firm's many employees are separated by oceans, languages, and cultures (www.suzlon.com).

In general, a new or emerging firm should choose one of the strategies for entering the global marketplace and then determine which foreign markets to enter and when. The determinants of entry, timing, and costs will lead to the selection of a sound strategy. Some industries are local in nature, and others are international or global. A new manufacturer of integrated circuits is immediately required to quickly build an international strategy since the competitive marketplace is global. For many firms, expansion from local to regional and then to national markets will follow a natural progression. The first step is to expand to selected nations, establishing appropriate distribution channels and supply chains.

International opportunities exist in many industries. Microsoft and Intel sell their products worldwide. London's hit plays, such as *Phantom of the Opera,* move to New York and eventually on worldwide tours. Global opportunities to reduce costs, improve capabilities, and match local needs call for a transnational or global strategy. The resources necessary to mount these strategies can be large, however.

Cisco Systems uses an international strategy that enables it to derive about 50 percent of its revenues from foreign sales in 150 countries. Many new ventures or emerging firms need to consider the development of a global strategy to access unique capabilities or advantages. Reasons for considering entering the global marketplace are listed in Table 15.5. In turn, the speed at which a new venture enters the global marketplace is dependent on competitors' actions, enabling technologies, and the venture's knowledge and perceptions of foreign markets [Oviatt and McDougall, 2005].

It is conventional for a start-up to fully develop and test its product in a local or national marketplace before taking it global. Since emerging firms have limited ability to acquire information and knowledge about foreign markets and to manage foreign activities [Julien and Pamangalahy, 2003], they often limit their international efforts. Over time, as firms identify international opportunities in the same region and around the world, they must balance the pursuit of these opportunities against their available resources [Kuemmerle, 2005]. Successful entrepreneurs generalize lessons learned in one context into simple rules that apply to the next context [Bingham, 2005].

A successful global start-up usually has an international vision from inception, a strong worldwide network, and a unique product in demand worldwide. Consumers throughout the world increasingly demand the same selection of consumer goods, particularly automobiles, clothing, many food and beverage products, and consumer durables such as appliances and electronics. For many businesses, this global consumerism has focused sharp attention on the development of global brands, which are rapidly creating brand equity positions for

TABLE 15.5 Reasons to develop a global strategy for a new venture.

■ Access low-cost labor or materials	■ Access attractive markets for the firm's products
■ Work around trade barriers	
■ Access unique capabilities and learning from others	■ The industry is global and all competitors are worldwide
■ Obtain economies of scale	■ Possession of a strong brand that is known worldwide

TABLE 15.6 Five forms of entry mode into international markets.

Mode	Description	Advantages	Disadvantages
1. **Exporting**	Send goods abroad for sale in another nation	Ability to sell elsewhere	Transportation costs Difficulties with the agent
2. **Licensing**	Legal permit to use knowledge or patent to make product	Low-cost entry	Weak control of the licensee
3. **Franchising**	Rights granted to a business to sell product using brand, name, and methods of operation	Low-cost entry	Lack of control over quality
4. **Joint venture**	Corporation held jointly with a local entity	Access to partner's capabilities Shared costs	Lack of control
5. **Wholly owned subsidiary**	Company incorporated in another nation	Ability to act directly	High costs

their companies. Nokia, Toyota, and Pfizer are global brands. New ventures need to plan for globalization in various ways. As their industry becomes global, they must plan for production, regulatory, and organizational factors [Farrell, 2004].

The mode of entry into another national or regional market offers five possibilities, as listed in Table 15.6. Exporting is an easy method to start up elsewhere. However, high transportation costs can be a disadvantage. Licensing to another party for a fee can be an inexpensive method and can provide some control of the marketing and manufacturing carried out by the licensee. Franchising is a form of licensing with an agreement to follow rules and procedures of operation. This method may result in loss of control over the quality of the product.

A joint venture with a foreign company affords access to the partner's capabilities. But, it also results in diminishing control. Thus, care must be taken to align the goals of both partners and to address the imbalance in bargaining power if one partner is much larger than the other [Lu and Beamish, 2006]. A wholly owned subsidiary enables the parent company to exercise full control but it can be a costly approach.

McDonald's and Hilton Hotels are usually set up as franchises in other nations. Intel and Hewlett-Packard act through wholly owned subsidiaries. Fuji-Xerox is a joint venture between Xerox-USA and Fuji Photo Film of Japan. Harley-Davidson exports about 25 percent of its motorcycles directly to dealers and distributors in other countries.

Eastman Kodak is the world's largest producer of photographic products. It has major plants in the United States and eight other countries. About one-half of its employees and one-half of its sales are outside the United States. To regain its worldwide market share, Kodak is moving aggressively into the worldwide filmless digital photography field.

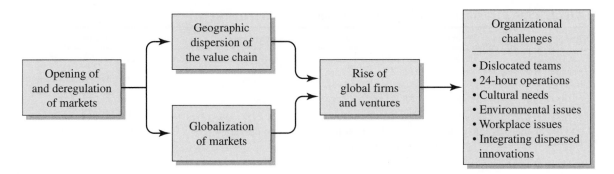

FIGURE 15.2 Forces and consequences of globalization.

Globalization is spreading as markets open and deregulate. As shown in Figure 15.2, the forces of globalization are powerful [Barkema, 2002]. Many new firms will need to consider the global factors in their industry and create a plan for responding. To be successful, global entrepreneurs need to articulate a global purpose, build alliances, create and manage supply chains, and learn how to build a trusting organization across different cultures and institutional frameworks [Isenberg, 2008].

As new firms grow, they need to develop global relationships and work with people from other cultures. Cultural intelligence is the capacity of a person to adapt to new cultural settings and to interact effectively with people from different cultures [Early, 2004]. The three components of cultural intelligence are cognitive, physical, and emotional intelligences. Global business depends on good cultural relationships and smart cultural intelligence. Successful members of global teams learn to cope with different national, corporate, and vocational cultures.

Yves Doz of INSEAD defines a **metanational** company as possessing three core capabilities: (1) being the first to identify and capture new knowledge emerging all over the world; (2) mobilizing this globally scattered knowledge to out-innovate competitors; and (3) turning this innovation into value by producing, marketing, and delivering efficiently on a global scale [Fisher, 2002]. Doz identifies one example of such a firm as Nestlé. Other examples are IBM and Hewlett-Packard. Perhaps the most challenging task is to access knowledge and innovations everywhere and then integrate them into the firm's competencies and products.

Competition is similar to a three-dimensional game of global chess. The moves that an organization makes in one market may achieve goals in another market in ways that are not immediately apparent to its rivals. Where this strategic interdependence between markets exists, the complexity of the competitive situation can quickly overwhelm ordinary analysis. A firm needs an analytic process for mapping the competitive landscape and anticipating how its moves in one market can influence its interactions in others.

Leading firms in technology-intensive industries are based on core capabilities that can be leveraged globally. Innovative companies strive to draw knowledge and coordinate capabilities worldwide. They build and discover new assets and capabilities, and work to achieve economies of scale and scope. As the firms grow, they move toward a global network of capabilities and assets [Tallman and Fladmol-Lindguist, 2002].

A truly global firm, such as IBM, HP, Cisco, or Intel from the information technology industry, operates as a single integrated organization. The firm will move people and projects anywhere in the world based upon the right set of costs, skills, and the business environment. These firms operate as a "global network enterprise"—a flexible assembly of firms that together strives to provide the best set of business processes. These networks are no longer transactional (contract-based), but rather rely on mutual cooperation and trust [Branscomb, 2008]. They rely on a cooperative network of innovation partners.

15.4 AgraQuest

AgraQuest has been looking for its first acquisition candidate so that it can grow revenues more quickly. Since it takes several years to develop a new biopesticide and obtain approval for its use, it is very attractive to acquire companies that have one or two approved products. AgraQuest is hoping to acquire one of these firms by issuing its common stock to the owners of the firm. Most likely, the owners of the acquired company would prefer to receive cash, since AgraQuest stock is not yet publicly held.

AgraQuest has a strategy for selling its products internationally through distributors in other countries. Serenade is sold via distributors in Japan, Mexico, Chile, and New Zealand. A distributor in each of those countries has an exclusive sales arrangement. Serenade is sold directly to the banana producers in Costa Rica. After receiving approval to distribute Serenade in Europe in 2009, AgraQuest reached a wide-ranging agreement with BASF. BASF gained rights to distribute Serenade for specific applications in many countries throughout Europe, Africa, the Middle East, Asia and Latin America that were not covered by AgraQuest's existing partnerships.

15.5 Summary

A new venture may start as a purchase of an existing company by a team of entrepreneurs. Alternatively, entrepreneurs can start their new venture and acquire other small firms to grow their own company. Most, if not all, acquisitions are justified on the basis of an expected synergy, which is the increased effectiveness of the combined firms. Acquisitions can be used for efficiency improvement, geographic expansion, product or market extension, and technology acquisition.

Most new firms start with a regional or national strategy and later develop a plan to export their product internationally. As these firms grow, they may shift from exporting to establishing a wholly owned subsidiary in other nations.

> **Principle 15**
> All new technology business ventures should formulate a clear acquisition and global strategy.

Video Resources

Visit http://techventures.stanford.edu to view experts discussing content from this chapter.

Challenges of a Global Company	Scott Kriens	Juniper Networks
Global Acquisition: Lessons Learned	Stephanie Tilenius	eBay

15.6 Exercises

15.1 An acquirer looks for a company with a good profit margin, a proven history, and a fair price. Choose an industry of interest and list five criteria for selecting candidates for acquisition.

15.2 A software firm is available for purchase, but it has experienced no growth for several years. The firm provides a cash return annually before taxes and owner's salary of $100,000. It has annual sales of $1 million. As a purchaser of this firm, you select a discount rate of 14 percent. Calculate the price you would offer for the firm. Assume that you can increase and maintain a growth rate of sales and cash flow of 2 percent annually.

15.3 Describe how the acquisition strategies of these firms have differed: Apple, Google, Microsoft, Oracle, and Qualcomm.

15.4 Identify a technology-based company that used acquisition to enter into an international market. Describe the strategic alignment and the executed deal terms. Was it a success? What were the key challenges that arose?

15.5 Interview a technology executive based in a country different from the company's headquarters. How are various company functions handled (e.g., marketing, sales, operations, development)? What are the key challenges of operating away from company headquarters?

15.6 Social networking is one of the most prevalent Internet user activities today in many countries. However, given the geographic virality and social aspect of social networks, it has been very challenging for any specific network to dominate internationally. Le Monde created a compelling graphic describing this dynamic in 2008 (radar.oreilly.com/archives/2008/04/worldwide-social-network-market-share.html).

Leveraging the concepts in Section 6.3, Section 13.3, and Section 15.3, (a) describe the international expansion strategies selected by top social networking companies (e.g., Facebook, MySpace, Friendster, Bebo, Cyworld, Orkut), (b) describe some of the reasons each social networking firm has developed the global presence it has, and (c) how could this impact the business models chosen by each company.

15.7   Neopets (www.neopets.com) offers its virtual world of virtual pets globally. The firm offers its service in two languages, English and Chinese. Study the Neopets site and determine if it should be offered in many languages. What are the advantages of establishing a Neopets site for each language?

15.8   When Google initially expanded search services into China, there was surprise over Google's decision to censor some content per requests by the Chinese government. Why did some view this as "evil"? How did Google view the concern? What are other challenges facing companies expanding into international markets?

VENTURE CHALLENGE

1.   Describe your venture's approach to acquisitions and mergers as a growth strategy.

2.   For your venture, discuss the plans for going global and describe your approach using Figure 15.1 and Table 15.6.

PART 4

Financing and Building the Venture

The venture should have a clear revenue model and a workable path to profitability. Furthermore, there should be a plan of how the wealth created will be harvested by the owners. A comprehensive financial plan will be designed to demonstrate the potential for growth and profitability that is based on accurate and reliable assumptions. With the financial plan in hand, sources of investment capital will be explored and tested. Then the terms and valuation of the deal with the investors will commence. The presentation of the total business plan will require a compelling story about the venture. Furthermore, skillful negotiations with all owners and partners will be required. When funded and launched, the venture team must continuously and ethically implement the business plan and adapt to changing conditions. ■

CHAPTER 16

Profit and Harvest

Profit is the product of labor plus capital multiplied by management. You can hire the first two. The last must be inspired.
Fost

How will a new venture generate revenue and achieve positive cash flow?

A new firm creates a sales model describing how it will generate revenues from its customers. Then it determines a cost model and how to generate profits from its revenues. The revenue and profit engines show how the firm will create powerful value for its customers and how customers will enable the new firm to profit. Many new ventures assume that profit will flow naturally from sales but discover that profits are not guaranteed. It is difficult to operate in a market that is chronically unprofitable.

A new firm seeks positive cash flow as soon as is feasible and acts to move to profitability early in its life. Managing revenue growth is important since uncontrolled growth can lead to negative cash flow and the need to constantly raise new funds from outside investors. Furthermore, a firm needs a plan to harvest the benefits of its growing venture for all owners. Entrepreneurs must also be realistic and accept that termination of the new venture is a possibility. ■

16.1 The Revenue Model

A firm's **revenues** are its sales after deducting all returns, rebates, and discounts. A firm's **revenue model** describes how the firm will generate revenue; five models are listed in Table 16.1. Most firms generate revenues by selling a product in units to a customer using a **product sales model.** For example, Dell sells its personal computers to one customer at a time, and Intel sells its chips to electronics companies.

In the **subscription revenue model,** a business offers content or a membership to its customers and charges a fee permitting access for a certain period of time. This model is used by magazines, information and data sources, and content websites. *Consumer Reports* offers its information to magazine subscribers (members) as well as to subscribers to its online service for a fee. This model is also used by clubs, cooperatives, or other member-based organizations. Technology firms sometimes license their technology for a fee.

The **advertising revenue model** is used by media companies such as magazines, newspapers, and television broadcasters that provide space or time for advertisements and collect revenues for each use. The media entities that are able to attract viewers or listeners to their ads will be able to collect the highest fees. MSN, Google, and Yahoo collect most of their revenues through the sale of advertising space.

Some firms receive a fee for enabling or executing a transaction. The **transaction fee revenue model** is based on providing a transaction source or activity for a fee. Examples of firms based on transaction fees are Charles Schwab, Visa, and eBay.

The **affiliate revenue model** is based on steering business to an affiliate firm and receiving a referral fee or percentage of revenues. For example, this revenue model is used in the real estate business and by companies that steer business to Amazon.

Magazines and newspapers such as the *New York Times* use both the subscription and the advertising models to generate revenues. Some new ventures use a mix of the five revenue models. Amazon.com, for example, uses the product sales, transaction fee, and affiliate revenue models.

Google is a good example of a firm with multiple sources of revenue. Google generates revenues through advertising and licensing its technology to others for a subscription fee. Other competitors collect fees from higher placement of a Web address on its list, but Google avoids that model.

Any new business needs to determine its revenue model and test it on potential customers. Who is the customer? Will the target customer pay for the

TABLE 16.1 Five revenue models.

■ Product sales	■ Transaction fee revenue
■ Subscription (member) fee	■ Affiliate revenue
■ Advertising revenue	

> **JBoss: Generates Revenue Through Service and Support**
>
> JBoss of Atlanta, Georgia, has an unusual revenue model and mission: "JBoss Inc.'s mission is to revolutionize the way enterprise middleware software is built, distributed, and supported through the Professional Open Source model. We are committed to delivering innovative and high quality technology and services that make JBoss the safe choice for enterprises and software providers." Revenue is generated by supporting software that is given away for free. By providing professional service and support for open source software, JBoss found a comfortable and profitable niche within its industry. It was acquired by Red Hat in 2006. (See http://www.jboss.com.)

offering? Consider the case of the Weather Channel (www.weather.com), which was initiated in 1982 when cable television was building across the United States, but lost money for several years. The Weather Channel assumed the TV viewer at home was its customer and tried to develop a revenue model. Perhaps, it thought, advertisers would pay to advertise on the channel. Having committed more than $30 million for satellites and the transmission system, the Weather Channel in a few years was on the brink of bankruptcy as a result of few advertisers signing on. At that point, the cable operators agreed to pay a subscriber fee to help keep the channel operating. It turned out that the customer was the cable operator, not the viewer, and while the Weather Channel is not glamorous television, it is profitable today [Batten, 2002].

Because they have yet to achieve profitability, most start-ups use a revenue metric such as revenue per employee to track their performance during their formative years. Their goal is to exceed $100,000 per employee as soon as possible. Most mature growth firms exceed $200,000 per employee, and technology companies sometimes exceed $400,000 per employee.

16.2 The Cost Model

A **cost driver** is any factor that affects total firm costs [Niraj et al., 2001]. Typically, costs vary over time or volume of output. The four primary types of cost drivers are: fixed, variable, semivariable, and nonrecurring. Fixed costs do not vary at all with volume. Examples include rent and management salaries. Variable costs change directly and proportionally with volume. Examples include sales commissions (which vary with sales) and materials costs (which vary with the amount of goods produced). Semivariable costs also change with volume, though not as directly and they contain elements of fixed costs. For example, a store may need to keep a minimum staff on hand (which is a fixed cost), but can add more staff as the number of customers increases (which is a variable cost). Finally, nonrecurring costs occur infrequently or irregularly. Examples include equipment and building purchases

[Hamermesh et al., 2002]. In order to maximize profits, entrepreneurs must be aware of their various cost drivers and of the types of costs.

16.3 The Profit Model

Profit is the net return after subtracting the costs from the revenues. The **profit model** is the mechanism a firm uses to reap profits from its revenues. Google makes most of its profits by auctioning text-based advertisements that appear in search results and participating websites. Microsoft makes most of its profits from licensing Windows and Office software. General Electric generates one-half of its profits from its financial arm, GE Capital. Newspapers make most of their profits from classified ads. Hewlett-Packard and Xerox make most of their profits from replacement toner cartridges.

Figure 16.1 shows the value of a product and the distribution of the value to the customers and the profit captured by the firm. To remain profitable, a firm strives to reduce its costs while maintaining or increasing the value of its product to the customer. To generate profits, a firm needs to examine all its activities on the value chain and determine if its cost versus value generated is in line.

Profit accrues to a company that maintains a competitive advantage as conditions change. When the PC industry took off in the 1980s, IBM ceded the operating system's rights to Microsoft when it incorrectly determined that profit would flow to the branded integrator of hardware and software components. Key to profit capture is ownership of the unique, value-added element of the value chain or the product makeup. Examples would be ownership of an essential pipeline, control of the customer interface, or ownership of unique locations for a retail operation. It pays to hold the largest "value-added" step in a value chain or the unique innovation that no one else can match.

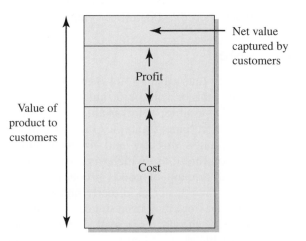

FIGURE 16.1 Value of a product and profit.

During the early years of a firm, the firm may be patient for growth but should be impatient for profitability. As a firm works to gain profitability, it is testing its assumptions that customers will pay for a profitable product [Christensen and Raynor, 2003].

The revenue and profit engines are driven by the firm's business model, strategy, resources, capabilities, operations, and processes, as shown in Figure 16.2.

The best conditions for profit occur when the perceived value of a product to a customer is high and the cost to produce the product is low. Figure 16.3 shows a value grid that enables a firm to determine its potential to reap large profits. Low cost to produce a product leads to low price per unit of perceived value to the customer. The upper-right quadrant is a high profit location that many firms seek to occupy [Chatterjee, 1998].

Most start-ups initially invest time and energy in learning about their customers. Then they use that knowledge to create improved solutions for them. They lose money initially but make money after a period we will call T, as shown in Figure 16.4. Of course, it is best to keep T relatively short and the peak negative profit (NP) small [Slywotzky, 2002]. The profit curve shown in Figure 16.4 is often called a "hockey stick" expectation.

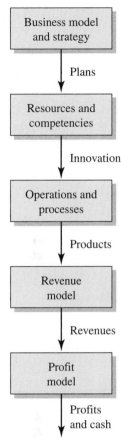

FIGURE 16.2
Revenue and profit flow from the firm's operations.

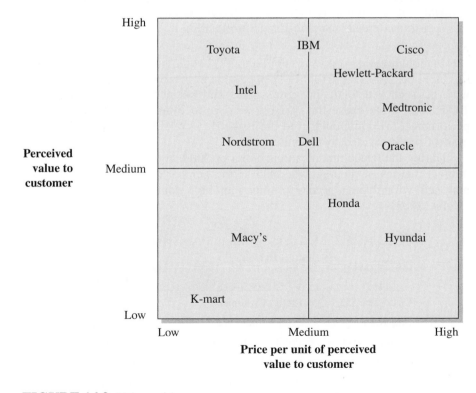

FIGURE 16.3 Value grid.

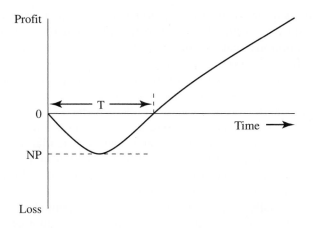

FIGURE 16.4 Early losses of a successful start-up turn profitable after a time, T. The peak negative profit is NP.

It is useful to try to estimate the attractiveness (the potential profitability) of a market segment. It is of little value to win market leadership in a market segment that is chronically unprofitable [Ryans et al., 2000].

A firm can create a metric for its profitability as

$$\text{metric} = \frac{\text{profit}}{x}$$

when x is chosen to fit the firm's goals and business model. As shown in Table 16.2, firms chose the variable x so as to illuminate their profitability performance. A commonly used profit metric is **profit margin,** which is the ratio of profit divided by revenues.

During the telecommunications boom of 2000, companies often used poor indirect metrics of their growth such as the number of building leases. They reasoned, wrongly, that access to office buildings would translate to customers [Malik, 2003].

TABLE 16.2 Metric for profitability performance for selected firms.

x	Metric	Example firm
Customer	Profit / Customer	Gillette
Employee	Profit / Employee	Abbott Labs
Customer visit	Profit / Customer visit	Walgreens
Tons of output	Profit / Tons	Nucor
Revenue	Profit / Revenue	ExxonMobil

A business model designed for high customer relevance that delivers high value will have the best chance of capturing profit. The best business model helps the customer in the difficult or time-consuming areas of their purchasing process. One of the most powerful profit models is the **installed base profit model.** The supplier builds a large installed base of users who then buy the supplier's consumable products. This is the model used by Gillette, which sells a razor at a modest price while building a large base of users who purchase consumable razor blades. Nine types of profit models are listed in Table 16.3 [Slywotzky et al., 1999]. A new venture will wisely select its needed profit model and work hard to build its strength and resiliency in the competitive marketplace.

Managers have always known that some customers are more profitable than others. For some emerging firms, 20 percent of the firms' customers may provide most of the firms' profits. Furthermore, the firms' worst customers may be costing more than they pay for products or services. Securing profitable customers while getting rid of unprofitable customers can help to double a firm's profits. It pays to be attentive to the best customers and ignore the worst [Selden and Colvin, 2002].

TABLE 16.3 Nine types of profit models.

Name	Description	Examples
1. **Installed base**	Build a large installed base of customers and sell consumables or upgrades	Gillette Hewlett-Packard printers
2. **Protected innovation**	Create a unique, innovative product and protect it using patents and copyrights	Merck Microsoft
3. **New business model**	Find unmet customer needs and build a new business model	Netflix Google
4. **Value chain specialization**	Specialize in one or two functions on a value chain	Nucor Intel
5. **Brand**	Create a valued brand for your product	Intel Coca-Cola
6. **Blockbuster**	Focus on creating a series of big winners	Pixar Schering Plough
7. **Profit multiplier**	Build a system that reuses a product in many forms	Disney Virgin Group
8. **Solution**	Shift from product to unique total solutions	General Electric Microsoft
9. **Low cost**	Create a low-cost product to offer low price per unit of value	Ryanair Dell

Recall that profit is

$$\text{Profit} = (P - VC)Q - FC \tag{16.1}$$

where P = price, VC = variable costs, Q = total number of units sold, and FC = fixed costs. Managing profitability can be achieved by lowering fixed or variable costs, raising units sold, or raising price. One may be able to find a customer segment that is willing to pay a higher price or purchase more units. Otherwise, costs need to be lowered.

Another important measure of performance is **cash flow,** which is the sum of retained earnings minus the depreciation provision made by the firm. Without positive cash flow, a firm may eventually use up all its cash and close its doors. A profit and cash flow model focuses attention on the nature of the driving forces of the revenue and profit models.

Amazon.com has used aggressive price discounting and free shipping to boost its revenues. However, its low operating margin (P − VC) made profitability elusive during its formative years.

Many mass-market retailers such as Wal-Mart continually lower prices by squeezing inefficiencies from their operations and sacrificing profit margins on products in favor of selling in high volumes. Hewlett-Packard tries to avoid lowering prices to keep its profit margin robust.

All entrepreneurs need to find a suitable profit model for their firm. If profitability appears to be highly elusive and at best in the distant future, it may be wise to not proceed with the venture. We discuss the matter of terminating a venture in Section 16.6.

Cemex is a Mexican cement company. In the past, it described its profit metric as yards of concrete sold each day. The customer, it realized, values the delivery of the right amount of concrete at the right time. Using that metric, profitability increased. The profit model led to the creation of a new business model [McGrath, 2005].

16.4 Managing Revenue Growth

New businesses normally strive to build up revenues and profits so that they can meet their goals. Most entrepreneur teams are naturally inclined to grow their business rapidly. Other entrepreneurs limit the growth of their firm for personal or lifestyle reasons. The degree of commitment of the entrepreneurial team to the growth of the firm can be called **entrepreneurial intensity** [Gundry and Welsch, 2001].

Commitment to growth leads to sacrifices an entrepreneur is willing to make. High growth can require significant financial resources, leading the entrepreneurs to seek outside capital and often give up majority ownership of the firm. Low growth would include firms growing revenues at a rate less than 10 percent per year, and high rates of growth would exceed 25 percent per year. Many high-growth firms grow at 50 percent or higher each year for several years after founding. Technology entrepreneurs who seek a high-growth

TABLE 16.4 Characteristics exhibited by entrepreneurial teams that seek high growth rates.

■ Strong entrepreneurial intensity	■ Emphasis on a team-based organizational structure
■ Willingness to incur the costs of growth	
■ Willingness to use a wide range of financing sources	■ Focus on innovation

strategy will usually select a team-based organizational structure and exhibit high entrepreneurial intensity. Furthermore, they are willing to endure the burdens associated with the demands of high growth. The characteristics of a high-growth entrepreneurial team are listed in Table 16.4. High-growth entrepreneurs are willing to put aside some of their personal or family goals and make sacrifices because they are committed to the growth of their ventures.

Leasing of a firm's products to customers can delay revenues into the future and provide long-term cash flow. Because IBM leased computers and office machines, customer service representatives remained close to customers and were able to keep the relationship strong.

A growing business requires cash for working capital, assets, and operating expenses. If a company grows too fast, it will need to continually raise additional cash from investors. The cash required will depend on the operations model of the firm, which depends on its accounts payable cycle, as well as assets and working capital required [Churchill and Mullins, 2001]. In general, most growing firms are unable to sustain a growth rate of sales exceeding 15 percent using their internally generated cash. Some service businesses are less asset-intense and may be able to self-finance a sales growth rate of 20 percent to 30 percent. Very few, if any, firms can self-finance if they plan to grow at 50 percent or more per year. For high growth, a financing plan for outside cash will be required.

Since service businesses are often less asset-intensive and more labor-intensive than production firms, growth typically adds to costs and may not produce economies of scale. Some companies with a heavy emphasis on service such as IBM, Dell Computer, and Southwest Airlines have managed to successfully combine growth and profitability while other companies have not. Successful service-oriented companies are able to design and implement the right strategies to keep costs low, strengthen customer loyalty, and gain competitive advantage.

The profitability of a firm may be a function of the growth rate of revenues, as shown in Figure 16.5. At a low growth rate such as G_1, the firm is unable to meet demand and loses sales to its competitors. At a very high growth rate such as G_2, the firm is unable to efficiently manage its operations and only achieves profitability P_2. Growth rate G_m maximizes a firm's profitability P_m. An emerging or new firm should try to estimate its growth rate G_m that would maximize profitability. For many new firms, G_m ranges between 20 percent and 40 percent.

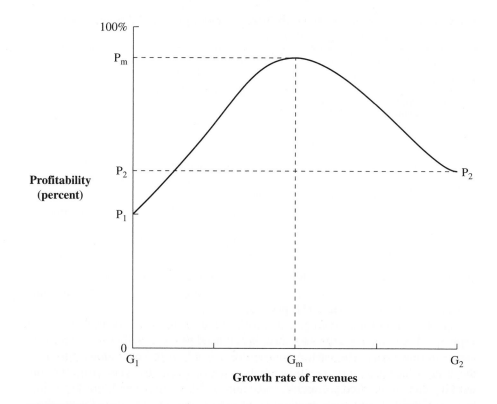

Profitability (percent)

Growth rate of revenues

FIGURE 16.5 Profitability of a firm as a function of the growth rate of revenues.

The profitability of a firm may be represented by its return on capital or return on equity (ROE). Thus, one quick estimate of a firm's ability to grow is to state that a firm may grow organically at a rate less than its return on equity without turning to outside financial sources. We define **organic growth** as growth enabled by internally generated funds.

A more complete equation for a sustainable change in sales-to-sales ratio, $\Delta S/S$, for a firm is [Ross and Jaffe, 2002]:

$$\frac{\Delta S}{S} = \frac{PM\,(1 + L)}{T - [PM(1 + L)]} \qquad (16.2)$$

where PM = profit-to-sales ratio, L = debt-to-equity ratio, and T = the ratio of total assets to sales. If a start-up firm has no debt (L = 0), we have

$$\frac{\Delta S}{S} = \frac{PM}{T - PM}$$

For example, if PM $= 0.10$ and T $= 0.5$, then

$$\frac{\Delta S}{S} = \frac{0.10}{0.5 - 0.10} = 0.25$$

or the sustainable sales growth rate is 25 percent. Consider an asset-intensive business with T $= 1.0$ and determine the sustainable growth rate when PM $= 0.10$. Then, we have

$$\frac{\Delta S}{S} = \frac{0.1}{1 - 0.1} = 0.11$$

and the sustainable growth rate is 11 percent. If this asset-intensive firm uses debt so that L $= 0.8$, then

$$\frac{\Delta S}{S} = \frac{0.1(1 + .8)}{1 - [0.1(1 + 0.8)]} = \frac{0.18}{1 - 0.18} = 0.22$$

and the sustainable growth rate is 22 percent.

A start-up will need to examine its expected growth rate and its financing needs carefully. As an example, consider the growth of a fictional green-tech venture that raised $68 million via an initial public offering in 2005. Its sales were $130 million in 2007 and $518 million in 2010, growing at an annualized rate of 44 percent. During the same period, the firm's long-term debt rose from $135 million to $947 million, and its debt-equity ratio rose from 0.16 in 2007 to 0.96 in 2010. This venture was able to grow sales at a significant rate by increasing its debt (and associated risk) significantly.

Service firms require less money to start and expand than asset-intensive industries. Building and growing a service business requires adding employees. A service business has low capital and asset intensity and requires little debt. Consider Robert Half International (www.rhi.com), which grew revenues from $220 million in 1992 to $2.45 billion in 2001 at an annual rate of 29 percent. The firm has negligible debt, thus L $= 0$. The profit-to-sales ratio averaged 0.06 during 1992 to 2001. The ratio of total assets to sales was approximately 0.15. Then, the sustainable growth in sales was

$$\frac{\Delta S}{S} = \frac{PM}{T - PM} = \frac{0.06}{0.15 - 0.06} = 0.67$$

or the sustainable growth rate was 66.7 percent.

To grow steadily and avoid stagnation, a company should learn how to scale up and extend its business, lengthen its expansion phase, and accumulate and apply new knowledge to new products and markets faster than competitors. Entrepreneurs should choose a plan that fits with the knowledge, learning skills, and assets that the organization possesses or is developing.

Rapid growth and good profitability can often cover up some underlying problems of an organization. They can provide a cushion for wasteful decisions

TABLE 16.5 Incentives for growth by new ventures.

■ Attracting capital for market expansion	■ Development of a reputation and brand
■ Attracting capable team members	■ Growing profitability and financial rewards for owners and employees
■ Achieving economies of scale	

regarding the allocation of financial, human, and other resources. The excitement of growth can also veil inadequacies in leadership or management skills. Growth can mask a lack of planning or an inadequate orientation toward long-term issues. Success can disguise a variety of shortcomings while breeding a dangerous form of arrogance.

The incentives for growth by a new venture are several, as summarized in Table 16.5. One incentive to grow is to attract capital investment to expand markets and product lines. Also, growth creates a sense of pride among the employees and provides opportunity for expanding financial reward.

A firm's ability to use and coordinate new assets and activities depends on its organization and managerial capabilities. Rapid growth can challenge those capabilities severely. It is important for a publicly held company to have consistent, predictable financial growth. Consistency requires controlled growth of assets and personnel additions. Growing a staff at a rate greater than 15 percent a year will challenge any organization. Paychex, Inc., is a $2 billion payroll-processing firm that has increased revenues at an average of 18 percent for many years. With economies of scale, it has increased profits by 20 percent each year.

Most companies employ a mix of organic growth using both inside resources and external sources of resources. A balanced approach to growth attempts to break down barriers to growth and improve the company's core competencies. Some firms, such as Linear Technology and Dell, have been successful in building revenues at 30 percent or more per year while maintaining a balanced mix of internally and externally financed growth.

Most highly innovative firms become high-growth firms, compared to low-innovation firms [Kirchhoff, 1994]. Microsoft, for example, grew its revenues at 29 percent per year for the decade 1991 to 2001 and grew its profits at 35 percent per year for the same period.

Microsoft's Quest for Growth

In 1975, two college students, Bill Gates and Paul Allen, began to develop and sell software that ran on the original personal computers (PCs). Soon they incorporated their firm as Microsoft. In 1981, industry giant IBM decided to enter the PC market and hired Microsoft to develop an operating system. Microsoft quickly created DOS, which became the dominant operating system in the industry. In March of 1986, Microsoft raised $61 million in its initial public offering of stock. In 1992, Microsoft announced the third version of Windows, marking the end of its collaboration with IBM on operating systems. Windows allowed Microsoft to expand its offerings of applications, greatly enhancing its revenue generation.

Applications made by Microsoft continue to generate significant revenue for the company. Products such as the Office Suite and Explorer continue to be ubiquitous on computers around the world, despite increasing challenges from open source competitors. In late 2001, Microsoft built on its strength in the computer gaming industry by entering the game console market. Eventually, Xbox 360 gained market share to become a popular console alongside Sony's PlayStation 3 and Nintendo's Wii. The Xbox allows Microsoft to gain additional revenue through the sale of games.

Microsoft has cleverly added revenue streams over the years, while continuing to play to its core competencies as a software development firm. Current forays, such as its Zune music products and smartphone technologies, seek to add revenue growth to the firm in the future. Microsoft is currently the #1 software company in the world, a position it has maintained for most of the last 25 years. In 2009, Microsoft's market capitalization was nearly $225 billion with revenues of $60 billion. This valuation reflects nearly 30 years of sustained profitability and continuous revenue growth.

The sources of revenue growth for a firm can include increasing brand recognition and international expansion, as shown in Table 16.6. Very few firms are able to increase prices in the face of tough competition. New offerings of valuable products are a good source of revenue growth. Successful companies active in new product offerings are Apple and Google.

The market value of a firm can be described as a result of three drivers, as shown in Figure 16.6 [Rappaport et al., 2001]. Changes in volume, price, and sales mix lead to changes in sales growth rate. Operating profit margin (profit before taxes divided by revenues) is driven by four factors, as shown in Figure 16.6. Incremental investment rate is the result of investment efficiencies. Operating leverage is the ratio of profit margin increases to upfront preproduction expenses for new product development and capacity expansion. Accounting for all these factors, a firm seeks to increase its

TABLE 16.6 Sources of revenue growth.

■ Increasing brand recognition	■ Acquisition of other firms
■ Intellectual property licensing	■ Price increases
■ International expansion	■ New product offerings

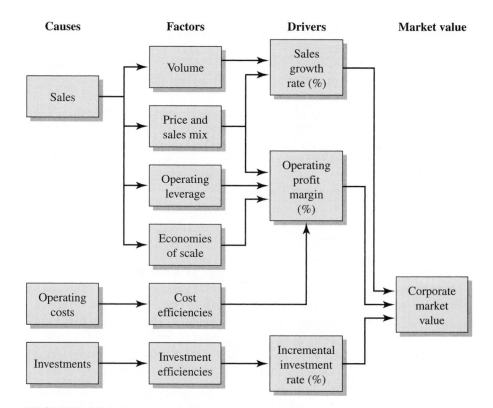

FIGURE 16.6 Causes and drivers of market value.

Note: Operating leverage $= \dfrac{\Delta \text{ Profit margin}}{\Delta \text{ Costs for product development manufacturing}}$

efficient use of investments and reduce its operating costs. Furthermore, increased volume, operating leverage, economies of scale, and an improved price and sales mix can lead to an improved sales growth rate and operating profit margin.

Millennium Cell: When Market Value Disappears

Wireless devices such as cell phones and laptop computers use batteries for power. The replacement for a battery could be a small hydrogen fuel cell. Stephen Tang and his team at Millennium Cell developed a miniature fuel cell and used the factors of Figure 16.6 to increase his investment and cost efficiencies and improve his product development capabilities. Unfortunately, this could not be achieved profitably. The firm ceased operations and filed for bankruptcy in 2009.

16.5 The Harvest Plan

Assuming a successful venture, any investor will want to know the plan for providing a cash return to all investors in a timely way. Assuming the venture has a favorable outcome in a few years, how will the investors reap or harvest their fair share of the wealth created by the venture? A **harvest plan** defines how and when the owners and investors will realize or attain an actual cash return on their investment. It delineates how and when they will extract some of the economic value from the investment. Professional investors will expect a return on their investments within five to seven years. Thus, investors will expect a plan for cash liquidity for themselves. Note that "harvest" does not mean the challenges and responsibility of the business are over.

For high-growth firms, the value created by an innovative venture can lead to significant returns over a four- to ten-year period, depending on the industry and market conditions. Both the founders and the investors will desire to access the financial return accrued by the new growing firm at the end of that period. This will mean some action will be necessary to yield a cash flow from the firm to the investors and owners. Table 16.7 lists five methods of harvesting a firm. The sale of the entrepreneurial firm to an acquiring firm is an attractive route for the founders and investors. Proceeds from the sale of a private company usually consist of cash, shares of the acquiring company, or a combination of shares and cash.

Fast-growing companies with annual revenues greater than $20 million may find a solution in the public stock market by using an initial public offering (IPO). If the investors are patient, the issuance of cash dividends to individuals can serve to provide cash to the investors. Of course, it is often possible to arrange a sale of the firm to the managers and employees of the firm. Finally, many owners of relatively small firms will consider passing the firm on to family successors.

The selection of a harvest strategy will depend on the interests of the founders and investors. Professional investors such as venture capitalists expect large annualized returns and normally seek the issuance of an initial public offering by year five or six. Alternately, the venture may be acquired by another, typically larger, firm that provides the liquidity sought by the professional investors.

TABLE 16.7 Five methods of harvesting the wealth created by a new firm.

- Sale of the firm to an acquiring firm
- Sale of the firm's stock on a public market through an initial public offering
- Issuance of cash dividends to the owners and investors
- Sale of the firm to the managers and employees
- Transfer of the firm through gifts and sales to family successors

The entrepreneurial team may describe an **exit strategy** for their venture after a specified period. This plan will be part of the negotiation with the investors.

A planned sale to employees and managers may be outlined in the business plan. This transfer can use an employee stock ownership plan (ESOP). The firm first establishes an ESOP and guarantees any debt borrowed by the ESOP for the purpose of buying the company's stock. Then the ESOP borrows money from a bank, and the cash is used to buy the owner's stock. The shares of the firm are held by a trust, and the company makes annual tax-deductible contributions to the trust so it can pay off the loan. As the loan is paid off, shares are released and allocated to the employees. While an ESOP benefits the owner by providing a market for selling stock, it also carries with it some tax advantages that make the approach attractive to owner and employees alike.

Few events in the life of the entrepreneur or the firm are more significant than the harvest. Without the opportunity to harvest, a firm's owners and investors will be denied a significant amount of the value that has been created over the firm's life. The founders may need a harvest strategy due to a desire to retire or diversify their portfolio of assets. Investors may need to realize their returns to invest them elsewhere or benefit in other ways. The timing of the harvest may be uncertain, but a harvest strategy does help the entrepreneur team plan together for the future.

A good time to sell a company is when it is very successful. When that time arrives, it may be best to review or exercise the harvest plan. At that time, a firm needs to determine a realistic valuation for the firm and obtain advice from its board of directors. Entrepreneurs often choose to sell for personal, nonfinancial reasons, such as burnout from the long hours and high stress of running their own businesses. Entrepreneurs who have raised money from family and friends may be especially eager to sell, since they feel a heightened pressure to return their investors' money. Moreover, entrepreneurs typically have much of their personal wealth tied up in a single company, making them eager to sell so they can diversify their holdings.

When entrepreneurs decide to sell, their choice of buyers is about more than price. Company leaders often choose buyers based on "soft" criteria such as strategic and organizational fit. They care about the fate of their employees as well as whether their strategic vision will be carried on. It is rare for entrepreneurs to hold a formal auction for their company; instead, they hold informal discussions with a small number of potential buyers with whom they see a good fit [Graebner, 2004].

First Solar, Inc., was founded in 1999 by Harold McMaster. The company manufactures cadmium telluride-based thin-film solar panels. In 2006, the company became profitable and raised about $400 million in its IPO. Subsequently, the founders sold some of their shares as a way to harvest a portion of their gains. They retained ownership of the majority of their stock, but had the alternative of liquidating shares in the future.

AIR: A Successful Harvest

In 1994, David Edwards, a postgraduate researcher in Robert Langer's lab at MIT, began working on a novel drug delivery system that used large, porous particles to deliver drugs directly to the lungs. Although the technology was entering a seemingly crowded market, Edwards's idea had the potential to perform significantly better than other forms of inhalation delivery systems. From the technology's initial stages, Langer, who had founded several successfully biotechnology firms, recognized the commercial potential of Edwards's idea, but an attempt in 1995 to license the technology to a public drug delivery firm was unsuccessful.

Reluctant to start his own company when the technology was still pre-liminary, Edwards left MIT in early 1995 for a faculty position at Penn-sylvania State University. While at Penn State, he continued to refine the delivery system and to visit Langer at MIT about once a month. By early 1997, Edwards and Langer were sufficiently pleased with their progress to approach Terry McGuire, a Harvard Business School graduate and recent founder of Polaris Ventures. Langer knew McGuire because he had invested in several of Langer's ideas. McGuire was initially hesitant to invest in the novel technology, which was entering a crowded market. His faith in Langer, however, as both a stellar scientist and entrepreneur, along with an influential *Science* article and external affirmation of the importance of Edwards's technology, convinced McGuire to put aside his uncertainty. In the summer of 1997, McGuire invested $250,000 in return for 11 percent of Advanced Inhalation Research (AIR) and an option to purchase an additional 9 percent. McGuire took on the role of temporary CEO, and Edwards took a leave of absence from Penn State to return to Boston to work on the idea full time. In January 1998, the company opened its headquarters in Cambridge, Massachusetts. The first three employees were all previously affiliated with the chemical engineering department at MIT.

Once established in their new offices, Langer and Edwards quickly went to work on their first human clinical trial. As their science progressed, so too did interest from outside parties. The founders decided on a two-tier strategy. For drugs that had gone off patent, they would manufacture generic versions of the drugs themselves. For newly developed drugs, AIR would partner with leading pharmaceutical firms to exclusively manufacture the drugs for the delivery system. McGuire focused on successfully closing the deals with minimal dilution of Polaris's stake in the firm.

For a nascent firm, AIR was extremely successful. It had raised capital for an early-stage technology, was making progress in its clinical trials, and was forming favorable partnerships. Nevertheless, when suitors began knocking at the door, AIR needed to consider their offers seriously.

To continue to grow, AIR would need to scale up its operations significantly and move from a small R & D firm to a larger manufacturing operation. This task was extremely difficult and one that only a small handful of biotechnology firms had achieved. In late 1998, just a year and a half after AIR was founded, Alkermes offered to acquire the firm. In February 1999, AIR agreed to Alkermes's offer of an all-stock deal valued around $125 million. The deal was considered a success on all sides. Polaris received a healthy return on its investment, AIR's technology was a strategic fit for Alkermes's existing drug delivery systems, and Alkermes's savvy management team and established reputation in the industry allowed it to find the most appropriate partner for AIR.

All three founders remain active in the biotechnology industry. Edwards, now a professor at Harvard University, cofounded Pulmatrix in the spring of 2003 and has been involved in the formation of a number of nonprofit organizations. McGuire remains a general partner with Polaris Ventures and has invested in other early-stage life science firms. Finally, Langer continues to be wildly prolific. He holds more than 600 granted and pending patents, has licensed his technology to more than 100 firms, and was named one of the 100 most important people by CNN and *Time* magazine.

Sources: Roberts and Gardner, 2000, and www.alkermes.com.

16.6 Exit and Failure

A large percentage of new ventures shut down within a few years of initiation. Some terminate their efforts when they fail to achieve the original goals; others terminate when they simply run out of cash. Most entrepreneurs and investors assert that new ventures fail because of inadequate management skills, a poor strategy, and inadequate capitalization as well as poor market conditions [Zacharakis et al., 1999]. For most entrepreneurs, an inadequate team with inadequate past experience leads to failure.

Many people learn from business failure by revising their knowledge and assumptions about their skills [Shepherd, 2003]. Learning can be measured in terms of an increased understanding of why the business failed, in an effort to prevent repeating the same mistakes. Often entrepreneurs are overconfident and unrealistically overrate their knowledge and skills. They discount the risks inherent in a new venture. Furthermore, they may exaggerate their ability to control events and people. Examples of highly visible failures in telecommunications are Helio and Amp'd Mobile.

Knowing when to stop or terminate a venture may be as important as knowing when to start. The concept of **sunk costs** is that a cost that has already incurred cannot be affected by any present or future decisions. In other words, funds and time invested on a new venture are already gone, regardless of any action you take today or later. If a new venture has not worked out as planned,

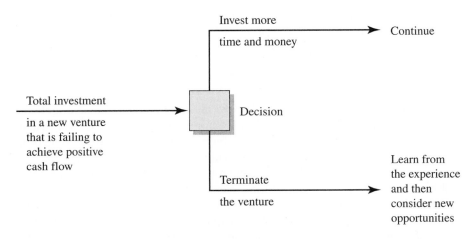

FIGURE 16.7 Decision tree for the sunk cost dilemma.

one should look at proceeding with the venture as a new decision, as shown in Figure 16.7. The decision to terminate or continue should, if possible, be looked at afresh with the information at hand. If a venture has run out of cash and the market has not responded to the new venture, it may be wise to exit the venture.

Pandesic was a joint venture of SAP and Intel designed to develop information architectures for Internet companies. Founded in 1997, Pandesic intended to create a unique e-business solution that would automate the entire business process for companies doing business on the Internet. By March 1997, Pandesic had grown to 100 employees. By the end of 1997, sales were slowly appearing, and by 1998, it was still struggling without many sales. By April 1999, the sales force was reorganized. As a result, sales started to grow, but the firm was still unprofitable. Pandesic had 400 employees and 100 customers by mid-2000, but it was still experiencing negative cash flow (said to be $80 million per year) and closed its doors in July 2000. The decision to terminate was based on large cash losses and recognition that profitability was not within reach [Girard, 2000].

If the decision is to continue, it may be wise to consider the next phase of the company as a turnaround and devise deliberate interventions that increase the level of communication, collaboration, and respect among all the participants [Kanter, 2003].

If the decision is to terminate the new venture, it is best to try to learn from it. Ideally, it is the venture that failed, not the people in it. Picking oneself up, learning from the venture, and then moving on is the best process. Every exit is an entry somewhere else.

The alternative of investing more time and money should be based on rational recognition of the sunk costs and the potential for recovery and success.

An entrepreneur can examine the situation afresh and decide if the opportunity still looks good enough to invest more time and money.

16.7 AgraQuest

AgraQuest has a sales revenue model based on selling units (pounds or gallons) of biopesticides. Its profit model is based on a protected innovation via patents and its emerging brand (reputation) in the biopesticide industry. To become profitable, AgraQuest needs to increase the quantity of units sold (Q) and reduce its variable costs (VC), as described in equation 16.1. It has high fixed costs (FC), and is underutilizing its manufacturing plant.

AgraQuest's growth plan calls for doubling sales revenues each year for the next several years. With unused capacity at its plant, this should be achievable but will require increased working capital. Therefore, several million dollars of working capital will be sought through a line of credit from its bank.

Getting accepted in the pesticide marketplace has been AgraQuest's toughest challenge. Conventional, nonorganic growers are wary of "greener" solutions because of a long history of untruthful claims made by companies peddling unreliable products.

The harvest plan for AgraQuest is either an initial public offering of its shares of common stock or acquisition by a large agricultural technology firm such as DuPont or Monsanto. This may be achievable now that AgraQuest is profitable and has sales exceeding $10 million dollars.

16.8 Summary

A new firm formulates its revenue model to clearly describe how it will generate and grow its revenues. Revenues are important, but positive cash flow and profitability are critical to ultimate success. Thus, cost and profit models that can be readily implemented must be created early in a firm's planning. Profit does not occur naturally but rather flows from value shared with the customer. Thus, enough customers must experience high value and find that the firm is the best, if not the only, provider of this value.

Managing revenue growth to match the growth of cash flow is important to achieve organic, internally funded growth. Otherwise, the new firm must constantly seek new financial resources from investors and lenders. Many firms find it necessary to terminate their activities when they are unable to access new sources of funds. With positive results and reasonable growth, a firm should consider a plan to harvest the rewards so that all owners receive a financial return on their investment.

> **Principle 16**
> A new firm with a powerful revenue and profit engine can achieve strong but manageable growth leading to a favorable harvest of the wealth for the owners.

Video Resources

Visit http://techventures.stanford.edu to view experts discussing content from this chapter.

Reducing Company Costs: PayPal and SpaceX	Elon Musk	SpaceX
Generating Revenue	Mark Zuckerberg	Facebook

16.9 Exercises

16.1 A company's value chain was discussed in Section 14.1. Figure 16.1 outlines the division between profit and cost. Extend Figure 16.1 to account for the profits and costs of partners and suppliers in a company's value chain.

16.2 Google, the search engine firm, uses a complex revenue model and a related profit engine. It has a large base of users and advertisers, and works to link its users to retailers. Describe its revenue and profit model. Contrast this with Yahoo's revenue model. Examining recent quarterly reports will give an indication of the "split" of revenue among business units.

16.3 Zipcar gives its members short-term, on-demand use of a fleet of cars (www.zipcar.com). Before launching this service, Zipcar proposed a pricing scheme of a $300 refundable security deposit, a $300 annual subscription fee, and $1.50 per hour plus 40 cents per mile. Create another pricing scheme that will be more profitable.

16.4 Skype offers Voice over Internet Protocol (VoIP) telephony services and PC to PC calling. The telephone service began as a free service. Describe the revenue and profit model for Skype specifically. Has its acquisition by eBay changed Skype's revenue model?

16.5 Salesforce.com sells software as a service delivered online. Corporations pay about $60 per month for each user (www.salesforce.com). Describe its revenue model and profit model.

16.6   Research and determine the most profitable industry sector in a country of interest (profitability on either a relative or absolute basis). What is driving this high profitability? Is it sustainable, and for how long?

16.7   Compare the most recent yearly income statement for Microsoft, Apple, Dell, Sony, and Qualcomm. Examine the gross margins for each of these companies (cost of goods sold divided by revenue). Complete a similar ratio analysis against Sales, General & Administrative (SG&A) and Research & Development (R&D). When normalized against revenues, what do these ratios tell us about the strategies and operations of these companies?

VENTURE CHALLENGE

1.   Describe the revenue model of your venture.

2.   Describe the profit model of your venture.

3.   Using Table 16.7, discuss your harvest plan.

CHAPTER 17

The Financial Plan

Budgets are not merely affairs of arithmetic, but in a thousand ways go to the root of prosperity of individuals, the relation of classes, and the strength of kingdoms.

William E. Gladstone

How do entrepreneurs describe the financial elements of their new venture?

Entrepreneurs build a financial plan to determine the economic potential for their venture. This plan provides an estimate of the potential of the venture. Of course, any estimate is based on a set of assumptions regarding sales revenues and costs. Using the best available information and their intuition, entrepreneurs calculate the potential profitability of the venture. Furthermore, they need to determine the flow of cash monthly to identify the cash investments that will be required over a two- or three-year period. An income statement and a balance sheet also are required to demonstrate profitability and liquidity.

Using the estimates of sales, the venture team can determine the number of units it needs to sell to break even. Furthermore, it can calculate several measures of profitability that demonstrate the return provided by its venture for investors. The best ventures grow sales consistently and provide positive cash flow and profit early in their life. ■

17.1 Building a Financial Plan

A sound business plan is based on a solid vision and a business design or concept. It is an expression of the theory of the business in the form of a story. This story also needs to make sense financially as a business model. The business model tells a story about the customer and the value proposition that leads to revenue and profit. To create this value for the customer, a new firm needs to build a financial plan that describes the expected revenues, cash flows, profits, and investments necessary to achieve them. The purpose of any business is to create value for its customers and to generate a return on investment for its owners. A financial plan provides an estimate of projected cash flow and return on investment.

To create a financial plan, entrepreneurs must clearly state their assumptions about sales and costs. What resources will it take, over what time frame, to achieve expected sales and profitability? The calculation of cash flows is based on a set of assumptions, which we will call the **base case,** that portrays the most likely outcomes. It may be prudent to also determine the outcome of a situation in which the expectations are not realized as expected, called the **pessimistic case. Cash flow** is the amount of cash flowing into or out of a firm during a specific period. It is arrived at by subtracting the amount paid out in dividends from the net profit and then adding back noncash expenses such as depreciation. See Table 17.12 at the end of this chapter for a glossary of accounting and financial terms.

The entrepreneurs' goal is to develop a solid set of financial projections that will include a pro forma income statement. **Pro forma** means provided in advance of actual data. Pro forma statements are forecasts of financial outcomes. The creation of a set of financial projections starts with a sales forecast based on a set of assumptions regarding the customer and sales growth. Then the calculation of projected sales over a two- or three-year period can be developed. This is step 1, as shown in Table 17.1. The second step is to state the assumed costs of doing business in the time frame described in step 1. In step 2, the costs associated with the projected sales can be calculated. Step 3 is to calculate the expected income and cash flow forecast over the time frame based on a set of assumptions regarding the timing of sales and receipts as well as payables to vendors and others. The final step is to calculate the balance sheet on an annual basis for the two- or three-year period. The balance sheet at the starting point of the new venture will need to be described by stating the assumed starting investments and required assets. Please visit this textbook's websites for tools and links useful towards building a financial plan.

The cash flows, assets, balance sheet, and revenue projections are all interconnected through linkages. Accounting items are classified into "accounts" according to their nature, translated into monetary units, and organized in statements. The basic accounting formula is

$$\text{Assets} = \text{liabilities} + \text{equity}$$

where assets are what the company owns and liabilities are the amounts it owes to other persons or entities. Equity is the company's net worth (book value) expressed as

TABLE 17.1 Four steps to building a financial plan.

1. Sales forecast

- Time frame—two or three years
- Assumptions about sales per customer, number of customers, and growth rate of sales
- Calculation of the sales forecast

2. Costs forecast

- Assumptions about the costs of doing business in the specified time frame
- Calculation of the costs associated with the projected sales of step 1

3. Income and cash flow forecast

- Assumptions about the timing of cash receivables and payables specified in the time frame
- Calculation of the income and cash flow associated with the projected sales and costs on a monthly basis over the time frame

4. Balance sheet

- Assumptions about the starting value of cash and assets
- Calculation based on the income and cash flows from step 3

$$\text{Equity} = \text{assets} - \text{liabilities}$$

Equity is the ownership of the firm, usually divided into certificates called common or preferred shares of stock. The assets and liabilities are linked to income and expenses, as shown in Figure 17.1. Assets are used to generate income, and liabilities require expenses such as rent, payments, or return of loans. Book value is the firm's net worth and is often called accounting value. Market value is share price times the number of issued shares. Note that book value is not equal to market value, which is the perceived value of the firm given its growth potential.

The financial plan is critical to the evaluation of the business model of the new venture. With sound assumptions, the projected results will help in evaluating the venture and its financial viability. The resulting financial plan is only as good as the quality of the assumptions. One reason forecasting models are so fallible is that they rely on the assumptions that the user chooses to input [Riggs, 2004].

New firms should select two or three parameters of the business that display the greatest impact on the cash flows of the business. Examples are sales growth rate and new customer acquisition rate. They should then test the changes in sales as each of these parameters is changed. For example, a software firm, which licenses its products, will examine the potential range for its growth rate of licenses sold.

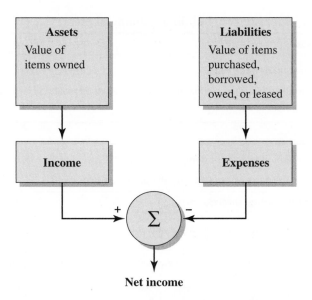

FIGURE 17.1 Assets generate income and liabilities lead to expenses. Net income is income minus expenses.

17.2 Sales Projections

Sales projections will normally be developed for a two- or three-year period on a monthly basis. The sales forecast for a new venture is often the weakest link in the financial plan. Since the new venture has not actually obtained sales, the firm can only work with assumptions based on inadequate information.

In this chapter, consider a fictitious new venture, named e-Travel. It seeks to sell travel guide books (e-books) directly to readers who will download the guides via the Internet and read them on their e-book reader, such as a laptop computer or a handheld device. Short books such as these travel guides operate on a pull model, since customers order them only when they wish to read them. This is opposite of the usual push model, in which the publisher prints the book and then tries to find a purchaser. For travelers, an e-reader is easier to use than several heavy books. Using keywords such as "pizza restaurant Denver," the reader can use the search function to get the information quickly.

The new venture, named e-Travel, needs to build a financial plan. The first step is to build the sales projections. e-Travel has created a network of the best authors of travel guides. These authors have signed publisher agreements and have provided e-Travel with electronic guides to over five hundred cities, regions, and leisure and recreation destinations throughout the world. All these guides are written following a common format, and key search words are identified.

e-Travel expects to sell 1,200 guides starting in the third month of operation. The price per guide is $15, paid by credit card when the guide is ordered online. Based on market research, the expected growth rate of sales is 10 percent per month. The pessimistic growth rate is 1 percent per month. The sales projections for the expected growth rate are shown in Table 17.2 for a three-year period. Assuming a 10 percent growth rate per month, sales amount to over $3 million in year three. Note the bolded ovals for key results in tables.

17.3 Costs Forecast

To determine the expected costs of doing business, the new venture team must examine its needs for facilities, equipment, and employees. Our example business, e-Travel, will need an office, computers, software, and office furniture. The authors will constantly update the information in their guides and will receive royalties at 12 percent of net revenues paid on the fifteenth day of the month following the sale generated by their guides. The costs of doing business will include salaries, marketing, and communication costs, and normal office utilities and supplies. The cost assumptions for e-Travel are summarized in Table 17.3. Every new venture needs to construct a set of assumptions at a similar level of detail.

17.4 Income Statement

The income statement reports the economic results of a firm over a time period. The income statement is calculated as shown in Figure 17.2 [Maher et al., 2004]. Note that operating expenses are often grouped into four general categories: sales and marketing, general and administrative, research and development, and depreciation. Other formats are acceptable.

The income (profit and loss) statement for e-Travel depicts the venture's expected performance over a period of time—in this case, three years. The sales, costs, and profits or losses are shown monthly. The purpose of the income statement is to show how much profit or loss is generated. Due to the online nature of e-Travel, it has no cost of goods sold. The profit and loss statement is shown in Table 17.4. The firm is profitable by the fifth month and shows a profit of $19,954 for the first year (for the base case).

17.5 Cash Flow Statement

The cash flow statement shows the actual flow of cash into and out of the venture. The cash flow statement tracks when the venture actually receives and spends the cash. A venture with positive cash flow can continue to operate without new debt or equity capital. If a cash flow statement reveals projected negative cash in some period, it will be necessary to plan for a new capital infusion. We define cash flow as the sum of retained earnings minus the depreciation provision made by a firm [Maher et al., 2004].

TABLE 17.2 Sales projections for the expected growth rate (10 percent per month).

Year 1

	1	2	3	4	5	6	7	8	9	10	11	12	Year total
Month	1	2	3	4	5	6	7	8	9	10	11	12	
Units	0	0	1,200	1,320	1,452	1,597	1,757	1,933	2,126	2,339	2,573	2,830	19,127
Price per unit	$15	$15	$15	$15	$15	$15	$15	$15	$15	$15	$15	$15	
Sales dollars	$0	$0	$18,000	$19,800	$21,780	$23,955	$26,355	$28,995	$31,890	$35,085	$38,595	$42,450	$286,905

Year 2

	1	2	3	4	5	6	7	8	9	10	11	12	Year total
Month	1	2	3	4	5	6	7	8	9	10	11	12	
Units	3,113	3,424	3,766	4,143	4,557	5,013	5,514	6,065	6,672	7,339	8,073	8,880	66,559
Price per unit	$15	$15	$15	$15	$15	$15	$15	$15	$15	$15	$15	$15	
Sales dollars	$46,695	$51,360	$56,490	$62,145	$68,355	$75,195	$82,710	$90,975	$100,080	$110,085	$121,095	$133,200	$998,385

Year 3

	1	2	3	4	5	6	7	8	9	10	11	12	Year total
Month	1	2	3	4	5	6	7	8	9	10	11	12	
Units	9,768	10,745	11,820	13,002	14,302	15,732	17,305	19,036	20,940	23,034	25,337	27,871	208,892
Price per unit	$15	$15	$15	$15	$15	$15	$15	$15	$15	$15	$15	$15	
Sales dollars	$146,520	$161,175	$177,300	$195,030	$214,530	$235,980	$259,575	$285,540	$314,100	$345,510	$380,055	$418,065	$3,133,380

TABLE 17.3 Cost assumptions for e-Travel.

■ Author royalties: 12 percent of net revenues paid on the fifteenth of the month following the sale

■ Credit card: 1 percent to credit card providers, paid electronically as sale is processed

■ Office rent: $1,500 per month

■ Physical assets (computers, furniture, etc.): $48,000 per year; four-year life

■ Depreciation of equipment (monthly):

Year 1	Year 2	Year 3
$1,000	$2,000	$3,000

■ Salaries (monthly):

	Year 1	Year 2	Year 3
President	$4,000	$ 5,500	$ 6,500
Vice president	4,000	5,000	6,000
Administrative manager	1,500	3,000	3,500
Total:	$9,500	$13,500	$16,000

■ Social Security taxes and other benefits: 15 percent of total salaries

■ Marketing (monthly):

Year 1	Year 2	Year 3
$2,000	$2,500	$3,000

■ Utilities, supplies, travel, communication (monthly):

Year 1	Year 2	Year 3
$2,000	$3,000	$4,000

■ Interest expense: $1,000 per month ($100,000 loan at 12 percent per year; interest paid monthly; principal to be paid at end of five years)

■ Income taxes: 30 percent of income before taxes

A growing business needs cash to operate. The detailed cash flow process is shown in Figure 17.3. Firms calculate their cash on hand at the end of each month. Therefore,

$$TC(N + 1) = (CF - Disbursements) + TC(N)$$

where $TC(N + 1)$ is the cash at the end of month $(N + 1)$, $TC(N)$ is the total cash at the end of month (N), and CF is the cash flow for the month.

The cash flow statement for e-Travel is provided in Table 17.5 (for the base case). It is assumed that the founders invest $140,000 in cash and obtain a bank loan of $100,000 secured by their personal assets. This $240,000 is for operations

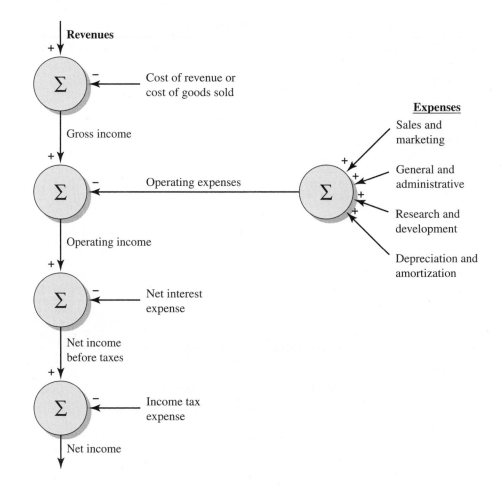

FIGURE 17.2 Calculation of the income statement.

and the initial purchase of long-term assets such as computers and furniture, as shown in month 1 of year 1 in Table 17.5. The initial investment of $240,000 can be considered an equity investment since the loan is personally guaranteed by the two founders. Under the base case assumption of 10 percent growth in revenues each month, the cash flow quickly becomes positive.

17.6 Balance Sheet

The new venture team should prepare a balance sheet at the opening of the business and for the end of each year. The balance sheet depicts the conditions of the business by displaying the assets, liabilities, and owners' equity of the business [Maher et al., 2004]. The format for a balance sheet is shown in Figure 17.4 for a business at the end of the year.

TABLE 17.4 Profit and loss statement.

Year 1

Month	1	2	3	4	5	6	7	8	9	10	11	12	Year total
Revenues	$0	$0	$18,000	$19,800	$21,780	$23,955	$26,355	$28,995	$31,890	$35,085	$38,595	$42,450	$286,905
Expenses:													
Author royalties	0	0	2,160	2,376	2,614	2,875	3,163	3,479	3,827	4,210	4,631	5,094	34,429
Credit card charges	0	0	180	198	218	240	264	290	319	351	386	425	2,871
Marketing	2,000	2,000	2,000	2,000	2,000	2,000	2,000	2,000	2,000	2,000	2,000	2,000	24,000
Depreciation	1,000	1,000	1,000	1,000	1,000	1,000	1,000	1,000	1,000	1,000	1,000	1,000	12,000
Interest	1,000	1,000	1,000	1,000	1,000	1,000	1,000	1,000	1,000	1,000	1,000	1,000	12,000
Office rent	1,500	1,500	1,500	1,500	1,500	1,500	1,500	1,500	1,500	1,500	1,500	1,500	18,000
Salaries	9,500	9,500	9,500	9,500	9,500	9,500	9,500	9,500	9,500	9,500	9,500	9,500	114,000
Social Security and benefits	1,425	1,425	1,425	1,425	1,425	1,425	1,425	1,425	1,425	1,425	1,425	1,425	17,100
Utilities, supplies, travel, communication	2,000	2,000	2,000	2,000	2,000	2,000	2,000	2,000	2,000	2,000	2,000	2,000	24,000
Profit (loss) before income tax	(18,425)	(18,425)	(2,765)	(1,199)	523	2,415	4,503	6,801	9,319	12,099	15,153	18,506	28,505
Income tax (credit)	(5,528)	(5,528)	(830)	(360)	157	725	1,351	2,040	2,796	3,630	4,546	5,552	8,551
Net profit (loss)	($12,897)	($12,897)	($1,935)	($839)	$366	$1,690	$3,152	$4,761	$6,523	$8,469	$10,607	$12,954	$19,954

Year 2

Month	1	2	3	4	5	6	7	8	9	10	11	12	Year total
Revenues	$46,695	$51,360	$56,490	$62,145	$68,355	$75,195	$82,710	$90,975	$100,080	$110,085	$121,095	$133,200	$998,385
Expenses:													
Author royalties	5,603	6,163	6,779	7,457	8,203	9,023	9,925	10,917	12,010	13,210	14,531	15,984	119,805
Credit card charges	467	514	565	621	684	752	827	910	1,001	1,101	1,211	1,332	9,985
Marketing	2,500	2,500	2,500	2,500	2,500	2,500	2,500	2,500	2,500	2,500	2,500	2,500	30,000
Depreciation	2,000	2,000	2,000	2,000	2,000	2,000	2,000	2,000	2,000	2,000	2,000	2,000	24,000
Interest	1,000	1,000	1,000	1,000	1,000	1,000	1,000	1,000	1,000	1,000	1,000	1,000	12,000
Office rent	1,500	1,500	1,500	1,500	1,500	1,500	1,500	1,500	1,500	1,500	1,500	1,500	18,000
Salaries	13,500	13,500	13,500	13,500	13,500	13,500	13,500	13,500	13,500	13,500	13,500	13,500	162,000
Social Security and benefits	2,025	2,025	2,025	2,025	2,025	2,025	2,025	2,025	2,025	2,025	2,025	2,025	24,300
Utilities, supplies, travel, communication	3,000	3,000	3,000	3,000	3,000	3,000	3,000	3,000	3,000	3,000	3,000	3,000	36,000
Profit before income tax	15,100	19,158	23,621	28,542	33,943	39,895	46,433	53,623	61,544	70,249	79,828	90,359	562,295
Income tax	4,530	5,747	7,086	8,563	10,183	11,969	13,930	16,087	18,463	21,075	23,948	27,108	168,689
Net profit	$10,570	$13,411	$16,535	$19,979	$23,760	$27,926	$32,503	$37,536	$43,081	$49,174	$55,880	$63,251	$393,606

(continued on next page)

TABLE 17.4 (continued)

Year 3

Month	1	2	3	4	5	6	7	8	9	10	11	12	Year total
Revenues	$146,520	$161,175	$177,300	$195,030	$214,530	$235,980	$259,575	$285,540	$314,100	$345,510	$380,055	$418,065	$3,133,380
Expenses:													
Author royalties	17,582	19,341	21,276	23,404	25,744	28,318	31,149	34,265	37,692	41,461	45,607	50,168	376,007
Credit card charges	1,465	1,612	1,773	1,950	2,145	2,360	2,596	2,855	3,141	3,455	3,801	4,181	31,334
Marketing	3,000	3,000	3,000	3,000	3,000	3,000	3,000	3,000	3,000	3,000	3,000	3,000	36,000
Depreciation	3,000	3,000	3,000	3,000	3,000	3,000	3,000	3,000	3,000	3,000	3,000	3,000	36,000
Interest	1,000	1,000	1,000	1,000	1,000	1,000	1,000	1,000	1,000	1,000	1,000	1,000	12,000
Office rent	1,500	1,500	1,500	1,500	1,500	1,500	1,500	1,500	1,500	1,500	1,500	1,500	18,000
Salaries	16,000	16,000	16,000	16,000	16,000	16,000	16,000	16,000	16,000	16,000	16,000	16,000	192,000
Social Security and benefits	2,400	2,400	2,400	2,400	2,400	2,400	2,400	2,400	2,400	2,400	2,400	2,400	28,800
Utilities, supplies, travel, communication	4,000	4,000	4,000	4,000	4,000	4,000	4,000	4,000	4,000	4,000	4,000	4,000	48,000
Profit before income tax	96,573	109,322	123,351	138,776	155,741	174,402	194,930	217,520	242,367	269,694	299,747	332,816	2,355,239
Income tax	28,972	32,797	37,005	41,633	46,722	52,321	58,479	65,256	72,710	80,908	89,924	99,845	706,572
Net profit	$67,601	$76,525	$86,346	$97,143	$109,019	$122,081	$136,451	$152,264	$169,657	$188,786	$209,823	$232,971	$1,648,667

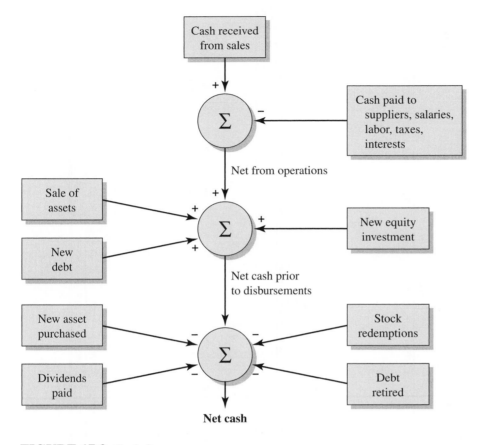

FIGURE 17.3 Cash flow process.

The balance sheet for e-Travel is shown in Table 17.6. The balance sheet shows the assets, such as cash, equipment, furniture, and accumulated depreciation. The liabilities are the loan payable and the royalties to the authors. Total owners' equity consists of contributions of $140,000 and retained earnings. Table 17.6 shows the balance sheet at the end of month 1, year 1, year 2, and year 3. The base case balance sheet provides evidence of the financial strength of e-Travel.

17.7 Results for a Pessimistic Growth Rate

Any new venture needs to plan for the likely case and prepare for the worst case. For e-Travel, we will assume the pessimistic case occurs when the sales grow at a rate of only 1 percent per month. The summary of the results for the pessimistic case is shown in Table 17.7. Table 17.7A shows the sales projections for the first three years. Sales for year 3 are $284,010 for the pessimistic case, while they were estimated at $3,133,380 for the expected case. Table 17.7B shows the profit and loss statement for the pessimistic case. Note that the firm is not profitable in the pessimistic case in any year.

TABLE 17.5 Cash flow statement.

Year 1

Month	1	2	3	4	5	6	7	8	9	10	11	12	Year total
Operating activities													
Net profit (loss)	($12,897)	($12,897)	($1,935)	($839)	$366	$1,690	$3,152	$4,761	$6,523	$8,469	$10,607	$12,954	$19,954
Add: Depreciation	1,000	1,000	1,000	1,000	1,000	1,000	1,000	1,000	1,000	1,000	1,000	1,000	12,000
Add: Increase in royalties payable			2,160	216	238	261	288	316	348	383	421	463	5,094
Cash flow from operations	(11,897)	(11,897)	1,225	377	1,604	2,951	4,440	6,077	7,871	9,852	12,028	14,417	37,048
Investing activities													
Purchase of long-term assets	(48,000)												(48,000)
Financing activities													
Bank loan	100,000												100,000
Owners' cash contributions	140,000												140,000
Increase (decrease) in cash	180,103	(11,897)	1,225	377	1,604	2,951	4,440	6,077	7,871	9,852	12,028	14,417	229,048
Beginning cash balance	0	180,103	168,206	169,431	169,808	171,412	174,363	178,803	184,880	192,751	202,603	214,631	0
Ending cash balance	$180,103	$168,206	$169,431	$169,808	$171,412	$174,363	$178,803	$184,880	$192,751	$202,603	$214,631	$229,048	$229,048

Year 2

Month	1	2	3	4	5	6	7	8	9	10	11	12	Year total
Operating activities													
Net profit	$10,570	$13,411	$16,535	$19,979	$23,760	$27,926	$32,503	$37,536	$43,081	$49,174	$55,880	$63,251	$393,606
Add: Depreciation	2,000	2,000	2,000	2,000	2,000	2,000	2,000	2,000	2,000	2,000	2,000	2,000	24,000
Add: Increase in royalties payable	509	560	616	678	746	820	902	992	1,093	1,200	1,321	1,453	10,890
Cash flow from operations	13,079	15,971	19,151	22,657	26,506	30,746	35,405	40,528	46,174	52,374	59,201	66,704	428,496
Investing activities													
Purchase of long-term assets	(48,000)												(48,000)
Increase (decrease) in cash	(34,921)	15,971	19,151	22,657	26,506	30,746	35,405	40,528	46,174	52,374	59,201	66,704	380,496
Beginning cash balance	229,048	194,127	210,098	229,249	251,906	278,412	309,158	344,563	385,091	431,265	483,639	542,840	229,048
Ending cash balance	$194,127	$210,098	$229,249	$251,906	$278,412	$309,158	$344,563	$385,091	$431,265	$483,639	$542,840	$609,544	$609,544

Year 3

Month	1	2	3	4	5	6	7	8	9	10	11	12	Year total
Operating activities													
Net profit	$67,601	$76,525	$86,346	$97,143	$109,019	$122,081	$136,451	$152,264	$169,657	$188,786	$209,823	$232,971	$1,648,667
Add: Depreciation	3,000	3,000	3,000	3,000	3,000	3,000	3,000	3,000	3,000	3,000	3,000	3,000	36,000
Add: Increase in royalties payable	1,598	1,759	1,935	2,128	2,340	2,574	2,831	3,116	3,427	3,769	4,146	4,561	34,184
Cash flow from operations	72,199	81,284	91,281	102,271	114,359	127,655	142,282	158,380	176,084	195,555	216,969	240,532	1,718,851
Investing activities													
Purchase of long-term assets	(48,000)												(48,000)
Increase (decrease) in cash	24,199	81,284	91,281	102,271	114,359	127,655	142,282	158,380	176,084	195,555	216,969	240,532	1,670,851
Beginning cash balance	609,544	633,743	715,027	806,308	908,579	1,022,938	1,150,593	1,292,875	1,451,255	1,627,339	1,822,894	2,039,863	609,544
Ending cash balance	$633,743	$715,027	$806,308	$908,579	$1,022,938	$1,150,593	$1,292,875	$1,451,255	$1,627,339	$1,822,894	$2,039,863	$2,280,395	$2,280,395

Balance sheet of ABC Corporation, 31 December 200X

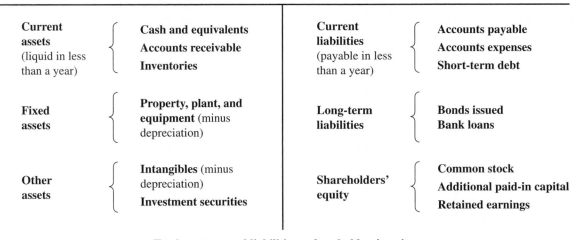

Total assets = total liabilities + shareholders' equity

FIGURE 17.4 Format for a balance sheet.

We show the cash flow statement for the pessimistic case in Table 17.7C. Notice that the ending cash balance turns negative in month 1 of year 3. The company would need a cash infusion in month 1 of year 3 to continue operating.

To be a profitable venture, e-Travel needs to attain a sales growth rate that exceeds 4 percent per month over the first two years. This figure can be determined by modifying the spreadsheet calculation with various growth rates.

17.8 Breakeven Analysis

In the initial stages of building a financial plan, it is useful to know when a profit may be achieved. **Breakeven** is defined as when the total sales equals the total costs. Total sales (R) are

$$R = Q \times P$$

where Q = number of units sold and P = price per unit. Total cost (TC) is

$$TC = FC + VC$$

where FC = total fixed costs and VC = variable costs. Thus, breakeven is the volume of sales (Q) at which the venture will neither make a profit nor incur a loss. Sales in excess of the volume of sales needed to cover costs will result in a profit.

Total fixed costs are $221,100 for e-Travel in year 1 for the base case, and variable costs are 13 percent of sales, since royalty and credit card costs are 12 percent and 1 percent, respectively. Then, to determine Q, we have

$$R = TC$$
$$R = \$221,100 + (0.13 \times R)$$

TABLE 17.6 Balance sheet.

End of month 1 of year 1			End of year 1		
Assets			**Assets**		
Cash		$180,103	Cash		$229,048
Equipment and furniture		48,000	Equipment and furniture		48,000
Accumulated depreciation		(1,000)	Accumulated depreciation		(12,000)
Total assets		$227,103	Total assets		$265,048
Liabilities			**Liabilities**		
Loan payable		$100,000	Royalties payable		$5,094
			Loan payable		100,000
Owners' equity			**Owners' equity**		
Owners' contributions		140,000	Owners' contributions		140,000
Retained earnings (deficit)		(12,897)	Retained earnings		19,954
Total owners' equity		127,103	Total owners' equity		159,954
Total liabilities and owners' equity		$227,103	Total liabilities and owners' equity		$265,048
End of year 2			**End of year 3**		
Assets			**Assets**		
Cash		$609,544	Cash		$2,280,395
Equipment and furniture		96,000	Equipment and furniture		144,000
Accumulated depreciation		(36,000)	Accumulated depreciation		(72,000)
Total assets		$669,544	Total assets		$2,352,395
Liabilities			**Liabilities**		
Royalties payable		$15,984	Royalties payable		$50,168
Loan payable		100,000	Loan payable		100,000
Owners' equity			**Owners' equity**		
Owners' contributions		140,000	Owners' contributions		140,000
Retained earnings		413,560	Retained earnings		2,062,227
Total owners' equity		(553,560)	Total owners' equity		2,202,227
Total liabilities and owners' equity		$669,544	Total liabilities and owners' equity		$2,352,395

TABLE 17.7A Sales projections for the pessimistic growth rate (1 percent per month).

Year 1		Year 2		Year 3	
	Year total		Year total		Year total
Units	12,550	Units	16,804	Units	18,934
Price per unit	$15	Price per unit	$15	Price per unit	$15
Sales dollars	$188,250	Sales dollars	$252,060	Sales dollars	$284,010

TABLE 17.7B **Profit and loss statement for the pessimistic growth rate (1 percent per month).**

Year 1		Year 2	
	Year total		Year total
Revenues	$188,250	Revenues	$252,060
Expenses:		Expenses:	
Author royalties	22,590	Author royalties	30,246
Credit card charges	1,883	Credit card charges	2,521
Marketing	24,000	Marketing	30,000
Depreciation	12,000	Depreciation	24,000
Interest	12,000	Interest	12,000
Office rent	18,000	Office rent	18,000
Salaries	114,000	Salaries	162,000
Social Security and benefits	17,100	Social Security and benefits	24,300
Utilities, supplies, travel, communication	24,000	Utilities, supplies, travel, communication	36,000
Profit (loss) before income tax	(57,323)	Profit before income tax	(87,007)
Income tax (credit)	0	Income tax	0
Net profit (loss)	($57,323)	Net profit (loss)	($87,007)

Year 3	
	Year total
Revenues	$284,010
Expenses:	
Author royalties	34,079
Credit card charges	2,840
Marketing	36,000
Depreciation	36,000
Interest	12,000
Office rent	18,000
Salaries	192,000
Social Security and benefits	28,800
Utilities, supplies, travel, communication	48,000
Profit before income tax	(123,709)
Income tax	0
Net profit (loss)	($123,709)

TABLE 17.7C Cash flow statement for the pessimistic growth rate (1 percent per month).

Year 1

Month	1	2	3	4	5	6	7	8	9	10	11	12	Year total
Operating activities													
Net profit (loss)	($18,425)	($18,425)	($2,765)	($2,609)	($2,452)	($2,295)	($2,138)	($1,982)	($1,812)	($1,643)	($1,473)	($1,304)	($57,323)
Add: Depreciation	1,000	1,000	1,000	1,000	1,000	1,000	1,000	1,000	1,000	1,000	1,000	1,000	12,000
Add: Increase in royalties payable			2,160	22	21	22	21	22	23	24	23	24	2,362
Cash flow from operations	(17,425)	(17,425)	395	(1,587)	(1,431)	(1,273)	(1,117)	(960)	(789)	(619)	(450)	(280)	(42,961)
Investing activities													
Purchase of long-term assets	(48,000)												(48,000)
Financing activities													
Bank loan	100,000												100,000
Owners' cash contributions	140,000												140,000
Increase (decrease) in cash	174,575	(17,425)	395	(1,587)	(1,431)	(1,273)	(1,117)	(960)	(789)	(619)	(450)	(280)	149,039
Beginning cash balance	0	174,575	157,150	157,545	155,958	154,527	153,254	152,137	151,177	150,388	149,769	149,319	0
Ending cash balance	$174,575	$157,150	$157,545	$155,958	$154,527	$153,254	$152,137	$151,177	$150,388	$149,769	$149,319	$149,039	$149,039

Year 2

Month	1	2	3	4	5	6	7	8	9	10	11	12	Year total
Operating activities													
Net profit (loss)	($8,234)	($8,064)	($7,895)	($7,712)	($7,529)	($7,346)	($7,164)	($6,981)	($6,798)	($6,615)	($6,432)	($6,237)	($87,007)
Add: Depreciation	2,000	2,000	2,000	2,000	2,000	2,000	2,000	2,000	2,000	2,000	2,000	2,000	24,000
Add: Increase in royalties payable	23	23	24	25	25	25	26	25	25	25	25	27	298
Cash flow from operations	(6,211)	(6,041)	(5,871)	(5,687)	(5,504)	(5,321)	(5,138)	(4,956)	(4,773)	(4,590)	(4,407)	(4,210)	(62,709)
Investing activities													
Purchase of long-term assets	(48,000)												(48,000)
Increase (decrease) in cash	(54,211)	(6,041)	(5,871)	(5,687)	(5,504)	(5,321)	(5,138)	(4,956)	(4,773)	(4,590)	(4,407)	(4,210)	(110,709)
Beginning cash balance	149,039	94,828	88,787	82,916	77,229	71,725	66,404	61,266	56,310	51,537	46,947	42,540	149,039
Ending cash balance	$94,828	$88,787	$82,916	$77,229	$71,725	$66,404	$61,266	$56,310	$51,537	$46,947	$42,540	$38,330	$38,330

(continued on next page)

TABLE 17.7C (continued)

Year 3

Month	1	2	3	4	5	6	7	8	9	10	11	12	Year total
Operating activities													
Net profit	($11,416)	($11,220)	($11,024)	($10,829)	($10,633)	($10,424)	($10,216)	($10,007)	($9,799)	($9,589)	($9,380)	($9,172)	($123,709)
Add: Depreciation	3,000	3,000	3,000	3,000	3,000	3,000	3,000	3,000	3,000	3,000	3,000	3,000	36,000
Add: Increase in royalties payable	27	27	27	27	27	29	29	29	29	28	29	29	337
Cash flow from operations	(8,389)	(8,193)	(7,997)	(7,802)	(7,606)	(7,395)	(7,187)	(6,978)	(6,770)	(6,561)	(6,351)	(6,143)	(87,372)
Investing activities													
Purchase of long-term assets	(48,000)												(48,000)
Increase (decrease) in cash	(56,389)	(8,193)	(7,997)	(7,802)	(7,606)	(7,395)	(7,187)	(6,978)	(6,770)	(6,561)	(6,351)	(6,143)	(135,372)
Beginning cash balance	38,330	(18,059)	(26,252)	(34,249)	(42,051)	(49,657)	(57,052)	(64,239)	(71,217)	(77,987)	(84,548)	(90,899)	38,330
Ending cash balance	($18,059)	($26,252)	($34,249)	($42,051)	($49,657)	($57,052)	($64,239)	($71,217)	($77,987)	($84,548)	($90,899)	($97,042)	($97,042)

or

$$0.87R = \$221,100$$

Therefore,

$$0.87 \, (Q \times \$15) = \$221,100$$

or

$$Q = 16,943$$

Therefore, after selling about 17,000 e-Travel guides, the firm is profitable.

17.9 Measures of Profitability

Investors in new ventures are interested in measures of annual return on their investment. The **return on invested capital** (ROIC) is the ratio of the net income earned each year expressed as a percentage of the total invested capital in the firm. The ROIC for Hewlett-Packard and IBM in 2008 was 19 percent and 27 percent, respectively. It is also called **return on investment** (ROI). The ROI for a venture is

$$ROI = \frac{net\ income}{investment}$$

where the income is distributed or allocated to the investor. As a firm grows, it may not actually distribute cash to its investors for some period. In that case, it is retaining earnings and using the retained cash earnings as investment capital. Then the retained earnings are added to the original equity investments to yield the owners' equity [Riggs, 2004].

Net income and owners' equity can provide the ratio called **return on equity** (ROE). ROE is calculated as

$$ROE = \frac{net\ income}{owners'\ equity}$$

Using Tables 17.4 and 17.6, the return on equity for e-Travel for year 2 is

$$ROE = \frac{\$393,606}{\$553,560} \times 100\% = 71.1\%$$

Investor (owner) returns can be calculated at the time of distribution of cash or when the equity is priced in a public market. Assuming the ownership held by the original investors in e-Travel can be sold at the end of year 3 for \$720,000, the multiple (M) achieved by the investor group is

$$M = \frac{\$720,000}{\$240,000} = 3.0$$

TABLE 17.8 Top 10 accounting principles for entrepreneurs.

1. The fundamental equation inherent in financial statements is: Assets = Liabilities + Owners' Equity.

2. The balance sheet is a *snapshot* at a moment in time of *financial position,* while the income statement reports on *financial performance* for a *period.*

3. An income statement details changes in retained earnings for the period.

4. Accounting disputes turn almost solely on valuation and timing.

5. Five key principles govern valuation: Realization (accrual), conservatism, consistency, materiality, and historic cost.

6. Since valuations require judgments, financial statements are necessarily only estimates.

7. "Book values" seldom equal "market values," particularly for long-term assets and owners' equity.

8. The lifeblood of any operation is cash.

9. Ratios are the key tool for drawing meaning from financial statements.

10. A company's ability to finance its growth internally is a function of its return on equity.

Source: Riggs, 2006.

The annual compound return over the three years is 44.2 percent, since $(1.442)^3 = 3.0$. Therefore, the annual return on investment (ROI) is

$$ROI = 44.2\%$$

Note that we designate the investment as \$240,000 since that original investment is the total investment by the founders. We consider a loan countersigned personally as equivalent to an equity investment by the founders.

Table 17.8 summarizes ten core accounting principles for entrepreneurs. Table 17.12 defines various financial terms and ratios at the end of this chapter. See appendix C for sources to compare financial performance metrics.

17.10 AgraQuest

The original business plan for AgraQuest, dated May 5, 1995, requested start-up financing of \$1.1 million for equipment, \$2.5 million for operations, and \$2.9 million for cash reserves—a total initial investment of \$6.5 million. It also projected an initial public offering after five years. The projected sales and income for the first five years is shown in Table 17.9. The projected investment needs for the five years are provided in Table 17.10.

AgraQuest was unable to meet the expected results, and sales grew to only \$6.4 million in 2003. The two assumptions that caused the projections to be unrealistic were (1) contract screening and natural molecules sold and (2) product development schedule. AgraQuest's plan assumed that screening of molecules for other firms and sale of molecules to other firms would result in \$2.4 million in 1997 and \$4.3 million in 1998. None of these revenues were

TABLE 17.9 **Profit and loss statement for AgraQuest as projected in its original business plan.**

	Profit and loss statement				
	1996	1997	1998	1999	2000
Sales	550	2,375	4,250	8,900	16,500
Income	(2,338)	(3,094)	(2,768)	(1,693)	4,284
Average shares outstanding	7,002	9,462	10,972	11,722	11,722

Note: All figures are in thousands of dollars or shares.

TABLE 17.10 **Projected investment requirements as provided in AgraQuest's original business plan.**

	Projected investment requirements				
	1996	1997	1998	1999	2000
Capital required	6,500	5,200	8,600	10,200	0

Note: All figures are in thousands of dollars.

realized. Furthermore, it projected sales of its natural products would be $2.9 million in 1999 and $6.5 million in 2000. These sales were severely delayed because the assumption was that products would be approved by the Environmental Protection Agency in 18 months, but it actually took 36 months. Furthermore, the pipeline of products developed more slowly than planned. These problems demonstrate the fragility of assumptions for any business plan.

17.11 Summary

The entrepreneurial team builds its financial plan to determine the economic potential for its venture and demonstrate it to potential investors. The plan uses projected figures based on the underlying assumptions of the business venture. This plan shows the profit and loss statement and the cash flow statement, which can be used to draw up the balance sheet. Monthly figures are used for the first year or two and quarterly figures for the next two or three years. Furthermore, a calculation of the sale of the required number of units for breakeven will be useful. The best new ventures are able to grow sales consistently and show positive cash flow and profit early in their life.

Principle 17
A sound financial plan demonstrates the potential for growth and profitability for a new venture and is based on the most accurate and reliable assumptions available.

Video Resources

Visit http://techventures.stanford.edu to view experts discussing content from this chapter.

Keeping a Financial Focus	Gajus Worthington	Fluidigm
Extreme Relevance of Cash Flow	Elizabeth Holmes	Theranos

17.12 Exercises

17.1   Define the interrelationships between an income statement, balance sheet, and cash flow statement. Which one is most important for a new venture and why?

17.2   Examples of revenue models are listed in Table 16.1. Select three of these revenue models and explain the challenges in creating sales projections for each.

17.3   Viscotech, Inc. is planning to enter the field of electro-optical systems for automated optical inspection and detection of defects in manufacturing components and modules. The projected financial revenues and income are shown in Table 17.11. The firm plans to start with an equity investment of $10,000 and a five-year loan of $500,000. Determine the cash on hand at the end of each year. Also, determine the return on equity and the return on investment for each year.

17.4   A new venture is launched with an initial investment (cash on hand) of $80,000. It generates sales of $40,000 each month. It has monthly operating costs of $36,000. The firm purchases equipment costing $30,000 each month for the first 4 months. Calculate the return on investment at the end of 12 months. Determine if the cash on hand remains positive at the end of each month. What, if any, investment is required and when?

17.5   A software firm has fixed costs of $800,000 and variable costs of $12 per unit. Calculate the breakeven quantity (Q) when each unit sells for

TABLE 17.11 Viscotech projections.

	Year 1	Year 2	Year 3	Year 4	Year 5
Revenues	1,500	3,400	5,900	10,600	15,400
Income after tax	(500)	(100)	200	400	600
Depreciation	250	300	350	400	400
Average shareholders' equity	1,000	700	600	800	1,400
Long-term debt	500	450	300	200	100

Note: All figures are in thousands of dollars.

$50. If the firm sells 50,000 units in a year, what is its profit for that year? Assume the tax rate for the firm is 20 percent.

17.6 A new firm, Sensor International, is preparing a plan based on its new device to be used in a security network. The cost of manufacturing, marketing, and distributing a package of six sensors is 14 cents, and the price to the distributor is 68 cents. The firm calculates its one-time fixed costs at $121,000. Determine the number of units required for breakeven.

17.7 Continuing exercise 17.6, Sensor International sells 300,000 packages in its first year and 400,000 in its second. If the investor's original investment was $100,000, determine the return on investment in year 1 and year 2 for the firm. Assume a tax rate of 20 percent.

17.8 Reconsider the firm described in exercise 17.6 when it is determined that fixed costs have declined to $30,000 and the firm has determined by studies that it can only expect to sell 60 units this year. What price should it select to ensure selling the 60 units profitably?

17.9 Superconductor Inc. is planning to start on January 1, 2010, with an initial investment of $200,000 from its founder team. It projects sales of $12,000 in February 2010 and a growth rate of sales of 10 percent per month for the foreseeable future (at least two years). It expects to receive payment upon delivery for its unique devices. Its costs are $18,000 for its first month of operation, and it plans for a growth rate of expenses at 2 percent per month. Prepare a cash flow statement for the first 24 months of the firm, and determine if and when an additional investment would be required.

17.10 A new clean technology company has fixed costs of $420,000 and variable costs per unit of $3,100. Competitor analysis shows that the price for a comparable product ranges from $6,500 to $9,300. The goal of the firm is to attain profitability this year while increasing its market share next year. What price should the firm select? What is the breakeven quantity for the selected price?

VENTURE CHALLENGE

1. Describe the assumptions for your venture that will be used to create the financial projections. Sketch out an income statement.

2. How long will it take to break even and become cash-flow positive?

3. How much cumulative cash is necessary to reach that point?

TABLE 17.12 Glossary of accounting and financial terms.

Assets—The value of what the company owns.

Asset velocity—The ratio of revenues to net assets, which include plant and equipment, inventories, and working capital.

Balance sheet—The financial statement that summarizes the assets, liabilities, and shareholders' equity at a specific point in time.

Book value—The net worth of the firm, calculated by total assets minus intangible assets (patents, goodwill) and liabilities.

Cash flow—The sum of retained earnings minus the depreciation provision made by the firm.

Depreciation—The allocation of the cost of a tangible, long-term asset over its useful life. The reduction of the value of asset from wear and tear.

Discount rate—The rate at which future earnings or cash flow are discounted because of the time value of money.

Dividends—A distribution of a portion of the net income of a business to its owners.

Earnings per share—The ratio of net income to shares of stock outstanding.

Equity—The firm's net worth (book value).

Financial statement—A report summarizing the financial condition of a business. It normally includes a balance sheet and an income statement.

Income statement—A financial statement that summarizes revenues and expenses.

Liabilities—The amounts owed to other entities.

Net income—Total income for the period less total expenses for the period.

Pro forma—Provided in advance of actual data.

Retained earnings—Represents the owner's claim on the earnings that have not been paid out in dividends.

Return on invested capital—The ratio of net income to investment.

Return on investment—The ratio of net income to investment.

Return on revenues—The ratio of net income to revenues.

Return on stockholders' equity—Net income divided by average stockholders' equity.

Revenues—Sales after deducting all returns, rebates, and discounts.

Statement of cash flows—The statement that summarizes the cash effects of the operating, investing, and financing activities for a period of time.

Working capital—Current assets minus current liabilities.

CHAPTER 18

Sources of Capital

Capital is to the progress of society what gas is to a car.
James Truslow Adams

What are the sources of capital that a new venture can use to finance the start and growth of its company?

Entrepreneurs can estimate the capital required for their new business by reviewing the financial projections they prepare using the methods detailed in Chapter 17. In examining the projections and the cash flow statement, it becomes clear how much capital will be needed and when. The entrepreneurs may provide some of the required capital, and friends and family may help with modest investments. Government grants and bootstrapping can also supply necessary capital to launch a venture. Most high-growth ventures that expect to grow to a significant scale will need outside capital from professional investors. Typically, several stages of investment will be required over the life of the business.

This chapter addresses the task of attracting investors to a new business and creating an investment offering that will meet the firm's needs and the investors' requirements for an attractive return. In this chapter, we describe the funds that may be available from various sources.

Several stages of investment may be required, and it must be determined what percentage ownership is offered to the investor. This determination is based on the valuation of the new business at each stage. With the mutual agreement of the investor and the firms, the terms of the deal will be recorded and the arrangement completed.

An alternative to an equity investment is debt financing from a bank or other financial institution. Furthermore, a line of credit can help finance short-term cash-flow requirements.

Mature ventures use an initial public offering (IPO) to raise additional growth capital and to offer early investors a means of harvesting the value created in an emerging firm. Preparation for an IPO can be an important milestone for a firm with solid growth potential. ∎

18.1 Financing the New Venture

The financial projections for a new venture, as described in Chapter 17, provide the entrepreneur with an estimate of the cash flow over the first two or three years. From the cash-flow statement, the entrepreneurial team can determine the requirement for financial capital. Some new firms are cash-flow positive almost immediately, and the entrepreneurs themselves can provide the necessary start-up funds. On the other hand, if it takes one year or more for the cash flow to become positive, a sizable investment may be necessary. Most high-tech firms take several years to become cash-flow positive.

A new venture may require capital to purchase fixed assets such as computers and manufacturing equipment. Furthermore, a new business needs capital to operate while building a customer base. **Working capital** is the capital used to support a firm's normal operations and is defined as current assets minus current liabilities. Financial capital is necessary to permit a new venture to purchase assets and provide working capital. As a firm grows, its needs for financial capital will normally increase. Choosing the right sources of capital for a business can be as important as choosing the team members and the location of the business. This decision will influence the future of the firm.

Finding the capital they need can be a difficult and time-consuming task for entrepreneurs. Getting and completing an investment agreement can take 3 to 12 months. For many entrepreneurs, it may be wise to first secure a few customers and then seek investors. Finding the right financial backer takes a good business plan and time. It will be necessary to continuously tell the venture's story and answer myriad questions. However, revealing proprietary information understandably makes entrepreneurs uneasy. The chance that important information will leak to competitors is real.

Many investors take a long time reviewing the plan and interviewing the team, only to turn down the proposal in the end. The entrepreneur must assume a tentative deal will not be consummated and keep looking for investors even when one investor is seriously interested. While it's tempting to end the hard work of finding money, continuing the search not only saves time if one deal falls through but also strengthens the negotiating position.

Financial capital for new ventures is available, but the key is knowing where to look. Entrepreneurs must do their homework before attempting to raise money for their ventures. Understanding which sources of funding are best suited to the various stages of a company's growth and then learning how those sources operate are essential to success.

The issue of how much money to seek is difficult to resolve. Entrepreneurs wish investors to provide all the money necessary before positive cash flow. However, most investors want to divide their investment into several milestone-based stages. Furthermore, most investors will be wary of the pro forma projections and tend to accept only the pessimistic projection or variations. Investors attempt to factor uncertainty into their calculations, while entrepreneurs are, by nature, more optimistic.

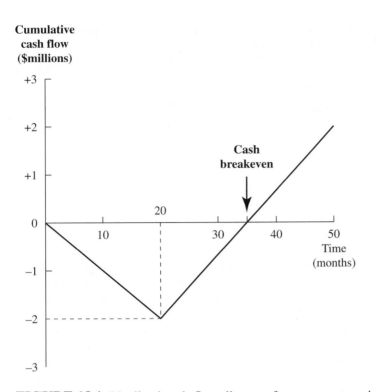

FIGURE 18.1 Idealized cash-flow diagram for a new enterprise.

An idealized cash-flow diagram is shown in Figure 18.1. This new enterprise has a burn rate of $100,000 per month in the first 20 months. It starts generating positive cash flow in the twenty-first month and reaches cumulative cash breakeven in the thirty-fifth month. This firm would require an investment of at least $2 million.

Uncertainty of venture outcomes can lead to a wide range of estimates of results. Breaking a firm's development into several stages can help investors build confidence in the firm over time. However, staged investments require the entrepreneur to raise funds several times—a potentially distracting and risky effort.

Often the two parties, investors and entrepreneurs, possess asymmetrical information. The investors may know more about the industry, while the entrepreneurs may know more about the venture. Furthermore, it may be difficult for the investors to determine an appropriate value for the intellectual property of the new venture. What is a patent worth? Often one finds out in court. Furthermore, how does one value the team's capabilities? Finally, the dynamics of the marketplace are difficult to assess. In addition, the marketplace provides variable valuation multiples as the market mood changes. These four factors that lead to different perceptions of the investors and the entrepreneurs are summarized in Table 18.1 [Gompers and Lerner, 2001].

TABLE 18.1 Five factors that lead to the different perceptions of investors and entrepreneurs.

■ Uncertainty of projected outcomes	■ Dynamics of the industry and the financial marketplace
■ Asymmetrical information	
■ Assigning a value to intellectual property and intangibles	■ Concentration of wealth risk in a venture by entrepreneurs while investors will have risk spread through a diversified portfolio

The process of securing investment capital requires that the effects of the first four factors shown in Table 18.1 be reduced in the negotiation stage. This can be done through a full discussion of the risks, clear goals, and value of intellectual property and the leadership team. Also, the investors must build confidence that the entrepreneurs can properly manage the firm within the dynamics of the marketplace. Finally, it should be recognized by both parties that the entrepreneurs will have their financial wealth concentrated in the venture, while the investors will have a set of diversified investments. The goal of the entrepreneurs is to find investors who examine the factors in Table 18.1 and eventually align their view of these factors with those of the entrepreneurs. These investors can be called *aligned investors*. Entrepreneurs are advised to find aligned investors since they will be responsive to changing needs and provide required flexibility. Identifying aligned investors who have industry, operating, and team-building expertise is a critical first step in successful fundraising.

Entrepreneurs who are seeking financing send a credible signal to potential investors through their willingness to invest in the business. Investors may be attracted to entrepreneurs who are willing to invest a significant part of their private wealth in the venture [Ogden et al., 2003].

18.2 Venture Investments as Real Options

Investors make capital investments in opportunities for future cash returns. Professional investors often think of an opportunity as an *option*. **Options** may be defined as rights but not obligations to take some action in the future. Investments in new ventures can be viewed as investments in opportunities with an uncertain outcome. The outcome of an investment in a new venture is highly uncertain with many risks. However, an investment in any new venture may create unforeseen opportunities that can be exploited in the future. A sound practice in the early stages of developing a business is to keep a number of options open by committing investments only in stages while exploring multiple business paths. Once uncertainty has been reduced to a tolerable level and widespread consensus exists within the organization on an appropriate path, the full commitment to that path can be made. Investors who hold options are given the right to make a decision now or at a later date. They can exercise that right to either proceed to the next step or cut their losses and decide to cease investing.

In financial terms, an option is simply the right to purchase an asset at some future date and at a predetermined price. A **real option** is the right to invest in (or purchase) a real asset—shares in a start-up at a future date.

Columbus and the Queen's Option

Christopher Columbus developed a plan to seek out exotic spices and develop a spice trade to Europe. He was an Italian sea captain who submitted, without success, a proposal to the king of Portugal in 1484 to reach Asia by sailing west. Over the next four years, he made inquiries of many European courts. Finally, in 1492, after years of fruitless effort, he received the support of Queen Isabella and the Court of Madrid. The court invested 1.4 million maravedis (currency of the time), and Columbus invested 250,000 maravedis, mostly obtained from friends and family. The contract called for his designation as "admiral" and receipt of one-eighth of the profits from of all gold, silver, gems, and spices produced or mined in "his dominions."

Columbus returned to Spain in March 1493 with some gold and a large amount of information about how to get to the "New World." Queen Isabella had purchased a real option on the discovery of gold and spices. As a result of Columbus's discovery, she decided to exercise the option and send Columbus, Pizzaro, and Cortez to the New World to find and bring back fabulous wealth to Spain. Centuries later, today's professional investors are often purchasing a real option in a start-up. For example, Kleiner Perkins and Sequoia Capital did so in Google in 1999.

Let's consider a simplified mathematical model of venture investments. Intellectual capital (knowledge) can be transformed to economic capital to increase or create cash flows as well as strategic capital to exploit new opportunities. We state that economic capital is the intrinsic value of the series of cash flows. The strategic capital is the option value (OV).

We may state broadly that the value of an investment in a new venture (V) is

$$V = IV + OV$$

where IV = intrinsic value and OV = option value.

Net present value (NPV) is the present value of the future cash flow of a venture discounted at an appropriate rate (r). The intrinsic value of a venture is the net present value (NPV) of the venture using a discount rate (r), equal to the expected return for the venture.

The net present value of a series of cash flow (c_n) is

$$NPV = \sum_{n=0}^{N} \frac{c_n}{(1 + r)^n}$$

(18.1)

where n = 0, 1, 2, · · · N. For example, the NPV for a new firm over the first two years might be

$$NPV = -100,000 + \frac{65,000}{(1 + r)} + \frac{35,000}{(1 + r)^2}$$

where r = 0.15, the discount rate for this firm. The initial cash flow is negative since an investment of $100,000 was required at n = 0. Then we may calculate NPV as

$$NPV = -100,000 + 65,000(0.870) + 35,000(0.756)$$
$$= -100,000 + 83,000$$
$$= \$ -17,000$$

The intrinsic value (IV) is equal to NPV for this case. However, the investor has an option to reinvest in the firm after the first two years. Typically, investors establish options by making an initial investment in a venture, which grants them an option to invest again later. As information flows over time, the uncertainty is reduced, and later-stage investments can be seen as less risky.

The value of an option (OV) is a function of four factors: the life of the option (T), the volatility (or uncertainty) of the price of the underlying asset (σ), the discount rate (r), and the level of the exercise price (X), relative to the current price of the venture (P). Clearly, having a longer period of time in which to decide whether or not to exercise an option increases the likelihood that the value of the firm will in the future exceed the exercise price. The higher the degree of uncertainty about the future value of the venture, the more one should be willing to pay for the option. The uncertainty can be represented by the standard deviation of the firm's price (σ) over the period (T). If the firm is new, use the standard deviation for comparable firms.

The value of an option increases with the relative value of the current stock price at the initial investment (P) to the exercise price of the option (X). The ratio P/X increases as X declines. The value of an option also increases with the discount rate (r), since higher discount rates make the option more valuable due to high discounting of future cash flows. The four factors contributing to the value of a real option are summarized in Table 18.2.

The value of the option is based on the four factors T, σ, P/X, and r. Then the option value is

$$OV = f (T, \sigma, P/X, r)$$

TABLE 18.2 Value of a real option based on four factors.

- Increases with the level of uncertainty measured by the standard deviation (σ)
- Increases with the length of time (T) the person holding the option has to decide whether or not to exercise it
- Increases with the ratio of the current stock price (P) to the exercise price (X) or P/X.
- Increases with the discount rate (r)

which increases as all four factors increase. The option value can be calculated using the Black-Scholes formula [Boer, 2002]. An option calculator is available at www.blobek.com/black-scholes.html. Another valuation model for options is the binomial model, which uses a decision tree format [Copeland and Tufano, 2004].

As a first approximation, we can use a linear approximation for OV as follows:

$$OV = k_1T + k_2\sigma + k_3(P/X) + k_4r$$

where k_i are unspecified constants. For high relative values of the four factors, the option value will be significant and may exceed the NPV for the first few years of a high-risk venture.

Let us again consider the hypothetical case discussed in the preceding paragraphs. An initial investment of $100,000 was made with an option to invest in a second round of investments after two years. Thus, consider the case where the period (T) is relatively long, the uncertainty (standard deviation) of the firm's value is high, and the discount rate is relatively high. Also, if the current stock price of the firm at the time of the initial investment is low, say $10, and if the exercise price is preset at $5, then the option value is quite high. Perhaps an investor could estimate an option value of $50,000. Then the value of the investment is

$$V = IV + OV$$
$$= -\$17,000 + 50,000 = +\$33,000$$

Often, the value of an investment in a start-up is largely its real option value.

Early Valuation of Genentech

Genentech was founded in 1976 by entrepreneur Robert Swanson and biochemist Herbert Boyer to use gene-splicing technology to develop new pharmaceuticals. Genentech, devoid of profit, went public in 1980 and raised $35 million. With little cash flow in sight, Genentech was valued solely on its potential to exploit its intellectual property and introduce important new pharmaceuticals. Genentech became profitable in 1993, and was acquired by Roche in 2009.

Investments that appear overly risky from a purely financial view may be viable once the opportunities for future action are taken into account. However, as soon as an option no longer promises to provide future value, it should be abandoned. Entrepreneurs sometimes allow unpromising ventures to drag on far too long. Entrepreneur teams can develop a collective belief that their venture will succeed, a conviction that overcomes skepticism and perhaps reality [Carr, 2002].

18.3 Sources of Capital

Many sources of financial capital exist for a new business, as listed in Table 18.3. There are two types of capital: equity and debt. **Equity capital** represents the investment by a person in ownership through purchase of the stock of the firm. The holders of equity shares are called stockholders. **Debt capital** is money that a business has borrowed and must repay in a specified time with interest. An example is a leased vehicle. Debt capital usually does not include any ownership interest in the new firm.

Equity financing of a venture at formation often involves funds provided by founders, friends, and family. Lenders and equity investors may expect the entrepreneur team to invest a significant amount of their own capital in the new business. If an entrepreneur will not invest in a risky venture, why should the outside investor? Usually, a new business can obtain debt capital only after some period in the marketplace and success evidenced by growing revenues and accounts receivable. Investments from family and friends are an excellent source of seed capital and can get a start-up far enough along to attract money from private investors or venture capital companies. Family and friends are often willing to invest because of their relationships with one or more of the founders. However, family and friends should receive and review all the financing documents. Furthermore, they should be able to afford losing their investment and be comfortable with the risk.

In 1995, Mike and Jackie Bezos invested $300,000 in their son Jeff's start-up, Amazon.com. Today, those shares are worth many times what they paid for them. Entrepreneurs can approach professional investors who may invest in very promising, high-growth ventures. Wealthy individuals will also invest in new ventures. They are often called **angels,** and it is estimated that angels personally invest in more than 50,000 firms annually in the United States. **Venture capitalists** are professional managers of investment funds. They normally invest in more than 5,000 firms annually in the United States. The number of investments made by these two groups increases in good economic conditions. Angels and venture capitalists have invested about the same total amount in recent years.

In the United States, the Small Business Administration (SBA) will help fund start-ups as well as provide advice on funding sources. Small Business

TABLE 18.3 Sources of capital.

■ Founders	■ Leasing companies
■ Family and friends	■ Established companies
■ Small business investment companies	■ Public stock offerings
■ Small Business Innovation Research grants (United States)	■ Government grants and credits
■ Wealthy individuals (angels)	■ Customer prepayments
■ Venture capitalists	■ Pension funds
■ Banks	■ Insurance companies

TABLE 18.4 Comparison of major selected sources of growth capital.

Source of capital	Amounts	Advantages	Disadvantages
1. **Individuals (Angels)**	$10,000 to $1 million with low to medium levels of patience and expertise	Create little dilution for the venture; can move fast because of minimal negotiation and due diligence requirements	Lack sufficient funds for capital-intensive opportunities; can lack long-term perspective; may not provide good advice
2. **Venture capital firms**	$1–$20 million with high levels of patience and expertise	Possess large sums of money to deploy; provide recruiting assistance and other services; enhance venture's reputation and credibility immediately	Require larger percentage ownership of the venture; expect significant role in making major decisions; play active role in building executive leadership team
3. **Corporations**	$5–$50 million with medium to high levels of patience and expertise	Generate moderate dilution for the venture; provide opportunity for distribution and product development assistance	Create problems with other potential relationships (e.g., corporation's competitors); can put the venture's intellectual property at risk

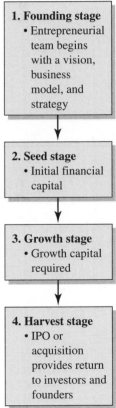

FIGURE 18.2
Four financial steps in building a successful firm.

Innovation Research grants are available to technology start-ups for the development of innovative technologies (see www.sba.gov/sbir).

New enterprises are not in a position to seek their early-stage financial capital through an initial public offering on the stock market. Only the most qualified and experienced teams with an outstanding opportunity are able to make an initial public offering in the early stages of building a business. A spin-off of an established company may be able to make an IPO quickly.

The four financial steps for building a successful firm are summarized in Figure 18.2. Often, a firm can start with seed capital from its founders and friends. However, most technology firms require significant capital in the growth phase and turn to professional or wealthy investors. Table 18.4 details the pros and cons of several different sources of equity capital for the growth phase.

18.4 Bootstrap and Seed Financing

The initial funds used to launch a new firm are usually called **seed capital**. The first round of capital needs may be limited, and funds may be readily available. Moreover, there are many steps that a small firm can take to reduce costs and to improve cash flow, such as obtaining advance payments from customers, delaying payments to others, and sharing employees, assets, or business space with other firms [Winborg and Landström, 2000].

Launching a start-up with modest funds from the entrepreneurial team, friends, and family is often called **bootstrap financing**. The term originated from the saying "they pulled themselves up by their bootstraps." To bootstrap a venture means to start a firm by one's own efforts. For many ventures with modest potential returns to investors, it is best to attract investors who are known to the entrepreneurs. Professional investors can only back a small fraction of the number of firms that start up each year. They seek to invest in large opportunities with high rates of return that have defensible competitive advantages, well-defined business plans, and well-known, proven founders.

An advantage of bootstrap financing is the ability to make some mistakes and yet keep going. Entrepreneurs who are unsure of their markets or who do not have the experience to deal with investor pressure are better off without other people's capital, even if they can somehow get investors to overlook their limited credentials and experience.

Bootstrap entrepreneurs often start with a modest business plan and a focused opportunity and look for a quick route to breakeven and positive cash flow. Many entrepreneurs underestimate the time and the marketing costs entailed in overcoming customer inertia and conservatism, especially with respect to new, unproven products.

Bootstrap companies start small and build their experience and know-how as they go. Eventually, these modest beginnings can turn into large successes as the firms find new opportunities. For example, Princeton Review was launched to compete with local private tutors of uneven quality. Eventually, the firm found its place on the national scene and competed with the Kaplan chain. Actual figures are difficult to obtain, but up to 75 percent of start-ups are bootstrap, self-financed firms. Self-financed start-ups concentrate on sales activity to bring cash flow into the business.

Many businesses fit the model of a bootstrap opportunity. They keep costs low, seek out markets that competitors are ignoring, and build the business one step at a time. Entrepreneurs often must fund a significant portion of their new business since investors and lenders are reluctant to provide the required capital [Quadrini, 2001].

Pierre Omidyar launched AuctionWeb on Labor Day, 1995, while he was still employed full-time at a software firm. As an experienced software developer, Omidyar was intrigued by the opportunities to build a business on the Internet, and his vision was to provide a "perfect market" for buyers and sellers on an Internet auction site [Cohen, 2002]. He wrote the program for the site over the Labor Day weekend using his personal website provided by his home Internet service provider. The best domain name he could find was eBay.com. Throughout the fall of 1995, AuctionWeb had hosted thousands of auctions. By February 1996, Omidyar's Internet service provider started charging him a commercial rate for his website, so he started charging a small fee for each sale. Starting in February, AuctionWeb was profitable. By April,

AuctionWeb took in $5,000 in sales. In June, when revenues doubled to $10,000, Omidyar decided to quit his day job. He quickly attracted a friend, Jeff Skoll, to join him, first as a consultant and then full time by August 1996. By late 1996, they moved out of Omidyar's house to an office building in Campbell, California. By October 1996, AuctionWeb had a total of four employees when it hosted 28,000 auctions that month. AuctionWeb was dedicated to thriftiness and controlling costs. By January 1997, the site had hosted 200,000 auctions, and AuctionWeb was projecting revenues of over $4 million for 1997. Early in 1997, Omidyar and Skoll wrote their first business plan and started looking for investors. By June 1997, Benchmark Capital, a venture capital firm, paid $5 million for 21.5 percent of the company, which then changed its name to eBay.

Bootstrap companies usually follow five rules: (1) start small and probe the market, (2) learn from your customer and adjust the business model, (3) adjust the revenue and profit engine, (4) keep costs to a minimum, and (5) start expanding the company, once the new venture starts growing, while keeping the cost curve below the revenue curve. Often, using bootstrap financing can instill long-term frugality and financial discipline in a new venture.

A useful indicator of the value of the quality of a new venture proposal for funding is the actual proportion of the lead entrepreneurs' wealth that is committed to the new venture [Prasad et al., 2001]. Typically, bootstrap start-ups can take care of the first (seed) round of the company. If the firm starts to grow, it may need investments from professional investors. The advantages and disadvantages of bootstrap financing are listed in Table 18.5.

Founding of Siebel

Tom Siebel earned an MBA and an MA in computer science from the University of Illinois. After graduation, he joined Oracle in 1982. By 1992, Siebel and Pat House had founded Siebel Systems. For the first 18 months, everyone worked for no salary but received equity shares. Siebel stated: "This was never about making money. It was never about going public; it was never about the creation of wealth. This was about an attempt to build an incredibly high-quality company." [Malone, 2002]. Siebel Systems was purchased by Oracle for $6 billion in 2005.

TABLE 18.5 Advantages and disadvantages of bootstrap financing.

Advantages	Disadvantages
■ Flexibility	■ Unable to fund growth phase
■ Maintain ownership	■ Lack of funding commitment for future
■ Operating control by founders	■ Loss of advice from professional investors
■ Little time spent on finding investors	

18.5 Debt Financing

New ventures with sales and cash flow can consider recurring short-term or long-term debt financing. Debt provides financial leverage to a firm and enables the firm to increase its return on equity. The principle of financial leverage works as long as a firm's earnings are consistent and are larger than the interest charged for the borrowed money. Of course, if the firm's net earnings should drop below the interest cost of borrowed money, the return on owners' equity will decrease. Thus, most new ventures avoid financial leverage until they achieve stable growth.

Debt financing can be easier to arrange, and often cheaper, than equity for profitable companies. Any profitable firm can borrow money if it's willing to pay high enough interest rates. Borrowers have the advantage of not giving up ownership or control of the firm—unless they cannot pay the interest. Also, the tax deduction for interest paid cuts the effective cost of debt. Using debt, however, exposes a company to another kind of risk. If it does not show enough profit to cover debt payments, its existence may be in peril.

Banks and the Small Business Administration will lend to qualified firms. Asset-backed borrowing may be available to new ventures. Leasing equipment is a form of borrowing. Venture leasing, while expensive, may be cheaper than giving away more equity. An asset leaseback is also a way to secure cash by giving up an asset such as a building or equipment.

Many new ventures arrange for a line of credit, a form of short-term borrowing, to be used as needed. The firm pays a fee for the right to access borrowed funds as needed. Sometimes a new venture can secure a bank loan, which will be guaranteed by the Small Business Administration. These asset-based loans can be favorable to the new venture, if needed.

Small Business Administration loans in 2008 amounted to $18.2 billion to about 78,000 borrowers. Under the program, the federal agency does not actually make the loan—banks do, mostly—but it guarantees about 75 percent of the loaned amounts. And it covers losses on the guaranteed portions from a combination of fees charged to banks and taxpayer funds.

18.6 Grants

A number of grants are available to support very early-stage and small-scale business efforts. The Small Business Innovation Research (SBIR) program is a U.S. government program, coordinated by the Small Business Administration, through which several government agencies make grants to small businesses. A "small business" is defined as an American-owned for-profit business with fewer than 500 employees. A similar program, the Small Business Technology Transfer (STTR) program, is focused on developing partnerships between small businesses and nonprofit U.S. research institutions such as universities. Together, SBIR and STTR grants to small businesses total $2 billion each year. Information about both programs can be obtained from the Small

Business Administration (www.sba.gov). Entrepreneurs must be aware, however, that start-ups that are 51 percent or more owned by venture capitalists are not eligible for these grants. This policy is under review.

The National Collegiate Inventors and Innovators Alliance (NCIIA), also based in the United States, offers Advanced E-Team Grants to move products or technologies from the idea stage to prototype and to help collegiate innovators secure intellectual property. Advanced E-Team Grants range in size from $1,000 to $20,000; the grant period is 12 to 18 months. More information can be obtained from NCIIA (www.nciia.org). Other examples in the United States include DARPA grants, loan guarantees, and international sources.

As high-growth ventures often are based upon cutting-edge research, including research conducted at universities, entrepreneurs are also wise to investigate research funding opportunities from national and regional government agencies and foundations. While these grants typically are intended to support research activities only, such funding can be critical for start-ups in areas such as energy, nanotechnology, and medicine, where the technical challenge is the primary risk faced by a new venture.

18.7 Angels

Angels are wealthy individuals, usually experienced entrepreneurs, who invest in business start-ups in exchange for equity in the new ventures. The term *angel* for an investor in a new firm was originally used for a person who backs a new theater production on Broadway. Angels are people who share the vision of the new venture and provide support, advice, and money. In a sense, angels can provide wings for lifting a new creation. Angels often have personal experience and interest in the industry that the start-up is entering. Angel investing is a fast-growing segment of the new business financing industry, and it is often ideal for start-ups that have outgrown the capacity of investments from bootstrap financing but are still too small to attract the interest of venture capital companies. For example, after raising the money to launch Amazon.com from family and friends, Jeff Bezos turned to angels, attracting $1.2 million from a dozen angels. Angels funded about 57,000 U.S. ventures for a total of about $26 billion in 2007.

Angels often invest because they understand the industry and are attracted by the opportunity as well as the potential return. They may place less emphasis on an early return strategy and enjoy working with new entrepreneurs. Angels serve as investors, advisers, and mentors for the new ventures they support. They may help new entrepreneurs create and refine a business model, find top talent, build business processes, test their ideas in the marketplace, and attract additional funding. Angels tend to invest close to home and limit their investments to early-stage companies. Most of the new ventures they fund were recommended to them by a business associate or an angel group. A summary list of criteria for investments by angels is provided in Table 18.6.

Angels can be helpful investors, but sometimes they can be overbearing and have a negative impact. Choosing the right angel is important, so entrepreneurs

TABLE 18.6 Criteria for angel investments.

■ Within the industry in which the angel has experience	■ Entrepreneurs with attractive personal characteristics such as integrity and coachability
■ Located within convenient driving distance of the angel	■ Good market and growth potential
■ Recommended by trusted business associates	■ Seeking an investment of $50,000 to $1 million that offers minority ownership of about 10 to 40 percent

should check references and capabilities of potential investors carefully. In some geographical regions, angels join together to form groups of angels. These groups work together to screen investment opportunities. For example, the Band of Angels meets monthly in Silicon Valley to hear a few presentations by start-ups (see www.bandangels.com). Angel groups are established in many cities (see appendix C for a partial list).

In 1976, Steve Wozniak and Steve Jobs designed the Apple I computer in Jobs's bedroom and built the prototype in Jobs's house. To start the company, Jobs sold his Volkswagen and Wozniak sold his Hewlett-Packard calculator, which raised $1,300. With that capital and credit from local electronics suppliers, they set up their first production line. Jobs met Mike Markkula, a former marketing manager at Intel and wealthy angel who invested $91,000 in cash and personally guaranteed a bank line of credit for $250,000. Jobs, Wozniak, and Markkula each held one-third ownership of Apple Computer [Young, 1988]. Markkula, the angel investor, became chairman of the company in 1977.

The founders of Google, Sergey Brin and Larry Page, approached Sun Microsystems cofounder Andy Bechtolsheim in 1998, who wrote a $100,000 check after a 15-minute pitch describing the new business. By 1999, the venture capital firms of Kleiner Perkins, Caufield and Byers, and Sequoia also invested.

Angels are usually available through referrals from their colleagues. However, a referral is not necessarily something you ask for. A referral is a favor that you earn.

18.8 Venture Capital

Venture capital is a source of funds for new ventures that is managed by investment professionals on behalf of the investors in the venture capital fund. The people who manage the venture capital fund are called venture capitalists. These funds typically invest in new ventures with high potential returns. The private venture capital firm is seeking equity participation in these high-potential firms. Each year in the United States, over 5,000 new ventures receive funding from venture capital. The venture capital firm engages in careful screening and due diligence before investing. It brings in other venture capital investors in a financing round and prepares contracts and restrictions (called term sheets) with the new venture. Venture capital firms are usually interested in technology ventures with high potential.

A typical venture capital firm will have enough investments so as to diversify its portfolio. Each venture capital firm will have several partners who are experienced, full-time investors. Their knowledge of finance and technology enables them to judge the potential investment. Normally, venture capital investing is staged financing. Thus, new information and risk reduction at each stage enables the venture capitalist to make better decisions. The new ventures are held to staged goals, milestones, and deadlines for achieving these milestones. A typical investment from a venture capital firm is $1 million to $5 million in the first financing round, with a total commitment of $10 million to $20 million. Over several stages with several venture funds involved, the combined funding potential can be $50 to $100 million.

Many entrepreneurs start up a new business to gain independence, only to find that they have a new set of partners—the venture capitalists acting as members of the board of directors. Most venture capitalists look for ventures that can become profitable and attain at least $100 million a year in revenues in the next five or six years. Entrepreneurs that envision high-growth ventures will likely turn to venture capitalists [Florin, 2005].

A venture capital portfolio of 20 to 30 new ventures may achieve an overall annual return of 30 percent. Perhaps one-half of the new ventures fail or provide a low return. Fortunately, two or three new ventures may provide an overall annual return of 50 to 100 percent. Therefore, venture capital firms are looking for new ventures that potentially can return at least 50 percent annually. Obviously, these candidates must be firms with high-potential growth.

The four investment requirements used by the venture capitalists in the selection process are (1) the industry is well-known to them; (2) the amount of the investment is greater than $1 million; (3) the company is at the appropriate stage of progress; and (4) the potential return is 40 percent or more annually. The track record of the team is also critically reviewed [Gompers and Sahlman, 2002].

Venture capitalists prefer staged or phased financing. The four stages are shown in Table 18.7. Venture capital is normally available in the development and growth stages, with greatest emphasis on the development stage. Venture capital money is limited-term money [Zider, 1998]. Typically, a venture capital firm wants to harvest (realize) its return on investment within five to 10 years. The harvest is usually facilitated through an initial public offering in the public equity markets or an acquisition by an established company.

Amazon: Raising Capital in Three Ways

New enterprises that can become large businesses can require large capital investments as they grow. Amazon raised $8 million from venture capitalists and $54 million from its 1997 IPO. The largest capital infusion for Amazon was a $1.25 billion debt offering in 1999.

TABLE 18.7 Investment stages.

1. Seed or start-up stage: Complete the team, formalize the plan, complete initial arrangements. Financial capital from angels, friends, and family.

2. Development stage (series A): Product development and prototype, ready for launch. Financial capital from venture capital funds.

3. Growth stage (series B or C, and others as required): Launch and growth phase. Financial capital from venture capital firms and corporations.

4. Competitive or maturity stage (initial public offering): Mature firm in a competitive context. Financial capital from offerings in the public equity markets.

Venture capitalists expect the new business to achieve certain outcomes, often called milestones, at the end of a funding stage before proceeding to the next stage. Examples of the milestones for each stage are shown in Table 18.8. Each milestone functions as a miniplan about each stage for the venture. For example, a milestone may state that a working prototype will be available in six months. Entrepreneurs should therefore develop a series of milestones and then tie their fundraising activities to the completion of them [Berkery, 2008].

A special type of venture capital is available for social benefit firms or those that promise to create products that lead to sustainable resources or environments. For example, the Silicon Valley Social Venture Fund, known as SV2, is dedicated to funding organizations in the San Francisco Bay area that facilitate social change (www.sv2.org). The Omidyar Network provides funding to for-profit ventures that attempt to make a positive impact on society (www.omidyar.net).

Venture capital funds concentrate on attractive, disruptive, and high-growth industries. They concentrated on computers and biotechnology in the 1980s and communications and the Internet in the 1990s. Areas of concentration in the 2000s include biomedical devices, genomics, energy, and mobile computing. By investing in emerging industries, venture capitalists hope to build great companies in important industries while reaping large rewards.

Venture capitalists carry out one or more of five functions: (1) they provide capital for start-ups, (2) they evaluate projects for other participants in the venture, (3) they provide expertise for the development of the firm, (4) they serve as the central coordinator for all the participants involved during a firm's infancy, and (5) they provide introductions to potential customers,

TABLE 18.8 Milestones for each stage of funding.

Stage	Milestone = expected outcome
Seed	Formation of initial team and completion of business plan
Series A	Product development completed and early customer interest
Series B	Product test and customer acceptance proven with some revenue
Series C	Rapid sales growth and international expansion

employees, collaborators, and future financiers. The importance of these non-financial contributions cannot be overstated. Venture-capital-backed firms demonstrate more responsive product development, greater collaborative activity, and more agile management decision making [Hsu, 2006; Arthurs and Busenitz, 2006].

In a typical deal at the development stage, a group of two or three venture capitalists will invest \$5 million to \$15 million in exchange for 40 percent to 60 percent preferred-equity ownership. The preferred class of stock provides the venture capitalists with preference over common stock held by founders, family, friends, and other first-stage investors. They will hold a liquidation preference on rights to assets. Furthermore, venture capitalists seek voting rights over key decisions such as the sale of the firm or the timing of an IPO. They usually require seats on the board of directors.

Venture capital is high-risk and high-return capital available to high-potential firms. Venture capitalists plan for a 50 percent or more annualized return on their investment with built-in protections and controls over the new venture.

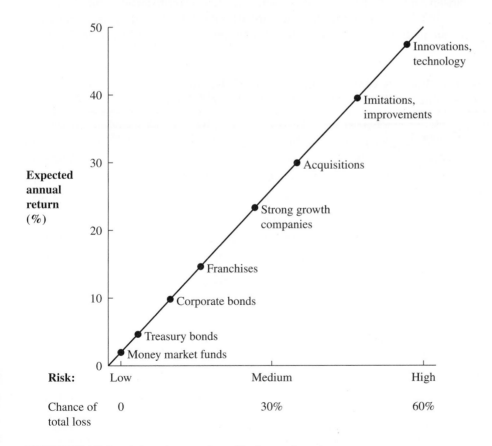

FIGURE 18.3 Risk and reward profile for various investments.

The structure of most venture capital contractual agreements (deals) favors the venture capitalist and may place the entrepreneur at a disadvantage if plans do not work out. However, the venture capital firms bring sizable money, industry knowledge and contacts for recruiting and customer assistance, and a pathway to a public stock offering. The risk-and-reward profiles for various types of investments are shown in Figure 18.3. Venture capitalists work at the high risk–high return end of the profile.

Venture capital investments in the United States in new emerging ventures averaged $25 billion per year over the decade of the 1990s and peaked at $106 billion in 2000 during the dot.com and telecommunications boom. Assuming $24 billion per year for new start-ups is available in the future and 3,000 ventures are funded, the average investment is $8 million per venture. Worldwide venture capital investments in technology ventures are shown in Table 18.9 for 1995 to 2008 along with IPOs and M&A deals.

Venture capitalists get good rates of return by buying shares of private companies early and then helping management use that cash to turn the start-ups into businesses with growing revenues, profitability, and cash flow. The big winners generally have to earn 10 times the original investment money in four to five years. A multiple of 10 times is equivalent to a 58.5 percent

TABLE 18.9 Worldwide technology venture capital investments, initial public offerings (IPOs), and merger and acquisition (M&A) deals, 1995–2008 ($ millions).

Year	Number of VC deals	VC investment	Number of IPOs	Offering amount	Post-IPO valuation	Number of M&A deals	M&A deal value
1995	1,935	$6,979	164	$6,223	$24,384	80	$2,892
1996	2,983	$11,355	211	$7,785	$36,559	92	$6,781
1997	3,502	$15,182	111	$3,678	$18,049	128	$6,263
1998	4,486	$24,704	64	$3,168	$14,949	177	$8,018
1999	7,035	$63,750	252	$19,451	$123,540	191	$38,134
2000	12,340	$132,133	250	$23,915	$127,207	286	$65,616
2001	7,595	$56,319	29	$2,347	$13,095	351	$16,205
2002	5,068	$32,748	18	$1,693	$5,844	312	$7,142
2003	5,550	$28,321	20	$1,382	$6,186	276	$6,785
2004	5,752	$32,218	82	$9,238	$53,195	323	$13,959
2005	5,646	$33,682	50	$3,757	$14,234	336	$21,061
2006	6,124	$39,119	51	$4,556	$20,148	354	$15,618
2007	5,870	$39,978	80	$8,652	$44,634	310	$22,478
2008	5,723	$41,690	6	$470	$2,646	218	$10,519

Note: M&A figures based only on deals with disclosed transaction values. See this textbook's websites for most recent data.
Source: Thomson Venture Economics.

TABLE 18.10 Characteristics of a company attractive to venture capital.

■ Potential to become a leading firm in a high-growth industry with few competitors	■ Outstanding opportunity
	■ Founders' capital invested in the venture
■ Highly competent and committed management team and high human capital (talent)	■ Recognizes competitors and has a solid competitive strategy
■ Strong competitive abilities and a sustainable competitive advantage	■ A sound business plan showing how cash flow turns positive within a few years
■ Viable exit or harvest strategy	■ Demonstrated progress on the product design and good sales potential
■ Reasonable valuation of the new venture	

annual return over the five years. If a venture capitalist invests $5 million in the early stage of a company, his or her ownership stake is hoped to be worth $50 million in five years. If the venture capitalist owns one-half of the company, the firm must be worth $100 million after five years for that return.

The characteristics of a good venture capital deal from the venture capitalists' view and the venture founders' view are summarized in Table 18.10. A large opportunity in a fast-growing industry led by a very competent, experienced team is attractive to venture capitalists. With a good venture capitalist partner, the founders of the new venture can realize their dream of a well-capitalized venture that can make a big difference in their industry. At the same time, the founding team must recognize that venture capital has a high cost of potential loss of ownership, control, and even changing the vision of the venture. A list of U.S. venture capital firms is available from the National Venture Capital Association (www.nvca.org). See appendix C for a partial list.

New and emerging businesses can use the five steps outlined in Table 18.11 to secure venture capital. Every one of these steps takes time, and the whole process may take 3 to 12 months. Given the extent of technology markets, there is no scarcity of good ideas. What is scarce is experienced, competent, and committed entrepreneurs. Investors strive to find the best entrepreneurial teams to invest in.

FedEx: Getting the Product Right

Fred Smith used his family money to found Federal Express (FedEx) in 1973. He leased several planes and built a 25-city network. He knew additional funding would depend on a solid start, so he tested the system for two weeks, shipping empty packages cross-country. In mid-April 1973, he opened for business. To expand his network, Smith turned to venture capitalists. He got his venture capital because he reduced the perceived risk with his own money in FedEx. By 1975, FedEx was profitable. FedEx sold stock to the public with an IPO in 1978.

TABLE 18.11 Five steps for a venture capital deal.

1. Determine the amount of cash needed and its use.

2. Locate appropriate venture capital investors and secure a referral to them.

3. Determine which risks are to be reduced in this financing round:

   ■ Team risk—recruiting great people

   ■ Capital risk—enough cash to achieve milestones

   ■ Technology risk—prove that the science works in a product

   ■ Market risk—deliver extraordinary customer experiences

4. Agree on valuation and ownership structure.

5. Agree on a contract (term sheet) describing the deal and its terms.

18.9 Corporate Venture Capital

Large companies such as Intel, Microsoft, and Cisco Systems have engaged in investing in external start-ups [Chesbrough, 2002]. **Corporate venture capital** is the investment of corporate funds in start-up firms that are not part of the corporation. In this case, established corporations are acting as venture capitalists. However, the corporate venture capitalist in many cases is looking for new ventures that can offer synergies between itself and the new venture. Another approach is for the corporate venture capitalist to make an investment to use its industry knowledge and then generate a high return on investment. Corporate investors often use external venturing programs as a mechanism for identifying and monitoring promising acquisition targets.

An example of a strategic investment is one of many made by Microsoft in start-up companies that could help advance its Internet businesses. Other firms, such as Intel, make corporate venture capital investments in firms making products complementary to their own. Another possible corporate investment strategy is in start-ups that may be valuable to the corporation's future operations. A summary of the four forms of corporate venture capital investments is provided in Figure 18.4.

Many start-ups can benefit from corporate venture capital, since corporations can make good partners and the cost of the capital can be less than that from regular venture capitalists. However, corporate venture capitalists can take longer to close a deal and may ask for distribution rights. Nevertheless, corporate venture capital may increase the credibility of the new venture and may not exert as much control as regular venture capitalists. Perhaps the most important benefit of corporate venture capital is the potential for a strong partnership between the corporate venture capitalists and the new venture that strategically couples the strength of the large corporation with the innovation of the start-up [Mason and Rohner, 2002].

Entrepreneurs need to be cautious. Corporate venture capitalists often invest to be aware of potential competitors and may not have their interests aligned with those of the start-up.

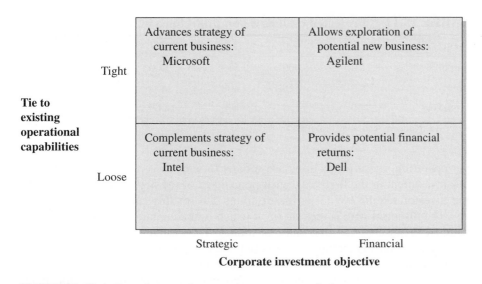

FIGURE 18.4 Four forms of corporate venture capital.

Source: Burgelman et al., 2004.

Intel's venture capital arm has been the most active investor of any kind in technology companies that received venture capital. For example, Intel invested $200 million in more than 100 companies in 2002 [Clark, 2003a]. Intel has made 1,000 investments since 1992, accounting for a total of over $7.5 billion. Intel Capital's main mission is to nurture technologies that could stimulate demand for Intel's chip businesses.

18.10 Valuation

The **valuation rule** is the algorithm by which an investor such as an angel or venture capitalist assigns a monetary value to a new venture. For many operating businesses, determining the net present value (NPV) is the best method for selecting alternative projects. A new enterprise, however, has uncertain pro forma cash flows, and investors find it difficult to use projected cash flows as reliable measures. Even if potential cash flows can be estimated, how does an investor decide what is a fair price for a share of ownership in a new firm?

Theoretically, the value of a company is equal to the present value of all dividends or cash disbursements paid now or later. A new firm has no historical results to use to project future cash flows. Furthermore, fads and social dynamics play a role in the determination of the value of a firm. Thus, determining the value of a start-up is difficult.

Discounted cash flow is a method of calculating the present value of a future stream of cash flow based on discounting back future flows at the end of a number of years using a discount rate (r) (see equation 18.1). Let us start with the discounted cash flow rules for the valuation of a new venture, ABC Inc., with projected cash flows as shown in Table 18.12. Since the firm is not

TABLE 18.12 Projected cash flow and profit for ABC Inc.

Year	1	2	3	4	5	6
Sales	0	1,000	2,500	5,000	8,000	10,000
Profit	−600	−10	400	650	1,000	1,200
Cash flow	−1,100	0	500	1,200	1,500	1,800

Note: All figures in thousands of dollars.

yet operating, these projections are subject to wide uncertainty. With a potential investment in the first year, what percentage of the firm's ownership should the investor require? What is the discount rate for this calculation?

The **discount rate** is the rate (r) at which future earnings or cash flow is discounted because of the time value of money. The discount rate (r) for a firm is its cost of capital. Therefore, the discount rate for investors will be their expected return on investment. Thus, the new enterprise may propose using a discount rate of 15 percent, while the investors may demand a return of 30 percent or more. McNulty and colleagues [2002] have shown the real cost of capital of a biotech start-up is about 35 percent. Also, for the cash flow calculation, we need to have an estimate of the cash flow for year 6 and later. However, it is unreasonable to project into year 6 and later since the estimates become less reliable for later years. Thus, investors will be reluctant to use the discounted cash flow method for valuation of a firm that is yet to provide any reliable cash flow.

Venture capitalists or angels will want a harvest of the value of the firm by an IPO, acquisition, or buyout of their share of ownership by year 5 to 10. Realistically, the IPO or acquisition may not happen until year 6 or later. Examining the projections for ABC Inc. shown in Table 18.12, a valuation can be determined for year 5 using a method favored by venture capitalists and angels.

The **new venture valuation rule** uses the projected sales, profit, and cash flow in a target year (N) and the projected earning growth rate (g) for five years after year N. The investment by the investor (I) is made at the beginning of year 1. The investor requires an annual return (gain) on investment of G for N years. Thus, investors expect a capital return after N years of $(1 + G)^N$ times their original investment I. Therefore,

$$CR = (1 + G)^N \times I$$
$$= M \times I \qquad (18.2)$$

where CR is the capital return and M is the multiple of the investment. Thus, if an investor invests $1.1 million in the series A stage and the expected annual return on investment, G, is 45 percent over a five-year period, then

$$M = (1 + 0.45)^5 = 6.41$$

Therefore,

$$CR = M \times I = 6.41 \times 1.1 = 7.05$$

The percentage ownership (PO) demanded by the investor will then be

$$PO = \frac{CR}{MV} \times 100\%$$

where MV is the expected market value of the new venture in year N.

To calculate the expected market value in year 5, we may use the price-to-earnings or **PE ratio** of comparative firms to estimate the market value in year N. Then,

$$MV = PE \times EN \tag{18.3}$$

where EN is the earnings in year N. In this case for year 5, EN = \$1,000,000 (Table 18.12). The comparative PE ratio is obtained by looking at the industry PE ratios while accounting for the expected growth rate of earnings over the ensuing years. In this case, we might expect a growth rate of earnings of 20 percent for several years following year 5. Then, examining the industry data, we estimate the appropriate PE ratio is 16. Therefore, we have

$$MV = 16 \times \$1,000,000$$
$$= \$16,000,000$$

Then the required percentage ownership is

$$PO = \frac{CR}{MV} \times 100\%$$
$$= \frac{7.05}{16.00} \times 100\% = 44.0\%$$

The market value can also be calculated using a price-to-sales ratio (PS) for comparative firms. If comparative firms have a PS = 2.3, then the market value would be MV = PS × S. The market value for ABC Inc. is

$$MV = PS \times S$$
$$= 2.3 \times \$8,000,000 = \$18,400,000$$

Then the percentage ownership required by the investor would be

$$PO = \frac{CR}{MV} \times 100\% = \frac{7.05}{18.40} \times 100\% = 38.3\%$$

Using these calculations, the investor may reasonably expect to receive 40 to 50 percent of the firm's ownership in a series A investment. Given the uncertain nature of the sales and profit projections, the valuation of a firm is a function of the potential of the firm to achieve a big success in a short time. This simple example assumes no more stages of investment are required in the five-year period.

Netscape was formed by Marc Andreesen, coauthor of the University of Illinois Mosiac Web browser, and Jim Clark, founder and former CEO of Silicon

Graphics. As an angel investor, Clark invested $3 million in Netscape in May 1994. Clark, a respected entrepreneur, then offered a $6.4 million investment opportunity (at $2.25 per share) for 15 percent of the firm to Kleiner Perkins, a premier venture capital firm. Netscape's sales rocketed to $16 million in the first six months of operations [Lewis, 2000]. Just before the IPO, Clark owned 30.0 percent of the firm and Andreesen owned 12.3 percent. The multiple (M) they achieved in 17 months on their seed shares was 37.3. Kleiner Perkins achieved a multiple of 12.4 on its first-stage investment in just 13 months. Netscape was the first high-profile venture of the Internet boom companies and reaped large rewards. A summary of the valuation of Netscape (price per share) is shown in Table 18.13.

When a firm makes an offering, it raises an investment (INV). The firm hopes to set its **pre-money** (before the investment) **value** as PREMV. Then

$$\text{post-money value} = \text{pre-money value} + \text{investment}$$

When **post-money value** is POSMV, we have

$$\text{POSMV} = \text{PREMV} + \text{INV}$$

The percentage of the company sold to the investors is

$$\frac{\text{INV}}{\text{POSMV}} \times 100\%$$

Consider a firm, EZY Inc., with a series of investments as shown in Table 18.14. A set of venture capital investors invests in the firm in the series A round at 90 cents per share and owns 40.0 percent of the firm after the purchase. At the next stage, series B, the firm has missed its milestones, and the investors offer a reduced share price of 50 cents per share. This round (or stage) of financing is called a *down round,* since the price per share goes down. As a result, the venture capitalists receive a significant increase in their ownership in the firm. If the firm meets or exceeds its milestones, subsequent rounds can be "up priced." Also, entrepreneurs are urged to seek new investment rounds from other investors to check market pricing conditions.

The experience of FedEx in its early days of venture capital investments is shown in Table 18.15. Note the down round in September 1974.

The timing of staged investments is a critical issue for the CEO and CFO of a new company that has not yet achieved breakeven. This company is using

TABLE 18.13 Netscape valuation at four stages.

Stage	Date	Price per share	New investment ($ millions)
Seed	4/94	$0.75	$3.1
Series A	7/94	$2.25	$6.4
Series B	6/96	$9.00	$18.0
IPO	8/96	$28.00	$160.0

TABLE 18.14 Financial stages of EZY Inc.

Stage	Investors	Price per share	Ownership by FFF*	Ownership by venture capitalists
Seed	Founders Friends Family	$0.10	100%	—
Series A	Venture capital group	$0.90	60.0%	40.0%
Series B	Venture capital group	$0.50	34.0%	66.0%

*Founders, friends, and family.

TABLE 18.15 Venture capital financing of FedEx.

	Date	Investment ($ millions)	Share price ($ per share)
Series A	September 1973	$12.25	—
Series B	March 1974	$6.40	$7.34
Series C	September 1974	$3.88	$0.63
IPO	1978	—	$6.00

Source: Gompers and Lerner, 2001.

cash from an earlier investment to cover its negative cash flow. We define **burn rate** as cash in minus cash out, on a monthly basis. Thus, if a firm has $800,000 cash in the bank and has a burn rate of $100,000 per month, it will run out of cash in eight months unless it can reduce its burn rate.

Some companies spend money unwisely, so a good rule is to keep the burn rate as low as possible. The alternative is to raise new rounds of new financing, if that is possible, to avoid dilution.

Amp'd Mobile's Bankruptcy

A big opportunity often leads to success, but not always. Founded by Peter Adderton in 2005, Amp'd Mobile was a Mobile Virtual Network Operator (MVNO) located in Los Angeles, California. As a mobile phone service provider targeting youth and young professionals, it sought to use emerging technologies like 3G wireless internet to provide phones capable of delivering media content and advanced social networking capabilities. In 2007, Amp'd filed for Chapter 11 bankruptcy. It had spent all $360 million it had acquired over five rounds of funding. The company had acquired only 175,000 subscribers, which did not produce enough revenue to cover its operating costs. Some worried that Amp'd did not perform proper credit checks on subscribers, perhaps contributing to its large percentage of nonpayers.

18.11 Terms of The Deal

The terms of the investment transaction are critical to the entrepreneur. The issues that concern professional investors are provided in Table 18.16. Clearly, trust and integrity are necessary between investors and the founding entrepreneurs. The venture capitalist willing to pay the highest price is not necessarily the person whom the entrepreneur will want most in the deal. Another venture capitalist who is not willing to pay quite as much may be a better partner in growing the business. Normally, founders are required to earn (vest) their stock over time.

Professional investors will normally want **preferred stock,** which has claims on dividends, if ever paid, and assets before common stock owners. With all the factors involved in completing a financing deal, the new venture needs to engage an attorney to review any terms of the deal and perhaps represent the new venture in negotiating the deal.

The terms of any deal should reflect the likelihood that the firm will require more capital at a later time. Many deals make it difficult to raise capital later at an attractive price. Often the deal is written on an assumption that everything works out as planned—an unlikely outcome. The plan for the future should include reasonable methods of obtaining new capital infusions. Often the provisions for protecting the investors in the current deal will be onerous if another capital infusion is required.

The terms of the deal should address the means of achieving potential return and the allocation of risk between investor and new venture. It may be better to lower the price to the investor and get the investor to share the future risk with the new venture. The investors are looking for protection of their investment, and the new venture is seeking a capital infusion but needs to retain the right to pursue future capital infusions. Excessive protection clauses need to be traded away for lower prices for stock purchased by the investor. If possible, the investor should pay less for ownership and share the risk with the new enterprise.

TABLE 18.16 Issues to be resolved within the terms of the deal.

■ Percent ownership for the investor group	■ Type of security
■ Timing of investment	■ Reservation of ownership for future employees (stock option pool)
■ Control exerted by investor	
■ Vesting periods for ownership by the entrepreneur team	■ Antidilution provisions
	■ Milestones of achievement, if there are multiple stages or steps to the investment ("traunches")
■ Rights to require an IPO and registration rights	
	■ Stock option plans

18.12 Initial Public Offering

The first public equity issue of stock made by a company is referred to as an **initial public offering** (IPO). The newly issued stock is sold to all interested investors of the general public in a cash offer. In the United States, the IPO is a sale of a portion of the company to the public by filing with the Securities and Exchange Commission (SEC) and listing its stock on one of the stock exchanges. The offer is managed via a financial intermediary, an investment bank, which aids in the sale of the securities. The investment bank performs such services as formulating the method of issuance, pricing, and selling the new securities.

Determining the offering price is a difficult task, but new issues are normally priced somewhat below their intrinsic value to ensure a sufficient number of new buyers. The total cost to the firm for issuance of new securities can be up to 10 percent of the funds raised in a typical offering raising $50 million.

The new venture firm has three possible reasons to issue an IPO: (1) raise new capital, (2) liquidity, and (3) image or brand. Many fast-growing firms will need large capital infusions of more than $30 million, and the public market is suitable for these larger amounts. Second, liquidity—the ability to easily convert ownership to cash—is facilitated by the IPO. The third reason is to help build brand reputation by allowing public ownership of the new venture firm. These advantages of issuing an IPO are summarized in Table 18.17.

Entrepreneurs and early investors can obtain a harvest of the value they have created by the liquidity resulting from an IPO. The timing of the IPO is critical since the IPO market can be very volatile. The IPO market was favorable in the period 1998 to 2000 and very unfavorable in the period 2001 to 2003. See Table 18.9 for IPO data for 1995 to 2008.

The process for an IPO is described in Table 18.18. The disadvantages of issuing on IPO include: (1) offering costs, (2) disclosure and scrutiny, and (3) perceived short-term pressures. These disadvantages are listed in Table 18.17. For small offerings, less than $25 million, the costs can amount to 15 percent of the offering. The time required to prepare all the documents can also be onerous to a small, emerging firm. The disclosure and scrutiny can be burdensome.

TABLE 18.17 Advantages and disadvantages of issuing an IPO.

Advantages	Disadvantages
▪ Raising new capital with the possibility of later, additional offerings	▪ Offering costs and effort required
▪ Liquidity: Ability to convert ownership to cash, potential of harvest for investors and founders	▪ Disclosure requirements and scrutiny of operations
▪ Visibility: Build brand and reputation	▪ Perceived pressures to achieve short-term results
	▪ Possible loss of control to a majority shareholder

TABLE 18.18 The process for an initial public offering (IPO) in the United States.

1. Examine the state of the stock market and the potential for an IPO. Consider Sarbanes-Oxley compliance costs.

2. Interview several investment banking firms and select two or three. Select a law firm with IPO experience.

3. Conduct organization meeting and set the schedule for preparation.

4. Draft the registration statement and conduct due diligence. Prepare the issuer to become a public company, including attention to board and committee composition, disclosure controls, and internal controls.

5. Complete the registration statement, provide it to a financial printer, and file the registration statement with the SEC (Securities and Exchange Commission). Gain clearance from NASDAQ or other stock exchange.

6. Receive initial comments from the SEC. Revise registration statement and file first amendment to registration statement. Submit comment response letter to the SEC.

7. Management prepares for "road show" presentations and travel.

8. Receive additional comments from the SEC. File second amendment to registration statement. Resolve SEC comments and print preliminary prospectus and begin marketing efforts.

9. Conduct "road show" of presentations to potential large investors.

10. Resolve final issues with SEC. File final amendment to registration statement and request that the SEC declare the registration statement effective. Price the offering and commence the sale of the stock.

Any misstep can cause havoc with the company's share price. Furthermore, many new firms may find the pressure for reports of quarterly earnings improvements difficult to satisfy.

The IPO That Ignited the Dot Com Boom

During the late 1990s, Silicon Valley became known for IPOs that "popped." High demand resulted in shares that were so sought after on the first day of public sale, the share price closed at much higher values than at the opening trades. In 1995, this was an uncommon occurrence. In August of that year, a start-up called Netscape went public at the young corporate age of 16 months. Demand for the shares was so high that trading was delayed for over two hours as a clearing price was sought. On that day, the stock eventually rose to nearly three times its initial price, closing at nearly double its initial offering price. Netscape had raised $140 million in its IPO, but its impact was even larger. Netscape is credited with making the Internet accessible to mainstream users around the world. By launching the hottest IPO market in history, it stimulated the appetite of investors for technology IPOs for several years. Google had a similar effect with its IPO in 2004.

Determining the appropriate offering price is the most important thing the lead investment bank must do for an initial public offering. The issuing firm faces a potential cost if the offering price is set too high or too low. If the issue is priced too high, it may be unsuccessful and be withdrawn. If the issue is priced below the true market price, the issuer's existing shareholders will experience an opportunity loss.

The IPO marketing process includes a road show. Road shows involve the lead underwriters and key firm managers marketing the firm to prospective investors (institutional investors) via presentations in major cities and one-on-one meetings with mutual fund managers.

Amazon.com was founded in July 1994 and opened for business in July 1995. It reached sales of $16 million for the quarter ending March 31, 1997, but was operating at a loss. Jeff Bezos hoped to take the company public to raise additional funds and to build public recognition for it. The total investment from Bezos, angels, and Kleiner Perkins amounted to approximately $9 million. Amazon selected Deutsche Morgan Grenfell (DMG) as its investment bankers in February 1997 and began the process of preparing the necessary documents for submission to the Securities and Exchange Commission (SEC). Amazon went public with an IPO selling three million shares at a price of $18 each. Its revenue per share, pre-IPO, was approximately $1.80. Thus, the firm went public at a price-to-sales ratio of 10. The firm increased its sales from $148 million in 1997 to $1.5 billion in 2002, but was unprofitable in those years.

If a firm intends to eventually issue an IPO when the market is favorable, it is wise to work from the beginning to be in position to qualify its IPO. Thus, a firm planning to go public needs to meet all the regulatory requirements in place when creating the prospectus. This means having audited financial statements, a complete management team, a sustainable competitive advantage, and an independent board of directors.

The prospectus or selling document is a part of the information provided to the SEC for approval. The information in the prospectus must be presented in an organized, logical sequence and an easy-to-read, understandable manner to obtain SEC approval. Some of the most common sections of a prospectus are: prospectus summary, description of the company and its business, risk factors, use of proceeds, dividend policy, capitalization, dilution, management, owners, and the financial statements.

The cover page of the prospectus for Netflix's IPO in 2002 is shown in Figure 18.5a. The table of contents for the Netflix prospectus is shown in Figure 18.5b. The summary page, the offering page, and the summary financial data are provided in Figures 18.5c, d, and e, respectively. At the time of the IPO, the business was not yet profitable, but the company was narrowing its losses. Netflix raised $82.5 million from the offering of 5.5 million shares. Netflix used the cash for operations and to pay off debt of $14 million.

Its 2001 revenues of $76 million reflected real demand, much of it in the San Francisco Bay area. Nationwide, subscribers totaled 456,000 in 2001, up from 292,000 the year before. Netflix's financial picture includes its relatively low costs.

5,500,000 Shares

NETFLIX®

Common Stock

This is Netflix, Inc.'s initial public offering of common stock. We are selling all of the shares.

Prior to the offering, no public market existed for the shares. The common stock has been approved for quotation on the Nasdaq National Market under the symbol "NFLX."

Investing in our common stock involves risks that are described in the "Risk Factors" section beginning on page 5 of this prospectus.

	Per Share	Total
Public offering price	$15.00	$82,500,000
Underwriting discount	$1.05	$5,775,000
Proceeds, before expenses, to Netflix, Inc.	$13.95	$76,725,000

The underwriters may also purchase up to an additional 825,000 shares from us at the public offering price, less the underwriting discount, within 30 days from the date of this prospectus to cover overallotments.

Neither the Securities and Exchange Commission nor any state securities commission has approved or disapproved of these securities or determined if this prospectus is truthful or complete. Any representation to the contrary is a criminal offense.

The shares will be ready for delivery on or about May 29, 2002.

Merrill Lynch & Co.

Thomas Weisel Partners LLC

U.S. Bancorp Piper Jaffray

The date of this prospectus is May, 2002.

FIGURE 18.5   **(a)**

TABLE OF CONTENTS

You should rely only on the information contained in this prospectus. We have not, and the underwriters have not, authorized any other person to provide you with different information. If anyone provides you with different or inconsistent information, you should not rely on it. We are not, and the underwriters are not, making an offer to sell these securities in any jurisdiction where the offer or sale is not permitted. You should assume that the information appearing in this prospectus is accurate only as of the date on the front cover of this prospectus or other date stated in this prospectus. Our business, financial condition, results of operations and prospects may have changed since that date.

Through and including June 16, 2002 (the 25th day after the date of this prospectus), all dealers effecting transactions in these securities, whether or not participating in this offering, may be required to deliver a prospectus. This is in addition to the dealers' obligation to deliver a prospectus when acting as underwriters with respect to their unsold allotments or subscriptions.

Netflix is a registered trademark and Netflix.com, CineMatch and Mr. DVD are trademarks of Netflix, Inc. Each trademark, trade name or service mark of any other company appearing in this prospectus belongs to its holder.

SUMMARY

This summary highlights information contained elsewhere in this prospectus. You should read the entire prospectus carefully, including "Risk Factors" and our financial statements and the notes to those financial statements appearing elsewhere in this prospectus before you decide to invest in our common stock.

Our Company

We are the largest online entertainment subscription service in the United States providing more than 600,000 subscribers access to a comprehensive library of more than 11,500 movie, television and other filmed entertainment titles. Our standard subscription plan allows subscribers to have three titles out at the same time with no due dates, late fees or shipping charges for $19.95 per month. Subscribers can view as many titles as they want in a month. Subscribers select titles at our Web site (www.netflix.com) aided by our proprietary CineMatch technology, receive them on DVD by first-class mail and return them to us at their convenience using our prepaid mailers. Once a title has been returned, we mail the next available title in a subscriber's queue. In 2001, our total revenues were $75.9 million, and our net loss was $38.6 million. For the three months ended March 31, 2002, our total revenues were $30.5 million, and our net loss was $4.5 million. As of March 31, 2002, we had an accumulated deficit of $141.8 million.

In 2001, domestic consumers spent more than $29 billion on in-home filmed entertainment, representing approximately 78% of total filmed entertainment expenditures, according to Adams Media Research. Consumer video rentals and purchases comprised the largest portion of in-home filmed entertainment, representing $21 billion, or 71% of the market in 2001, according to Adams Media Research.

The home video segment of the in-home filmed entertainment market is undergoing a rapid technology transition away from VHS to DVD. The DVD player is the fastest selling consumer electronics device in history, according to DVD Entertainment Group. In September 2001, standalone set-top DVD player shipments outpaced VCR shipments for the first time in history, and this trend continued throughout the remainder of 2001. At the end of 2001, approximately 25 million U.S. households had a standalone set-top DVD player, representing an increase of 91% in 2001, according to Adams Media Research. Adams Media Research estimates that the number of U.S. households with a DVD player will grow to 69 million in 2006, representing approximately 62% of U.S. television households in 2006.

Our subscription service has grown rapidly since its launch in September 1999. We believe our growth has been driven primarily by our unrivalled selection, consistently high levels of customer satisfaction, rapid customer adoption of DVD players and our increasingly effective marketing strategy. We primarily use pay-for-performance marketing programs and free trial offers to acquire new subscribers. In the San Francisco Bay area, where the U.S. Postal Service can make one- or two-day deliveries from our San Jose distribution center, approximately 2.8% of all households subscribe to Netflix.

Our proprietary CineMatch technology enables us to create a customized store for each subscriber and to generate personalized recommendations which effectively merchandise our comprehensive library of titles. We provide more than 18 million personal recommendations daily. In April 2002, more than 11,000 of our more than 11,500 titles were selected by our subscribers.

We currently provide titles on DVD only. We are focused on rapidly growing our subscriber base and revenues and utilizing our proprietary technology to minimize operating costs. Our technology is extensively employed to manage and integrate our business, including our Web site interface, order processing, fulfillment operations and customer service. We believe our technology also allows us to maximize our library utilization and to run our fulfillment operations in a flexible manner with minimal capital requirements.

The Offering

Common stock offered by Netflix .... 5,500,000 shares

Common stock to be outstanding after the offering ...... 20,648,074 shares

Use of proceeds .... We estimate that our net proceeds from this offering will be approximately $74.7 million. We intend to use the net proceeds for:

- repayment of approximately $14.1 million of indebtedness under our subordinated promissory notes, including accrued interest as of May 22, 2002; and

- general corporate purposes, including working capital.

Risk factors .... See "Risk Factors" and other information included in this prospectus for a discussion of factors you should carefully consider before deciding to invest in shares of our common stock.

Nasdaq National Market symbol.................................... NFLX

Unless we indicate otherwise, all information in this prospectus: (1) assumes no exercise of the overallotment option granted to the underwriters; (2) assumes the conversion into common stock of each outstanding share of our preferred stock, which will occur automatically upon the completion of this offering; (3) is based upon 15,148,074 shares outstanding as of May 22, 2002, including shares to be issued to certain studios immediately prior to this offering based on our capitalization as of May 22, 2002; (4) gives effect to a one-for-three reverse stock split effected in May 2002; and (5) excludes:

- 4,352,472 shares of common stock issuable upon the exercise of stock options outstanding as of May 22, 2002, with a weighted average exercise price of $3.11 per share, of which 1,162,022 were vested as of May 22, 2002, and 988,608 shares of common stock available for future option grants under our 1997 Stock Plan and 2002 Stock Plan, as of May 22, 2002;

- 7,017,962 shares of common stock issuable upon exercise of warrants with a weighted average exercise price of $3.20 per share; and

- 583,333 shares of common stock reserved for issuance under our 2002 Employee Stock Purchase Plan.

Summary Financial and Other Data

The summary financial data below should be read together with "Management's Discussion and Analysis of Financial Condition and Results of Operations" and the consolidated financial statements and the related notes included elsewhere in this prospectus.

	Year Ended December 31,			Three Months Ended March 31,	
	1999	2000	2001	2001	2002
		(in thousands, except per share data)			
Statement of Operations Data:					
Total revenues	$ 5,006	$ 35,894	$ 75,912	$ 17,057	$30,527
Gross profit (loss)	633	11,033	26,005	(1,120)	15,369
Operating loss	(30,031)	(57,557)	(37,227)	(20,417)	(4,054)
Net loss	(29,845)	(57,363)	(38,618)	(20,598)	(4,508)
Net loss per common share:					
Basic and diluted	$ (21.41)	$ (40.57)	$ (21.15)	$ (12.26)	$ (2.20)
Pro forma—basic and and diluted[1]			$ (2.74)		$ (0.30)
Supplemental pro forma[2]			$ (2.60)		$ (0.26)
Number of shares used in computing per common share amounts:					
Basic and diluted	1,394	1,414	1,086	1,680	2,047
Pro forma—basic and and diluted[1]			14, 099		14, 834
Supplemental pro forma[2]			14,532		15,701

	As of March 31, 2002		
	Actual	Pro Forma[1]	Pro Forma As Adjusted[3]
		(in thousands)	
Balance Sheet Data:			
Cash and cash equivalents	$ 15,671	$ 15,671	$ 76,471
Working capital (deficit)	(9,547)	(9,547)	51,253
Total assets	44,740	44,740	105,540
Long-term debt, less current portion	4,117	4,117	959
Redeemable convertible preferred stock	101,830	—	—
Stockholders' equity (deficit)	(90,872)	10,958	74,916

	Year Ended December 31,			Three Months Ended March 31,	
	1999	2000	2001	2001	2002
		(in thousands)			
Other Data:					
EBITDA[4] (unaudited)	$(21,223)	$(28,179)	$ (1,716)	$(3,600)	$ 3,583
Adjusted EBITDA[5] (unaudited)	(24,405)	(43,860)	(13,722)[6]	(8,012)[6]	666
Number of subscribers (unaudited)	107	292	456	303	603
Net cash provided by (used in):					
Operating activities	$(16,529)	$(22,706)	$ 4,847	$(2,805)	$ 6,505
Investing activities	(19,742)	(24,972)	(12,670)	(4,087)	(5,798)
Financing activities	49,408	48,375	9,059	(927)	(1,167)

(1) The pro forma balance sheet data, pro forma net loss per share—basic and diluted, and pro forma number of shares—basic and diluted give effect to the conversion of all outstanding shares of our preferred stock into shares of common stock automatically upon completion of this offering.

(2) The supplemental pro forma net loss per share—basic and diluted gives effect to the assumed repayment of our subordinated promissory notes as of July 11, 2001 with the proceeds from the offering for the shares solely sold to repay these subordinated promissory notes.

(3) The pro forma as adjusted column gives effect to the sale of 5,500,000 shares of common stock offered by us at the initial public offering price of $15.00 per share and the application of the net proceeds from the offering, after deducting underwriting discounts and commissions and estimated offering expenses, including repayment of our subordinated promissory notes.

(4) EBITDA consists of operating loss before depreciation, amortization of intangible assets, amortization of DVD library, non-cash charges for equity instruments granted to non-employees, gains or losses on disposal of assets and stock-based compensation. EBITDA provides an alternative measure of cash flow from operations. You should not consider EBITDA as a substitute for operating loss, as an indicator of our operating performance or as an alternative to cash flows from operating activities as a measure of liquidity. We may calculate EBITDA differently from other companies.

(5) Adjusted EBITDA consists of EBITDA less amortization of DVD library. Adjusted EBITDA provides an alternative measure of cash flow from operations. You should not consider Adjusted EBITDA as a substitue for operating loss, as an indicator of our operating performance or as an alternative to cash flows from operating activities as a measure of liquidity. We may calculate Adjusted EBITDA differently from other companies.

(6) Adjusted EBITDA for the year ended December 31, 2001 and for the three months ended March 31, 2001 has been "normalized" to reflect DVD library amortization as if a one-year amortizable life had been used beginning as of January 1, 2000 instead of July 1, 2001. As more fully discussed in Note 1 to the Notes to Financial Statements, on January 1, 2001, we revised our DVD library amortization policy from an accelerated method using a three-year life to the same accelerated method over a one-year life.

TABLE 18.19 Conditions that favor an IPO.

■ Market value greater than $200 million	■ Profit margins greater than 14 percent
■ Sales greater than $100 million	■ Return on capital greater than 14 percent

The market for IPO issuance is cyclical. Thus, the ability of a company to go public is a function of timing as well as its financial performance. There are several reasons for going public. First, it is one way to obtain the financial resources needed to grow. Second, it may do great things for the firm's reputation. Third, it's an effective tool for recruiting employees. Fourth, and probably most important, it provides liquidity and an avenue for eventually cashing out of the company. However, only a few firms are able to go public, even in good times. A firm needs to be in an industry with favor in the market and be able to show a good financial story of growth and profitability. The firm also needs to be able to sell a large enough number of shares to net at least $30 million so that its total costs of the IPO process remain reasonable.

While a fast-growing, high-impact company may benefit from an IPO, it is wise to consider the burdens of public ownership. Aggressive regulation, stringent record keeping requirements, and a fickle public market can be significant to smaller firms. Firms with a market value of less than $200 million may wish to avoid these burdens and remain private. Furthermore, firms with a market value of less than $300 million are often thinly traded and illiquid. The conditions required for a U.S. technology venture to go public are summarized in Table 18.19.

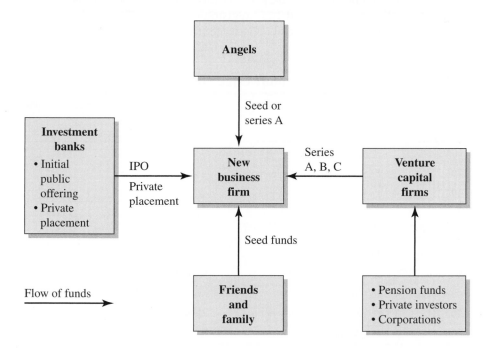

FIGURE 18.6 Potential for the flow of funds to a new business firm.

The overall investment process for a new firm can be portrayed as shown in Figure 18.6. The seed round of financing may be achieved with friends and family. Angels may supply the series A round of financing. Venture capitalists will often supply the funds in series A, B, and C. Then investment bankers will facilitate an IPO or large private placement.

18.13 AgraQuest

The original AgraQuest business plan dated May 5, 1995, provided the profit and loss statement shown in Table 17.9. It also provided a table of projected investment requirements, shown in Table 17.10. These projections were optimistic but consistent with the growth in interest in biotech firms and their potential to become large, important companies. The original business plan called for an IPO after five years.

Pam Marrone showed the plan to numerous venture capitalists over an 18-month period. Eventually, in February 1997, she found a team of investors interested in opportunities in sustainable agriculture. These venture capitalists as a group invested $3.2 million in the firm for about 46 percent of the firm. Subsequent stages of venture capital were attracted over the years, as shown in Table 18.20. Note the down rounds starting with round F.

As a result, after round H, the venture capital firms owned about 90 percent of the firm. Marrone, the original founder, owned only 3 percent of the firm as a result of all the venture capital rounds.

The initial public offering market was favorable in 2000, and AgraQuest prepared an IPO document and SEC registration to raise $75 million with a share price of $11 to $13. The lead underwriter (investment banker) was Merrill

TABLE 18.20 Sequence of venture capital rounds for AgraQuest.

Year	Round	Amount raised ($ thousands)	Share price ($)	Percent owned by venture capital firms
1995	Seed—Founders	50	0.025	—
1995	Seed—Family and friends	450	0.35	—
1997	A—Venture capital	3,200	1.05	46.1%
1998	B—Venture capital	6,000	1.70	59.6%
1999	C—Venture capital	7,100	2.35	66.4%
April 2000	D—Venture capital	7,200	3.20	72.3%
December 2000	E—Venture capital	15,000	5.00	76.1%
2003	F—Venture capital	11,000	1.37	88.0%
2004	G—Debt with equity options	3,000	1.25	—
2005	H—Venture capital	14,000	0.07	90.0%

Lynch. The prospectus was completed and filed in August 2001. However, the market for U.S. IPOs disappeared with the September 11, 2001, terrorist attacks. As a result, an expenditure of $1 million on underwriting costs was lost as the IPO filing was withdrawn in April 2002.

18.14 Summary

The entrepreneur leaders of new ventures create a set of financial projections that can be used to estimate the cash investments that will be needed as well as when they will be needed. Using that information, the entrepreneurs seek out investment capital. For the first or seed round, they may rely on their own funds and investments from friends and family. Eventually, most technology firms will need a significant investment from others such as wealthy individuals called angels or professional investors called venture capitalists.

Several stages of investment may be the best means of acquiring investments based on performance milestones for each stage. A real option is the right to purchase an asset at a future date. Thus, venture capital investors often use staged investing with milestones to exercise their investment opportunity with a start-up. For big-impact start-ups, it is the future value of the potential of the new firm that is most attractive.

Using venture capital valuation methods, a start-up's value is established and an agreement on the division of ownership may be obtained. As the expected growth of revenues and profitability are achieved, the firm and the investors may wish to exercise an option to sell. This could be an initial public offering (IPO) or the acquisition of the start-up by a larger firm in order to harvest the value that has been created by the partnership of investors, founders, and employees.

Principle 18
Many kinds of sources for investment capital for a new and growing enterprise exist and should be compared and managed carefully.

Video Resources

Visit http://techventures.stanford.edu to view experts discussing content from this chapter.

The Benefit of Picking the Right VC	Marc Fleury	JBoss
Venture Capital vs. Customer Funding	Vic Verma	Savi
What Is the Average Size of a VC Investment?	Steve Jurvetson	DFJ
Flying with Angel Investors	Ron Conway and Mike Maples	Angel investors

18.15 Exercises

18.1   Viscotech Inc. is described in exercise 17.3. Determine the percentage ownership an angel group may demand for investing the $1 million sought by Viscotech at the start of year 1. If Viscotech is unable to obtain the bank loan of $500,000, it will need an equity investment of $1.5 million. What ownership percentage will the angel group demand for this investment? Assume the annual interest payment on the loan was planned to be 10 percent of the principal.

18.2   Glenn Owens's attractive technology start-up requires $10 million to launch. Projections show earnings of $10 million and sales of $80 million in the fifth year. The venture capital firm expects a return of 50 percent per year for the five-year period prior to an IPO. What valuation would you assign to the new venture? What ownership portion should the venture capitalist expect to receive? Perform a sensitivity analysis on the valuation and rate of return.

18.3   DGI, a new firm in formation, has developed a set of projections shown in Table 18.21. The expected and pessimistic cases are shown. The harvest of the firm is planned for the fifth year. The firm is seeking an initial investment of $1 million before launching in year 1 as well as a commitment for $1 million at the end of year 2 for expansion. Acting as an adviser to a venture capital firm, prepare an offer of investment to submit to DGI. Assume the PE (price earnings) ratio in DGI's industry is 15.

18.4   The CEO of an early-stage software company is seeking $5 million from venture capitalists. The reasonable projected net income of $5 million in year 5 can be valued at a PE of 20. Furthermore, the sales in year 5 are projected at $25 million. What share of the company would the venture capitalists require if their anticipated rate of return were 50 percent? The company has one million shares outstanding before the venture capitalists purchase shares. What price per share should the venture capitalists pay?

TABLE 18.21   Revenues and net income for the expected case and the pessimistic case for DGI.

Year	1	2	3	4	5
Expected revenues	0.84	2.82	5.44	8.35	11.55
Expected net income	0.18	1.25	2.67	4.17	5.86
Pessimistic revenues	0.42	1.41	2.72	4.18	5.78
Pessimistic net income	(0.11)	0.26	0.77	1.25	1.81

Note: All figures in millions of dollars.

18.5 Consider a new firm in the nanotechnology field that seeks a second round of financing. This year, it has revenues of $2 million and projects profitability of $200,000 next year on revenues of $3 million. It is raising $1 million from a new set of investors. What share of the company should it offer to the new investors? Assume it can increase profits at a rate of 25 percent per year over the next five years.

18.6 Rackspace is a Web hosting company that went public in 2008 (www.rackspace.com). The firm had sales of $362 million in 2007. The firm went public via an OpenIPO auction process using bids for shares (www.wrhambrecht.com/ind/auctions/openipo/index.html). Examine the OpenIPO auction process and the Rackspace offering. What was the offer price and offering size? What was the post-money valuation of the firm? Was this valuation reasonable?

18.7 Consider the Netflix IPO prospectus in Figure 18.5. Calculate the percentage of Netflix that is being sold to the public. Assume all issuable stock options (including unvested shares) and warrants were exercised before the IPO. What percentage of the company would then be sold?

18.8 Explain the purpose and value of staged financing (a) for investors and (b) for the entrepreneur(s).

18.9 In what circumstances should grants be viewed as an attractive source of capital? When would a grant be less appropriate?

VENTURE CHALLENGE

1. What sources of capital will you use for your venture?

2. Why did you select these sources?

3. How much capital is needed initially and for what purpose?

4. What percentage of your venture do you plan to offer to outside investors?

CHAPTER 19

Presentations and Deal Negotiations

Leadership involves remembering past mistakes, an analysis of today's achievements, and a well-grounded imagination in visualizing the problem of the future.
Stanley C. Allyn

How does the new venture present its vision and story and negotiate a deal with investors?

The creators of a new enterprise need to tell their story about the future of their business. Establishing credibility and trust through presentations of the new venture's plan for a novel solution to an important problem can lead to an investment.

A short presentation of the plan (called an elevator pitch) can help interest investors in seeing a more complete presentation. The presentation, often delivered in an investor's office or at a venture fair, may create the necessary interest from an investor.

The negotiation of a deal with an investor is an important part of the process. One can cement the relationship or destroy it through the negotiation process. Often agreements with terms contingent upon performance may be appropriate. The integrated story and the business plan should show how the business solution would be profitable within a reasonable period. The investors are interested in a favorable return. They also want to sense that they will be partners with trustworthy and capable entrepreneurs. ■

19.1 The Elevator Pitch

Often entrepreneurs have a chance to make their case to a potential investor or ally. The importance of a compelling story was discussed in Chapters 2 and 7. A short version of the new venture story is often called the **elevator pitch,** which gets its name from the two-minute opportunity to tell a story during an elevator ride. Chance meetings (in places such as elevators) will offer opportunities to entrepreneurs to make a brief case for their venture. A prepared short version of the venture story can be a powerful door opener.

The goal of the short story is to get approval to proceed to the next step, where the entrepreneur can tell the longer version of the story and secure new colleagues, allies, and investors. Thus, the entrepreneur must recognize that there isn't time to elaborate on many details. Instead he or she must quickly convey the essence of the opportunity in a way that invites a longer conversation. A short version of the venture story demonstrates that the entrepreneurs know their business and can communicate it effectively.

The short version of the venture story starts with an introduction, moves into a description of the opportunity, and then describes the potential benefits of the new venture. The secret of strong short stories is in grabbing the attention of listeners, convincing them with the promise of mutual benefit, and setting the stage for follow-up. The story can start with a captivating question such as: "If IKEA can provide baby-sitting, why can't movie theaters?"

The short version of the story should convey the vision of the venture. In the case of Genentech, it is: "We discover and make biotechnology pharmaceutical products to reduce or overcome the effects of cardiovascular, pulmonary, and cancer diseases." An important vision can provide the inspiration for the venture and its short story.

There are techniques that a storyteller can employ to make the pitch memorable. First, the storyteller should speak in terms the listeners can understand and should focus on one simple message. Second, the storyteller should use concrete images and examples rather than ambiguous or abstract phrases. Third, the storyteller should generate interest and curiosity by exposing gaps in current solutions and then filling these gaps through the proposed new venture. Ideally, this will cause an emotional response from the listeners. Finally, the storyteller must be passionate about the idea, demonstrating a genuine attachment to the problem and solution [Heath and Heath, 2007].

Intuit's Elevator Pitch

During the founding of Intuit, Scott Cook described his venture in this way: "Homemakers need to pay the family's bills. They hate the hassle of bill collection and payment. They need a software computer program to quickly and easily pay their bills. Other programs are too slow and too hard to learn. Our solution is a fast, easy-to-use program with no instruction books needed. So the bill payer needs Quicken!"

19.2 The Presentation

The new venture team will be expected to verbally present its business plan to investor groups, angels, potential employees, allies, and suppliers. The purpose of these meetings will be to persuade them to cooperate, support, and participate in the new venture. Effective persuasion is a negotiating and learning process that leads to a shared vision. Formerly, people thought of persuasion as a simple process: state the position or plan, outline the supporting arguments, and then ask for the action or deal being sought. Today, most would-be investors and allies want to engage in a dialogue with the new venture team. The team outlines the venture and invites feedback and alternative solutions. Persuasion involves compromise and the development of relationships with the investors and other participants.

New enterprises must be able to sell their story to potential investors and clients. Selling ideas is intrinsically difficult. Clients and investors are naturally risk-averse when it comes to large projects that call for large investments with payoffs that are many years in the future. They are even more risk-averse when the projects do not originate from within their own organization.

Effective persuaders first establish credibility. Second, they frame their goals in a way that establishes common ground with their listener. Third, they offer solid evidence to support their plan. Finally, they build a good relationship with their potential investors or allies. This four-step method of persuasion is summarized in Table 19.1 [Conger, 1998].

Credibility and trust are established through experience over time. Thus, the shortest path to credibility is with someone the entrepreneurs already know. Otherwise, the entrepreneurs' track record and references will be particularly important. The expertise required for the venture must always exist on the venture team. It may be useful to review the concept of influence and persuasion as described in Section 13.2.

The business plan must appeal strongly to the potential investor or ally. Framing the unique benefits of the new venture so that they match the goals of the investor or ally is critical. The next step is to provide solid evidence supporting the business plan. Here, a vivid story or analogy will help to bring the plan alive. Finally, it is important to build a good relationship with the investor or ally. In this step, the entrepreneurs demonstrate their commitment to the plan and show some of their passion for the project.

TABLE 19.1 Four-step method of persuasion.

■ Establish credibility with the investor, ally, client, or talent.	■ Offer solid, compelling evidence to support the plan.
■ Frame the goals of a new venture to be consistent with the goals of the investors or allies; describe the unique benefits of the venture.	■ Build a good relationship with the investor or ally.

TABLE 19.2 Four-part pitch.

1. This is a painful problem that customers need to solve.	3. This company has a profitable and proven solution for this problem.
2. Many customers who have this painful problem have money to spend on alleviating the pain.	4. Management has implemented and planned effectively in the past, and they can execute well in the future.

A good presentation captures the listeners and lets them respond to the problem the entrepreneurs propose to solve. A four-part "pitch" is described in Table 19.2.

In presenting the business plan, it is also useful to answer the nine questions provided in Table 19.3. Of course, there will be other issues, but these nine are almost always part of a presentation. It is helpful to rehearse the presentation with an audience of a few trusted colleagues or friends who can respond with suggestions.

Trust is the basis of nearly every enduring business relationship. The presentation of a business plan is a vehicle for winning trust from potential investors and allies. Trust, confidence, and relationships are built over time. Speaking about the venture's goals needs to invoke a positive response from the listener. What difference will it make if the venture is successful? Will people live better lives or enjoy new alternatives?

A successful pitch often follows the 10/20/30 rule: 10 slides, 20 minutes, and 30-point-font text. Most presentations to investors or allies should be based on about 10 slides. Each slide should use at least 30-point font text. This will keep the listener interested and the presenters focused on their key points. A sample 10-slide presentation is outlined in Table 19.4. Ten slides can be used for a 20-minute presentation with time for discussion [Kawasaki, 2004].

In any presentation, the speaker should convey a sense of urgency about the problem and a strong commitment to make the solution robust. Good speakers highlight the unique benefits on slides 3 and 4.

TABLE 19.3 Nine questions to answer in the business plan presentation.

1. What is the product and what problem is it solving?
2. What are the unique benefits of the product?
3. Who is the customer?
4. How will it be distributed and sold?
5. How many people will buy it in the first and second years?
6. How much will it cost to design and build the product?
7. What is the sales price?
8. When will you break even?
9. Who are the key team members and how are they qualified to build this business?

TABLE 19.4 A sample 10-slide presentation.

1. Company name, presenter name, contact information
2. Description of the problem: the need and the market
3. Solution: the product and its key benefits
4. Business model and profitability
5. Competition and strategy
6. Technology and related processes
7. Marketing and sales plans
8. Leadership team and prior experience
9. Financial projections summary
10. Current status and funds required

Listeners are swayed by the quality of the ideas but even more by evidence that the presenters are creative and innovative. The investors or new team members seek to be part of a creative collaboration. The listeners look for passion and evidence that the proposed solution is a big change or discontinuity. Furthermore, the goal is to engage the listeners so that they become part of a creative collaboration. The best outcome is that the presenters successfully project themselves as creative types and get their listeners to view themselves as creative collaborators in the process of building the new venture [Elsbach, 2003].

19.3 Negotiating the Deal

After the presentation of the business plan and follow-on discussions, an investor or ally may be confident about the potential venture but hold different expectations about it. Thus, the investor and the venture team may have different views regarding the valuation for the firm and the appropriate terms of the deal. A good deal is one that fairly meets the needs of the new venture while enabling the relationship between the firm and the investor to flourish in the future. Thus, the pricing and terms of the deal must be balanced with the future of the relationship. If possible, the new venture needs to have alternatives to closing a bad deal. With a good alternative, the new venture can walk away from a bad deal. A new venture should understand its own interests and its own no-deal alternatives or options. Price, control, and ownership percentage are usually key factors for negotiation.

Investors will normally engage in a process of checking the backgrounds of the venture team, the market data, and the key elements of the business plan. This process, called **due diligence,** consists of verifying facts and data provided in a business plan before making a commitment to the terms of an investment deal.

Negotiating a fair deal is a skill that can be learned. Most entrepreneurs have limited experience negotiating a fair deal with investors. **Negotiation** may

TABLE 19.5 Four principles of negotiations.

1. Focus on describing the problem (task or deal) and take the people out of the discussion.

    Goal: All participants are solving the problem.

2. Focus on the interests of the parties, not their original positions.

    Goal: Each party states what they seek and the associated goals.

3. Generate a variety of options or possibilities that advance the interests of the parties.

    Goal: Several real solutions.

4. Create a final deal based on fair and objective standards.

    Goal: Real, measurable standards.

Source: Fisher and Ury, 1991.

be defined as a decision-making process among interdependent parties who do not share identical preferences. Consider an example of a manager and an employee negotiating a pay raise for the employee. They are interdependent but have different preferences for the outcome. The employee wants a raise, and the manager wants improved performance.

The best negotiation should produce an efficient, wise agreement and not damage the relationship between the parties. A wise or good agreement is one that meets the legitimate interests of both parties, resolves conflicts fairly, and is durable [Fisher and Ury, 1991]. One should try to avoid parties locking into a position, but rather reach for common ground.

The process of negotiating a good deal for all concerned can be based on four principles, as summarized in Table 19.5. Try to take personalities out of the discussion and avoid locking into positions. Try to get everyone working for a fair deal. Talk about the interests and goals of each party and avoid taking rigid positions, then generate several possible solutions that advance the interests of all parties. Finally, select the best solution with measurable outcomes.

Often negotiations will stall, and it may be best to try to reshape the scope and sequence of the negotiation. One or more of the parties can scan widely to identify elements outside of the deal currently on the table that might create a more favorable structure. For example, they may introduce new parties and terms to the deal and try to satisfy all parties [Lax and Sebenius, 2003].

Investors tend to have goals based on return on investment and time horizons to receiving this return. Entrepreneurs tend to have goals based on growth, success, and achievement, as well as return on investment. Both parties need to help generate some good options so that there is room to adjust the deal. Finally, the parties select a deal that gives them a fair solution to their needs. This deal should include measurable outcomes and adjustments in ownership or other factors if the agreed-to outcomes are not realized. Also, the deal should be a good start on a long, cooperative alliance between the investor and entrepreneur [Ertel, 2004].

In any negotiation, each side must choose between two options: accepting a deal or taking its best no-deal option. Typically, the best no-deal option is to move on to find and negotiate with a new investor. Thus, the new venture team evaluates the deal offered versus seeking another potential investor. Similarly, the other side is also considering losing the deal versus losing the deal's advantages. The investor will accept the deal if it meets its own interests better than its own no-deal option, which is to lose a good opportunity. Thus, the negotiation problem for the venture is to understand and shape the investors' perceptions and help them choose in their own interest what you want [Sebenius, 2001].

When making an investment, a venture capitalist considers three forms of risk: the market risk (establishing customers for the product), the technological risk (the extent to which the technologies or concepts are well developed and not threatened by potential competitors), and the management risk (the team's technical and leadership competencies to develop the new firm). The venture capitalist wants to know which risks will be reduced by its investment in this stage.

Differences regarding valuation and ownership usually reflect differences in estimations of the future performance of the new venture. Therefore, it may be wise to develop an agreement that includes terms that are contingent on the outcome of designated measures or events. Using contingent terms in a contract enables the parties to bet on the future rather than argue about it. An agreement with contingences could include terms on such measures as revenues, profits, or number of customers achieved in an agreed-upon period. The actual outcome could lead to an agreed-upon ownership percentage. For example, the investor agrees to an ownership split of 70-30 (firm versus investor) if the goals and milestones are met but will reserve shares for a readjusted split of 60-40 if they are not met. Investors and new ventures can come to very different conclusions about many kinds of future events, such as sales and market shares as well as competitors' moves. Whenever such a difference exists, so does an opportunity to craft a contingent contract that both sides believe to be in their best interests. A contingent contract results in the two parties sharing the risk.

A common way to reach agreement about ownership is to offer a deal with warrants to tie ownership to actual performance. A warrant is a long-term option to acquire common shares, usually at a nominal price. For example, an investor may receive a warrant to receive an agreed-upon number of shares if certain performance levels are not achieved by a certain date [Smith and Smith, 2004].

A factor that can complicate negotiations is the matter of dilution of ownership by the founder group. Investors will usually seek an antidilution clause to protect themselves from dilution of their ownership percentage. This antidilution clause will usually be triggered by a lower price (down round) in a subsequent financing round. All the terms of an investment agreement, often called a **term sheet,** should be reviewed by the new venture firm's attorney. A term sheet is a funding offer from a capital provider. It lays out the amount of an investment and the conditions under which the investors expect the entrepreneur to work using their money. The key is to remember that it's just an offer, and the entrepreneur can counter that offer and negotiate all the terms before finally accepting the funds. An excellent reference on these matters is *The Entrepreneur's Guide to Business Law* [Bagley and Dauchy, 2007].

19.4 Critical Issues for the Business Plan

We discussed the development of the business plan in Chapter 7 and the presentation of the business plan in Section 19.2. After the business plan is presented to a few potential investors, their suggestions and criticisms may require adjusting the business model of the firm or other parts of the business plan. Perhaps it is necessary to revise the product to make it more compelling. Offering a product that is "nice to have" for a customer is different than offering a product that the customer "must have." It is nice to have vitamins available, but when one has a headache, an aspirin or Advil is a "must have." Is the product a complete solution to the customer's problem, or is it only a part of the solution?

Investors will ask: Are the people committed to the business, and is the opportunity large in potential? Can we identify and reduce the risks? Will there be a way we can harvest our return on investment? Is the estimated growth rate attractive?

The plan in written form and the verbal presentation should hold together as an integrated whole. Use of outdated or incorrect data will leave doubts for investors. Unsubstantiated assumptions can also hurt a plan. For example, does the plan include an honest recognition of the competitors?

The business plan and its associated story can be viewed as the keystone of the business arch, as shown in Figure 19.1. All the elements of the business come together to form a business whole.

It will help to validate a business plan if the new venture has paying customers at the time of presentation. It is even better if the firm is about to become profitable. Investors need reasons to invest. They want to see advance orders, a letter of intent, or a customer list. That shows proof of customers and how they value the firm's product.

In 1968, Gordon Moore and Robert Noyce left Fairchild Semiconductor to found Intel. They brought along Andy Grove and several other colleagues. At Intel, they saw an opportunity to make a silicon transistor and later an integrated circuit. Moore and Noyce were leaders in their field and knew Arthur Rock, a San Francisco venture capitalist. They asked him one day if he could raise $3 million to start Intel. Rock had secured commitments for the $3 million investment by that evening. Moore and Noyce were well-known and could command a good deal for their new business. Most entrepreneurs are not as fortunate.

If the leaders of a new venture do not have a big, proven network and background, they will need to work hard to find and satisfy investors. They must convince them that their business is a one-time chance to get involved with something important that will exploit a big opportunity.

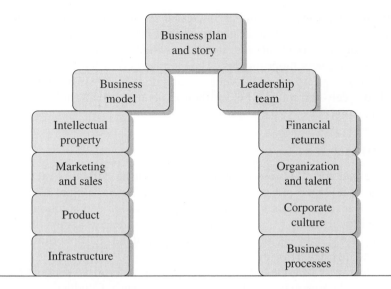

FIGURE 19.1 Integral nature of the business plan and the firm.

19.5 AgraQuest

The search for venture capital was long and hard for Pam Marrone. She pitched her business plan to over two hundred venture capital firms. Her plan called for a new firm, AgraQuest, entering a big market and creating a new solution. The market for pesticides is $28 billion and is dominated by large chemical companies. AgraQuest had a solution that used natural pesticides. Her story was:

1. Pesticides and herbicides help farmers to keep their yields up and avoid disastrous pestilence.
2. Chemical pesticides are harmful to people and the environment.
3. AgraQuest can readily develop natural biopesticides that protect people and the environment, and allow farmers to use them right up to the harvest.

Eventually, Marrone was introduced to the Investors Circle (<u>www.investorscircle.net</u>) by Calvert Social Ventures, and she presented at a venture fair in Chicago in May 1996. This led to interest from Rockefeller Ventures and its eventual investment in AgraQuest.

Marrone learned to give a passionate presentation and capture the listener's attention. Most investors, however, were wary of the agricultural market and declined to invest. The all-natural solution had been tried before and had failed.

One of her angel investors in the seed round, James Schlindwein (former CEO of Sara Lee and Sysco), started introducing her to agricultural venture investors. Marrone found her angel investor to be her best mentor and coach.

Eventually, Marrone found an interested set of social responsibility venture capital firms. They were dedicated to the natural solution and saw AgraQuest as a company that could succeed. It was a well-led company with a good process for discovering new biopesticides. AgraQuest received $3.2 million in venture capital from a group of social venture capitalists in 1997. By 2005, the company had received a total of $62 million in venture capital and $9.5 million in debt financing.

19.6 Summary

The potential for a successful new business can be communicated through a short pitch, a formal presentation, and a written business plan. Most entrepreneurs will need the skills to communicate their vision and solution in all three forms. Through these presentations of the entrepreneurs' story, the potential investor, new employee, or ally learns to understand the opportunity and recognize the competencies of the team members.

As investors become interested in the new enterprise, negotiations about valuation and performance milestones will commence. Conducting negotiations that retain and enhance the rapport with the investor is essential. The entrepreneur is negotiating until the moment of execution of an agreement and a transfer of funds. All the negotiations continuously address issues surrounding product, team, processes, business model, and intellectual property.

> **Principle 19**
> The presentation of a compelling story about a venture and the resulting skillful negotiations to close a deal with investors are critical to all new enterprises.

Video Resources

Visit http://techventures.stanford.edu to view experts discussing content from this chapter.

Tips for a Good Pitch	Heidi Roizen	Mobius
Make a Great Pitch	Guy Kawasaki	Garage

19.7 Exercises

19.1 Twitter allows customers to send (tweet) and read short messages to and from friends, colleagues, and others to answer the simple question "What are you doing?". Write an elevator pitch for this venture.

19.2 Business plan contests offer an opportunity for entrepreneurs to present their business plans. The Clean Tech Open in the United States offers one such contest (www.cleantechopen.com). Visit the site, study the description of the winners, and prepare a brief report on an enterprise that most interests you.

19.3 As the CEO of a new technology venture, you and your team have set a valuation for your firm of $10 million (pre-money) and found a willing venture capital firm. The venture capital firm, however, has set a valuation of $6 million. Revenues next year are projected to be about $6 million, and the firm will be profitable next year. Identify a negotiation approach for achieving a reasonable compromise valuation.

19.4 What is a term sheet? Specify three items in the term sheet that you would want as an entrepreneur. Specify item clauses in the term sheet that you would want as a venture capitalist. Why is it important for both parties to be happy with the resulting deal?

19.5 What are the key factors a venture capitalist uses to value a new venture? Describe what value parameters are most likely different between an investor and an entrepreneur.

19.6 What deal terms can a venture capitalist suggest to ensure entrepreneur incentives are aligned in both good and bad times for the firm? What are deals terms an entrepreneur would suggest to ensure investor incentive alignment in both good and bad times for the firm?

19.7 Why is it critical that the investor and the entrepreneur are both happy with the final deal? Who loses if this is not the case?

VENTURE CHALLENGE

1. State your venture's elevator pitch.

2. Provide an outline of the presentation for describing your venture to investors.

3. Sketch out a term sheet outlining your venture's capital needs, the amount of the company you are interested in selling (e.g., number of shares or what percentage of the total shares), and any other negotiation terms you consider important.

CHAPTER 20

Leading Ventures to Success

Well done is better than well said.
Benjamin Franklin

How do successful entrepreneurs transition from a solid business plan to an operating enterprise?

Creating a business plan for a new enterprise is important, but implementing the plan successfully is essential. Execution of a plan is a discipline for connecting strategy with reality by aligning goals and the firm's people to achieve the desired results. Execution is about turning a concept into a great business.

New businesses move from start-up to growth to maturity in stages. Managing a new business through these stages requires different skills and organizational arrangements. Start-ups need to plan for having the right people in the right positions as they grow.

Organizations, like people, need to learn and adapt to change. Organizing for recognizing and responding to challenges can build resilience in a start-up firm. The ability to adapt to change may be a firm's only truly sustainable advantage. Furthermore, to achieve long-term success, a firm needs an ethical base for action. ■

20.1 Execution of the Business Plan

Once the new venture has secured the necessary resources, the firm proceeds to the **implementation** phase by carrying out or putting into effect the elements of the business plan. Another common term for the implementation is execution, which is a system of getting things done. **Execution** is a discipline for meshing strategy with reality, aligning the firm's people with goals, and achieving the results promised [Bossidy and Charan, 2002]. Often, the unique difference between a successful company and its competitors is the ability to execute its plan.

Execution is not just tactics; rather, it is competency and an associated system built into a firm's goals and culture. Both Dell Computer and Gateway sell personal computers directly to customers, but Dell requires one-fifth the working capital needed to generate a million dollars of sales than does Gateway. Both companies sell build-to-order PCs, but Dell's asset velocity is five times that of Gateway. From Chapter 17, asset velocity is defined as the ratio of sales to net assets.

Execution is also the process of determining how well a firm is performing, acting to improve performance, following through, and ensuring accountability. Execution is following through on the strategy of the business plan. Any firm with a sound business model and strategy is still only as good as the implementation of the strengths of the model and strategy. In preparing the plan, all the team members have built their expectations and strategies into an agreed-upon, coherent road map. Thus, the team knows what needs to be done. The next step is to agree on how it is going to happen. Who does what and by when? The team sets short-term goals and priorities, and then assigns tasks to individuals. Rewards and recognition are linked to on-time performance. Missing deadlines is costly. Therefore, setting realistic deadlines is important. The success of a new enterprise can be attributed to the execution skills of many team players rather than the decision-making skills of an omniscient entrepreneur. Six questions that can be effectively used to achieve solid implementation are provided in Table 20.1.

The new venture needs team members with a wide set of capabilities so that one person can be responsible for several tasks. If necessary, additional people may be engaged to accomplish unique or difficult tasks. These tasks flow seamlessly from the firm's strategy. The team describes where it wants to be by a certain time and then divides the tasks among its members. It is critical to have a realistic assessment of the effort and time required for major tasks. The elements of an operating plan are tasks, milestones, and objectives. The leadership team will need to make trade-offs between tasks and goals so that the operating plan is realistic. A specific, written operating plan will help the firm to move forward

TABLE 20.1 Six questions for implementation.

1. **Why** is the objective a priority?	4. **Who** is on the team and will be accountable?
2. **What** is the action and the expected outcome?	
	5. **When** will the activity be completed?
3. **How** will the action be achieved?	6. **Where** will it be accomplished?

efficiently. Also, reviews of accomplishments and the operating plan will help keep the team on task. Good execution is based on clear priorities, good assumptions, and constantly monitored performance [Mankins and Steele, 2005].

Execution is fundamentally about turning a concept into a business. For example, after a short period, the firm may find its prototype has defects or the sales channel is not as attractive as originally thought. Every new venture runs into trouble before reaching fruition. Then it is time to reconfirm the vision and recommit to the execution task. This requires persistence and follow-through. While admitting mistakes, the team's focus must stay on the long-term strategy. For example, in 1998, when fraud started to increase on eBay, the online auction firm created an antifraud campaign. Nonpaying bidders would receive only one warning before receiving a 30-day suspension from bidding. At the same time, eBay offered free insurance through Lloyd's of London. As a result, eBay became more successful.

Execution is hard work. The setting of goals and deadlines should be the task of those who need to accomplish the work. The establishment of priorities is a big part of sound execution. Tasks can be divided into "must do," "should do," and "nice to do," with the priority kept on the "must do" activities as much as possible. The use of measurable goals and lists of necessary tasks and deadlines is very helpful. Cyrus Field tried four times to lay a telegraph cable across the Atlantic Ocean to link the United States and Britain. Using a new steamship, the *Great Eastern,* he succeeded in 1866 after nine years of effort. The saga was a dramatic example of Thomas Edison's maxim that "genius is 1 percent inspiration, 99 percent perspiration."

Consider the highly competitive computer-aided design software business and two solid competitors, Mentor Graphics and Cadence Design. Mentor and Cadence offer software for chip design at competitive prices. With aggressive competition and demanding customers, execution and details are everything. A firm's survival depends on repeat business. Knowing which details are more important than others can make a big difference. Few competitive advantages cannot be quickly imitated. Therefore, a new firm needs to out-execute its competitors. It must continually underpromise and overdeliver. Execution can go wrong for several reasons such as (1) allowing the strategy to shift over time and (2) synchronization, which is getting the right product to the right customer at the right time [Hrebiniak, 2005].

An emerging new venture often needs help on operational issues as it moves toward growth. One source of help is the new firm's suppliers and customers. They have in-depth capabilities that can often be accessed by a new firm. Often these large firms want the new firm to succeed and will lend a hand on tough issues.

During the initial period following launch of the firm, one of the primary goals will be to build and grow revenues. One key measure that can help in the early stages of a business is the ratio of revenues to expenditures plus assets employed, called a business index (BI), where

$$BI = \frac{revenues}{expenses + assets}$$

TABLE 20.2 Seven steps to building great companies.

1. **Leadership:** Be ambitious about the firm, possess strong will and resolve, desire sustained results, and retain personal humility.

2. **People:** Choose the right people, put them in the right positions, create a road map for success, and communicate it to everyone.

3. **Success:** Show unwavering faith that the firm will prevail in the long run, confront the realities and facts, and respond.

4. **Organizing principle:** Act on your passion, competencies, and economic engine to create a core principle for the business.

5. **Culture:** Build a culture of discipline where everyone is responsible for results; stay focused.

6. **Technology:** Select a technology application that will accelerate the firm's momentum.

7. **Momentum:** Build momentum slowly and consistently for the long run by constantly creating entrepreneurial projects.

Source: Collins, 2001.

The goal of a business is to steadily increase the BI ratio by growing revenues faster than expenses and assets.

As a new firm grows, one useful measure of sound execution is the sales-per-employee ratio. A successful technology company will have at least $200,000 sales per employee. For example, the emerging firm Network Appliance has about $500,000 sales per employee (www.netapp.com).

Great companies execute flawlessly. They deliver products that *consistently* meet customer expectations. Furthermore, they empower customer representatives on the frontline to respond to varying customers' needs. The goal is to achieve almost perfect operational execution by constantly improving processes, training staff, and eliminating inefficiencies, as evidenced by such firms as United Technologies [Joyce et al., 2003]. In contrast, often companies make a chain of mistakes that undermine the firm [Mittelstaedt, 2005]. Failing to recognize mistakes and fix them can be fatal.

Jim Collins [2001] describes seven steps to building a great business, as summarized in Table 20.2. If the new enterprise can implement most of these very well, it will have a good chance for success.

GE: Entrepreneurial Leadership

Jack Welch, retired CEO of General Electric, is perhaps the best-known U.S. executive of 1985 to 2000. He advocated many principles and methods of execution, such as "work-out" and "bullet-train speed." The key to Welch's popularity was his plain but powerful rhetoric about the critical issues of the firm [Lowe, 2001]. He insisted that each operating unit be number one or two in its market. While GE is a very large company, it retains today the atmosphere of entrepreneurial activity. General Electric meets all seven criteria for Collins's great companies.

A new venture can be said to execute well if its performance for its customers exceeds their expectations through all stages of its business. It is organized around a prevailing vision and a well understood long-term strategy. It keeps investing in new ideas, products, and people [McFarland, 2008].

20.2 Stages of a Business

A new business venture is expected to grow over time, normally following an S-curve, as shown in Figure 20.1. The five stages of a firm are: start-up, take-off, growth, slowing growth, and maturity. During the start-up period, the firm is organizing itself, accumulating the necessary resources, and launching its product. The second stage is take-off, when revenues start to grow. The growth stage is often the most profitable and is a time when management must look toward expanding its product offerings and/or serving new types of customers and/or geographic areas. This stage also entails high risk since the firm has little experience with these new goals. Figure 20.2 illustrates alternative growth strategies [Roberts, 2003].

Eventually, the firm's growth slows in the slow-growth period. Finally, the firm reaches the mature phase. Figure 20.1 shows a high-growth trajectory and

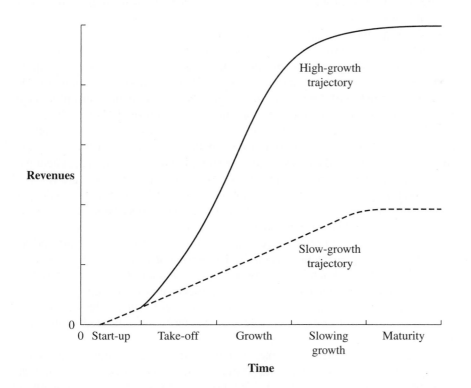

FIGURE 20.1 Growth trajectories for two businesses.

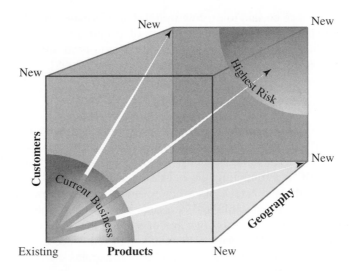

FIGURE 20.2 Alternative growth strategies for a venture.
(Source: Roberts, 2003.)

a slow-growth trajectory for two businesses. The high-growth firm experiences a growth rate of 40 percent per year or better. The slow-growth firm will have a growth rate of 10 percent per year.

Technology ventures needing a long start-up period for developing their product have to show success and raise their funds in stages. They continually must demonstrate credibility and engender trust in the investor community. The CEO and leadership of the venture must execute a creative and truthful strategy that unites the interests of the investors and the employees [Kleiner, 2003].

Moving from start-up to take-off may engender serious stresses as management practices that were appropriate for a smaller size and earlier time no longer work and are scrutinized by frustrated managers. High-growth ventures encounter fluid situations where rapid change and chaos occur. New products are released and marketing plans are in flux. New hires join the firm and decision-making is too slow. With high growth rates, new capital infusions may be necessary. Long working days become common and the potential for employee burnout increases. As ventures progress through stages, the original founders may depart and are replaced by management professionals. Fast-growing firms are more likely to replace the founders. Founders are more likely to remain, however, if they hold a sizable percentage ownership of the firm [Boeker and Karichalil, 2002].

A new venture on a slow-growth trajectory will enjoy a less-demanding workweek and fewer competitors but may be faced with limited profitability and access to capital.

As a new firm enters the take-off phase, the need for additional capital, resources, and employees will lead to more regularized processes and increasingly formalized communication. The take-off phase requires management skills and budgeting, accounting, and purchasing capabilities. By the time a firm enters the growth phase, the company moves toward decentralization and delegation of tasks. At this time, the firm may add midlevel managers for such tasks as purchasing, fulfillment, and sales. In the phase called "slowing growth," the challenge of flat revenues calls for new innovation and entrepreneurial leadership with an emphasis on renewal. The stages and their respective goals are summarized in Table 20.3.

When the business grows, the founding team is incredibly busy. Rapid growth puts an enormous strain on them. The business outgrows its production facilities and management capabilities. Typically, the management crunch hits in the third or the fourth year. That is when firms tend to outgrow their management base with quality falling, delivery dates missed, and customers not

TABLE 20.3 Stages and respective goals of a business.

Start-up	Design, make and sell; exploit the opportunity
	Create the product and go to market
	Build the core business
	Emphasis on creativity
	Informal communication
Take-off	Efficiency of operations
	Refine and strengthen the business model
	New procedures introduced
	Invest in quality and customer service
	Communication becomes formal
	Emphasis on direction from leaders
Growth	Expansion of revenues and market share
	Expand product lines
	Hierarchy in place
	Move to decentralized structure
	Emphasis on delegation of tasks and responsibilities
Slowing growth	Consolidation and renewal of the firm
	Leadership required
	New initiatives needed
	Manage working capital
	Emphasis on coordination and renewal
Maturity	New innovation
	Strong culture and history
	Development of personnel seeking new opportunities
	Emphasis on collaboration and renewal

paying on time. As the firm enters the growth phase, the leadership team needs to ask: what does the business need at this stage? Founding CEOs tend to depart when their firm reaches a rapid growth phase and needs strong managerial expertise [Boeker and Karichalil, 2002].

During the growth phase, competition heats up. Most products that are technology-based experience competition driven by technology, as shown in Figure 20.3 [Hirsh et al., 2003]. As technologies like fuel cells and hybrid engines improve, customers demand better performance with lower life-cycle cost of ownership. Firms that respond to these demands will continue to succeed.

During the growth phase, it is also important to have financial leadership from a chief financial officer who manages the operating cash (cash flow), the capital expenditures, and any increases in working capital. The timing and form of capital investments can have a salutary effect on the profitability of the firm.

Entrepreneurs often are unable to make the transition from start-up mode and consequently struggle to become effective managers of a high-growth firm. The habits and skills that make entrepreneurs successful can undermine their ability to lead larger organizations. Entrepreneurs tend to focus on details and tasks in the early stage of a firm, as they should. As the firm grows, a leader needs to work on leading a larger, more complex organization. However, entrepreneurs can learn to grow with their firm by developing their relationship, networking, and strategic capabilities and by moving from a task orientation to a coordinating approach [Hamm, 2002].

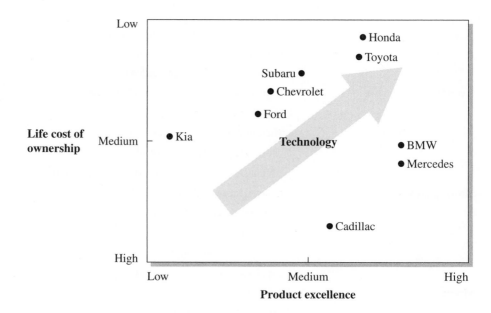

FIGURE 20.3 Technology drives most competition toward excellence and lower life-cycle cost of ownership. In this figure, automobiles are used to illustrate the principle.

As the firm approaches the slowing growth phase, a move toward hierarchy may be appropriate. Hierarchy helps us to handle the complexity of large organizations. Furthermore, people envision career ladders, readily understand the system, and identify with one subunit. Hierarchical structures provide rewards of power and status, and may be the best for managing complex activities in large firms. The leader strives to keep the best of the small firm—empowerment, teamwork, and shared leadership—while accepting the benefits of hierarchy in the large firm [Levitt, 2003].

Often an analogy helps us understand the management of transitions. We can envision the early-stage company in the start-up and take-off as a jazz band playing in a jam session with wonderful improvisations. The jazz band has fewer than 20 members, and all the members know each other and seamlessly take their turn leading. Several players can play several instruments. When a firm grows to more than 50 people, it starts to shift to acting like an orchestra with its separate sections—strings, wind instruments, and percussion. The orchestra needs one coordinating leader, called a conductor. He or she has a score they follow and describes a strategy the orchestra will follow for each piece of music. They act as one—as should a growing firm.

Cisco Systems's revenue growth between 1995 and 2000 was over 50 percent per year, primarily through acquisitions of small companies in exchange for Cisco stock. By 2002, Cisco's growth stalled, and the goal became managing expenses. The motto for the old Cisco was: faster, more sales. The motto for the new Cisco became: slower, better, profitable. From mid-1999 to late 2000, Cisco doubled its payroll from 22,000 to 44,000 employees. In 2001, the growth abruptly ended when businesses stopped buying. Telecommunications companies discovered they had massively overbuilt and ordered too much equipment from Cisco. Revenue fell for the first time in Cisco's history. By the summer of 2001, sales plunged one-third from their level six months earlier. By 2003, Cisco was cutting 8,500 workers [Thurm, 2003].

With a talented leadership team, a new venture can navigate the challenges of growth and the demand for initiatives as needs change. Throughout the evolution of the firm, the challenge is to contain and reduce costs, improve operating margins, manage demand and capacity, and continuously innovate.

eBay is an example of a successful move from take off to the growth stage. With a successful site and venture capital firm investment, eBay began looking in 1997 for a capable manager to take it into the growth phase. It found Meg Whitman, an experienced manager with solid marketing credentials. Whitman arrived in 1998 and immediately worked on the execution of an IPO for September 1998. In the first quarter of 1998, more than $100 million in goods changed hands on eBay, and the company's revenues exceeded $3 million a month [Cohen, 2002]. eBay sold about 9 percent of the company and raised $62.8 million. The IPO price was $18 per share, which jumped to $53 on the opening of the market. Whitman built a brand that promised trustworthy trading for buyer and seller. She turned eBay into a powerful auction firm with about $8.5 billion in revenues and an operating margin of 25 percent in 2008.

> ### Iridium: Getting It Wrong
>
> In 1991, Motorola founded a spin-off called Iridium, LLC, to build a set of 66 low-orbit satellites that would allow subscribers to make phone calls from any global location. In 1998, the company launched its service, charging $3,000 for a handset and $3 per minute for calls. By 1999, Iridium filed for bankruptcy. As the CEO of Iridium said [Finkelstein and Mooney, 2003]: "We're a classic MBA case study in how not to introduce a product. First, we created a marvelous technological achievement. Then we asked how to make money on it." Not all ventures successfully reach the growth stage and beyond.

Downturns and recessions happen in every industry. The response of an emerging business to these tests can lead to renewal and success or disarray and failure. Few companies have soared as high, sunk as low, and struggled to survive as the software maker Novell. Founded in 1983, it has a formidable competitor, Microsoft, and has experienced large swings in revenues and profitability. In the early 1980s, Novell pioneered the market for network operating systems. In the early 1990s, Novell missed the shift to the Internet and lost market share. Drifting through the 1990s, Novell tried for a renewal with Eric Schmidt as leader. Trying to move out of the slowing growth phase, Schmidt attempted to renew the innovation of the firm. By 1999, Novell launched new software for networks and the Internet. However, Novell revenues remained flat and profitability remained elusive. Schmidt became CEO of Google in 2001.

By the time a company reaches maturity with more than seven thousand stores and over two million employees, there is a good chance it will have lost its entrepreneurial zeal. Not so with Wal-Mart, with its mission of goods for the masses. Jim Collins [2003] states that Wal-Mart has built a consistent, growing company through its cultlike culture as well as its commitment to everyday low prices. Wal-Mart's discipline is: "Never think of your company as great, no matter how successful it becomes."

Managing a downturn is as challenging as managing a period of fast growth. In a recession, customers are slow to pay their bills, and suppliers become weak. Furthermore, the availability of new capital dries up. If possible, an emerging new venture can reformulate a positive agenda for managing through the down period by renewing and tightening its strategy while avoiding overreacting. An economic downturn is an opportunity to clean the slate and get back to economic reality. Every downturn is a chance to rebuild the core business. Contrary to conventional wisdom, downturn winners avoid diversification. Focus makes sense, along with renewal of the business [Rigby, 2001]. With a focus on the core business and a renewed strategy, the firm can see beyond the bad times. While managing costs, the firm prepares for the next upturn. If a firm has the resources, selected acquisitions may be a wise step for the future.

Making it through a downturn is not easy, and there's no ready path to success. Companies that successfully handle a downturn refocus on their core business and renew their strategy. They maintain a long-term view and strive to earn the loyalty of employees, suppliers, and customers. Coming out of the downturn, they maintain momentum in their business to stay ahead of the competition [Rigby, 2001].

What Ventures Need in CEOs

As an author and partner at Kleiner Perkins, Randy Komisar [2000] describes a start-up as requiring three types of CEOs at successive stages of its development. He cleverly uses descriptors in terms of dogs. The first CEO of a start-up is the "retriever." This CEO assembles the core team and the product to fit the original vision, and proceeds to access the necessary resources. The second CEO is the "bloodhound," who must sniff out a trail and find the right market and profitable customers. The third CEO is the "husky," who executes well and pulls the established firm steadily forward.

Another organizational issue flowing through the stages of a company's life is the matter of executive succession. As the firm moves through the stages, it often must change its CEO, particularly during periods of either very low or very high growth [Boeker and Wiltbank, 2005]. As needs change, the board of directors and the investors ask whether the incumbent has the skills to manage the firm through and into the next stage. Fewer than 40 percent of founder CEOs make it past the second round of venture capital financing [Bailey, 2003].

The founder of the firm that addresses a solid opportunity and grows rapidly faces a dilemma. On the one hand, the founders find it necessary to cede control and important decisions to their investor group. On the other hand, the company can grow faster with the necessary funds from the investors and subsequently increase the monetary value of the firm. If the founders want to get rich, they will probably need to cede control. If the founders want to retain control, they should consider a slower growth path and require less investment. Most founding CEOs initially want both wealth and control. Choosing between money and power forces entrepreneurs to decide what success really means to them [Wasserman, 2008].

Successful executive succession can lead to superior organizational performance [Dyck et al., 2002]. Succession planning is necessary for any new venture as it moves from one stage to another. The four factors of a good succession are sequence, timing, technique, and communication, as described in Table 20.4. Using the analogy of a relay race, there is a positive relationship between successful passing of the leadership baton and organizational performance. Succession is facilitated when the incumbent and the successor have a shared understanding of the timing and technique of the hand-off.

TABLE 20.4   Four factors of executive succession and the relay race analogy.

1. **Sequence**	Ensuring the successor has the appropriate skills and experience to lead the organization in the next stage. The executive in the start-up stage should have strong entrepreneurial skills, while the executive in the take-off phase should have good organizational skills.
2. **Timing**	Ensuring the leadership baton is passed in a timely and expeditious way from incumbent to successor.
3. **Baton-passing technique**	Methods used for passing the baton. Ensuring that the baton is handed off as expected and the incumbent lets go of the power.
4. **Communication**	Harmonious cooperation and clear communication between incumbent and successor.

20.3 The Adaptive Enterprise

Successful entrepreneurs know a great deal about what kind of priorities matter to their customers at particular historical junctures. Fashion, self-expression, status, community, and control are important factors in different periods of a customer's life. Entrepreneurs have a deep knowledge of their customers and products, and create meaningful brands and a range of organizational capabilities that consistently deliver on the promises of their brands. Furthermore, they learn from their experience and make rapid adjustments [Koehn, 2001].

No business plan survives its ultimate collision with reality. Changes in the marketplace and competition require any firm to react to change. Entrepreneurs must determine whether the assumptions on which their organization was built match the current reality.

One of the biggest tasks for leaders in growing firms is to repeatedly mobilize their companies to change to meet new opportunities and competitive challenges. The leadership team needs to reinvent its strategy through a process of continuous renewal in a time of constant change; this process is called strategic learning [Pietersen, 2002]. A key capability of leaders in adaptive organizations is the capacity to adapt. These leaders do not get stopped by tough challenges they encounter but learn new lessons and go on. Aldous Huxley [1990] stated it as: "Experience is not what happens to a man. It is what a man does with what happens to him." **Strategic learning** is a cyclical process of adaptive learning using four steps: learn, focus, align, and execute. This adaptive process of learning and executing, if done well, may be one of a company's sustainable competitive advantages. Challenging discontinuities are new technologies, globalization, the Internet, deregulation, convergence, and channel disintermediation. A firm's strategy defines how it will respond to its challenges. Thus, the leadership team needs to focus the firm's resources on the best opportunities in the shifting context of the business world.

TABLE 20.5 Characteristics of successful CEOs.

Prone to failure	More likely to succeed
1. Arrogant, hubris	1. Humble, open-minded
2. Overconfident with their answers to problems	2. Realistic, learning, always challenging answers
3. Underestimate major obstacles and risks	3. Examine carefully potential negative consequences and all risks
4. Rely on what worked in the past	4. Challenge every decision and look for wise changes and opportunities to learn

Source: Finkelstein and Mooney, 2003.

A **learning organization** captures, generates, shares, and acts on knowledge by revising its strategy as new knowledge becomes available. This type of firm is an **adaptive enterprise**—one that changes its strategy or business model as the conditions of the marketplace require.

In a small start-up consisting of 10 to 20 people with shared values and objectives, an informal process of developing renewed strategies will suffice. As the firm grows, it needs to continue to renew its strategy as conditions in the competitive marketplace require. At that time, the ability to adapt becomes a required organizational capability. The leadership team needs to learn from its experiences, adjust its strategy, and execute those changes. Learning to deal with change, discontinuity, and uncertainty and adapt in a timely way needs to become a skill for leaders of new ventures [Buchanan, 2004].

The effective management of risk is critical to success. Assessing the risks associated with new initiatives can help managers make adjustments to mitigate these risks. Dell Computer is always asking what could go wrong and considering ways to mitigate the downside. The characteristics of successful CEOs are shown in Table 20.5.

The learning organization, as described in Section 9.3, uses a learning process or cycle as shown in Figure 20.4. The goal of the learning process is to generate new strategies in a cycle of renewal. The first step consists of a situation analysis (see Section 4.3) of the competitive marketplace, industry dynamics, and the firm's strengths and weaknesses. The outcome of this step is insight into the issue and alternative responses. The second step is to redefine the vision, mission, strategy, and adjusted business model. The outcome of this step is a statement of the performance, resource, and capability gaps in terms of the desired strategy and the actual reality. The third step is to adjust the structure, process, people, and culture of the firm to work toward the new strategy. The outcome of this step is an adjusted business plan. The fourth step is to execute the newly adjusted business plan. The outcome of the fourth step is the actual adjusted performance. After a period of actual performance, the learning cycle starts again. The firm continues to learn and adjust as it repeats the strategic learning cycle.

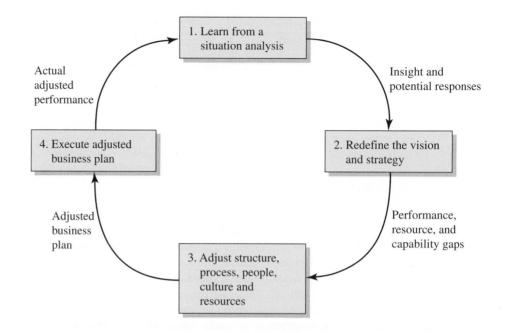

FIGURE 20.4 Strategic learning cycle of a learning organization. (Adapted from Pietersen, 2002.)

Effective learning involves continuing to ask the key questions about customers, competitors, capabilities, resources, and profitability. Entrepreneurs should not rely on excessive overconfidence and overcommit up-front resources, hindering their willingness to learn and adjust. The best degree of confidence lies at the level of willingness to decide and move ahead with the expectation that new knowledge will help the firm learn and adjust [Simon and Houghton, 2003].

Corporations formerly built to last like pyramids are now more like temporary arrangements. With a changing landscape, the adaptive organization has become a reality. CEOs and their firms fail when they fail to execute their strategy. The culture of the firm, if widely understood and shared, can help the firm's people to execute the strategy. With a strong culture and less formal direction, employees can take ownership over their actions and execute well. New firms can recruit, train, and reward people to take responsibility for their actions [Chatman and Cha, 2003].

Most organizations face all kinds of unpredictable challenges that collectively place huge demands on people's creativity and imaginations. Resilience is the ability to recover quickly from setbacks. **Resilience** is a skill that can be learned and increased. A resilient organization acts on its learning and possesses a staunch acceptance of reality, a set of strongly held positive values, and a powerful ability to adapt [Coutu, 2002]. The firm has a clear, undistorted sense of reality about its competitive position. Then it needs the ability to make meaning out of difficult challenges. Resilient leaders build a new, improved vision of the future. Value systems of resilient firms evoke meaning and noble purpose. The third factor is the ability to make a future from what is available—a kind of inventiveness or ability to improvise a solution. Companies that survive regard improvisation as a core skill [Coutu, 2002]. Highly resilient organizations have the ability to read a weak signal of a problem and respond to it. Action tempered by reflection is the best method of responding to change and information [Coutu, 2003]. One way to identify threats and weaknesses is to bring in outsiders to test the resilience of the firm and its processes.

George Eastman and Henry Strong created the Eastman Kodak firm in 1880 based on a new photographic technology process for preparing film negatives. The company sold shares to the public in 1884. Eastman sought to create a mass-market camera and developed a strategy based on continuous innovation and learning. The first Kodak, placed on the market in 1888, was a simple, handheld box camera containing a 100-exposure roll of paper stripping film. The entire camera was sent back to the manufacturer for developing, printing, and reloading when the film was used up. By 1900, Eastman introduced the Kodak Brownie with the motto: "You push the button and we do the rest." By 1927, Eastman Kodak had a virtual monopoly on the photographic industry in the United States. For half a century, Kodak, a "learning company," continuously adapted its strategy and products to build the most powerful brand in photography [Tedlow, 2001].

Amyris Biotechnologies: Adapting to New Industry

Amyris Biotechnologies is an excellent example of an adaptive company. Amyris was founded in 2003 by Dr. Jay Keasling, at the time a professor at U.C. Berkeley. In 2004, it received a $42.6MM grant from the Bill and Melinda Gates Foundation for the manufacture of a bacteria-produced antimalarial drug, artemisinin. Under the terms of the grant, Amyris would sell the drug at cost.

By 2005, Amyris had succeeded in producing the drug microbially. The company now had both R&D funding and significant experience with the methodologies of synthetic biology. It was on the cutting edge of engineering microorganisms to become highly effective chemical factories.

In 2006, Amyris saw a new opportunity. With rising fuel prices, biofuels were increasingly looked at as a potential solution to the world's energy needs. With Amyris's unique competency in bioengineering, it was well positioned to take a leadership position in the biofuels industry. Amyris expects to produce synthetic diesel and jet fuel from sugarcane by 2010. The company effectively adapted to market needs and expanded into this second industry.

Successful start-up teams have a business plan, but they are willing to adapt it as needed. Flexcar (www.flexcar.com) was started by Neil Peterson in Seattle to provide a time-sharing plan for automobile users. He thought that customers would use cars provided by his time-sharing company rather than buy their own car. This approach was popular in Europe, but Peterson found that people in Seattle still wanted their own car. He quickly learned to switch his marketing campaign to steer it toward universities and businesses, where the business model and sales pitch made more sense [Thomas, 2003].

Philippe Kahn, cofounder of Borland and Lightsurf, described adaptivity as improvisation [Malone, 2002]:

> I don't know what an entrepreneur is, but to me it's the difference between a jazz musician and a classical musician. I think a classical musician is a kind of guy who's gonna work for a big company. A jazz musician's gotta work in a small band and know how to improvise. I think that's really the analogy, that's really the difference.

It is wise to build an adaptive learning enterprise from a new venture's inception. Competitive advantage depends, in large part, on the ability of organizations to constantly change and reinvent themselves. They accomplish this through building and rebuilding a shared vision and team learning [Gabor, 2000]. Adaptive organizations strive to learn to manage within complex marketplaces. Complex, nonlinear, unpredictable markets challenge venture leaders to learn to manage continuous challenges. To a great extent, the iterative, adaptive process of Figure 20.4 is the only sustainable advantage for a firm in a dynamic economy.

20.4 Ethics

Life is filled with difficult ethical challenges. **Ethics** are a set of moral principles for good human behavior. Ethics provides the rules for conducting activities in a manner acceptable to society. **Moral principles** are concerned with goodness (or badness) of human behavior and usually are provided as rules and standards of human behavior. Thus, a common moral rule would be: do not lie. Of course, such rules are subject to interpretation, thus, the concept of the white lie.

Ethics is concerned with doing the right (moral) thing. Society also establishes laws to guide actions. For example, U.S. law states that bribes and kickbacks are illegal. Laws are subject to interpretation, and paying a fee for sales help may be legal, while paying a bribe is illegal.

The success of new ventures, either profit or nonprofit, depends on winning against competitors. The competitive marketplace can put pressure on the entrepreneur to act unethically. The business leader finds it difficult to be fair to others without sacrificing customers or profits. The troubles of Enron and WorldCom show the poor practices that arise when competitive pressures win out over ethical principles.

Ethical conduct may reach beyond the law, since the law is inadequate for every task. Doing the right thing is an undefined but helpful standard. One moral goal would be to tell the truth. Thus, a businessperson would try to provide full and truthful information about his or her product or service. Telling the truth is a critical part of integrity, and integrity is the basis for reputation. Thus, firms, at the least, find it in their interest to be truthful. Fortunately, good ethics and self-interest usually coincide, since most firms want to develop and maintain a high reputation [Beauchamp and Bouré, 2001].

Integrity can be defined as truthfulness, wholeness, and soundness. It can be described as a consistency of our words and our actions or our character and our conduct. A corporate model of integrity is based on ethical principles embedded in the corporate culture so that all stakeholders can conduct business to attain mutual benefits [Kaptein and Wempe, 2002]. A key task of a business leader is to establish an ethical culture [McCoy, 2007].

While firms have clear-cut business objectives such as profitability, they must consider them subordinate to ethical values. A firm's integrity cannot be sacrificed to short-term gain. The firm's moral compass points the way. The spotlight is on the CEO and his or her integrity.

People can have great values and still give way to error. One needs the competence and character to implement one's values. Often, major corruption begins with a single small misstep. As executives or employees take further actions that build upon or attempt to cover up this misstep, unethical behaviors can grow to massive proportions [McCoy, 2007]. Firms and individuals face a number of obstacles to ethical decision making, as indicated in Table 20.6.

As one becomes an entrepreneurial leader, the pressures to win at any cost will become powerful. To be a good team player, one may be asked to cut corners. We know that the lack of truth and the collapse of integrity can lead to terrible outcomes, as illustrated in the Enron case of 2002. On Enron, Robert Bryce and M. Ivins write [2002]:

> Enron failed because its leadership was morally, ethically, and financially corrupt. Whether the question was accounting or marital fidelity, the executives who inhabited the 50th floor at Enron's headquarters became incapable of telling the truth, to the Securities and Exchange Commission, to their spouses, or to their employees. That corruption permeated everything they did, and it spread through the company like wildfire.

The challenges of Enron extended beyond the company, too. Enron's rise and fall was based on a partnership between its financial division and the investment bankers who put together the deals. "Enron loves these deals,"

TABLE 20.6 Obstacles to ethical decision making.

Complacency: We believe that "it can't happen here."

Self-Delusion: We judge ourselves by our intentions, while others judge us by our actions

Rationalizations: We construct justifications and excuses for ethical missteps

Survival Mentality: We convince ourselves that ethical missteps are necessary

Source: McCoy, 2007.

wrote a Chase banker in 1998, "as they are able to hide funded debt from their equity analysts" [McLean and Elkind, 2003]. On Wall Street, investment bankers call their innovative structured-finance arrangements their "technology." In investment banking, the ethic for many was: "Can you get the deal? If you can, and you're not likely to be sued or jailed, it's a good deal."

MiniScribe: Cooking the Books

MiniScribe was a Longmont, Colorado, producer of disk drives that found itself in trouble when IBM canceled major purchasing contracts. When actual sales failed to materialize, the CEO badgered and bullied MiniScribe executives to meet quarterly revenue targets, no matter what it took. Executives turned to cooking the books. Their "cooking" activities included counting raw inventory as finished goods, creating false inventory, and grossly overstating actual shipments.

As MiniScribe's sales and profits continued to tumble, the pressures on the firm increased. At one point, executives rented a private warehouse. Over a weekend, staffers and spouses packed bricks in disk drive shipping boxes, then shipped pallet-loads of them to "BW," a fake customer. To pull off the brick shipping plan, they created a custom computer program called "Cook Book." Over $4 million worth of "very hard" disks were shipped utilizing Cook Book.

Once the fraud was exposed, MiniScribe's stock plummeted, and investors lost hundreds of millions of dollars, but not before the same executives, in an example of insider trading, sold most of their shares at a healthy profit. The CEO and CFO served time in prison. Unethical behavior most often gets exposed with severe consequences.

Fortunately, there are a number of tools to help one act ethically. McLemore [2003] suggests two tests for difficult actions: (1) are things so questionable that you lose sleep over them and (2) could you live with the newspaper reporting your actions tomorrow? If difficult issues arise, perhaps it is best to discuss the issues with a knowledgeable and trustworthy friend. Can you live with the action? Ernest Hemingway wrote in his novel, *Death in the Afternoon:* "What is moral is what you feel good after, and what is immoral is what you feel bad after."

Several tools for acting ethically are provided in Table 20.7. For individuals, it is helpful to have a "personal business plan" that records the importance of various activities and relationships and that sets goals for the coming year. Actions that deviate from this plan should be evaluated carefully. Individuals should also identify ethical "partners" who can serve as sounding boards and supporters. When confronted with unethical situations, McLemore [2003] wisely suggests you say, "I don't feel comfortable doing

TABLE 20.7   Tools for acting ethically in tough situations.

- Maintain involvement in a variety of activities and with a variety of people. This will help avoid being pressured to go along with what everybody around you says is acceptable behavior.

- Create a personal board of directors consisting of people you admire and who possess admirable values. If difficult situations arise, call on them for advice.

- Keep a cash reserve of six months to a year of salary. This will allow you to escape further involvement with an unethical firm and seek other options.

- Increase your defenses to negative forces of influence and persuasion by reading a book on the subject such as Cialdini, [2001].

- Apply the "front, left-hand-side page of the Wall Street Journal" test. Would your actions change if you knew they would be exposed publicly someday?

- Write down your personal core values for later use. They can provide a helpful reference in difficult times to remind you of what is important in your life.

- Take a break before making a decision. When feeling pressured, ask for time to leave the room and gather your thoughts.

that." Of course, the risk of loss of your position is real, but you can repeat without judgment, "I am uncomfortable doing that."

For firms, it is important to have a "Credo" that makes values explicit to all employees. At J&J, for example, all employees must read, accept, and sign the Credo, which aligns everyone along a clear set of principles. When trouble does occur, an open and transparent approach is essential for maintaining trust and facilitating ethical responses [McCoy, 2007].

A person with an ethical mind asks herself, "What kind of a person, worker and citizen do I want to be?" [Gardner, 2006]. When circumstances are tempting you to drop your standards, life becomes difficult. With a firm belief in personal integrity, a person needs to be willing to resign or be fired for what he or she believes is right.

Building an Ethical Venture the Intel Way

Entrepreneurs should plan an ethical foundation for the firm so that they can build integrity and reputation. Maintenance of integrity is critical to the long-run success of any firm. Former Intel CEO Craig Barrett promotes a "3M" philosophy to help leaders make decisions with the right amount of integrity and ethics. The three Ms—manager, media, and mother—represent the three constituencies with whom leaders should be comfortable sharing their decision. Only if the leaders expect that their manager, the media, and their mother will all approve of their decision should they proceed with their chosen course of action.

Source: Barrett, 2003.

20.5 AgraQuest

AgraQuest prepared a business plan that stated that the U.S. Environmental Protection Agency would certify and permit its natural biopesticides within one year. The actual time to approve Serenade was two years, and AgraQuest's second product, Sonata, took 30 months for approval. Furthermore, the firm developed a marketing campaign for its first product that was fundamentally flawed. As a result of poor execution and slow adaptation to challenges and difficulties, AgraQuest was late meeting its milestones until 2005.

The implementation of the research and development strategy and processes has been relatively flawless. It is the golden goose, but it is slow in gestating the products. AgraQuest has been a science company with a mixed execution record in product certification, marketing, and sales. What could have AgraQuest achieved if it followed the iterative process of Figure 20.4 early in its life?

By 2004, AgraQuest hired Mike Mille as CEO while Pam Marrone remained as president and board member. The emphasis shifted to execution, and gross margins rose from 25 percent to 45 percent in 2005. AgraQuest was on the path to increased revenues and profitability.

Pam Marrone left AgraQuest in April 2006 to found a new biotech company, Marrone BioOrganic Innovations, that focuses on natural weed control products. She remained on the AgraQuest board until 2007. Marrone found new energy and opportunity within the natural pesticides industry. Marrone BioOrganic Innovations raised $1 million in the seed round with family friends and angels. The firm raised $3 million in the series B round in 2008. Pam Marrone has learned to build a firm with a strategic learning cycle as illustrated in Figure 20.4.

Both AgraQuest and Marrone BioOrganic Innovations value social responsibility. In 2003, AgraQuest won the Green Chemistry Award from the U.S. Environmental Protection Agency and the best presentation award at the CleanTech Venture Conference. It has a code of ethics and a commitment to sustainability that all employees are required to read and sign.

20.6 Summary

The implementation of a creative, well-defined business plan is essential to the success of a new enterprise. Good execution depends on the logical alignment of the firm's strategy with its goals and the efforts of its people. Turning a concept into a successful reality depends on goals, deadlines, teamwork, and focus on achieving the desired outcomes. Choosing the right people for the right jobs and helping them see where to go and how to get there are critical elements of building a great company. With the right people, a great strategy, and a sound road map, a start-up can strive to achieve an outstanding execution of the plan.

A new business grows from fledgling start-up to growth to maturity in stages. Managing a new business through the stages requires varying skills and organizational arrangements. Start-ups need to have talented, multiskilled people in place as they move through the stages of growth.

Emerging firms are constantly subject to challenge and change. Organizing a firm for resilient response to these challenges calls for an adaptive corporation. The ability to adapt to change may be a firm's only sustainable advantage. Furthermore, a firm needs to sustain its ethical principles through difficult times.

Principle 20

The ability to continuously and ethically execute a business plan and adapt that plan to changing conditions provides long-term success.

Video Resources

Visit http://techventures.stanford.edu to view experts discussing content from this chapter.

Measuring Success: You Measure What Matters	John Thompson	Symantec
Why Are Ethics Important?	Frank Levinson	Finisar
Learning from Failure	Tina Seelig	Stanford University

20.7 Exercises

20.1   In May 2003, Zipcar of Boston decided it was time to bring in new funding to reach profitability (www.zipcar.com). However, the willing investors insisted on replacing the CEO and the board of directors. Examine Zipcar's subsequent progress in terms of execution and need at different stages of the life of this company.

20.2   The global financial crisis in 2007–2010 drastically impacted the spending patterns of many businesses and industries. Select an industry, list specific spending changes that occurred during this period, and describe how a start-up selling into this industry should adapt its strategies to account for the broader market conditions.

20.3   Your emerging new company is selling a high-priced software system to the oil and gas industry. Each sale amounts to $100,000 or more. Your firm is scheduled to deliver a system next week to one of your best customers. However, your chief technical officer has just told you that they have found a major software error that will take two to three weeks to fix. You are counting on the sale within this month so that you can meet payroll and pay all your delinquent bills. Your CFO suggests you ship the system now and send in a team later to fix the error. Your CTO wants to fix it first and then ship. What should you do?

20.4   Southeby's and Christie's are the two largest upscale auction houses. Both enjoyed a growing business in the boom years of the late 1990s. In 2000, both firms were accused of price fixing. The Sherman Antitrust Act was passed in 1890 to control the power of trusts and monopolists. In 1995, both firms announced they would charge a fixed, nonnegotiable sliding-scale commission on the sales price. Is this the age-old tactic of price fixing? What constitutes legal pricing policies versus illegal price fixing?

20.5   Your cash-strapped company is bidding for a badly needed contract. As the bid deadline nears, an employee of your nearest competitor pays you a visit. He says he will provide details of your competitor's bid in return for the promise of a job in six months, after the dust has settled. You know your competitor can survive losing this contract, but you cannot. Unfortunately, hiring a new employee will mean someone who currently works for you will have to go. Even so, is this an offer you cannot refuse?

20.6   Your new firm is considering offering one of two health benefit options. One is more complete but also more costly than the other. Should you ask your employees to accept the lower-cost option? Should you explain the benefits of both plans? If you do, most people will prefer the better plan. What should you do?

20.7   You attend a critical partner meeting with your CEO. After the meeting, your CEO misrepresents the results of the meeting to the broader management team to further a different agenda. How do you handle this situation with your CEO? With your other team members?

20.8   Select an example of a white-collar (business) crime in the technology industry and describe what happened. How could this crime have been avoided?

VENTURE CHALLENGE

1.   Briefly describe your plan for executing your business plan after you receive the resources.

2.   Describe your venture's plans to act as an adaptive organization.

3.   What mechanisms will you use to instill ethical behavior in your venture?

REFERENCES

Aaker, David. 2001. *Developing Business Strategies.* 6th ed. New York: Wiley and Sons.

Aaker, David, and E. Joachimsthaler. 2000. *Brand Leadership.* New York: Free Press.

Aaker, David, V. Kumar, and G. S. Day. 2001. *Marketing Research.* New York: Wiley & Sons.

Abate, Tom. 2008. "Who Is Doing What with Technology?" *San Francisco Chronicle* (28 July).

Adner, Ron, and D. A. Levinthal. 2002. "The Emergence of Emerging Technologies." *California Management Review* (Fall), pp. 50–66.

Agarwal, Rajshree, M. B. Sarkar, and R. Echambadi. 2002. "The Conditioning Effect of Time on Firm Survival." *Academy of Management Journal* 5:971–94.

Ahuja, Gautam, and C. M. Lampert. 2001. "Entrepreneurship in the Large Corporation." *Strategic Management Journal* 22:521–43.

Aiello, Robert, and M. Watkins. 2000. "The Fine Art of Friendly Acquisition." *Harvard Business Review* (December), pp. 101–16.

Albrinck, Jill, J. Hornery, D. Kletter, and G. Neilson. 2002. "Adventures in Corporate Venturing." *Strategy and Business* 22:119–29.

Allen, Thomas. 1995. *Managing the Flow of Technology.* Cambridge, MA: MIT Press.

Anand, Bharat, and A. Galetovic. 2004. "How Market Smarts Can Protect Property Rights." *Harvard Business Review* (December), pp. 73–79.

Anders, George. 2003. *Perfect Enough.* New York: Penguin Putnam.

Anderson, Chris. 2006. *The Long Tail.* New York: Free Press.

Anthony, Robert, David Hawkins, and Kenneth Merchant. 1999. *Accounting: Text & Cases.* New York: McGraw Hill.

Arthurs, Jonathan, and Lowell Busenitz. 2006. "Dynamic Capabilities and Venture Performance: The Effects of Venture Capitalists." *Journal of Business Venturing* 21:195–215.

Astebro, Thomas. 1998. "Basic Statistics on the Success Rate and Profits for Independent Inventors." *Entrepreneurship Theory and Practice* (Winter):41–48.

Audretsch, David, and Erik Lehmann. 2005. "Does the Knowledge Spillover Theory of Entrepreneurship Hold for Regions?" *Research Policy* 34:1191–1202.

Audretsch, David, Erik Lehmann, and Susanne Warning. 2005. "University Spillovers and New Firm Location." *Research Policy* 34:1113–22.

Austin, James, Roberto Gutierrez, Enrique Ogliastri, and Ezequiel Reficco. 2007. "Capitalizing on Convergence." *Stanford Social Innovation Review* (Winter): 24–31.

Bagley, Constance, and Craig Dauchy. 2007. *The Entrepreneur's Guide to Business Law.* Cincinnati: West/Thomson.

Bailey, Jeff. 2003. "For Investors, Founders Are Short-Term CEOs." *Wall Street Journal* (21 October), p. A24.

Baker, Wayne. 2000. *Achieving Success Through Social Capital.* San Francisco: Jossey-Bass.

Bakke, Dennis W. 2005. *Joy at Work.* Seattle: PVG Publishers.

Balachandra, R., M. Goldschmitt, and J. Friar, 2004, "The Evolution of Technology Generations," IEEE Trans. on Engineering Management, February, 3–12.

Barkema, Harry, J. Baum, and E. Mannix. 2002. "Management Challenges in a New Time." *Academy of Management Journal* 5:916–30.

Barney, Jay. 2002. *Gaining and Sustaining Competitive Advantage,* 2d ed. Upper Saddle River, N.J.: Prentice-Hall.

Barney, Jay. 2001. "Is the Resource Based View a Useful Perspective for Strategic Management Research? Yes." *Academy of Management Review* (January): 41–56.

Baron, James, and M. T. Hannan. 2002. "Organizational Blueprints for Success in High-Tech Start-Ups." *California Management Review* 3:18–24.

Baron, Robert, and G. D. Markman. 2003. "Beyond Social Capital." *Journal of Business Venturing* 18:41–60.

Barrett, Craig. 2003. Address at the AEA/Stanford Executive Institute, Stanford University, Palo Alto, Calif. (August 14).

Barthelemy, Jerome. 2003. "The Seven Deadly Sins of Outsourcing." *Academy of Management Executive* 2:87–100.

Bartlett, Christopher, and S. Ghoshal. 2002. "Building Competitive Advantage through People." *MIT Sloan Management Review* (Winter), pp. 34–41.

Batten, Frank. 2002. *The Weather Channel.* Boston: Harvard Business School Press.

Baumol, William. 2002. *The Free Market Innovation Machine.* Princeton, N.J.: Princeton University Press.

Baumol, William, Robert Litan, and Carl Schramm. 2007. *Good Capitalism, Bad Capitalism, and the Economics of Growth and Prosperity.* New Haven: Yale University Press.

Beatty, Jack. 2001. *Colossus—How the Corporation Changed America.* New York: Broadway Books.

Beauchamp, Tom, and N. E. Bowie. 2001. *Ethical Theory and Business,* 6th ed. Upper Saddle River, N.J.: Prentice-Hall.

Bechky, Beth. 2006. "Gaffers, Gofers, and Grips: Role-based Coordination in Temporary Organizations." *Organization Science* 17(1):3–21.

Bechky, Beth, and S. O'Mahony. 2006. "Stretchwork: Managing the career progression paradox in external labor markets." *Academy of Management Journal* 49(5):918–941.

Beckman, Christine. 2006. "The Influence of Founding Team Company Affiliations on Firm Behavior." *Academy of Management Journal* 49(4):741–58.

Beckman, Christine, M. Diane Burton, and Charles O'Reilly. 2007. "Early Teams: The Impact of Team Demography on VC Financing and Going Public." *Journal of Business Venturing* 22:147–73.

Bennis, Warren, and R. J. Thomas. 2002. "Crucibles of Leadership." *Harvard Business Review* (September), pp. 39–45.

Berkery, Dermot. 2007. *Raising Venture Capital for the Serious Entrepreneur.* New York: McGraw-Hill.

Bernstein, Peter. 1996. *Against the Gods.* New York: Wiley & Sons.

Bhardwaj, Gaurab, John Camillus, and David Hounshell. 2006. "Continual Coporate Entrepreneurial Search for Long-Term Growth." *Management Science* 52(2):248–61.

Bhargava, Hement. 2003. "Contingency Pricing for Information Goods." *Journal of Management Information Systems* (Fall): 115–38.

Bhatia, Vinit, and Gib Carey. 2007. "Patenting for Profits." *MIT Sloan Management Review* 48(4):15–16.

Bhattacharya, C.B., Sankar Sen, and Daniel Korschun. 2008. "Using Corporate Social Responsibility to Win the War for Talent." *MIT Sloan Management Review* 49(2):37–44.

Bhidé, Amar. 1994. "How Entrepreneurs Craft Strategies That Work." *Harvard Business Review* 72:150–161.

Bhidé, Amar. 2000. *The Origin and Evolution of New Business.* New York: Oxford University Press.

Bhidé, Amar. 2008. *The Venturesome Economy.* Princeton, N.J.: Princeton University Press.

Bingham, Christopher B. 2005. "Learning from Heterogeneous Experience: The Internationalization of Entrepreneurial Firms." Dissertation: Stanford University Department of Management Science and Engineering.

Birley, Sue. 2002. "Universities, Academics, and Spinout Companies: Lessons from Imperial." *International Journal of Entrepreneurship Education* 1:133–53.

Black, J. Stewart, and H. B. Gregersen. 2002. *Leading Strategic Change.* Upper Saddle River, N.J: Prentice-Hall.

Blank, Steven Gary. 2006. *The Four Steps to the Epiphany.* Palo Alto: Cafepress.

Boeker, Warren, and R. Karichalil. 2002. "Entrepreneurial Transitions." *Academy of Management Journal* 3:818–26.

Boeker, Warren, and Robert Wiltbank. 2005. "New Venture Evolution and Managerial Capabilities." *Organization Science* 16(2):123–33.

Boer, Peter. 2002. *The Real Options Solution.* New York: Wiley & Sons.

Bolino, Mark, W. Turnley, and J. Bloodgood. 2002. "Citizenship Behavior and the Creation of Social

Capital." *Academy of Management Review* 4:505–22.

Bossidy, Larry, and Ram Charan. 2002. *Execution.* New York: Crown.

Bosworth, Michael. 1995. *Solution Selling.* New York: McGraw-Hill.

Boulding, William, and M. Christen. 2001. "First Mover Disadvantage." *Harvard Business Review* (October), pp. 20–21.

Bower, Joseph. 2001. "Not All M&As Are Alike and That Matters." *Harvard Business Review* (March), pp. 93–101.

Bradach, Jeffrey. 1997. "Using the Plural Form in the Management of Restaurant Chains." *Administrative Science Quarterly* 42:276–303.

Bradley, Bill, P. Jansen, and L. Silverman, 2003. "The Nonprofit Sector's $100 Billion Opportunity." *Harvard Business Review* (May), pp. 94–103.

Brandenburger, Adam, and Barry Nalebuff. 1997. *Co-opetition.* New York: Currency Doubleday.

Branscomb, Lewis M. 2008. "Research Alone Is Not Enough." *Science* (15 August), pp. 915–916.

Branscomb, Lewis, F. Kodama, and R. Florida. 1999. *Industrializing Knowledge.* Cambridge: MIT Press.

Brown, David. 2002. *Inventing Modern America.* Cambridge: MIT Press.

Brown, John Seely, and Paul Duguid. 2001. "Creativity versus Structure: A Useful Tension." *MIT Sloan Management Review* (Summer), pp. 93–94.

Brown, Rex. 2005. *Rational Choice and Judgment.* New York: Wiley.

Brown, Shona, and K. M. Eisenhardt. 1998. *Competing on the Edge.* Boston: Harvard Business School Press.

Brown, Tim. 2008. "Design Thinking." *Harvard Business Review* (June), pp. 84–92.

Bryce, Robert, and M. Ivins. 2002. *Pipe Dreams: Greed, Ego and the Death of Enron.* New York: Public Affairs–Perseus.

Buchanan, Mark, 2004. "Power Laws and the New Science of Complexity Management," Strategy and Business, 34:71–79.

Buckingham, Marcus. 2005. *The One Thing You Need to Know.* New York: Free Press.

Bunnell, David. 2000. *Making the Cisco Connection.* New York: Wiley & Sons.

Burgelman, Robert A. 2002. *Strategy as Destiny.* New York: Free Press.

Burgelman, Robert A., and L. Valikangas. 2005. "Managing Internal Corporate Venturing Cycles." *Sloan Management Review* (Summer), pp. 26–34.

Burgelman, Robert A., C. M. Christensen, and S. C. Wheelwright. 2004. *Strategic Management of Technology and Innovation.* Burr Ridge, Ill.: McGraw-Hill Irwin.

Burton, Thomas. 2005. "Medtronic Settles Patent Fight." *Wall Street Journal* (25 April), p. B4.

Carl, Fred. 2007. "The Best Advice I Ever Got." *Harvard Business Review* (February), pp.____.

Carman, James, and Eric Langeard. 1980. "Growth Strategies for Service Firms." *Strategic Management Journal* 1:7–22.

Carr, Geoffrey. 2008. "The Power and the Glory." *Economist* (19 June).

Carr, Nicholas. 2002. "Unreal Options." *Harvard Business Review* (December), p. 22.

Castrogiovanni, Gary, and Robert Justis. 1998. "Franchising Configurations and Transitions." *Journal of Consumer Marketing* 15(2):170–90.

Chakravorti, Bhaskar, 2004. "The New Rules for Bringing Innovations to Market," *Harvard Business Review* (March), pp. 58–67.

Chandler, Alfred, and J. W. Cortada. 2003. *A Nation Transformed.* New York: Oxford University Press.

Charan, Ram. 2007. *What the Customer Wants You to Know.* New York: Portfolio.

Chatman, Jennifer, and S. E. Cha. 2003. "Leading by Leveraging Culture." *California Management Review* (Summer), pp. 20–33.

Chatterjee, Sayan. 1998. "Delivering Desired Outcomes Efficiently." *California Management Review* (Winter), pp. 78–94.

Chatterjee, Sayan. 2005. "Core Objectives: Clarity in Designing Strategy." *California Management Review* (Winter), pp. 33–49.

Chesbrough, Henry. 2002. "Making Sense of Corporate Venture Capital." *Harvard Business Review* (March), pp. 90–99.

Choi, Young Rok, Moren Lévesque, and Dean Shepherd. 2008. "When Should Entrepreneurs Expedite or Delay Opportunity Exploitation?" *Journal of Business Venturing* 23:333–55.

Chowdhury, Sanjib. 2005. "Demographic Diversity for Building an Effective Entrepreneurial Team: Is It Important?" *Journal of Business Venturing* 20:727–46.

Chrisman, James J., Ed McMullan, and J. Hall. 2004. "The Influence of Guided Preparation on the Long-term Performance of New Ventures." *Journal of Business Venturing* (Fall), pp. 18–26.

Christensen, Clayton. 1999. *Innovation and the General Manager.* Burr Ridge, Ill.: McGraw-Hill Irwin.

Christensen, Clayton. 2002. "The Rules of Innovation." *Technology Review* (June), pp. 21–28.

Christensen, Clayton, and M. E. Raynor. 2003. *The Innovator's Solution.* Boston: Harvard Business School Press.

Christensen, Clayton, and Richard Tedlow. 2000. "Patterns of Disruption in Retailing." *Harvard Business Review* (January), pp. 36–42.

Christensen, Clayton, M. Raynor, and M. Verlinden. 2001. "Skate to Where the Money Will Be." *Harvard Business Review* (November), pp. 73–81.

Christensen, Clayton, S. D. Anthony, and E. A. Roth. 2004. *Seeing What's Next: Using the Theories of Imovation to Predict Industry Change.* Boston: Harvard Business School Press.

Churchill, Neil, and J. W. Mullins. 2001. "How Fast Can Your Company Afford to Grow?" *Harvard Business Review* (May), pp. 135–42.

Cialdini, Robert. 2008. *Influence: Science and Practice.* Boston: Allyn and Bacon.

Cialdini, Robert. 1993. *Influence.* New York: Morrow.

Clark, Don. 2003a. "Intel's John Miner to Be President of Investment Unit." *Wall Street Journal* (18 April), p. C5.

Clark, Don. 2003b. "Renting Software Online." *Wall Street Journal* (3 June), p. B1.

Clemons, Eric, and J. A. Santamaria. 2002. "Maneuver Warfare." *Harvard Business Review* (April), pp. 57–65.

Coburn, Pip. 2006. *The Change Function.* New York: Portfolio.

Cohen, Adam. 2002. *The Perfect Store.* Boston: Little, Brown and Company.

Cohen, Wesley, and Dan Levinthal. 1990. "Absorptive Capacity: A New Perspective on Learning and Motivation." *Administrative Science Quarterly* 35:128–53.

Coleman, James. 1990. *Foundations of Social Theory.* Cambridge, MA: Belknap Press.

Collins, James, and William Lazier. 1992. *Beyond Entrepreneurship.* Upper Saddle River, N.J.: Prentice-Hall.

Collins, James, and J. Porras. 1996. "Building Your Company's Vision." *Harvard Business Review* (September), pp. 65–77.

Collins, James. 2001. *Good to Great.* New York: Harper Collins.

Combs, James, Steven Michael, and Gary Castrogiovanni. 2004. "Franchising: A Review and Avenues to Greater Theoretical Diversity." *Journal of Management* 30(6):907–31.

Conger, Jay. 1998. "The Necessary Art of Persuasion." *Harvard Business Review* (May), pp. 85–95.

Copeland, Tom, and Peter Tufano, 2004. "A Real-World Way to Manage Real Options," *Harvard Business Review* (March), pp. 90–99.

Corstijens, Marcel, and J. Merrihue. 2003. "Optimal Marketing." *Harvard Business Review* (October), pp. 114–21.

Courtney, Hugh. 2001. *20–20 Foresight.* Boston: Harvard Business School Press.

Coutu, Diane. 2003. "Sense and Reliability." *Harvard Business Review* (April), pp. 84–90.

Coutu, Diane. 2002. "How Resilience Works." *Harvard Business Review* (May), pp. 46–55.

Covey, Stephen. 2006. The Speed of Trust. New York: Free Press.

Covey, Stephen, and A. R. Merrill. 1996. First Things First. New York: Free Press.

Crawford, Fred, and Ryan Matthews. 2001. *The Myth of Excellence.* New York: Crown Business.

Crois, Rob, J. Liedtka, and L. Weiss. 2005. "A Practical Guide to Social Networks. "*Harvard Business Review* (March), pp. 124–32.

Cross, Rob, and L. Prusak. 2002. "The People Who Make Organizations Go—or Stop." *Harvard Business Review* (June), pp. 105–12.

Dahan, Ely, and V. Srinivasan. 2000. "The Predictive Power of Internet-Based Product Concept Testing." *Journal of Product Innovation Management* 17:99–109.

Dahle, Cheryl. 2005. "The Change Masters." *Fast Company* (January).

Davenport, Thomas, and J. Glaser. 2002. "Just in Time Delivery Comes to Knowledge Management." *Harvard Business Review* (July), pp. 107–11.

Davenport, Thomas, and L. Prusak. 1998. *Working Knowledge.* Boston: Harvard Business School Press.

Davidsson, Per. 2002. "What Entrepreneurship Research Can Do for Business and Policy Practice." *International Journal of Entrepreneurship Education* 1:5–24.

Davis, Jason, Kathy Eisenhardt, and Christopher Bingham. 2009. "Optimal Structure, Market Dynamism, and the Strategy of Simple Rules." *Administrative Science Quarterly* 54(3):415–452.

Davis, Julie L., and S. S. Harrison. 2001. *Edison in the Boardroom.* New York: Wiley & Sons.

Davis, Stan, and Christopher Meyer. 2000. *Future Wealth.* Boston: Harvard Business School Press.

Dean, Thomas, and Jeffrey McMullen. 2007. "Toward a Theory of Sustainable Entrepreneurship: Reducing Environmental Degradation Through Entrepreneurial Action." *Journal of Business Venturing* 22:50–76.

de Bettignies, Jean-Etienne, and James Brander. 2007. "Financing Entrepreneurship: Bank Finance Versus Venture Capital." *Journal of Business Venturing* 22:808–32.

De Carolis, Donna Marie, and Patrick Saparito. 2006. "Social Capital, Cognition, and Entrepreneurial Opportunities: A Theoretical Framework." *Entrepreneurship Theory and Practice* (January), pp. 41–56.

Dees, J. Gregory, J. Emerson, and P. Economy. 2002. *Strategic Tools for Social Entrepreneurs.* New York: Wiley & Sons.

DeLong, Thomas, and V. Vijayaraghavan. 2003. "Let's Hear It for B Players." *Harvard Business Review* (June), pp. 96–102.

Dembo, Ron, and Andrew Freeman. 1998. *Seeing Tomorrow.* New York: Wiley & Sons.

DeMeyer, Arnold, C. H. Loch, and M. T. Pich. 2002. "Managing Project Uncertainty." *MIT Sloan Management Review* (Winter), pp. 60–67.

Demos, Nick, S. Chung, and M. Beck. 2002. "The New Strategy and Why It Is New." *Strategy and Business* 25:15–19.

Dhar, Ravi. 2003. "Hedging Customers." *Harvard Business Review* (May), pp. 86–92.

Diamond, Jared. 2000. "The Ideal Form of Organization." *Wall Street Journal* (12 December), p. A17.

Dobrev, Stanislav, and William Barnett. 2005. "Organizational Roles and Transition to Entrepreneurship." *Academy of Management Journal* 48(3):433–49.

Dorf, Richard C. 2001. *Technology, Humans and Society: Toward a Sustainable World.* San Diego: Academic Press.

Dorf, Richard C. 2010. *Introduction to Electronic Circuits.* New York: Wiley.

Douglas, Evan, and Dean Shepherd. 1999. "Entrepreneurship as a Utility Maximizing Response." *Journal of Business Venturing* 15:231–51.

Douglas, Evan, and D. Shepherd. 2002. "Self-Employment as a Career Choice." *Entrepreneurship, Theory and Practice* (Spring): 81–89.

Doz, Yves, and Gary Hamel. 1998. *Alliance Advantage.* Boston: Harvard Business School Press.

Drucker, Peter. 2002. *Managing in the Next Society.* New York: St. Martin's Press.

Drucker, Peter F. 1995. *Managing in a Time of Great Change.* New York: Penguin Books.

Dushnitsky, Gary, and Michael Lenox. 2005. "When Do Incumbents Learn from Entrepreneurial Ventures? Corporate Venture Capital and Investing Firm Innovation Rates." *Research Policy* 34:615–39.

Dyck, Bruno, et al. 2002. "Passing the Baton: The Importance of Sequence, Timing, Technique and Communication in Executive Succession." *Journal of Business Venturing* 17:143–62.

Dye, Renee. 2000. "The Buzz of Buzz." *Harvard Business Review* (November), pp. 139–46.

Dyer, Jeffrey, P. Kale, and H. Singh. 2004. "When to Ally and When to Acquire." *Harvard Business Review* (July), pp. 109–15.

Earley, P. Christopher, and Elaine Mosakowski. 2004. "Cultural Intelligence." *Harvard Business Review* (October), pp. 139–146.

Ebben, Jay, and Alec Johnson. 2006. "Bootstrapping in Small Firms: An Empirical Analysis of Change over Time." *Journal of Business Venturing* 21:851–65.

Economist. 2005b. "Less Is More." (9 July), p. 11.

Economist. 2005c. "Profiting from Obscurity." (7 May), p. 72.

Economist. 2006. "Behind the Bleeding Edge" (23 September), p. 16.

Economist. 2008. "Beneath Your Feet—Geothermal Could Be Hot" (21 June), p. 15.

Eisenhardt, Kathleen, and D. N. Sull. 2001. "Strategy as Simple Rules." *Harvard Business Review* (January), pp. 106–16.

El-Haik, Basem, and D. M. Roy. 2005. *Service Design for Six Sigma.* New York: Wiley.

Elias, Stephen, and R. Stim. 2003. *Patent, Copyright and Trademark,* 6th ed. Berkeley: Nolo Press.

Elsbach, Kimberly. 2003. "How to Pitch a Brilliant Idea." *Harvard Business Review* (September), pp. 40–48.

Ensley, Michael, and Keith Hmieleski. 2005. "A Comparative Study of New Venture Top Management Team Composition, Dynamics and Performance Between University-Based and Independent Startups." *Research Policy* 34:1091–05.

Ertel, Danny. 1999. "Turning Negotiation into a Corporate Capability." *Harvard Business Review* (May-June), pp. 55–70.

Erikson, Truls. 2002. "Entrepreneurial Capital: The Emerging Venture's Most Important Asset and Competitive Advantage." *Journal of Business Venturing* 17:275–290.

Ertel, Danny. 2004. "Getting Past Yes." *Harvard Business Review* (November), pp. 60–68.

Estrin, Judy. 2009. *Closing the Innovation Gap.* New York: McGraw-Hill.

Ettenberg, Elliott. 2002. *The Next Economy.* New York: McGraw-Hill.

Ettenson, Richard, and J. Knowles. 2008. "Don't Confuse Reputation with Brand." *MIT Sloan Management Review* (Winter), pp. 19–21.

Fahey, Liam, and R. M. Randall. 1998. *Learning from the Future.* New York: Wiley & Sons.

Farrell, Diana. 2004. "Beyond Offshoring." *Harvard Business Review* (December), pp. 82–90.

Fast Company, 1997, "John Doerr's Startup Manual," (March), p. 82.

Ferguson, Glover, S. Mathur, and B. Shah. 2005. "Evolving from Information to Insight." *Sloan Management Review* (Winter), pp. 51–58.

Fernandez-Araoz, Claudio. 2005. "Getting the Right People at the Top." *Sloan Management Review* (Summer), pp. 67–72.

Fine, Charles, et al. 2002. "Rapid Response Capability in Value Chain Design." *MIT Sloan Management Review* (Winter), pp. 69–75.

Finkelstein, Sydney. 2003. *Why Smart Executives Fail.* New York: Portfolio Penguin.

Finkelstein, Sydney, and A. Mooney. 2003. "Not the Usual Suspects." *Academy of Management Executive,* 2:101–12.

Fischer, Bill, and Andy Boynton. 2005. "Virtuoso Teams." *Harvard Business Review* (July), pp. 117–123.

Fisher, Lawrence. 2002. "Yves Doz: The Thought Leader Interview." *Strategy and Business* 29:115–23.

Fisher, Robert, and W. Ury. 1991. *Getting to Yes.* New York: Penguin.

Fleming, John H., C. Coffman, and J. K. Harter. 2005. "Manage Your Human Sigma." *Harvard Business Review* (July), pp. 107–14.

Fleming, Lee, and Olar Sorenson. 2001. "The Dangers of Modularity." *Harvard Business Review* (September), pp. 20–21.

Fleming, Lee, Santiago Mingo, and David Chen. 2007. "Collaborative Brokerage, Generative Creativity, and Creative Success." *Administrative Science Quarterly* 52:443–75.

Florida, Richard. 2002. *The Rise of the Creative Class.* New York: Basic Books.

Florida, Richard, and J. Goodnight. 2005. "Managing for Creativity." *Harvard Business Review* (July), pp. 125–32.

Florin, Juan. 2005. "Is Venture Capital Worth It?" *Journal of Business Venturing* 20:113–35.

Foster, William, and J. Bradach. 2005. "Should Non-Profits Seek Profits?" *Harvard Business Review* (February), pp. 92–100.

Freeman, John, and Jerome Engel. 2007. "Models of Innovation: Startups and Mature Corporations." *California Management Review* 50(1):94–119.

Freiberg, Kevin, and Jackie Freiberg. 1997. *Nuts!* New York: Broadway Books.

Friedman, Thomas L. 1999. *The Lexus and the Olive Tree.* New York: Farrar, Straus, and Giroux.

Friedman, Thomas. 2008. *Hot, Flat, and Crowded.* New York: Farrar, Straus and Giroux.

Gabor, Andrea. 2000. *The Capitalist Philosophers.* New York: New York Times Books.

Gaffney, John. 2001. "How Do You Feel About a $44 Tooth Bleaching Kit?" *Business 2.0* (October), pp. 126–27.

Galor, Oden, and O. Moav. 2002. "National Selection and the Origin of Economic Growth." *Quarterly Journal of Economics* (November): 1133–91.

Gardner, Howard. 2006. *Five Minds for the Future.* Boston: Harvard Business School Press.

Gargiulo, Terrence, 2002. *Making Stories*, Westport, CT:Quorum.

Garrett, E. M. 1992. "Branson the Bold." *Success* (November), p. 22.

Garvin, David. 1993. "Building a Learning Organization." *Harvard Business Review* (July), pp. 78–91.

Garvin, David, and Lynee Levesque. 2006. "Meeting the Challenge of Corporate Entrepreneurship." *Harvard Business Review* (October), pp. 102–12.

Gatewood, Elizabeth. 2001. "Busting the Stereotype." *Kelley School Business Magazine* (Summer), pp. 14–15.

Gavetti, Giovanni, and J. W. Rivkin. 2005. "How Strategists Really Think." *Harvard Business Review* (April), pp. 54–63.

Gawer, Annabelle, and M. A. Cusumano. 2008. "How Companies Become Platform Leaders." *MIT Sloan Management Review* (Winter), pp. 28–35.

GEM Executive Report. 2008. (November 18).

Gershenfeld, Neil. 2005. *FAB: The Coming Revolution on Your Desktop.* New York: Basic Books.

Gersick, Kelin, and John Davis. 1997. *Life Cycles of a Family Business.* Boston: Harvard Business School Press.

Gerstner, Louis. 2002. *Who Says Elephants Can't Dance?* New York: Harper Collins.

Gilbert, Brett, Patricia McDougall, and David Audretsch. 2008. "Clusters, Knowledge Spillovers and New Venture Performance: An Empirical Examination." *Journal of Business Venturing* 23:405–22.

Girard, Kim. 2000. "Pandesic's Failed Union." *Business 2.0* (September), pp. 16–18.

Gittell, Jody H. 2003. *The Southwest Airlines Way.* New York: McGraw-Hill.

Gladwell, Malcolm. 2000. *The Tipping Point.* Boston: Little, Brown.

Goldenberg, Jacob, Roni Horowitz, Amnon Levav, and David Mazursky. 2003. "Finding Your Innovation Sweet Spot." *Harvard Business Review* (March), pp. 120–29.

Goleman, Daniel, R. Boyatzis, and A. McKee. 2002. *Primal Leadership.* Boston: Harvard Business School Press.

Goleman, Daniel, R. Boyatzis, and A. McKee. 2001. "Primal Leadership: The Hidden Driver of Great Performance." *Harvard Business Review* (December), pp. 44–51.

Gompers, Paul, and J. Lerner. 2001. *The Money of Invention.* Boston: Harvard Business School Press.

Gompers, Paul, and W. A. Sahlman. 2002. *Entrepreneurial Finance.* New York: Wiley & Sons.

Gosling, Jonathan, and H. Mintzberg. 2003. "The Five Minds of a Manager." *Harvard Business Review* (November), pp. 54–63.

Gottfredson, Mark, R. Puryear, and S. Phillips. 2005. "Strategic Sourcing." *Harvard Business Review* (February), pp. 135–39.

Govindarajan, Vijay, and Chris Trimble. 2005a. "Building Breakthrough Businesses Within Established Organizations." *Harvard Business Review* (May), pp. 58–68.

Govindarajan, Vijay, and Chris Trimble. 2005b. "Organizational DNA for Strategic Innovation." *California Management Review* (Spring), pp. 47–76.

Grabowski, Robert, L. Navarro-Serment, and P. K. Khosla. 2003. "An Army of Small Robots." *Scientific American* (November), pp. 63–67.

Graebner, Melissa. 2004. "Momentum and Serendipity: How Acquired Leaders Create Value in the Integration of Firms." 25:*Strategic Management Journal.*

Graham, Paul. 2008. *www.paulgraham.com blog* (October).

Green, Heather. 2003. "Companies That Really Get It." *Business Week* (25 August), p. 144.

Greene, Patricia, C. G. Brush, and M. M. Hart. 1999. "The Corporate Venture Champion." *Entrepreneurship Theory and Practice* (Spring): 103–22.

Grove, Andy. 2003. "Churning Things Up." *Fortune* (11 August), pp. 115–18.

Gruber, Marc. 2007. "Uncovering the Value of Planning in New Venture Creation: A Process and Contingency Perspective." *Journal of Business Venturing* 22:782–807.

Gundry, Lisa, and H. Welsch. 2001. "The Ambitious Entrepreneur." *Journal of Business Venturing* 16:453–70.

Hamel, Gary. 2001. "Revolution versus Evolution: You Need Both." *Harvard Business Review* (May), pp. 150–56.

Hamel, Gary. 2000. *Leading the Revolution.* Boston: Harvard Business School Press.

Hamermesh, Richard, Paul Marshall, and Taz Pirmohamed. 2002. "Note on Business Model Analysis for the Entrepreneur." *Harvard Business School Case* 9–802–048.

Hamm, John. 2002. "Why Entrepreneurs Don't Scale." *Harvard Business Review* (December), pp. 110–15.

Hammer, Michael. 2001. *The Agenda.* New York: Crown Business.

Hardy, Quentin. 2003. "All Eyes on Google." *Forbes* (26 May), pp. 100–10.

Hargadon, Andrew, and Y. Douglas. 2001. "When Innovations Meet Institutions." *Administrative Science Quarterly* 46 (September): 476–501.

Hargadon, Andrew, and R. I. Sutton. 2000. "Building an Innovation Factory." *Harvard Business Review* (June), pp. 157–66.

Harzberg, Friderick. 2003. "How Do You Motivate Employees?" *Harvard Business Review* (January), pp. 87–92.

Hastie, Reid, and R. M. Dawes. 2001. *Rational Choice in an Uncertain World.* Thousand Oaks, Calif.: Sage.

Hayward, Mathew, Dean Shepherd, and Dale Griffin. 2006. "A Hubris Theory of Entrepreneurship." *Management Science* 52(2):160–72.

Heath, Chip, and Dan Heath. 2007. *Made to Stick.* New York: Random House.

Heifetz, Ronald, and D. Laurie. 2001. "The Work of Leadership." *Harvard Business Review* (December), pp. 131–40.

Helft, Miguel. 2002. "Fashion Fast Forward." *Business 2.0* (May), pp. 61–66.

Henderson, James, Benoit Leleux, and Ian White. 2006. "Service-for-Equity Arrangements: Untangling Motives and Conflicts." *Journal of Business Venturing* 21:886–909.

Henderson, Rebecca, and Kim Clark. 1990. "Architectural Innovation." *Administrative Science Quarterly* 35:9–30.

Hill, Charles W., and Gareth R. Jones. 2001. *Strategic Management,* 5th ed. Boston: Houghton Mifflin.

Hill, Charles, and F. Rothaermel. 2003. "The Performance of Incumbent Firms in the Face of Radical Technological Innovation." *Academy of Management Review* 28:257–74.

Hill, Michael, R. Ireland, S. Camp, and D. Sexton. 2002. *Strategic Entrepreneurship.* Malden, Mass.: Blackwell.

Hinds, Pam, and Sara Kiesler. 2002. *Distributed Work.* Cambridge, MA: MIT Press.

Hirsh, Evan, S. Hedlund, and M. Schweizer. 2003. "Reality Is Perception—The Truth about Car Brands." *Strategy and Business* (Fall): 20–25.

Hitt, Michael A., R. D. Ireland, S. M. Camp, and D. L. Sexton. 2001. "Entrepreneurial Strategies for Wealth Creation." *Strategic Management Journal* 22:479–91.

Ho, Yew Kee, H. T. Keh, and J. M. Ong. 2005. "The Effects of R&D and Advertising on Firm Value."

IEEE Transactions on Engineering Management (February), pp. 3–14.

Holt, Douglas. 2003. "What Becomes an Icon Most?" *Harvard Business Review* (March), pp. 43–49.

Hoover, Gary. 2001. *Hoover's Vision.* New York: Texere.

Hounshell, David. 1985. *From the American System to Mass Production.* Baltimore: Johns Hopkins University Press.

Jane M. Howell, Christine M. Shea, Christopher A. Higgins. 2005. "Champions of product innovations: Defining, developing, and validating a measure of champion behavior." *Journal of Business Venturing* 20(5):641–661.

Hoy, Frank. 2006. "The Complicating Factor of Life Cycles in Corporate Venturing." *Entrepreneurship Theory and Practice* (November), pp. 831–36.

Hrebiniak, Lawrence G. 2005. *Making Strategy Work.* Upper Saddle River, N.J.: Pearson.

Hsu, David. 2004. "What Do Entrepreneurs Pay for Venture Capital Affiliation?" *Journal of Finance* (August), pp. 1805–36.

Hsu, David. 2006. "Venture Capitalists and Cooperative Start-up Commercialization Strategy." *Management Science* 52(2):204–19.

Hsu, David, Edward Roberts, and Charles Eesley. 2007. "Entrepreneurs from Technology-Based Universities: Evidence from MIT." *Research Policy* 36:768–88.

Hughes, Jonathan, and Jeff Weiss. 2007. "Simple Rules for Making Alliances Work." *Harvard Business Review* (November), pp. 122–31.

Huse, Morten. 2000. "Boards of Directors in SMEs: A Review and Research Agenda." *Entrepreneurship & Regional Development* 12:271–90.

Huntington, Tom. 2003. "The Gimmick That Ate Hollywood." *Invention and Technology* (Spring): 34–45.

Huxley, Aldous. 1990. *The Perennial Philosophy.* New York: Harper Collins.

Iansiti, Marco, and Roy Levien, 2004. "Strategy as Ecology."*Harvard Business Review* (March), pp. 68–78.

Ibarra, Hermina. 2002. "How to Stay Stuck in the Wrong Career." *Harvard Business Review* (December), pp. 40–47.

Ibarra, Hermina, and Kent Lineback. 2005. "What's Your Story?" *Harvard Business Review* (January), pp. 65–71.

Isenberg, Daniel. 2008. "The Global Entrepreneur." *Harvard Business Review* (December).

Jackman, Jay, and M. H. Strober. 2003. "Fear of Feedback." *Harvard Business Review* (April), pp. 101–7.

Jackson, Ira, and J. Nelson. 2004. *Profits with Principles.* New York: Doubleday.

Jakle, John A., K. A. Sculle, and J. S. Rogers. 1996. *The Motel in America.* Baltimore: Johns Hopkins University Press.

Jassawalla, Avan, and H. C. Sashittal. 2002. "Cultures that Support Product Innovation Processes." *Academy of Management Executive* (August): 42–54.

Jensen, Richard, and Marie Thursby. 2001. "Proofs and Prototypes for Sale: The Licensing of University Inventions." *American Economic Review* 91:240–259.

Joyce, William, N. Nohria, and B. Roberson. 2003. *What Really Works.* New York: Harper Collins.

Julien, Pierre Andre, and C. Pamangalahy. 2003. "Competitive Strategy and Performance of Exporting SMEs." *Entrepreneurship Theory and Practice* (Spring): 227–45.

Kahneman, Daniel, and Dan Lovallo. 1993. "Timid Choices and Bold Forecasts: A Cognitive Perspective on Risk Taking." *Management Science* 39(1):17–31.

Kalnins, Arturs, and Wilbur Chung. 2006. "Social Capital, Geography, and Survival: Gujarti Immigrant Entrepreneurs in the U.S. Lodging Industry." *Management Science* 52(2):233–47.

Kanter, Rosabeth Moss. 2003. "Leadership and the Psychology of Turnarounds." *Harvard Business Review* (June), pp. 58–67.

Kaplan, Robert, and D. Norton. 2000. *The Strategy Focused Organization.* Boston: Harvard Business School Press.

Kaplan, Robert S., and David P. Norton. 2004. *Strategy Maps.* Boston: Harvard Business School Press.

Kaptein, Muel, and J. Wempe. 2002. *The Balanced Company.* New York: Oxford University Press.

Karlgaard, Rich. 2005. "Flying Ellipse's Pocket Jet." *Forbes* (August 15), p. 27.

Karra, Neri, Paul Tracey, and Nelson Phillips. 2006. "Altruism and Agency in the Family Firm: Exploring the Role of Family, Kinship, and Ethnicity." *Entrepreneurship Theory and Practice* (November), pp. 861–877.

Katila, Riitta, Jeff Rosenberger, and Kathy Eisenhardt. 2008. "Swimming with Sharks: Technology Ventures, Defense Mechanisms, and Corporate Relationships." *Administrative Science Quarterly* 53(2):295–332.

Kauffman Foundation. 2009. "Start-ups Critical for Job Creation in U.S." *Business Dynamic Statistics Report,* US Census Bureau (January).

Kawasaki, Guy. 2004. *The Art of the Start.* New York: Penguin.

Kelley, T. 2001. *The Art of Innovation: Lessons in Creativity from IDEO, America's Leading Design Firm.* New York: Doubleday.

Kelley, Thomas, and Jonathan Littman. 2005. *The Ten Faces of Innovation: IDEO's Strategies for Defeating the Devil's Advocate and Driving Creativity throughout Your Organization.* New York: Doubleday.

Kellner, Tomas. 2002. "One Man's Trash." *Forbes* (4 March), pp. 96–98.

Kessler, Eric, and Paul Bierly. 2002. "Is Faster Really Better? An Empirical Test of the Implications of Innovation Speed." *IEEE Transactions on Engineering Management* (February): 2–12.

Khurana, Rakesh. 2002. *Searching for a Corporate Savior.* Princeton, N.J.: Princeton University Press.

Kidwell, Roland, Arne Nygaard, and Ragnhild Silkoset. 2007. "Antecedents and Effects of Free Riding in the Franchisor-Franchisee Relationship." *Journal of Business Venturing* 22:522–44.

Kim, W. Chan, and R. Mauborgne. 2005. "Blue Ocean Strategy. "*California Management Review* (Spring), pp. 105–21.

King, Jr., Martin Luther. 1963. *Strength to Love.* Minneapolis: Fortress Press.

Kingsbury, Kathleen. 2008. "Oregon Software Company Helps Adopt Open Source to Collect Public Data." *Time* (July 14).

Kirchhoff, Bruce A. 1994. Entrepreneurship and Dynamic Capitalism. New York: Praeger.

Kirkman, Bradley, et al. 2002. "Five Challenges to Virtual Team Success." *Academy of Management Executive* 3:67–79.

Kirkpatrick, David. 2003. "Brainstorm 2003." *Fortune* (27 October), pp. 187–90.

Klein, Alec. 2003. *Stealing Time.* New York: Simon and Schuster.

Klein, Mark, and A. Einstein. 2003. "The Myth of Customer Satisfaction." *Strategy and Business* 30:8–9.

Kleiner, Art. 2003. "Making Patient Capital Pay off." *Strategy and Business* (Fall): 26–30.

Knopper, Steve. 2009. *Appetite for Self Destruction.* New York: Free Press.

Koehn, Nancy. 2001. *Brand New.* Cambridge: Harvard Business School Press.

Komisar, Randy. 2000. *The Monk and the Riddle.* Boston: Harvard Business School Press.

Kotter, John P., and D. S. Cohen. 2002. *The Heart of Change.* Boston: Harvard Business School Press.

Krajewski, Lee, and L. Ritzman. 2002. *Operations Management,* 6th ed. Upper Saddle River, N.J.: Prentice-Hall.

Krupp, Fred, and M. Horn. 2008. *Earth: The Sequel.* New York: Norton.

Kuemmerle, Walter. 2005. "The Entrepreneur's Path to Global Expansion." *Sloan Management Review* (Winter), pp. 42–49.

Kuemmerle, Walter. 2002. "A Test for the Fainthearted." *Harvard Business Review* (May), pp. 4–8.

Langerak, Fred, and E. J. Hultink. 2005. "The Impact of New Product Development Acceleration Approaches on Speed and Profitability." *IEEE Transactions on Engineering Management* (February), pp. 30–41.

Lapre, Michael, and L. N. Van Wassenhove. 2002. "Learning across Lines." *Harvard Business Review* (October), pp. 107–11.

LaSalle, Diana, and T. A. Britton. 2003. *Priceless.* Boston: Harvard Business School Press.

Lassiter, Joseph B., III. 2002. "Entrepreneurial Marketing: Learning from High-Potential Ventures." *Harvard Business School Case 9–803–036.*

Lawrence, Thomas B., E. A. Morse, and S. W. Fowler. 2005. "Managing Your Portfolio of Connections." *Sloan Management Review* (Winter), pp. 59–65.

Lax, David A., and J.K. Sebenius. 2003. "3-D Negotiation: Playing the Whole Game." *Harvard Business Review* (November), pp. 65–74.

Lax, Eric. 1985. "Banking on Biotech Business." *New York Times* (22 December).

Le Breton-Miller, Isabelle, and Danny Miller. 2006. "Why Do Some Family Businesses Out-Compete? Governance, Long-Term Orientations, and Sustainable Capability." *Entrepreneurship Theory and Practice* (November), pp. 731–46.

Le Breton-Miller, Isabelle, and Danny Miller. 2006. "Why Do Some Family Businesses Out-Compete? Governance, Long-Term Orientations, and Sustainable Capability." *Entrepreneurship Theory and Practice* 30:731-746.

Lechler, Thomas. 2001. "Social Interaction: A Determinant of Entrepreneurial Team Venture Success." *Small Business Economics* 16:263–78.

Lee, Jae-Nam, and Y. G. Kim. 2005. "Understanding Outsourcing Partnership." *IEEE Transactions on Engineering Management* (February), pp. 43–58.

Lee, Soo Hoon, P. K. Wong, and C. L. Chong. 2005. "Human and Social Capital Explanations for R&D Outcomes." *IEEE Transactions on Engineering Management* (February), pp. 59–67.

Leibovich, Mark. 2002. *The New Imperialists.* Paramus, N.J.: Prentice-Hall.

Leifer, Richard, Christopher M. McDermott, Gina Colarelli O'Connor, and Lois S. Peters. 2000. *Radical Innovation.* Boston: Harvard Business School Press.

Leonard-Barton, Dorothy. 1995. *Wellsprings of Knowledge.* Boston: Harvard Business School Press.

Leslie, Mark, and Charles Holloway. 2006. "The Sales Learning Curve." *Harvard Business Review* (July-August).

Lévesque, Moren, and Dean A. Shepherd. 2004. "Entrepreneurs' Choice of Entry Strategy in Emerging and Developed Markets." *Journal of Business Venturing* 19:29–54.

Lévesque, Moren, Dean Shepherd, and Evon Douglas. 2002. "Employment or Self-Employment: A Dynamic Utility-Maximizing Model." *Journal of Business Venturing* 17:189–210.

Levitt, Harold. 2003. "Why Hierarchies Thrive." *Harvard Business Review* (March), pp. 96–102.

Lewis, Michael. 2003. *Moneyball.* New York: Norton.

Lewis, Michael. 2000. *The New, New Thing.* New York: Norton.

Liker, J. K. 2004. *The Toyota Way.* New York: McGraw-Hill.

Lohr, Steve. 2003. "Silicon Valley Hikes Wireless Frontier." *New York Times* (7 April).

Lojacono, Gabriella, and G. Zaccai. 2004. "The Evolution of the Design-Inspired Enterprise." *Sloan Management Review* (Spring), pp. 75–79.

Lord, Michael, S. W. Mandel, and J. D. Wager. 2002. "Spinning out a Star." *Harvard Business Review* (June), pp. 115–21.

Lounsbury, Michael, and M. Glynn. 2001. "Cultural Entrepreneurship: Stories, Legitimacy and the Acquisition of Resources." *Strategic Management Journal* 22:545–64.

Low, Murray B., and E. Abrahamson. 1997. "Movements, Bandwagons and Clones: Industry Evolution and the Entrepreneurial Process." *Journal of Business Venturing* 12:435–57.

Lowe, Robert A. 2001. "The Role and Experience of Start-ups in Commercializing University Inventions." *Entrepreneurial Inputs and Outcomes* 13:189–222.

Lu, Jane, and Paul Beamish. 2006. "Partnering Strategies and Performance of SMEs' International NointVentures." *Journal of Business Venturing* 21:461–86.

Lucier, Chuck, and J. Dyer. 2003. "Creating Chaos for Fun and Profit." *Strategy and Business* 30:14–20.

Luenberger, David. 2006. *Information Science.* Princeton: Princeton University Press.

Lumpkin, G. T., and B. B. Lichtenstein. 2005. "The Role of Organizational Learning in the Opportunity Recognition Process." *Entrepreneurship Theory and Practice* (July), pp. 451–72.

Lynn, Gary, and Richard Reilly. 2002. *Blockbusters.* New York: Harper Collins.

Magretta, Joan. 2002. *What Management Is.* New York: Free Press.

Maher, Michael, C. P. Stickney, and R. L. Weil. 2004. *Managerial Accounting,* 8th ed. Cincinnati: Southwestern.

Majumdar, Sumit. 1999. "Sluggish Giants, Sticky Cultures and Dynamic Capability Transformation." *Journal of Business Venturing* 15:59–78.

Malik, Om. 2003. *Broadbandits.* Hoboken, N.J.: Wiley & Sons.

Malone, Michael. 2002. *Betting It All: The Entrepreneurs of Technology.* New York: Wiley & Sons.

Mandel, Michael J. 2004. *Rational Exuberance.* New York: Harper Collins.

Mangalindan, Mylene, and S. L. Hwang. 2001. "Insular Culture Helped Yahoo! Grow, But Has Now Hurt It in the Long Run." *Wall Street Journal* (9 March), p. A1.

Mangelsdorf, Martha. 2008. "How Hard Times Can Drive Innovation." *Wall Street Journal* (15 December).

Mankins, Michael, and R. Steele. 2005. "Turning Strategy into Great Performance." *Harvard Business Review* (July), pp. 65–72.

Markides, Costas, and Paul Geroski. 2004. "The Art of Scale." *Strategy and Business* 35:51–59.

Markides, Constantinos, and Paul Geroski. 2005. *Fast Second: How Smart Companies Bypass Radical Innovation to Enter and Dominate New Markets.* San Francisco: Josey-Bass.

Markman, Gideon, D. Balkin, and R. A. Baron. 2002. "Inventors and New Venture Formation." *Entrepreneurship Theory and Practice* (Winter): 149–65.

Marn, Michael, Eric Roegner, and Craig Zawada. 2004. *The Price Advantage.* New York: John Wiley & Sons.

Martens, Martin, Jennifer Jennings, and P. Devereaux Jennings. 2007. "Do the Stories They Tell Get Them the Money They Need? The Role of Entrepreneurial Narratives in Resource Acquisition." *Academy of Management Journal* 50(5):1107–32.

Martin, Roger L. 2002. "The Virtue Matrix." *Harvard Business Review* (March), pp. 69–75.

Martin, Roger, and Sally Osberg. 2007. "Social Entrepreneurship: The Case for Definition." *Stanford Social Innovation Review* (Spring), pp. 28–39.

Marvel, Matthew, Abbie Griffin, John Hebda, and Bruce Vojak. 2007. "Examining the Technical Corporate Entrepreneurs' Motivation: Voices from the Field." *Entrepreneurship Theory and Practice* (September), pp. 753–68.

Mason, Heidi, and T. Rohner. 2002. *The Venture Imperative.* Boston: Harvard Business School Press.

Maurer, Indre, and Mark Ebers. 2006. "Dynamics of Social Capital and Their Performance Implications: Lessons from Biotechnology Start-ups." *Administrative Science Quarterly* 51:262–92.

McCall, Morgan. 1998. *High Flyers.* Boston: Harvard Business School Press.

McCoy, Bowen H. 2007. *Living into Leadership.* Stanford: Stanford University Press.

McElroy, Mark. 2003. *The New Knowledge Management.* Boston: Elsevier.

McFarland, Keith. 2008. *The Breakthrough Company.* New York: Crown.

McGrath, James, F. Kroeger, M. Traem, and J. Rocken Haeuser. 2001. *The Value Growers.* New York: McGraw-Hill.

McGrath, Rita G. 2005. "Market Busting: Strategies for Exceptional Business Growth." *Harvard Business Review* (March), pp. 81–89.

McGrath, Rita Gunther, Thomas Keil, and Taina Tuki- ainen. 2006. "Extracting Value from Corporate Venturing." *MIT Sloan Management Review* 48(1):50–56.

McKee, Robert. 2003. "Storytelling That Moves People." *Harvard Business Review* (June), pp. 51–55.

McKenzie, Ray. 2001. *The Relationship-Based Enterprise.* New York: McGraw-Hill.

McLean, Bethany, and P. Elkind. 2003. *The Smartest Guys in the Room.* New York: Portfolio.

McLemore, Clinton. 2003. *Street-Smart Ethics.* Louisville, Ky.: Westminster John Knox Press.

McMullen, Jeffrey, and Dean Shepard. 2002. "Regulatory Focus and Entrepreneurial Intention." Presentation at the Academy of Management Meeting (August).

McMullen, Jeffrey, and Dean Shepherd. 2006. "Entrepreneurial Action and the Role of Uncertainty in the Theory of the Entrepreneur." *Academy of Management Review* 31:132–152.

McNulty, James, et al. 2002. "What's Your Real Cost of Capital?" *Harvard Business Review* (October), pp. 114–21.

Melnyk, Steven, and M. Swink. 2002. *Value-Driven Operations Management.* New York: McGraw-Hill.

Mendonca, L. T. and H. Rao. 2008. "Lessons from Innovation's Front Lines: An Interview with IDEO's CEO" *McKinsey Quarterly* (November).

Mezias, John, and W. H. Starbuck. 2003. "What Do Managers Know, Anyway?" *Harvard Business Review* (May), pp. 16–17.

Michael, Steven C. 2003. "First Mover Advantage through Franchising." *Journal of Business Venturing* 18:61–80.

Miles, Morgan, and J. Covin. 2002. "Exploring the Practice of Corporate Venturing." *Entrepreneurship, Theory and Practice* (Spring): 21–40.

Minniti, Maria, and W. Bygrave. 2001. "A Dynamic Model of Entrepreneurial Learning." *Entrepreneurship Theory and Practice* (Spring): 5–16.

Mintzberg, Henry, B. Ahlstrand, and J. Lampel. 1998. *Strategy Safari.* New York: Free Press.

Mittelstaedt, Robert. 2005. *Will Your Next Mistake Be Fatal?* Upper Saddle River, N.J.: Pearson.

Mokyr, Joel. 2003. *The Gifts of Athena—Historical Origins of the Knowledge Economy.* Princeton, N.J.: Princeton University Press.

Moore, Geoffrey. 2002. *Crossing the Chasm.* New York: Harper Collins.

Moore, Geoffrey. 2000. *Living on the Fault Line.* New York: Harper Collins.

Moore, Geoffrey A. 2004. "Darwin and the Demon: Innovating Within Established Enterprises." *Harvard Business Review* (July), pp. 86–92.

Mullins, John. 2006. *The New Business Road Test.* Harlow, England: Prentice Hall.

Mullins, John, and Randy Komisar. 2009. *Getting to Plan B: Breaking Through to a Better Business Model.* Boston: Harvard Business School Press.

Muna, Farid A., and Ned Mansour. 2005. "Leadership lessons from Canada geese." *Team Performance Management* 11:316–326.

Murray, Fiona, and S. O'Mahony. 2007. "Exploring the Foundations of Cumulative Innovation: Implications for Organization Science" *Organization Science* 18(6):1006–1021.

Nagle, Thomas, and John Hogan. 2006. *The Strategy and Tactics of Pricing.* Upper Saddle River, N.J.: Prentice-Hall.

Nalebuff, Barry, and Ian Ayres. 2003. *Why Not?* Boston: Harvard Business School Press.

Nambisan, Satish. 2002. "Designing Virtual Customer Environments for New Product Development." *Academy of Management Review* 27(3):392–413.

Nicolaou, Nicos, Scott Shane, Lynn Cherkas, Janice Hunkin, and Tim Spector. 2008. "Is the Tendency to Engage in Entrepreneurship Genetic?" *Management Science* 54(1):167–79.

Niraj, Rakesh, Mahendra Gupta, and Chakravarthi Narasimhan. 2001. "Customer Profitability in a Supply Chain." *Journal of Marketing* 65:1–16.

Norman, Patricia. 2001. "Are Your Secrets Safe? Knowledge Protection in Strategic Alliances." *Business Horizons* (November), pp. 51–60.

Northouse, Peter G. 2001. *Leadership,* 2d ed. Thousand Oaks, Calif.: Sage.

Ogawa, Susumu, and Frank Piller. 2006. "Reducing the Risks of New Product Development." *MIT Sloan Management Review* 47(2):65–71.

Ogden, Joseph, F. Jen, and P. O'Connor. 2003. *Advanced Corporate Finance.* Upper Saddle River, N.J.: Prentice-Hall.

O'Reilly, Charles, and M. L. Tushman. 2004. "The Ambidextrous Organization." *Harvard Business Review* (April), pp. 22–30.

Oviatt, Benjamin, and Patricia McDougall. 2005. "Defining International Entrepreneurship and Modeling the Speed of Internationalization." *Entrepreneurship Theory and Practice* (September), pp. 537–53.

Ozgen, Eren, and Robert Baron. 2007. "Social Sources of Information in Opportunity Recognition: Effects of Mentors, Industry Networks, and Professional Forums." *Journal of Business Venturing* 22:174–92.

Packalen, Kelley. 2007. "Complementing Capital: The Role of Status, Demographic Features, and Social Capital in Founding Teams' Abilities to Obtain

Resources." *Entrepreneurship Theory and Practice* (November), pp. 873–91.

Packard, David. 1995. *The HP Way.* New York: Harper Collins.

Pascale, Richard T., and J. Sternin. 2005. "Your Company's Secret Change Agents." *Harvard Business Review* (May), pp. 73–81.

Paydarfar, David, and William Schwartz. 2001. "An Algorithm for Discovery." *Science* (6 April), p. 13.

Perlow, Leslie, and S. Williams. 2003. "Is Silence Killing Your Company?" *Harvard Business Review* (May), pp. 52–58.

Perlow, Leslie, G. Okhuysen, and N. P. Repenning. 2002. "The Speed Trap." *Academy of Management Journal* 5:931–55.

Peters, Tom, and R. Waterman. 1982. *In Search of Excellence.* New York: Harper & Row.

Petroski, Henry. 2003. *Small Things Considered: Why There Is No Perfect Design.* New York: Knopf.

Pfeffer, J., and R. Sutton. 2006. "Evidence-based Management." *Harvard Business Review* (January), pp. 51–60.

Pfeffer, Jeffrey, and R. Sutton. 2000. *The Knowing-Doing Gap.* Boston: Harvard Business School Press.

Phinisee, Ivory, I. Elaine Allen, Edward Rogoff, Joseph Onochie, and Monica Dean. 2008. 2006–2007 *National Entrepreneurial Assessment for the United States of America.* Global Entrepreneurship Monitor.

Pietersen, Willie. 2002. *Reinventing Strategy.* New York: Wiley & Sons.

Porter, Michael. 2001. "Strategy and the Internet." *Harvard Business Review* (March), pp. 63–78.

Porter, Michael E. 1998. *On Competition.* Boston: Harvard Business School Press.

Post, James, L. E. Preston, and S. Sachs. 2002. "Managing the Extended Enterprise." *California Management Review* (Fall), pp. 6–20.

Prahalad, C. K. 2005. *The Fortune at the Bottom of the Pyramid.* Upper Saddle River, N.J.: Pearson.

Prahalad, C. K., and A. Hammond. 2002. "Serving the World's Poor, Profitably." *Harvard Business Review* (September), pp. 48–57.

Prahalad, C. K., and M. S. Krishnan. 2008. *The New Age of Innovation.* New York: McGraw-Hill.

Prahalad, C. K., and V. Ramaswamy. 2000. "Co-opting Customer Competence." *Harvard Business Review* (January), pp. 79–87.

Prahalad, C. K., and V. Ramaswamy, 2004. *The Future of Competition: Co-creating Unique Value wth Customers.* Boston: Harvard Business School Press.

Prasad, Dev, G. Vozikis, and G. Bruton. 2001. "Commitment Signals in the Interaction between Business Angels and Entrepreneurs." *Entrepreneurial Inputs and Outcomes* 13:45–69.

Prestowitz, Clyde. 2005. *Three Billion New Capitalists.* New York: Basic Books.

Prusak, Laurence, and D. Cohen. 2001. "How to Invest in Social Capital." *Harvard Business Review* (June), pp. 86–94.

Quadrini, Vincerizo. 2001. "Entrepreneurial Financing, Savings and Mobility." *Entrepreneurial Inputs and Outcomes* 13:71–94.

Rappaport, Alfred, Michael J. Mauboussin, and Peter L. Bernstein. 2001. *Expectations Investing.* Boston: Harvard Business School Press.

Ratner, Mark, and D. Ratner. 2004. *Nanotechnology: A Gentle Introduction to the Next Big Idea.* Upper Saddle River, N.J.: Prentice Hall.

Reichheld, Frederick. 2001. "Lead for Loyalty." *Harvard Business Review* (July), pp. 76–84.

Ridgway, Nicole. 2003. "Something to Sneeze at." *Forbes* (21 July), pp. 102–4.

Ries, Al, and Jack Trout. 2001. *Positioning: The Battle for Your Mind.* New York: McGraw-Hill.

Rigby, Darrell. 2001. "Moving Upward in a Downturn." *Harvard Business Review* (June), pp. 99–105.

Rigby, Darrell, and C. Zook. 2002. "Open Market Innovation." *Harvard Business Review* (October), pp. 80–89.

Riggs, Henry. 2006. *Understanding the Financial Score.* San Rafael, CA: Morgan and Clayppol.

Riggs, Henry. 2004. *Financial and Economic Analysis for Engineering and Technology Management,* 2d ed. Hoboken, N.J.: Wiley & Sons.

Robbins-Roth, Cynthia. 2000. *From Alchemy to IPO.* Cambridge, Mass.: Perseus.

Roberts, John. 2004. *The Modern Firm*. New York: Oxford University Press.

Roberts, Michael. 2003. "Managing the Growing Venture." *Harvard Business School Note 9–803–137*.

Roberts, Michael, and Diana Gardner, 2000. Advanced Inhalation Research. Harvard Business School, Case 899292.

Rogers, Everett. 2003. *Diffusion of Innovations,* 5th ed. New York: Free Press.

Rohlfs, Jeffrey. 2001. *Bandwagon Effects in High Technology Industries.* Cambridge: MIT Press.

Roman, Kenneth. 2003. How to Advertise, 3rd ed. New York: Thomas Dunne.

Rosen, Corey, J. Case, and M. Staubus. 2005. "Every Employee an Owner. Really." *Harvard Business Review* (June), pp. 122–30.

Rosenberger, J. D. 2005. "The Flip Side of the Coin: Nascent Technology Ventures and Corporate Venture Funding." Dissertation: Stanford University Department of Management Science and Engineering.

Ross, Stephen, R. Westerfield, and J. Jaffe. 2002. *Corporate Finance.* New York: McGraw-Hill Irwin.

Rothaermel, Frank, and D. Deeds. 2004. "Exploration and Exploitation Alliances in Biotechnology." *Strategic Management Journal* (Winter): 100–21.

Rothaermel, Frank, and David Deeds. 2006. "Alliance Type, Alliance Experience and Alliance Management Capability in High-Technology Ventures." *Journal of Business Venturing* 21:429–60.

Ruef, Martin. 2002. "Strong Ties, Weak Ties and Islands." *Industrial and Corporate Change* 3:427–49.

Ryans, Adrian, R. More, D. Barclay, and T. Deutscher. 2000. *Winning Market Leadership.* New York: Wiley & Sons.

Sachs, Jeffrey 2008. *Common Wealth.* New York: Penguin Press.

Sahlman, William. 1999. *The Entrepreneurial Venture.* Boston: Harvard Business School Press.

Santos, Filipe, and Kathy Eisenhardt. 2009. "Constructing Markets and Shaping Boundaries:

Entrepreneurial Action in Nascent Markets." *Academy of Management Journal* 52:643–671.

Sathe, Vijay. 2003. *Corporate Entrepreneurship.* New York: Cambridge University Press.

Sawhney, Mohan, and J. Zabin. 2001. *The Seven Steps to Nirvana.* New York: McGraw-Hill.

Schrage, Michael. 2002. "Ease of Learning." *Technology Review* (December), p. 23.

Schrage, Michael. 2001. "Playing around with Brainstorming." *Harvard Business Review* (March), pp. 149–54.

Schrage, Michael. 2000. *Serious Play.* Boston: Harvard Business School Press.

Schramm, Carl. 2004. "Building Entrepreneurial Economies." *Foreign Affairs* (July), pp. 104–15.

Schultz, Howard. 1997. *Pour Your Heart Into It: How Starbucks Built a Company One Cup at a Time* New York: Hyperion.

Schumpeter, Joseph. 1984. *Capitalism, Socialism and Democracy.* New York: Harper Torchbooks.

Schwartz, Evan. 2002. *The Last Lone Inventor.* New York: Harper Collins.

Schwartz, Barry. 2004. *The Paradox of Choice.* New York: Echo Press.

Schwienbacher, Armin. 2007. "A Theoretical Analysis of Optimal Financing Strategies for Different Types of Capital-Constrained Entrepreneurs." *Journal of Business Venturing* 22:753–81.

Sebenius, James. 2001. "Six Habits of Merely Effective Negotiators." *Harvard Business Review* (April), pp. 87–95.

Selden, Larry, and G. Colvin. 2003. "What Customers Want." *Fortune* (7 July), pp. 122–25.

Selden, Larry, and G. Colvin. 2002. "Will This Customer Sink Your Stock?" *Fortune* (30 September), pp. 127–32.

Shane, Scott. 1996. "Hybrid Organizational Arrangements and Their Implications for Firm Growth and Survival: A Study of New Franchisors." *Academy of Management Journal* 39(1):216–34.

Shane, Scott. 2001. "Technological Opportunities and New Firm Creation." *Management Science* (February): 205–20.

Shane, Scott, and S. Venkataraman. 2000. "The Promise of Entrepreneurship as a Field of Research." *Academy of Management Review* 25(1):217–26.

Shane, Scott A. 2005. *Finding Fertile Ground.* Upper Saddle River, N.J.: Pearson.

Shapiro, Hal, and H. Varian. 1998. *Innovation Rules.* Cambridge, Mass: Harvard Business School Press.

Shaw, Gordon, R. Brown, and P. Bromiley. 1998. "Strategic Stories: How 3M Is Rewriting Business Planning." *Harvard Business Review* (May), pp. 41–50.

Shepherd, 2002

Shepherd, Dean. 2003. "Learning from Business Failure." *Academy of Management Review* 2:318–28.

Shepherd, Dean, Evan Douglas, and Mark Shanley. 2000. "New Venture Survival." *Journal of Business Venturing* 15:393–410.

Shepherd, Dean, R. Ettenson, and A. Crouch. 2000. "New Venture Strategy and Profitability." *Journal of Business Venturing* 15:449–67.

Shepherd, Dean, and N. F. Krueger. 2002. "An Intentions-Based Model of Entrepreneurial Teams." *Entrepreneurship Theory and Practice* (Winter): 167–85.

Shepherd, Dean, and M. Levesque. 2002. "A Search Strategy for Assessing a Business Opportunity." *IEEE Transactions on Engineering Management* (May): 140–54.

Shepherd, Dean, and Mark Shanley. 1998. *New Venture Strategy.* Thousand Oaks, Calif.: Sage Publications.

Sheth, Jagdish, and R. Sisodia. 2002. *The Rule of Three.* New York: Free Press.

Shiller, Robert J. 2003. *The New Financial Order.* Princeton, N.J.: Princeton University Press.

Shrader, Rodney, and Mark Simon. 1997. "Corporate versus Independent New Ventures." *Journal of Business Venturing* 12:47–66.

Siegel, Robin, Eric Siegel, and Ian MacMillan. 1988. "Corporate Venture Capitalists: Autonomy, Obstacles, and Performance." *Journal of Business Venturing* 3:233–47.

Simon, Mark, and S. M. Houghton. 2003. "The Relationship between Overconfidence and the Introduction of Risky Products." *Academy of Management Journal* 2:139–49.

Simons, Robert. 2005. "Designing High-Performance Jobs." *Harvard Business Review* (July), pp. 55–62.

Sine, Wesley, Heather Haveman, and Pamela Tolbert. 2005. "Risky Business? Entrepreneurship in the New Independent-Power Sector." *Administrative Science Quarterly* 50:200–232.

Sloan, Paul. 2005. "What's Next for Apple?" *Business 2.0* (April), pp. 69–78.

Slywotzky, Adrian. 1996. *Value Migration.* Boston: Harvard Business School Press.

Slywotzky, Adrian. 2002. *The Art of Profitability.* New York: Warner Books.

Slywotzky, Adrian. 2007. *The Upside.* New York: Crown.

Slywotzky, Adrian, and J. Drzik. 2005. "Countering the Biggest Risk of All." *Harvard Business Review* (April), pp. 78–88.

Slywotzky, Adrian, and David Morrison. 2000. *How Digital Is Your Business?* New York: Crown Business.

Slywotzsky, Adrian, David J. Morrison, Ted Moser, Kevin A. Mundt, and James A. Quella. 1999. *Profit Patterns.* New York: Random House.

Smith, Janet, and R. L. Smith. 2004. *Entrepreneurial Finance,* 3d ed. Hoboken, N.J.: Wiley & Sons.

Sonnenfeld, Jeffrey. 2002. "What Makes Boards Great." *Harvard Business Review* (September), pp. 106–12.

Sørensen, Jesper. 2007. "Bureaucracy and Entrepreneurship: Workplace Effects on Entrepreneurial Entry." *Administrative Science Quarterly* 52:387–412.

Spekman, Robert, and L. Isabella. 2000. *Alliance Competence.* New York: Wiley & Sons.

Spulker, Daniel. 2004. *Management Strategy.* Burr Ridge, Ill.: McGraw-Hill.

Sternberg, Robert, L. A. O'Hara, and T. I. Lubart. 1997. "Creativity as Investment." *California Management Review* (Fall), pp. 8–21.

Stevenson, Howard, et al. 1999. *New Business Ventures and the Entrepreneur,* 5th ed. Burr Ridge, Ill.: McGraw-Hill Irwin.

Stipp, David. 2003. "Speed Reading Your Genes." *Fortune* (1 September), pp. 150–54.

Stangler, Dale. 2009. *The Economic Future Just Happened.* Ewing Marion Kauffman Foundation.

Stringer, Kortney. 2003. "How Do You Change Consumer Behavior?" *Wall Street Journal* (17 March), p. R6.

Stuart, Toby, and Waverly Ding. 2006. "When Do Scientists Become Entrepreneurs? The Social Structural Antecedents of Commercial Activity in the Academic Life Sciences." *American Journal of Sociology* 112(1):97–144.

Suarez, Fernando, and G. Lanzolla. 2005. "The Half Truth of First Mover Advantage." *Harvard Business Review* (April), pp. 121–27.

Sull, Don. 2004. "Disciplined Entrepreneurship." *MIT Sloan Management Review* (Fall), pp. 71–77.

Sull, Donald, and C. Spinosa. 2005. "Using Commitments to Manage Across Units." *MIT Sloan Management Review* (Fall), pp. 73–81.

Sutton, Robert. 2002. *Weird Ideas That Work.* New York: Free Press.

Sutton, Robert. 2001. "The Weird Rules of Creativity." *Harvard Business Review* (September), pp. 94–103.

Swanson, Robert A. 2007. "Co-founder, CEO, and Chairman of Genentech, Inc., 1976–1996," an oral history conducted in 1996 and 1997 by Sally Smith Hughes, Regional Oral History Office, The Bancroft Library, University of California, Berkeley.

Szulanski, Gabriel, and S. Winter. 2002. "Getting It Right the Second Time." *Harvard Business Review* (January), pp. 62–69.

Tallman, Stephen, and K. Fladmoe-Lindquist. 2002. "Internationalization Globalization and Capability Strategy." *California Management Review* (Fall), pp. 110–34.

Tapscott, Don, and A. D. Williams. 2008. *Wikinomics.* New York: Portfolio.

Tapscott, Don, D. Ticoll, and A. Lowy. 2000. *Digital Capital.* Boston: Harvard Business School Press.

Taylor, Suzanne, and K. Schroeder. 2003. *Inside Intuit.* Boston: Harvard Business School Press.

Tedeschi, Bob. 2003. "End of the Paper Chase." *Business 2.0* (March), p. 64.

Tedlow, Richard S. 2001. *Giants of Enterprise.* New York: Harper Collins.

Teitelman, Robert. 1989. *Gene Dreams.* New York: Basic Books.

Thomas, Paulette. 2003. "Entrepreneur's Biggest Problems." *Wall Street Journal* (17 March), p. R1.

Thompke, Stefan. 2001. "Enlightened Experimentation." *Harvard Business Review* (February), pp. 48–52.

Thompke, Stefan, and D. Reinertsen. 1998. "Agile Product Development." *California Management Review* (Fall), pp. 8–28.

Thompke, Stefan, and Eric von Hippel. 2002. "Customers as Innovators." *Harvard Business Review* (April), pp. 74–81.

Thompson, Leigh. 2003. "Improving the Creativity of Organizational Work Groups." *Academy of Management Executive* (February): 96–109.

Thurm, Scott. 2003. "A Go-Go Giant of the Internet Age, Cisco Is Learning to Go Slow." *Wall Street Journal* (7 May), p. A1.

Thurm, Scott. 2002. "Cisco Details the Financing for Its Start-Up." *Wall Street Journal* (12 March), p. A3.

Thursby, Jerry G., and Marie C. Thursby. 2004. "Are Faculty Critical? Their Role in University-Industry Licensing." *Contemporary Economic Policy* 22:162–178.

Tichy, Noel, and Warren Bennis. 2007. *Judgment.* New York: Portfolio.

Tiwana, Amrit, and A. A. Bush. 2005. "Continuance in Expertise-Sharing Networks." *IEEE Transactions on Engineering Management* (February), pp. 85–100.

Treacy, Michael. 2004. "Innovation as a Last Resort." *Harvard Business Review* (July), pp. 29–30.

Tsai, Kuen-Hung, and Jiann-Chyuan Wang. 2008. "External Technology Acquisition and Firm Performance: A Longitudinal Study." *Journal of Business Venturing* 23:91–112.

Ullman, David. 2003. *The Mechanical Design Process.* New York: McGraw-Hill.

Ulrich, Karl, and S. Eppinger. 2004. *Product Design and Development,* 3d ed. New York: McGraw-Hill.

Ulwick, Anthony. 2002. "Turn Customers Input into Innovation." *Harvard Business Review* (January), pp. 91–97.

Van den Ende, Jan, and N. Wijaberg. 2003. "The Organization of Innovation and Market Dynamics." *IEEE Transactions on Engineering Management* (August): 374–82.

Van Praag, Miriam. 2006. *Successful Entrepreneurship: Confronting Economic Theory with Empirical Practice.* Cheltenham: Edward Elgar.

Venkataraman, Sankaran. 2004. "Regional Transformation Through Technological Entrepreneurship." *Journal of Business Venturing* 19:153–67.

Vermeulen, Freek. 2005. "How Acquisitions Can Revitalize Companies." *Sloan Management Review* (Summer), pp. 45–51.

Vogelstein, Fred. 2003. "24/7 Customer." *Fortune* (24 November), p. 212.

Vogelstein, Fred. 2005. "Yahoo's Brilliant Solution." *Fortune* (8 August), pp. 42–54.

Vonderembse, Mark, and G. White. 2004. *Operations Management.* Hoboken, N.J.: Wiley & Sons.

Von Hippel, Eric. 2005. *Democratizing Innovation.* Cambridge, MA: MIT Press.

Waaser, Ernest et al., 2004. "How You Slice It: Smarter Segmentation for Your Sales Force." *Harvard Business Review* (March), pp. 105–10.

Wadha, Vivek, Richard Freeman, and Ben Rissing. 2008. "Education and Tech Entrepreneurship." *Ewing Marion Kauffman Foundation.*

Wadhwa, Anu, and Suresh Kotha. 2006. "Knowledge Creation Through External Venturing: Evidence from the Telecommunications Equipment Manufacturing Industry." *Academy of Management Journal* 49(4):819–35.

Wasserman, Noam. 2006. "Stewards, Agents, and the Founder Discount: Executive Compensation in New Ventures." *Academy of Management Journal* 49(5):960–76.

Wasserman, Noam. 2008. "The Founder's Dilemma." *Harvard Business Review* (February), pp. 103–109.

Welch, Jack. 2002. *Jack: Straight from the Gut.* New York: Warner.

Wellman, Barry. 2001. "Computer Networks as Social Networks." *Science* (14 September), pp. 2031–33.

Wennekers, Sander, L. M. Uhlaner, and R. Thurik. 2002. "Entrepreneurship and Its Conditions." *International Journal of Entrepreneurship Education* 1:25–64.

West, Joel, and S. O'Mahony. 2008. "The role of participation architecture in growing sponsored open source communities." *Industry and Innovation* 15(2):145–168.

Williamson, Ian, and Daniel Cable. 2003. "Organizational Hiring Patterns, Interfirm Network Ties, and Interorganizational Imitation." *Academy of Management Journal* 46(3):349–58.

Winborg, Joakim, and Hans Landström. 2000. "Financial Bootstrapping in Small Businesses: Examining Small Business Managers' Resource Acquisition Behaviors." *Journal of Business Venturing* 16:235–54.

Winer, Russell. 2000. *Marketing Management.* Upper Saddle River, N.J.: Prentice-Hall.

Winer, Russell. 2001. "A Framework for Customer Relationship Management." *California Management Review* (Summer), pp. 89–104.

Wingfield, Nick, and K. Lundegaard. 2003. "eBay Is Emerging as Unlikely Giant in Used-Car Sales." *Wall Street Journal* (7 February), p. A1.

Wolcott, Robert, and Michael Lippitz. 2007. "The Four Models of Corporate Entrepreneurship." *MIT Sloan Management Review* 49(1):75–82.

Wood, Robert C., and G. Hamel. 2002. "The World Bank's Innovation Market." *Harvard Business Review* (November), pp. 104–12.

Wright, Randall. 2008. "How to Get the Most from University Relationships." *MIT Sloan Management Review* 49(3):75–80.

Yakura, Elaine. 2002. "Charting Time: Timelines as Temporal Boundary Objects." *Academy of Management Journal* 5:956–70.

Yankelovich, Daniel, and David Meer. 2006. "Rediscovering Market Segmentation." *Harvard Business Review* (February): 122–31.

Young, Jeffrey S. 1988. *Steve Jobs—The Journey Is the Reward.* Glenview, Ill.: Scott-Foresman.

Zacharakis, Andrew, G. D. Meyer, and J. DeCastro. 1999. "Differing Perceptions of New Venture Failure." *Journal of Small Business Management* (July):1–14.

Zahra, Shaker, D. O. Neubaum, and M. Huse. 2000. "Entrepreneurship in Medium Size Companies." *Journal of Management* 5:947–76.

Zider, Bob. 1998. "How Venture Capital Works." *Harvard Business Review* (December), pp. 131–39.

Zimmerman, Monica, and G. J. Zeitz. 2002. "Beyond Survival: Achieving New Venture Growth by Building Legitimacy." *Academy of Management Review* 3:414–31.

Zott, Christopher, and Raphael Amit. 2007. "Business Model Design and the Performance of Entrepreneurial Firms." *Organization Science* 18(2):181–99.

Sample Business Plan

I-MOS SEMICONDUCTORS

I. Executive Summary

Introduction

I-MOS Inc. (I-MOS), a semiconductor intellectual property start-up, has developed a disruptive transistor technology that offers a 1000x reduction in static power dissipation and a 30 percent increase in chip performance. This technology solves a problem that has plagued the semiconductor industry for the past 30 years: how to increase the density of transistors on a chip without exponentially increasing the heat generated. The I-MOS solution not only dramatically reduces static power consumption and increases chip performance, but it does so without increasing semiconductor fabrication costs. In fact, this solution is fully compatible with all the existing tools and processes currently used in semiconductor fabrication.

I-MOS will commercialize this technology by licensing it to integrated device manufacturers (e.g., Intel, IBM, Motorola), fabless semiconductor companies (e.g., Xilinx, Qualcomm, Nvidia), and semiconductor foundries (e.g., TSMC, UMC, Chartered). I-MOS has already initiated discussions with Intel, Xilinx, and TSMC, and all three have expressed an interest in licensing this technology.

Market Opportunity and Solution

The semiconductor industry is expected to grow from $140 billion in 2002 to $240 billion by 2005.[1] The industry is currently plagued with the problem of how to increase the density of transistors on a chip without exponentially increasing the heat generated. This exponential increase in the "Static Leakage Problem" occurs because at the same time transistor threshold voltages have been scaled down substantially, escalating the leakage per transistor, the number of transistors per chip has dramatically increased. A significant opportunity exists for a high-performance, cost-effective, and scalable solution at the transistor level that addresses the Static Leakage Problem.

I-MOS has developed a disruptive transistor technology that reduces the static power dissipation by 1000x and provides a 30 percent increase in chip performance without increasing semiconductor manufacturing costs. The I-MOS solution is cost-effective, by working in conjunction with standard CMOS processes, and increases manufacturing efficiencies, due to lower device

---

[1] Goldman Sachs, *Technology: Semiconductors Industry Primer,* 2002.
Note: This business plan was prepard by Kailash Gopalakrishnan, Adam Wegel, Rajit Marwah, and Tod Sacerdoti for class credit while students at Stanford University in 2002.

variations. Two-thirds of the semiconductor industry, $93 billion in 2002 or $160 billion by 2005, can benefit from the I-MOS technology. However, I-MOS will initially target the segments of the industry that are particularly power and performance sensitive, including wireless devices, consumer electronics, graphics, and networking products. These segments currently represent approximately $50 billion today or $86 billion by 2005.

Business Model

The semiconductor intellectual property (IP) industry was an $890 million industry in 2002 and is growing at 25 percent annually.[2] I-MOS will contribute to this growth rate by utilizing a fabless nonexclusive IP-licensing business model in which we outsource all manufacturing and sales of the semiconductors incorporating our technology. Along with our IP, we also provide manufacturers with design libraries, SPICE models, and layout modifier software tools that seamlessly port existing chip designs to incorporate the new I-MOS technology.

We will work with our foundry partners to get our technology up and running in mainstream processes, and optimized for performance and static power, for every chip generation. We license our technology, on a nonexclusive basis, to integrated device manufacturers, fabless semiconductor companies and semiconductor foundries. We derive multiple revenue streams from each licensing arrangement: the manufacturers pay I-MOS an upfront license fee and a per chip royalty fee, and the foundries pay a per wafer fee.

Technology

The I-MOS technology was invented and patented by I-MOS Chief Technology Officer Kailash Gopalakrishnan under the guidance of Dr. James Plummer, Dean of the Stanford School of Engineering. I-MOS will have an exclusive license on the technology from the Stanford Office of Technology Licensing. Our technology has two applications: (1) it reduces static power dissipation by 1000X, has comparable dynamic power and has up to 30 percent increase in performance, or (2) it can reduce static power dissipation by 1000X, have comparable performance and reduce dynamic power by 20 percent over existing CMOS technology. These technological advancements are achieved by using breakdown voltage modulation, which reduces the subthreshold slope, resulting in much higher ON currents for performance and much lower OFF currents for lower static power than CMOS.

Current Status and Milestones

We have extensive device modeling and simulation results, and have demonstrated proof of concept with a successful initial run in silicon. Silicon-based prototypes were fabricated in the Stanford Nanofabrication Facility and have validated the I-MOS design.

---

[2] UBS Warburg, Global Equity Research, *At the Core of IT,* May 2002.

- **Milestone One:** We are planning to build silicon, germanium, and strained silicon, submicron devices to validate the scaling behavior of the I-MOS device shown in our simulations. We expect to have these experiments completed by June of 2003, coinciding with our first round of funding.

- **Milestone Two:** Our funding will allow us to demonstrate the I-MOS design in simple circuits, understand trade-offs, determine device characterization, study scaling properties, and sign up our first fab customer. We anticipate completing this milestone and raising our second round of funding by September of 2004.

- **Milestone Three:** We intend to demonstrate the I-MOS design on large SRAM test vehicle circuits, showing reliability, optimizing for performance and power, and signing up our first beta customers.

Financial Projections (millions)

	Yr 1	Yr 2	Yr 3	Yr 4	Yr 5	Yr 6	Yr 7
Revenue	$0	$0	$8	$19	$37	$73	$142
Net Income (AT)	($5)	($9)	($6)	($1)	$14	$18	$37
After Tax Margin	NA	NA	−68%	−6%	23%	25%	26%
Revenue Growth	NA	NA	NA	125%	98%	96%	95%

II. Market Opportunity

Summary:

- The semiconductor industry is enormous and growing.
- The semiconductor IP industry is experiencing rapid growth.
- The Static Leakage Problem has become a major concern for semiconductor companies.

The Semiconductor Industry is Enormous and Growing.

The worldwide market for semiconductor devices is large and growing, driven by the demand for electronic systems that are dependent on semiconductors. According to Gartner Research, the 2002 worldwide semiconductor market is $140 billion and is expected to grow to $240 billion by 2005.

Historically, the demand for semiconductor devices was met by vertically integrated semiconductor manufacturers, who designed, manufactured and tested their own products, in their own facilities, with their own tools. In the 1990s, the process for design and manufacturing grew in complexity and the cost of developing manufacturing facilities reached unmanageable levels, driving a disaggregation of the semiconductor industry. This disaggregation created a wave of growth of new semiconductor companies including fabless semiconductor chip designers, semiconductor equipment and tools vendors, and third-party semiconductor manufacturers, or foundries.

The Semiconductor IP Industry is Experiencing Rapid Growth.

In recent years, semiconductor design has increased further in complexity and geometries have rapidly decreased to meet market demand. Access to cutting-edge technology has become a competitive weapon for semiconductor manufacturers who are now forced to compete in an increasingly dynamic market. In the last year alone, the top 10 semiconductor companies spent over $12B on internal research and development. As semiconductor research has become more specialized, companies have begun to look to the outside for the latest technologies as a predictable way to generate R and D. These industry trends have accelerated the disaggregation of the semiconductor industry, fueling a new wave of growth in the market for semiconductor IP licensing companies. In 2001, semiconductor companies spent nearly $1 billion on IP licensing, and this is expected to grow at over 40 percent.[3] The total available market for IP licensing includes memory, microprocessors, microcontrollers, microperipherals, DSPs, ASICs and custom chips, representing two-thirds of the semiconductor industry, or $93 billion in 2002 and $160 billion by 2005.

The Static Leakage Problem has become a Major Concern for Semiconductor Companies.

For the past 30 years, semiconductor companies have been plagued with the problem of how to increase the density of transistors on a chip without exponentially increasing the heat generated. In the past, this "Static Leakage Problem" has been addressed by creating distance between the individual transistors in order to remove the concentration of heat or by finding ways to rapidly cool the chip when the heat is generated. In recent years, however, new geometries of semiconductors have been reduced at an increasing rate and the transistor threshold voltages, the voltage between the on and off positions at which the transistor is just considered on, have been scaled down substantially. As a result, the Static Leakage Problem per transistor has increased exponentially. In addition, the number of transistors per chip has dramatically increased. These factors are contributing to an exponentially increasing Static Leakage Problem.

> Shekhar Borkar, director of the Circuit Research Lab at Intel, said, "The [static leakage problem] is so bad that if current trends continue, future chips could reach 2,000 watts/cm^2, equivalent to a nuclear reactor . . . designers will have to trade off some performance to reduce power."

The industry has long been aware of the Static Leakage Problem, but it has been perceived to be the result of a fundamental principle of thermodynamics

---

[3] UBS Warburg, Global Equity Research, *At the Core of IT*, May 2002.

and, therefore, an impossible problem to solve. Semiconductor companies have addressed the problem by employing a variety of circuit workarounds, with varying degrees of success. One workaround used by circuit designers is to switch the power supply off from the circuits so that no static power may be dissipated when computing is done. This method is fundamentally inefficient due to the difficulty in determining which circuits are computing and which are not computing in a semiconductor. In addition, in a lot of high-density memory circuits such as SRAM and DRAM, it is not physically possible to switch the circuits off because these memories lose information when the power supply is switched off. In addition, the design complexity and cost of development increase significantly.

A more modern workaround proposed for future generations of chips is to design transistors with different threshold voltages on the same chip. For critical paths on the circuit, the high-speed (low threshold voltage) transistor may be used, but for the rest of the circuit the low speed (high threshold voltage) transistors would be used. Although this appears to be a convenient solution, it still results in a substantial reduction in performance. In addition, this solution introduces design complexity because circuit and layout designers have to design their circuits keeping two or more different transistors in mind and placing high-performance transistors judiciously in critical paths only. Every new transistor with a new threshold voltage necessitates additional process steps and new masks, which can increase the cost of making chips significantly. The design tools needed to accomplish these steps simply do not exist at this point.

> Intel Corp. plans to incorporate "sleep transistors" onto future-generation microprocessors to push clock frequencies higher and help tame the worsening leakage current that threatens high-speed processor designs (Article in the *EE Times* on June 14, 2002).

Several major market trends will exacerbate the Static Leakage Problem in the future. Static leakage per transistor has increased from 1 pA/μm in the 1990s to tens of nA/μm today and is projected to increase to many μA/μm by 2010 for high performance transistors. In addition, the number of transistors per chip is projected to increase well beyond the billion-transistor mark within the next few years. Coinciding with the technology market trends is the rapidly increasing demand for low power and high performance wireless devices with long battery lives, multimedia applications functionality, and many additional capabilities. Businesses and consumers are consistently demanding performance increases in laptops, PDAs, digital cellular phones, digital cameras, MP3 players, and other wireless products. Semiconductor companies are limited in their ability to deliver these performance improvements unless the Static Leakage Problem is addressed. I-MOS believes that the current market trends have created a significant opportunity for a high-performance, cost-effective, and scalable solution at the transistor level that addresses the exponentially increasing Static Leakage Problem.

Jeffrey Welser, IBM project manager for advanced CMOS, said, "We're scaling our gate thickness on our devices to less than 10 angstroms for future generations. The leakage to that is just unbelievably high. So we're looking furiously for new materials throughout the industry."

III. Solution

Summary:
- Dramatically reduced static power dissipation
- Increased chip performance
- Cost-effectiveness
- Increased manufacturing efficiencies

I-MOS has developed a disruptive transistor technology that reduces the static power dissipation by 1000x and provides a 30 percent increase in chip performance without increasing semiconductor manufacturing costs. I-MOS is focusing the application of its revolutionary technology on the Static Leakage Problem and, specifically, in chips manufactured in strained silicon and germanium, where the technology is most effective.

Since the transistor is the fundamental building block of a semiconductor, the I-MOS technology has a wide range of applications. However, the semiconductor markets that drive wireless devices such as laptops, PDAs, and cell phones, are most affected by the Static Leakage Problem. Imagine laptop batteries that lasted through multiple plane flights or PDAs that could play movies—I-MOS technology will push semiconductor driven products to new levels of performance. The key benefits of the I-MOS solution are:

Dramatically Reduced Static Power Dissipation

I-MOS technology reduces the overall static power dissipation of the chip by solving the problem at the fundamental level—by reducing the static leakage current per transistor. The I-MOS technology reduces the static leakage of the transistor to a level 1000x less than standard CMOS transistors.

Increased Chip Performance

The I-MOS technology enables customers to achieve up to a 30 percent increase in performance over CMOS semiconductors with comparable levels of dynamic power. Alternatively, if dynamic power is the main concern of our customer, the I-MOS technology can deliver a 20 percent reduction in dynamic power over CMOS with a comparable level of performance.

Cost-Effectiveness

Application of the I-MOS technology does not significantly impact manufacturing costs. The I-MOS technology is designed to work in conjunction with standard CMOS processes and the only necessary manufacturing modification is to selectively replace standard transistors from the design with the I-MOS transistors.

Increased Manufacturing Efficiencies

Typically, design changes that are necessary for higher performance result in a reduction in semiconductor yields. Due to our unique technology, our customers can achieve both higher performance and a higher yield, thereby generating a lower cost of goods sold. This manufacturing efficiency stems from the fact that there are certain variations from wafer to wafer, die to die and within the die itself. These variations cause certain transistors to be leakage prone and others to be slower. I-MOS devices have much lower variations as compared to standard CMOS devices, and this guarantees significantly lower development time, a shorter learning curve and higher yields.

IV. Business Model and Strategy

Summary:
- Complete technology development
- Develop strategic relationships with foundries
- Enable products in large and rapidly growing wireless markets
- Generate revenue through a combination of upfront license fees and royalties
- Build out our patent portfolio

Our goal is to establish the I-MOS transistor technology as the standard for high-performance semiconductors in the market. The broad application of this technology has led us to pursue a capital-efficient, IP-licensing business model in which we neither manufacture nor sell semiconductors incorporating our technology. We license the I-MOS technology, on a nonexclusive basis, to integrated semiconductor manufacturers, foundries and fabless semiconductor companies. This business model is scalable, has low fixed and variable costs, has high switching costs for our customers and is synergistic as we grow, in that it allows us to make interproduct sales and upgrade customers on relative products.

The disruptive transistor technology that I-MOS has developed can add value to approximately two-thirds of manufactured semiconductors. We will initially target the segments of the industry that are particularly power and performance sensitive, including: wireless devices, consumer electronics, graphics, and networking products. These segments represent approximately $50 billion in 2002 or $86 billion by 2005.

The I-MOS business model and strategy are designed to establish the I-MOS transistor as the standard transistor for high-performance semiconductors in the market and is based on achieving the following milestones.

Complete Technology Development

I-MOS has developed a revolutionary transistor technology that solves the Static Leakage Problem in the lab. We have extensive device modeling and simulation results that show this technology works and have done an initial run in silicon to show the proof of concept. In order to demonstrate this technology and get it up and running in the marketplace, we must first study scaling properties, demonstrate circuits, understand trade-offs and device characterization, optimize for speed and get our LMII layout tool completed.

Develop Strategic Relationships with Foundries

I-MOS must develop strong relationships with foundries today in order to incorporate our technology into their mainstream processing when technology development is complete. Foundries benefit from this relationship because they will be able to provide the most current transistor technology to their customers and therefore increase margins. Our value proposition to the foundries is powerful due to the minimal costs associated with bringing our technology on board. The strength of our relationship with the foundry will influence our ability to derive royalty fees from their customers as well, a common practice in the industry. An important element of our strategy may be developing strong relationships with the fabless semiconductor companies in order to create pull-through demand to influence the adoption of the I-MOS technology.

Enable Products in Large and Rapidly Growing Wireless Markets

I-MOS is initially targeting the largest and most rapidly growing segments of the wireless device market for our transistor technology. We believe these segments are the most likely to adopt this new technology because we enable performance advantages that can help them compete in the marketplace. These large and high-growth wireless markets include cell phones, PDAs, portable MP3 devices, digital cameras and handheld video games. Target customers in these markets include Broadcom, Texas Instruments, Qualcomm, SST Corp., Cirrus Logic, and others.

Generate Revenue Through a Combination of Upfront License Fees and Royalties

We intend on generating revenue from three types of companies—foundries, semiconductor manufacturers, and fabless semiconductor companies. Initially, we will work with foundries to bring our technology into their mainstream processes as we prove our technology and demonstrate scalability. After completing development in the foundries, we will license our technology to both fabless and integrated semiconductor companies who will generally pay a multimillion dollar upfront license fee to I-MOS for the broad use of our technology in their semiconductors. These fees may hinge on specific milestones for deliverables from I-MOS or the production of chips by the company. Initially, most engineering costs associated with bringing a customer on board will be consumed by I-MOS, but as we develop our customer base this will become a revenue generator for the Company. We hope to use this up-front fee structure as a capital efficient way to fund future development.

Both foundries and licensees will be responsible for paying I-MOS royalties on sales occurring throughout the life of the I-MOS technology being licensed. Foundries will pay I-MOS a per wafer fee for all wafers used in production. Fabless licensees will pay I-MOS a per chip fee, usually corresponding to a percentage of the cost of the chip. Integrated manufacturers will pay a per chip fee as a percentage of cost as well, but the fee structure will be slightly higher. Royalty rates will typically range from 2.5 percent–5 percent of the average selling price (ASP) of the chip; however, a wide variety of factors will impact the royalty rate including size of the up-front fee, number of chips in production, and many others.

Build Out Our Patent Portfolio

A key element of the I-MOS strategy is the development and protection of our intellectual property. We will aggressively pursue both offensive and defensive patents surrounding the core I-MOS transistor patent and will maintain a focus on building out our patent portfolio.

Case Study #1: Xilinx Inc.

Xilinx, Inc., founded in 1984 and headquartered in San Jose, California, is the leading supplier of complete programmable logic solutions, including advanced integrated circuits, software design tools, predefined system functions delivered as cores, and unparalleled field engineering support. Xilinx is the world-leader in FPGA-based products and its PLD solutions have more than 50 percent market share. Xilinx currently uses the most advanced 130 nm UMC process and it has 250M transistors in its high-end product 2VP125 (about 5X greater than Pentium IV). In an FPGA, a good fraction of transistors are unused in the design process but all the transistors contribute to static leakage. We talked to an advanced circuit researcher from Xilinx, who told us, "Static power dissipation is our biggest problem scaling forward because our customers always want higher performances. We would be very interested in looking at new technologies that can help us alleviate this problem. Reducing static power would also help us to expand our market by replacing ASIC's in cellphones."

Case Study #2: Intel Inc.

Intel, Inc., is the world leader in semiconductor manufacturing technology and supplies the computing and communications industries with chips, boards, systems, and software building blocks that are the

(continued on next page)

(continued from page 518)

"ingredients" of computers, servers and networking and communications products. Intel's products are always at the cutting edge of technology: its highest-end Pentium IV processors contain more than 50M transistors, and Intel has devoted a significant fraction of its R and D budget in device and circuit related solutions to the problems of static power dissipation. The director of the Circuit Research Lab at Intel noted, "The static leakage problem is so bad that if current trends continue, future chips could reach 2,000 watts/cm^2, equivalent to a nuclear reactor." We talked to an advanced process-integration group manager at Intel who informed us that "Undoubtedly, the static leakage problem seems to be one of the biggest roadblocks for deep submicron transistor scaling. We are constantly looking for new solutions that can push us to a better position on the static power versus performance curve."

V. Sales and Marketing Plan

Summary:

- Use direct technical sales force to target fabs and end users
- Offer support as well as technology

Our sales and marketing activities will be primarily focused on establishing and maintaining licensing arrangements with foundries, semiconductor manufacturers, and fabless semiconductor companies. Our sales strategy is to pursue targeted customers through the combination of our direct sales force and our strategic alliances. Since we generally license our technology to foundries that, in turn, sell products incorporating these technologies to fabless semiconductor companies, the foundry alliances we form will be an integral part of our sales strategy. In establishing these alliances, we seek partners who we believe will allow us to grow the overall market for our I-MOS transistor technology.

Use Direct Technical Sales Force to Target Fabs and End Users

Our marketing activities will also be directed toward the end product manufacturers. Through targeted advertising and co-marketing efforts, we will be focused on increasing the awareness of the I-MOS technology and generating interest from potential end customers. We believe these efforts will create demand for our product from the product manufacturers, which will create pull-through demand to the semiconductor manufacturers who will end up licensing our technology.

Offer Support As Well As Technology

Our goal will be to establish ourselves as an important partner to our customers and to develop relationships both at the executive level and at the engineering level. We anticipate dedicating substantial resources toward marketing to customers, implementing our technologies, and supporting our customers. This will include on-site engineering support, other technical support, trade shows, and more traditional marketing activities. We believe a close working relationship with our customers will allow us to identify new product areas and technologies to focus on for our future research and development efforts.

VI. Technology Overview

Summary:
- CMOS technology is thermodynamically limited to 60 mV/decade.
- I-MOS transistor physics achieves 5 mV/decade.
- No other solution offers comparable performance.
- Technology roadmap developed for future products and services.

CMOS Technology is Thermodynamically Limited to 60 mV/decade

The scaling of transistor feature sizes has demanded that chip supply voltages be continually shrunk in order to keep the dynamic power dissipation within tolerable limits for reliable operation. Reducing this supply voltage, while maintaining transistor performance, requires the threshold voltage of the transistors to be reduced in accordance with the supply voltage. One good metric of how much a transistor leaks in the "off" state is the subthreshold slope which describes the rate of change of the current with respect to the voltage below the threshold. In conventional transistors, this subthreshold slope is limited by basic thermodynamic principles to 60 mV/decade, which means that the transistor current changes by one order of magnitude for every 60 mV change in the voltage. Therefore, reducing the threshold voltage increases the leakage current of the transistor exponentially. In addition, chip temperatures have steadily increased in the past decade, thereby dramatically increasing the leakage current as well. The number of transistors per chip has doubled every two years for the past 30 years, and all these factors contribute to a dramatic rise in the static power dissipation in chips. These levels of static power dissipation are unacceptable in most applications and may impose fundamental limitations on transistor scaling itself.

I-MOS Transistor Physics Achieves 5 mV/decade

I-MOS has developed a disruptive transistor technology (Impact Ionization MOS, or I-MOS) that reduces static power dissipation by 1000X over conventional transistors and delivers up to 30 percent enhancement in performance. This dramatic reduction in static power dissipation stems from a *fundamental breakthrough in transistor physics.* Conventional CMOS transistors work by modulating the charge in the channel by using a process called drift-diffusion controlled by a gate terminal. The subthreshold slope in CMOS-based systems is thus thermodynamically limited to kT/q or 60 mV/decade at room temperature. The I-MOS works by avalanche breakdown voltage modulation in specifically designed p-n junction diodes. The avalanche breakdown process is an extremely abrupt and fast process. This built-in gain mechanism amplifies the nonlinearity of the system resulting in a subthreshold slope much lower than kT/q (~5 mV/ decade or lower) (fig. 1). We believe that I-MOS is the only technology that offers this huge reduction in static power and is the only semiconductor device in the world that has reduced the subthreshold slope below the thermodynamic limit of kT/q.

The 30 percent enhancement in performance is obtained by a combination of a number of factors. As mentioned earlier, the impact ionization process is an extremely steep and fast process and enhances the drive current in a transistor by about 20 percent over CMOS. In addition, the avalanche process causes a reduction in the swing at the output of the transistors by about 20 percent, which increases transistor speed. This enhancement in transistor performance is significant considering that it takes the industry 2–3 years to scale to a new transistor technology to enhance performance by 30 percent.

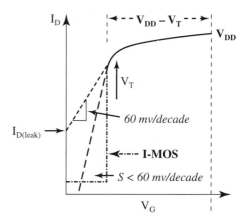

FIGURE 1 Comparison of the subthreshold slope of conventional transistors and the I-MOS and its implications on leakage.

No Other Known Future Technology Offers Comparable Performance

GaAs and InP based transistors provide high performance but also suffer from high static leakage. I-MOS is the only solution that can provide both higher levels of performance and dramatically lower static power. In addition, GaAs and InP are new materials and have more complex and expensive fabrication processes, which would likely require an expensive redesign of all current fabrication facilities. In contrast, the basic I-MOS concept works in any material. Our modeling work has shown that the best material for the I-MOS transistor would be strained silicon or germanium. IBM, Intel, AMD, TSMC, UMC, and the rest of the semiconductor industry, have announced their intentions to introduce strained silicon in mass production before 2005.

The static power dissipation problem is a lot worse especially if one considers all the process corners under which one would like the chip to operate. The worst-case process corner typically operates with 100 times more static power dissipation than the typical case. Since these levels of static power are unacceptable, this would end up directly affecting the yield. In addition to reducing this static power, the I-MOS also has about three times lower variability as compared to CMOS. This can improve the yield significantly for chips incorporating the I-MOS technology. A comparison is illustrated in fig. 2.

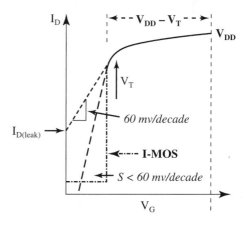

FIGURE 2 Comparison between the static power and performance trade-offs for CMOS, GaAs, and I-MOS transistors over all process corners.

Technology Roadmap Developed for Future Products and Services

Our core competency is the transistor technology related IP and the design tools that we offer to incorporate this I-MOS technology in any chip design. Typically, any transistor technology has three phases in its lifetime. The initial phase lasts for a couple of years and requires substantial research and development on the technology, process optimization and reliability testing of the transistor. After this phase, the technology is used in high performance chips and other designs that really need the extra computing power and reduced static power. In the final phase, which lasts for about three years, the technology is used in lower-end chips that are designed for reduced cost. We estimate that the I-MOS transistors that we introduce at every technology node would go through similar phases.

After doing extensive research and development from 2003–2005 on the 0.080 μm technology node, we plan to introduce the I-MOS design into production around 2005. We also plan to introduce advanced generations of I-MOS transistors in 2007, 2010, 2013, and 2016 for the 0.065 μm, 0.045 μm, 0.032 μm, and 0.022 μm technology nodes corresponding to the roadmap outlined by the International Technology Roadmap for Semiconductors (ITRS).

In addition, we envision expanding to have direct products in the future because the fundamental technology has wide ranging implications in a number of diverse fields. After having proven the technology in a number of diverse markets, we plan to introduce SRAM- and DRAM-based memory products along the way in parallel to pursuing licensing to other semiconductor markets. Furthermore, the I-MOS technology also has significant applications in the power semiconductor industry including power amplifiers, synchronous rectifiers, ESD protection, and we plan to introduce products that satisfy the needs of those markets as well.

One more significant breakthrough in computing stems from the optoelectronic-based applications of the I-MOS transistor. I-MOS provides a good device template in photodetector applications and can provide a gain-bandwidth product that is about a 1,000 times higher than conventional CMOS/GaAs based detectors. This can solve a very critical problem associated with clock distribution on chip. Optics, which has been proposed as an alternative to conventional techniques of electrical clock distribution in any ASIC or microprocessor chip, are becoming increasingly difficult because of increasing chip areas, rising clock frequency, and increased numbers of latches. The proprietary optoelectronic I-MOS technology is the only solution that can provide optoelectronic detectors for optically clocking integrated circuits. We plan to license our proprietary optical I-MOS solutions for on-chip clocking and chip-to-chip signalling to all partners, fabs and customers who need this technology for increased scalability in their high frequency designs. Fig. 3 summarizes all the products that I-MOS plans to offer and their rough time frames. We believe that this rich variety of products in both the IP licensing and product space will strongly position the company to scale upwards in accordance with Moore's law well into the twenty-first century.

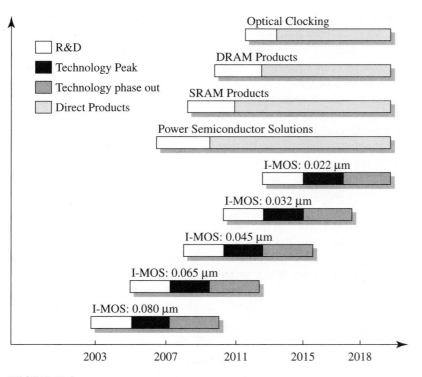

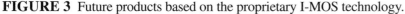

FIGURE 3 Future products based on the proprietary I-MOS technology.

VII. Technology Implementation

Summary:
- Compatible with existing fab processes
- Partner with fabs to implement technology
- Provide tools and support in addition to technology

Compatible with Existing Fab Processes

I-MOS is fully compatible with all of the current manufacturing tools and processes used for making strained silicon transistors. The footprint (or the area) occupied by the I-MOS transistor is exactly the same as the area occupied by conventional CMOS transistors for the same generation. In other words, there is no extra cost to be incurred in implementing the I-MOS technology. Most of the differences in the I-MOS device and structure are absorbed in the lithography masks and in the implant and anneal steps that are used in processing MOS transistors. We expect that the worst-case scenario would be the addition of a couple more masks for extra implant / spacer steps in the process of optimizing the device. One more distinct advantage of the I-MOS transistor is that its higher speed over CMOS permits us to go backward one technology generation for the same level of performance and much reduced static power. This permits us to cut down on the ever-increasing costs of manufacturing transistors and makes I-MOS chips much cheaper to fabricate than current CMOS chips for lower power and comparable or higher levels of performance and integration.

In low power chips for portable applications, in addition to reducing static power, the I-MOS could also be tweaked so that higher performance may be traded off for reduced dynamic or active power. We envision that we may be able to reduce the dynamic power by 20 percent or more for comparable levels of performance over CMOS transistors.

Circuit design using the I-MOS technology is only marginally different from conventional circuit design using CMOS transistors, and the slight difference stems from the fact that the output voltage range of a logic block containing the I-MOS transistor does not span the entire supply voltage range. This reduced swing at the output helps enhance performance and reduce dynamic power dissipation but comes at the expense of reduced noise margin, which is used to differentiate logic levels "1" and "0" from each other. We believe that this penalty can be easily accomodated because of lower spread in the I-MOS transistor thresholds. I-MOS also holds intellectual property rights for the circuit design methodology using the I-MOS transistor.

Partner with Fabs to Implement Technology

The implementation strategy would be to do joint development with the foundries and have the technology in the base-line process of the fabs. I-MOS will work with the foundries to understand the scalability of the transistor and develop simple circuit modules incorporating the I-MOS transistor as part of their first year milestones. We will then develop high-density and large-array SRAM test vehicles to understand the performance and static-power trade-off in these devices. We will also be optimizing the device specifications to maximize performance and minimize standby power. As stated earlier, the I-MOS technology is fundamentally compatible with all the existing tools and processes in IC fabrication. The I-MOS fabrication would use the conventional mask sets and would lithographically modify the ion-implantion and spacer steps that the I-MOS device would receive. This is evidenced by the fact that conventional transistors are n-p-n while the I-MOS is a specially designed p-i-n. This is necessary for implementing the breakthrough physics that fundamentally alters the device characteristics. The conventional transistors, also fabricated simultaneously on the chip, would be processed as they are normally done. Therefore, it is possible to fabricate I-MOS devices and conventional CMOS transistors on the same chip. In the process of the optimizing device characteristics for performance and reliability, we anticipate the need to add up to two more mask steps. This is relatively inexpensive, considering that most conventional IC processes today require about 33–40 mask steps. In addition, we believe that with the wide-spread acceptance of this technology

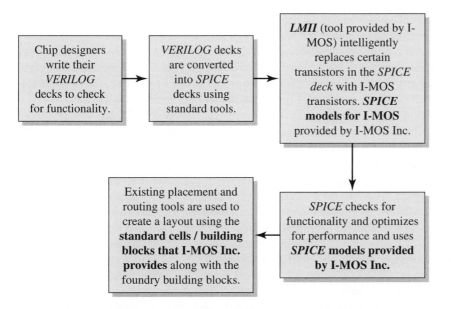

FIGURE 4 Schematic showing how semiconductor companies would use the I-MOS technology for the products.

and with process optimizations, we may be able to eliminate the low power (high V_T) and the low standby power (high V_{DD}, high V_T) transistors on the layout and cut down on the number of mask sets by about 5–10. Since a new scaled technology is introduced every 2–3 years, I-MOS would work with the foundries for developing a scaled I-MOS technology for every generation. I-MOS would then jointly work with these foundries to license this technology to their customers.

Provide Tools and Support in Addition to Technology

In addition to the I-MOS transistor related intellectual property, we also offer design tools, standard SPICE models for these transistors, and layout related tools for incorporating the I-MOS technology within new and existing designs. These tools would selectively replace certain transistors in the layout with I-MOS transistors, perform circuit-level optimizations, and help in seamlessly porting existing chip designs to take advantage of the I-MOS technology. A general flow chart showing how companies would use this technology is shown in fig. 4. All the procedural steps shown in fig. 4 are completely automated so that it is extremely easy to exploit the benefits of this technology.

VIII. Research and Development

I-MOS believes that its future success will depend critically on the continued development and introduction of new, scaled, and improved transistor technologies in accordance with or faster than Moore's law. To this end, company engineers are involved in developing new versions of the I-MOS technology that will allow higher performance, lower power, and higher levels of integration. In addition, we plan to have a highly-qualified technical advisory board comprised of professors and industry stalwarts from Stanford University, MIT, and SUN Microsystems, and we believe that this multidisciplinary expertise of our team of scientists and engineers will continue to advance our technological leadership and market.

I-MOS would intitally develop its transistor technology in one foundry (e.g., TSMC, TI, UMC) before expanding to satisfy the needs of customers in different foundries. The company plans to do extensive research and development on the scalability and the reliability of the transistor, gauging the power and performance in circuit frameworks and then optimizing the transistor for maximum performance. This would be carried out after extensive presentations and negotiations with the technology group in the foundries. We anticipate that the I-MOS process would require minimal process changes over the standard base-line strained-Si CMOS process in the foundry. After developing the technology and doing extensive reliability testing by March 2005, we plan to sign on beta customers (e.g., Xilinx, Nvidia) who use the foundry for their own silicon processing needs. These beta customers would be invaluable in providing direct feedback between technology development and client needs. After having demonstrated our value proposition to the above customers, we then anticipate that we will be able to turn some of them into permanent customers who would need the I-MOS technology to satisfy the emerging needs for low power and higher performance. We believe that I-MOS is strategically positioned to address the evolving and highly demanding needs of these markets.

IX. Intellectual Property

The I-MOS technology was originally developed as part of a DARPA-funded project at Stanford, and the patent for the I-MOS transistor is owned jointly by Stanford University, I-MOS CTO Kailash Gopalakrishnan, and Prof. James D. Plummer (Dean of the Stanford School of Engineering). Currently, the Office of Technology Licensing at Stanford University holds exclusive licensing rights to this technology. I-MOS believes that we can license this technology on an exclusive basis from Stanford University. Recent conversations with OTL have justified this expectation, and there are many examples of this licensing structure including T-RAM, Inc., Pixim, Inc., and Google.

As an IP licensing company, we believe that our patents, copyrights, mask work rights, trademarks, and trade secret laws are critical for our success and to that end the company will have an active program to protect its proprietary technology through the various types of filings mentioned above. In addition, the company would attempt to protect its trade secrets and other proprietary information through agreements with licensees and foundries, proprietary information agreements with employees and consultants, and other security measures. We estimate that we may have to file up to two U.S. patents a month and would have about a $500,000 or more in legal expenses in the initial years of development including the initial setup charges. We also seek to protect our software, documentation, and other written materials under trade secret and copyright laws.

X. Competition

I-MOS's competitors include integrated device manufacturers (IDMs) such as Intel and IBM, fabless semiconductor companies such as Sun and Transmeta, and foundries such as TSMC and UMC. Of these competitors, Intel, IBM, and Sun pose the largest competitive threat. These three companies have a combined annual R and D budget of over $10 billion and devote a significant portion of it to creating technologies that will make semiconductors dissipate less power, run faster, and cost less.

Specifically, the problem of reducing static leakage has long been recognized by the semiconductor industry. However, it has been perceived as an impossible problem to solve. Some of these companies claim that any potential solution is limited by the basic principles of thermodynamics itself. Therefore, many of these companies have focused their efforts on circuit workarounds, each with some degree of success.

The following is a summary of the primary competitors and a description of the competing solutions that each is pursuing:

- **Intel Corporation**—Intel has been researching "sleep transistor" technology in hopes of solving the issue of static leakage current. With this solution, the power supply itself is switched off when the transistor is not being used. Sleep transistor technology can provide a thousand-fold reduction in leakage power. The primary advantage of this technique is that it does not affect the basic transistor design and functionality but rather focuses on the circuit core instead. However, the technique is very hard to implement due to the fact that it is very difficult to determine which circuits are computing and which are not computing on a microprocessor. Hence, this kind of approach may have to be applied to larger blocks, which reduces the efficiency of the process in reducing static power. In addition, the design complexity goes up as well, which translates directly into an area and cost penalty. Also there is a loss of performance because the circuits which are switched off need a finite time to switch back on. It should also be noted that it may actually not be possible to switch the power off from certain circuits such as SRAM and DRAM because these memories lose information when the power supply is switched back on.

- **International Business Machines**—IBM has been researching silicon-on-insulator (SOI) which offers lower capacitance and, hence, increased speed. SOI is a technique that increases both dynamic power and performance by reducing the diffusion capacitance of transistors. In its native form, however, it does not affect static power. This technology is compatible with I-MOS, meaning that a wafer utilizing both SOI and I-MOS will experience performance improvements that are a summation of the improvements that would be realized from each technology independently.

■ **Advanced Micro Devices**—AMD, in conjunction with IBM, has been developing the Fin Field Effect Transistor (FINFET). This transistor utilizes a thin vertical "fin" to help control leakage of current through the transistor when it is in the "off" stage.[4] The FINFET is the only device solution that allows CMOS to scale to the end of the perceived roadmap. FINFET utilizes the same device physics as conventional CMOS but is nonplanar and hence requires different fabrication steps. Since FINFET is based on CMOS physics, it is still limited to 60mV/dec. The Intel DST (Depleted Substrate Solution) is also limited by the same principles to 60 mV /dec.

XI. Financials

Projected income statement

	Year 1 Aug-2004	Year 2 Aug-2005	Year 3 Aug-2006	Year 4 Aug-2007	Year 5 Aug-2008
Revenue	—	—	8,370,000	18,825,000	37,260,000
Cost of Revenue	—	—	—	—	—
Gross Profit	—	—	8,370,000	18,825,000	37,260,000
OPERATING EXPENSES					
Sales	—	674,000	1,988,900	3,105,094	3,651,840
Marketing	—	108,000	150,480	159,509	169,079
Engineering	4,029,907	6,884,609	10,749,401	15,464,000	18,501,490
General & Admin.	877,000	1,137,000	1,155,540	1,175,192	1,196,024
Total Oper. Exp.	4,906,907	8,803,609	14,044,321	19,903,795	23,518,433
Total Non-Op. Inc. (Exp.)	—	—	—	—	—
Pre-Tax Profits	(4,906,907)	(8,803,609)	(5,674,321)	(1,078,795)	13,741,567
Taxes	—	—	—	—	—[5]
Net Income	(4,906,907)	(8,803,609)	(5,674,321)	(1,078,795)	13,741,567
Net Margin	n/a	n/a	–68%	–6%	36.9%

Key Revenue Drivers

Licensing Structure, Average Selling Price, and Volume.
We model revenues based on a licensing structure that consists of (1) an average, one-time $1.5M license fee for access to the I-MOS patent portfolio for a

---

[4] Advanced Micro Device website.

[5] Although we expect to generate positive profits in Year 5, we do not anticipate paying taxes on the profit since the cumulative losses carried over from previous years exceed the expected Year 5 profit. As a reality check, our net margin in Year 5 would be 22.9 percent, consistent with the net profit margins of ARM and Rambus near 25 percent.

5-year term, and (2) an average royalty fee of 3 percent of the average selling price of all chips incorporating the I-MOS technology. We believe these are reasonable assumptions given the licensing structure of similar semiconductor IP licensing companies (see table in Appendix). We model an average price per chip at $3.50 based on ARM's average royalty per chip ($0.07) divided by the average royalty percentage per chip (2%). Since ARM licenses IP to many industries including mobile phones, storage, and consumer electronics, we believe $3.50 represents a reasonable "average" selling price per chip for us.[6]

Timing and Frequency of Customer Acquisition.

We do not anticipate receiving royalty revenue from our first customers until Q1 of Year 3, when the technology will have been developed for high scalability in commercial applications and the sales team will have had 3–6 months to land our first customer. The timing and frequency of customer acquisition reflects our strategy of first convincing fabs to integrate our technology into their processes and then having them walk us into their customers. We project multiple fabless customers relative to fabs because we expect each fab to have more than one customer. Supporting fabs is very costly and requires 8–10 dedicated engineers per fab. A table of our customer acquisition plan is included in the Appendix.

Key Expense Drivers

Headcount.

The chart below summarizes headcount growth for the first five years.

	Year 1	Year 2	Year 3	Year 4	Year 5
Sales	–	5	9	13	13
Marketing	–	1	1	1	1
Engineering	10	30	44	56	68
Genl. & Admin.	3	3	3	3	3
Total	13	39	57	73	85

Sales.

We anticipate hiring our first salesperson in Year 2 once technical risk has been sufficiently removed, which is also 1 year prior to when we expect our first customer to ship. We then plan on hiring one salesperson every 3 months (one a quarter) thereafter.

Engineering.

Based on conversations with the founder of a comparable early stage semiconductor company, we have estimated major development costs of:

■ A mask set per fab per year at $1.5M. Even after signing a customer, developing mask sets are essential to work on the next generation of chip and I-MOS technology.

---

[6] UBS Warburg, Global Equity Research Report, *At the Core of IT,* May 2002.

- Wafer costs of $1M in the first year, increasing thereafter. In the first month we estimate needing 30 wafers at a cost of $2500 per wafer and allocated a 2 percent increase in the cost per month to reflect rising costs of development over time.
- Leased testing equipment and software at $1.2M a year.
- Beginning in the second year, we plan to ramp development aggressively and hire two engineers a month for nine months, followed by hiring one additional engineer a month thereafter to support continued technology development.

Legal Expenses.
See Section VIII, "Intellectual Property," for a breakdown of these expenses.

Proposed Company Offering:

- Series A Financing—$6.5 million by September 2003
- Series B Financing—$8.5 million by September 2004
- Series C Financing—$9.0 million by September 2005

Total Anticipated Capital Requirements = $24.0 million

Series A:
I-MOS is currently seeking $6.5 million in Series A financing by September 2003. The key milestone we intend to achieve with the capital in this round is:

- Advancing the I-MOS technology from its current state to demonstrate proof of concept in simple devices and simple circuits that consist of less than 100,000 transistors.

The majority of this capital will be spent on technology development. We anticipate that the proceeds from this round will last us 15 months until November 2004 and will enable us to grow to 13 employees: 10 engineers and three G&A (two founders, one office manager).

Series B:
By the end of September 2004 (month 13), we intend on raising $8.5 million in Series B financing. In case securing the Series B financing takes longer than expected, we have built in a three-month cushion from our Series A financing. The milestones we intend to achieve with the capital in this round are:

- Demonstrating the I-MOS technology in bigger circuits and optimizing the technology for power and/or performance.
- Demonstrating reliability and verifying customer specifications.
- Signing first customer.

The majority of this capital will be used to triple the number of engineers and recruit our first salespeople. We anticipate that the proceeds from this round will carry us through month 27 (November 2005) and will enable us to grow to 39 employees.

Series C:

By the end of September 2005 (month 25), we intend on raising $9.0 million in Series C financing. In the event that securing Series C financing takes longer than expected, we have built in a three-month cushion from our Series B financing. The milestones we intend to achieve with the capital in this round are:

■ Revenue from first fab and fabless customer.
■ Cash flow breakeven and profitability.

The majority of this capital will be used to finance the expansion of the organization as we ramp up our sales and engineering teams in anticipation of rapid customer growth. We anticipate that this will be our last private round of financing and that the proceeds from this round will carry us through to cash flow breakeven and profitability in Q4 of Year 4.

Capitalization and Investor Return

In forecasting the future capitalization structure of the company and investor return, we assume that the company goes public at the end of Year 5 and offers 20 percent of its shares to the public. Assuming we meet our projections and value the company using the current industry median price to earnings multiple of 33, then I-MOS would be valued at the end of year 5 at $13.7 MM net income * 33 = $450 MM (see Appendix for public company comparison metrics).

Capitalization structure

Premoney Valuation	$6.5	$35.0	$100.0	
Invested Capital	$6.5	$8.5	$9.0	
Postmoney Valuation	$13.0	$43.5	$109.0	$450.0

Stakeholders	Post Financing Ownership Levels				
	Founding	Series A	Series B	Series C	Exit
Founders	100.0%	35.0%	27.0%	24.3%	19.5%
Employees		15.0%	15.0%	15.0%	12.0%
Series A Preferred		50.0%	38.5%	34.8%	27.8%
Series B Preferred			19.5%	17.6%	14.1%
Series C Preferred				8.3%	6.6%
Public Market					20.0%
Total Ownership	100.0%	100.0%	100.0%	100.0%	100.0%

Projected investor return

	Series A	Series B	Series C	Total
Years until Exit	5	4	3	
Invested Capital	$6.5	$8.5	$9.0	$24.0
Invested Capital Value upon Exit	$125.2	$63.5	$29.7	$218.4
IRR	80.7%	65.3%	48.9%	72.1%

XII. Management

Adam Wegel

Chief Executive Officer

Adam Wegel comes from Delphi Automotive Systems with over six years of varied experience in operations, manufacturing engineering, product development, finance, and sales and marketing. Most recently at Delphi, Adam developed a corporate strategy for selling telematics to the commercial fleet vehicle industry.

Adam is a Master in Business Administration candidate (March 2003) at the Stanford Graduate School of Business and a Master of Science in Mechanical Engineering candidate (March 2003) at the Stanford School of Engineering. Adam received in Bachelor of Science degree in Mechanical Engineering from North Carolina State University.

Kailash Gopalakrishnan

Chief Technology Officer

Kailash Gopalakrishnan comes from T-RAM, where he worked on various device and circuit issues for a novel memory product. At T-RAM, Kailash also invented two modified memory cells that were later patented. These key inventions helped the company solve the static power dissipation problem in their memory cells.

Kailash is a Stanford Graduate Fellow in the Semiconductor Devices Group at Stanford University, where he is pursuing his PhD in Electrical Engineering (expected September 2003). Kailash and his advisor, Dean James D. Plummer, have filed co-patents for various aspects of I-MOS technology.

Rajit Marwah

Chief Financial Officer

Rajit Marwah most recently comes from Echelon Corporation, where he researched worldwide vendors, technology, competition, international frequency regulations, and solution costs for the emerging low power, low data rate, medium range wireless data market (ZigBee, UltraWideband) for use in automatic meter reading (AMR) and other applications. Prior to Echelon, Rajit worked as an Associate for two summers at TL Ventures, a

$1.4 billion early stage technology venture capital firm, assisting partners in due diligence and structuring deals.

Rajit is a Master of Science in Management Science and Engineering and Bachelor of Arts in Economics candidate (June 2003) at Stanford University. Rajit is also a Vice President of BASES, Stanford's largest entrepreneurship organization, with over 4,500 members.

Tod Sacerdoti
Chief Marketing Officer

Tod Sacerdoti comes from Robertson Stephens, where he was a Corporate Finance Investment Banker focused on the Communication Infrastructure and Internet sectors. Working closely with research analysts, he managed two deals and developed strategic road maps, business model positioning and road show presentations for private technology companies. He was also the founder and CEO of DK Entertainment, a successful event production and marketing company.

Tod is a Master of Business Administration candidate (June 2003) at the Stanford Graduate School of Business. He received his Bachelor of Arts in Economics at Yale University, where he was President of Sigma Alpha Epsilon fraternity.

XIII. Risks and Mitigations

Risk	*Mitigation*
Market Risk	
The primary market risk is that we are unable to establish licensing contracts for our IP. While there are examples of very successful semiconductor IP companies, the number of such companies is fairly limited.	If we are unable to secure a licensing deal within six months of meeting our second milestone (proving the technology), we will consider taking a product approach and selling a relatively simple semiconductor product such as DRAM, SRAM, and/or FPGA.
Technology/Product Risk	
The primary technology risk is that I-MOS technology may not scale. Currently, the technology has been proven at the micron level. For the technology to be commercially feasible it must scale to submicron dimensions. In addition, to realize full power savings and performance increases, strained silicon must be utilized. The technology may not perform as well in strained silicon as simulations indicate.	Currently, extensive simulations and modeling have been run which validate I-MOS technology at submicron scale in strained silicon. To mitigate this risk, we plan to run I-MOS devices at a submicron scale in strained silicon as early as possible in the development process. Currently, we anticipate being able to complete this testing in nine months (September 2003).
Another major technology and market risk is that the semiconductor industry may not move toward a strained silicon based technology at all. Strained silicon is imperative in order to realize all the benefits and advantages of the I-MOS technology.	IBM, which pioneered strained silicon, has committed to introducing strained silicon in mass production by 2004 in its foundries. If we sense that the other major foundries are not moving in the direction of implementing strained silicon, we will shift our business strategy from a IP licensing business model to being a more product-oriented (SRAM or DRAM) company.
Team Risk	
The I-MOS management team has not been previously involved with a start-up, nor do they have significant management experience.	The management team will reduce this risk by hiring an experienced CEO after the first milestone is achieved (feasibility of I-MOS technology is proven). Furthermore, the management team will select a board of directors based not only on their relevant backgrounds but also on their willingness to be active mentors for the management team.
Financing Risk	
A significant financial risk is obtaining the capital required to launch a semiconductor IP company. Since there are relatively few successful semiconductor IP companies and since the level of required investment is high, we may experience difficulty in convincing the venture capital community to invest.	To reduce this risk, we will continue to establish relationships with potential customers early on in the development process. If we are unable to obtain the required financing, we may reconsider our licensing approach and investigate producing a relatively simple product, e.g., DRAM, SRAM, FPGA, to generate revenues sooner.

Appendix

FIGURE A1 Projected cash flow statement

	Year 1 Aug-2004	Year 2 Aug-2005	Year 3 Aug-2006	Year 4 Aug-2007	Year 5 Aug-2008
BEGINNING CASH BALANCE	—	1,419,593	772,984	3,859,663	2,572,868
Sources of Cash					
Net Income (Loss)	(4,906,907)	(8,803,609)	(5,674,321)	(1,078,795)	13,741,567
Incr. (Decr.) in A/P	—	—	—	—	—
Incr. (Decr.) in Accrued Exp.	—	—	—	—	—
Incr. (Decr.) in S.T. Notes Payable	—	—	—	—	—
Incr. (Decr.) in Other ST Liabilities	—	—	—	—	—
Incr. (Decr.) in LT Notes Payable	—	—	—	—	—
Incr. (Decr.) in Other LT Liabilities	—	—	—	—	—
Subtotal	(4,906,907)	(8,803,609)	(5,674,321)	(1,078,795)	13,741,567
Uses of Cash					
Incr. (Decr.) in A/R	—	—	—	—	—
Incr. (Decr.) in Inventory	—	—	—	—	—
Incr. (Decr.) in Other Current Assets	—	—	—	—	—
Incr. (Decr.) in Fixed Assets	(173,500)	(343,000)	(239,000)	(208,000)	(168,000)
Incr. (Decr.) in Other LT Assets	—	—	—	—	—
Subtotal	(173,500)	(343,000)	(239,000)	(208,000)	(168,000)
Operational Cash Flow	(5,080,407)	(9,146,609)	(5,913,321)	(1,286,795)	13,573,567
Financing Activity					
Preferred Stock	6,500,000	8,500,000	9,000,000	—	—
Common Stock	—	—	—	—	—
Subtotal	6,500,000	8,500,000	9,000,000	—	—
Net Cash Flow	1,419,593	(646,609)	3,086,679	(1,286,795)	13,573,567
ENDING CASH BALANCE	1,419,593	772,984	3,859,663	2,572,868	16,146,435

FIGURE A2 Projected balance sheet

	Year 1 Aug-2004	Year 2 Aug-2005	Year 3 Aug-2006	Year 4 Aug-2007	Year 5 Aug-2008
BALANCE SHEET					
Cash	1,419,593	772,984	3,859,663	2,572,868	16,146,435
Accounts Receivable	—	—	—	—	—
Inventory	—	—	—	—	—
Other Current Assets	—	—	—	—	—
Subtotal	1,419,593	772,984	3,859,663	2,572,868	16,146,435
Fixed Assets	173,500	516,500	755,500	963,500	1,131,500
Other LT Assets	—	—	—	—	—
Total Long-Term Assets	173,500	516,500	755,500	963,500	1,131,500
TOTAL ASSETS	1,593,093	1,289,484	4,615,163	3,536,368	17,277,935
Accounts Payable	—	—	—	—	—
Accrued Expenses	—	—	—	—	—
ST Notes Payable	—	—	—	—	—
Other Current Liab.	—	—	—	—	—
Subtotal	—	—	—	—	—
LT Notes Payable	—	—	—	—	—
Other LT Liabilities	—	—	—	—	—
Subtotal	—	—	—	—	—
Pfd. A Stock	6,500,000	6,500,000	6,500,000	6,500,000	6,500,000
Pfd. B Stock	—	8,500,000	8,500,000	8,500,000	8,500,000
Pfd. C Stock	—	—	9,000,000	9,000,000	9,000,000
Prior Yr. Ret. Earnings	—	(4,906,907)	(13,710,516)	(19,384,837)	(20,463,632)
YTD Earnings	(4,906,907)	(8,803,609)	(5,674,321)	(1,078,795)	13,741,567
Total Equity	1,593,093	1,289,484	4,615,163	3,536,368	17,277,935
TOTAL LIAB. & EQUITY	1,593,093	1,289,484	4,615,163	3,536,368	17,277,935

FIGURE A3 IP Semiconductor licensing structures

Company	Initial License Fee	Avg Royalty Rate	% of ASP
ARC	0.2+	0.51	8.3
ARM	5	0.07	1–3
MIPS	0.2–5.0	0.5	5–10
Parthus	<1.0	0.4–1.0	3–5

FIGURE A4 Customer acquisition plan

	Yr 1	Yr 2	Yr 3	Yr 4	Yr 5
Total Fab Customers	0	0	1	2	3
Total Chip Customers w/ own fab	0	0	1	2	3
Total Fabless Chip Customers	0	0	3	8	12
Total Customers	0	0	5	12	18

FIGURE A5 Industry comparables

Company	Market Cap ($MM)	Trailing Twelve Months				
		Sales ($MM)	Net Income ($MM)	Net Margin	Sales Multiple	Earnings Multiple
ARM	1,100	251	58.9	23.5%	4.4	18.7
Rambus	867.8	96.6	24.7	25.6%	9.0	35.1
MIPS	96.5	44.7	–12.3	–27.5%	2.2	–7.8
Synopsys						
TTP Communications						
Virage Logic	278.2	45.6	0.23	0.5%	6.1	1209.6
Mentor Graphics						
Parthus Technologies	137.7	42.5	–32.5	–76.5%	3.2	–4.2
Artisan Components	314.4	38	2.1	5.5%	8.3	149.7
DSP Group	433.8	126.2	13.1	10.4%	3.4	33.1
MEDIAN	314	46	2	5.5%	4.4	33.1

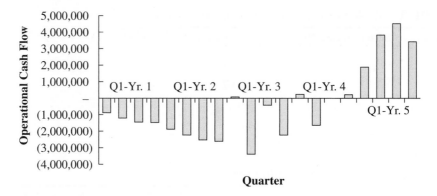

FIGURE 6 Projected operational cash flow by quarter

Cases

TREXEL

We've never met a customer who wasn't interested in our technology.
—David Bernstein, CEO of Trexel

David Bernstein hung up the phone with Alex d'Arbeloff, Trexel's largest investor, and contemplated an upcoming Board of Directors meeting scheduled for June 25, 1998. The meeting was only 10 weeks away and Bernstein, Trexel's president and chief executive officer, needed to present a coherent vision of the company's new strategy. Bernstein believed that Trexel's patented technology for manufacturing foamed plastics had the potential to revolutionize much of the worldwide plastics industry. His innovative process technology, known as MuCell, allowed the Woburn, Massachusetts company to produce foamed plastic utilizing 25% to 50% less material than traditional solid plastics without a significant decrease in the strength of the plastic. Bernstein believed the market for products produced via this technology could be in excess of 50 billion pounds of material per year representing potential worldwide annual revenues of over $100 billion. To date, Trexel had entered into numerous development partnerships with manufacturers, but no commercial products had made it to market. Bernstein was torn between his desire to pursue a variety of applications for the technology and the view of d'Arbeloff, and others, that he needed to limit Trexel's focus to one specific application. Bernstein and his investors had been confident that they could make sizeable inroads into the plastics business through a variety of potential applications, but time, money and human resource constraints had hindered Trexel's ability to fully capitalize on the opportunity. Bernstein was excited by the technology, but unsure of the best way to exploit its potential:

> The sheer size of the market for this platform technology is incredible. Unfortunately, it takes a lot of work to bring a technology from the laboratory to

Entrepreneurial Studies Fellow Matthew C. Lieb prepared this case under the supervision of Lecturer Michael J. Roberts as the basis for class discussion rather than to illustrate either effective or ineffective handling of an administrative situation.

the marketplace. I understand our investors' desire to focus on a single area given our inability to get any of our products into commercial production so far, but I'm just not convinced that committing to a single product is in our best interest right now. I would be more comfortable hedging our bet by pursuing multiple applications.

Bernstein knew that he needed to carefully analyze the many potential markets for MuCell in order to choose the best application(s) on which to focus. A wholehearted commitment to a specific market segment had the potential for earning substantial returns, but also the possibility of committing the company's limited resources to a single area that might never pay off.

PLASTIC FOAMING TECHNOLOGY

Traditional Technology

Traditional foaming of plastic had been in existence for almost 30 years and was used in the manufacturing of a variety of products such as meat trays, dinnerware products and disposable cups (see **Exhibit 1** for sample of products using traditional foaming technology). In the classic foaming process, plastic "pellets" were mixed with a gas (e.g., butane) under pressure. This mixing created cells (a cell was one air bubble and the plastic material around it) as the gas continually reacted with the plastic in a closed environment. This process required an extruder (a large machine that mixed the chemicals and plastic) to churn out foamed plastic products. The traditional methodology had several limiting characteristics arising from the fact that the technology produced relatively large and unevenly sized and shaped cells that could not be distributed uniformly within the plastic. The large size and lack of uniformity among the cells resulted in suboptimal strength and fatigue properties. Further limiting the effectiveness of traditional foaming was the inconsistency in product quality that arose from the difficulty in controlling the process. Finally, the "blowing" agents (chemicals used to create the air in the cells) in the traditional process also presented a regulatory challenge. Most of the agents were flammable, required special handling and regulatory approval for their use and release.

Trexel's Foamed Plastic Technology

Trexel's technology was originally developed at the Massachusetts Institute of Technology's (MIT) Polymer Processing Laboratory. In the early 1980s, Dr. Nam P. Suh, who headed the Mechanical Engineering Department at MIT, invented a microcellular foam process for thermoplastic polymers based on a precise process that utilized carefully controlled thermodynamic reactions within plastic raw material to create foam with small, evenly distributed and uniformly sized cells. MIT scientists developed a technique that utilized non-flammable gases—such as carbon dioxide and nitrogen—which were mixed with the plastic raw material under a carefully controlled set of temperature and high pressure conditions (see **Exhibit 2** for a process system diagram). Once the pressure was released, the evenly distributed gas "vaporized" and uniform air distribution in the plastic was achieved instantaneously (see **Exhibit 3** for key differences between Trexel's process and the traditional foaming process). Bernstein described the theory and advantages of Trexel's MuCell foaming technique:

> The primary motivation for all foaming is to use less material—you can use air instead of $1.35 per pound plastic. The trick, however, is to arrange the cells—the tiny bubbles—in such a way that they preserve the properties of the original solid material. Our technology permits a perfectly controlled approach to foaming. Indeed, it is so precise that we can create plastic that is significantly lighter than traditional plastics while preserving a high proportion of the key properties of the material. For instance, we can produce some products with a 30% weight reduction while sacrificing only a 10% reduction in stiffness. If the 10% strength reduction does not affect the performance of the product, then we have created a process that can save manufacturers a lot of money. This weight to strength tradeoff allows us to do things with plastics that have never been done before. To do this, we mix a gas, like nitrogen, into the solution of liquid plastic under high pressure. Then we remove the pressure—like popping off a champagne cork—and the gas expands, instantly becoming embedded in the plastic as tiny, uniformly sized bubbles.

The benefits of the MuCell process were threefold. First, the process allowed for the use of foamed plastics in applications that had previously relied on solid plastics because of the inherent limitations of traditional foaming. The use of MuCell plastics in place of solid material plastic reduced production costs by 20% to 25% as a result of decreasing both the amount of raw material used and the volume of production waste. MuCell technology typically produced products with a density and weight reduction relative to those products made from traditional foaming technologies, without a proportional drop in the strength of the materials. The second distinct advantage of the MuCell process was the improved mechanical properties of the foamed plastic (see **Exhibit 4** for mechanical property differences) relative to conventional foamed plastics. Specifically, the tensile (breaking or tearing) and compressive strengths of MuCell materials were greater than that of conventional foams. MuCell products also demonstrated improved performance at cold temperatures. The third key advantage of Trexel's technology relative to the traditional methods was the reduced environmental impact. The nonflammable "blowing agents" in the MuCell process were more environmentally friendly than the ozone-depleting chemicals required for traditional foaming. MuCell had received the Environmental Protection Agency's approval as a "safe alternative foam technology."

BACKGROUND

The Company

Dr. Suh founded Trexel in 1982 in an effort to capitalize on the commercial viability of numerous breakthroughs he had made in the polymer arena. Suh licensed various technologies from MIT and began engaging in ad hoc development efforts across a wide range of applications and products for the polymer industry. Bernstein described the philosophy of Trexel during this early phase:

> The eighties were a time when the company was struggling to find its focus. There were six or seven people here who were constantly chasing the next great idea. They were producing gauges and other devices for the polymer industry prior to 1993 when they first began working with microcellular foam. The process had been invented at MIT in 1982, but was not patented as a continuous—rather than batch—process until 1992. It was not until 1995 that Trexel fully licensed all of the necessary MIT patents required to proceed with commercializing the MuCell process. As soon as the licensing agreement was in place, the company began to focus more on MuCell. Although the scientists could only produce MuCell using a batch process, which is completely incompatible with commercial production, they were able to find development partners. The quality of the batch produced products was so good that people were eager to sign on with Trexel despite the fact that the company was not even close to being able to produce MuCell at commercial scale. Trexel signed MuCell-related deals for a variety of foamed plastic products—plates, pipe insulation, paper coating—all sorts of things. Unfortunately, the company was overly optimistic regarding what could really be accomplished given the state of the technology. The truth was that at this time, Trexel's theory for producing microcellular foam was really impressive, but the actual foam produced was a long way from being commercially viable. Lab conditions are drastically different from commercial manufacturing conditions. It's one thing to produce nice sheets of MuCell plastic in a laboratory and quite another to control the process so that you can make plastic products of different dimensions at a large scale in a manufacturing plant. The company had been unsuccessful in trying to raise outside equity, but even without external capital, strong customer interest in forming partnerships allowed them to use these development deals to "bootstrap" the technology development.

By 1994, the company had still not achieved commercial production of any products, but the progress of the technology was evident. Advances had been made in adapting the foamed plastic to different shapes at various levels of production. By 1995, Trexel's development efforts began to attract a great deal of attention from customers as companies saw the potential cost savings and improved product characteristics that MuCell made possible in the laboratory. The significant level of interest from major producers of plastic products prompted Dr. Suh to pursue a venture capital investment that would enable Trexel to more rapidly commercialize the technology. Dr. Suh needed to look no further than MIT's Mechanical Engineering Department for funding. The chairman of the Visiting Committee for the department was fabled Boston entrepreneur and angel investor Alex d'Arbeloff. d'Arbeloff, the chairman and CEO of Teradyne, Inc., was extremely interested in the commercial viability of the MuCell technology. d'Arbeloff assembled a group of investors and, in November of 1995, purchased 30% of Trexel for $2.2 million with the condition that a new CEO would be hired to run the business.

Bernstein Joins the Trexel Team

Concurrent with the equity infusion, d'Arbeloff recruited Bernstein (see **Exhibit 5** for management biographies) to be the president and CEO of the company. Bernstein, a 1976 graduate of Harvard Business School, had worked in a variety of managerial roles at Teradyne and Thermo Electron where he specialized in commercializing and marketing advanced technologies. Bernstein reflected on the opportunity:

> Alex d'Arbeloff, whom I really respected, was intrigued by the technology and wanted to bring me in to turn it into a viable business. I was immediately impressed by the technology. The company had been able to create some remarkable foamed plastics in the lab at small production volumes, so the potential of MuCell seemed enormous. Personally, I was excited to work in an environment that allowed me to implement decisions quickly. I had grown weary of the numerous layers of approval required to make something happen in larger organizations and was looking for something entrepreneurial. I liked the technology and I had a great deal of experience constructing licensing deals, so this opportunity was both exciting personally and a good fit with my previous experiences.

Strategically, Bernstein saw an opportunity to move Trexel's technology from the laboratory to commercial production by bringing in more skilled engineers, instilling a disciplined product management approach and by changing the fundamental business model. He commented on his vision of the business model:

> My predecessor had formed a number of development agreements with manufacturers to get the ball rolling towards commercialization of the technology. Strategically, I saw an opportunity to shift away from complete reliance on development partners to a more self-sustaining model. My view was that we could use the cash flow from the development agreements to fund our own internal projects. Under this scenario, we would take the knowledge and cash from our development programs and apply them to

internally developed products that were technologically similar to the development partnership projects. As our knowledge increased, we would eventually be able to capture the full value of the technology by manufacturing certain products in-house. Development contracts allowed us to spend other people's money to learn more about the technology.

Bernstein made significant changes almost immediately. His first move was to strengthen the company's intellectual property protection. Bernstein felt that the potential of Trexel's technology was so great that it was only a matter of time before other companies attempted to copy the process and beat Trexel to key market segments. The original MIT patents covered supercritical fluid, but it was clear to Bernstein that each application of the core technology to a specific plastic might also be patentable. Bernstein saw an opportunity to bolster the company's patent portfolio by implementing a formal process by which engineers documented their efforts and submitted patent applications in a routine fashion. This process, though often administratively cumbersome, was a critical step in building a base of protection for Trexel's long-term intellectual capital interests. To further shore up the company's position in the intellectual property arena, Bernstein retained the services of Wolf, Greenfield and Sachs, a Boston law firm specializing in patent law. Bernstein agreed to pay his patent attorneys close to $250,000 per year to protect Trexel's intellectual property interests. Though expensive, Bernstein felt the legal fees were well worth the money:

> Without those patents, our company had little to go on. When you are in the business of licensing technology, you must have patents to protect that technology; otherwise your work turns into nothing more than a consulting arrangement. Wolf, Greenfield and Sachs were expensive, but good. They filed close to one thousand claims and were able to get us broad protection for those claims entered into the patent process, including coverage in Europe and Asia (see **Exhibit 6** for sample patent).

Three-pronged Business Strategy

Confident that Trexel's intellectual property would be sufficiently protected, Bernstein turned his attention to the company's business strategy. Bernstein's goal was to implement a plan that would produce royalty revenues sufficient to cover the company's operating expenses within two years. He envisioned a three-part strategy where Trexel would continue to engage in large-scale development partnerships in an effort to generate cash as well as technological improvements. The partnerships would then allow Trexel to take the money and know-how from the partnership deals and quickly apply them to simple products that would be developed in-house. Production capability developed through internal development projects would eventually lead to Trexel's own full-scale production of high value-added product lines. Bernstein described his view of the situation:

When I joined the company, it was clear to me that these development agreements were hard to manage. "Handing off" the technology to a partner was very difficult because they were worried about their business today, not the potential of our technology down the road. Most partners simply could not afford to pull skilled employees off of revenue generating projects to work on MuCell development deals that held great future potential, but limited short-term benefits. We would propose next steps that needed to be taken to push the technology ahead and our partners would almost always agree with us. Unfortunately, we would come back to the same customers the following month only to find out that they hadn't even run their machine for weeks because the primary engineer for the project had been sent to Taiwan to work on something they deemed more urgent. Customers were—understandably—concerned with present revenue more than "the future." Nevertheless, while I was somewhat ambivalent about the development projects as they were structured now—even going so far as to cancel over $1 million in development deals—I believed that I couldn't afford to cut them all off until I had a working revenue model that could effectively replace the development partner revenue.

Development projects

The goal of the development projects was to demonstrate commercial feasibility for specific MuCell-enhanced products that would quickly lead to scale production and long-term royalty revenue. Development project deals would give customers an exclusive license in exchange for an up front development payment and a multi-year royalty agreement. Exclusivity agreements typically had a number of common characteristics. They offered exclusive use of the MuCell process for a specific product application (i.e., low-density polystyrene meat trays) over a three-year to five-year period assuming the customer achieved production levels sufficient to generate a minimum royalty payment and garner a minimum share of the specific product market.

Turnkey licensing

The knowledge gained from the development partnerships would allow Trexel to quickly license the technology for related products in a turnkey manner. Bernstein believed that Trexel's engineers would be able to rapidly transfer the technology developments from the partnership projects to similar products that were outside the bounds of the development projects' exclusivity agreements. Trexel would only target large-scale makers of technologically "simple" products with the turnkey licensing model.

Identification and retention of right to high value-added products

Bernstein saw enormous potential in retaining the right to develop certain products in-house. The commercial viability and market potential of the MuCell technology would only come to light through development projects and turnkey licensing. As the technology evolved, Bernstein hoped to identify specific products that represented extremely attractive cost/performance characteristics. Once these products were identified, Trexel would make strategic acquisitions enabling the company to manufacture products internally.

Implementing the Strategy

In early 1996, Bernstein set out to implement the three-pronged strategy by first focusing on what he believed were the "right" kind of development projects. Trexel focused its marketing efforts on billion-dollar companies that maintained significant research and development budgets. These companies could see the potential of Trexel's technology and wanted to get involved early. They had the resources to invest money in development and the patience to wait for the technology to mature to commercial viability. By September of 1996, Trexel had already entered into 11 development partnerships. These partnerships included MuCell products such as polystyrene sheets for arts and crafts applications, polypropylene pipe systems, building insulation foams made from recycled bottles, PVC tubing, and polystyrene meat trays. These development agreements accounted for $5 million in revenue to Trexel, in addition to agreements on future royalties that, if the projects were successful, could generate over $20 million dollars per year.

Each partnership project entailed slightly different technical challenges and varying degrees of partner participation. In some cases, Trexel allowed customers to design their own MuCell facility while Trexel served as a consultant. Other arrangements called for Trexel to design and install the equipment in Trexel's own facility and conduct all experimentation in-house. Bernstein commented on the development projects:

> Customers were so eager to reap the benefits of the MuCell technology that they were willing to pay anywhere from $300,000 to $400,000 up front—in addition to signing royalty agreements—to become development partners. I figured that if customers were paying us good money under development agreements they would have an incentive to get the technology into production. We signed lots of development contracts that essentially made us a technology job shop where we focused our attention on whatever products the development partners specified. Our arrangement with Sarto Plastics was pretty typical. They paid us $300,000 in development fees and agreed to pay a future royalty on sales. We, in turn, gave them exclusive rights to use the technology for disposable food service items.

While the development revenue earned under these agreements was clearly a positive aspect of the early business model, the actual development results were disappointing. It became, in Bernstein's words, "addictive to accept $300,000 even if we weren't sure we could deliver results." The promise of the technology was clear, but the shift from laboratory success to market acceptance was a long way off. Bernstein described the technological obstacles confronting Trexel engineers:

Our technology produces extremely small and perfectly sized cells because the gas uniformly and instantaneously comes out of the solution the instant pressure is dropped. It is extremely difficult to control the results of this "mini-explosion" because there is no room to adjust the outcome once the plastic has been formed—the outcome is instantaneous and permanent. The traditional technique provides a larger margin for error because the cells are formed over a period of time ranging from 30 to 40 seconds. As a result, the products can be shaped and adjusted somewhat with dies and other tooling devices before the final product is completely formed.

MuCell's instantaneous cell creation was especially problematic in the foaming of extruded plastic products with a thickness of greater than 1 mm. This problem was exacerbated when creating products with varying thickness. For instance, creating a container that was 1 mm thick in certain areas and 4 mm thick in others posed an enormous engineering challenge because of uneven pressure relationships at the time the material exited the die. Essentially, the desired characteristics of the products simply surpassed the initial capabilities of the technology. In the end, none of these development contracts appeared close to yielding products ready for commercial production. Bernstein offered his view:

On the positive side, the revenue from development partners gave us the opportunity to hold our venture capital financing in reserve—which was great from a cash management perspective. Unfortunately, our inability to actually produce market-ready products resulted in the alienation of some big customers. We simply were not able to quickly match the technology's performance to the expectations of our development partners. For example, we were working with a garden hose manufacturer on a project that appeared to have great promise. We developed a hose that used 45% less plastic than the old process. Unfortunately, the foamed plastic wasn't good under this kind of pressure and the hoses leaked. To make this application viable, we needed the manufacturer to use a different material formulation. Regrettably, the manufacturer's hose division was a small part of their overall business which relied on PVC materials. They were using hoses as a way to utilize the excess PVC material generated from their other business and, as a result, were reluctant to use any other material in the manufacturing of their hoses. In many cases, both Trexel and our development partners had unrealistic expectations of the benefits that the technology could deliver for particular applications—many of the applications simply weren't well suited to the MuCell process without significant modifications.

Trexel had learned a great deal about the technology and its limitations through development programs despite the lack of commercial production. Trexel engineers had successfully developed an assortment of new dies and tools that could be used with existing production machinery in the application of the MuCell process. Strategically, this was an important development because equipment modifications were more agreeable to potential users of the

MuCell technology than full-scale capital expenditures for MuCell-specific equipment. As a result, achieving commitments from potential partners would be easier going forward as capital equipment modifications became less of an issue. Another key learning point for Bernstein and his team was in their understanding of certain types of materials. The plastic industry was comprised of a variety of different plastic materials that exhibited a wide range of characteristics. Initially, Bernstein and his engineers believed that MuCell would work with almost any plastic compound. The reality was that certain materials (i.e., "rubbery" compounds) simply were not well suited to the MuCell process while others, such as polypropylene, worked well with Trexel's technology.

In April of 1996, Bernstein began to think about focusing more proactively on specific products rather than responding to the broad array of development partner interests. Though the development projects were moving ahead—six companies were in the process of installing MuCell production lines in their facilities—commercial results were not yet being realized. Trexel's customer-sponsored development focus was not producing the short development cycles that Bernstein and his investors had anticipated. Bernstein was convinced that the company's focus needed to be more on Trexel's internally driven development efforts. To that end, he embarked on a search for a specific product that met Trexel's objective of getting a product to market. Analysis by Trexel's scientists, marketers and Bernstein himself led to the decision to focus on wire insulation. The thin shape of wire allowed for easier application of the Trexel technology at its current state of evolution. Additionally, the inherent shortcomings of existing insulation products provided Trexel with an opportunity to significantly improve upon the current products in this market. Bernstein described the rationale behind focusing on wire insulation:

> I finally had the insight that we needed to get something—anything—into production quickly. We had to get one project working. We picked wire insulation because the product is thin—which typically makes things easier for us. In addition, the air bubbles make foamed plastic a better insulator than solid plastic. Finally, the existing materials are extremely expensive: customers were using Teflon, which costs $12 per pound. Reducing raw material costs by 40% for a $12 per pound material is much more valuable than saving 40% on a material like polyolefin which only costs $.40 per pound. Thus, MuCell was potentially very valuable in this application.

Bernstein signed a deal with a $1 billion wire and cable supplier to exchange processing knowledge and to set up a pilot wire and cable line at the customer's facility. Trexel's engineers, working closely with representatives from the wire and cable manufacturer, were able to produce a wire insulation product that represented significant cost savings for the customer relative to their old production process. Unfortunately, the savings were not sufficient to overcome a more practical problem—the insulation did not stick to the wire. The plastic insulation would swell up when exiting the extruding machine, causing a separation between the wire and the insulation. Trexel engineers were not able to resolve the problem and the partnership ended with no marketable product to speak of.

Modifying the Strategy: "Fast Track Development"

Though the wire insulation deal did not produce a product, Bernstein was intrigued by the shorter development cycle that the engineers had achieved. In the spring of 1997, Bernstein attempted to bring the lessons from the wire insulation partnership into a formal marketing program dubbed "Fast Track Development."

The financial performance of the company at this time was still lagging Bernstein's expectations despite the previously described development contracts and earnings (see **Exhibit 7** for financial information). The product development results were simply not coming fast enough. Bernstein knew that the success of the company would eventually be a result of tangible products succeeding in the marketplace. He felt that in order to implement a more market-oriented development process, Trexel would have to be the key driver in the move towards commercialization. In Bernstein's words, "development efforts needed to be more directed by Trexel than by Trexel development partners—we needed to be more pragmatic than visionary at this stage of the process."

Up until April of 1997, Trexel's development activity had been focused on commercializing the MuCell process in order to obtain royalty revenue and to confirm the adaptability of the process to the rigors of commercial scale production. Unfortunately, 1997 year-to-date royalty revenue was nonexistent. Bernstein hoped to remedy the current situation by engaging in fewer customer sponsored initiatives and focusing more on internally directed development. Bernstein described his decision to move even more of the development effort in-house:

> At this point in time, we had greatly improved our own understanding of the technological feasibility of MuCell. Unfortunately, the time to market for everything we were doing was simply too long. I made the decision to focus on the internal development of specific products. My goal was to shift our customer focus away from companies willing to spend several hundred thousand dollars on "high risk" development projects. We wanted customers who were more committed to investing their time and money in transferring Trexel-developed products to their own factories and producing products quickly.

To make this shift, Trexel would require an additional $2 million in financing. The funding would be used to purchase production equipment and to hire additional engineers and marketing staff. Bernstein went back to d'Arbeloff and the other initial investors to raise the capital required for the further development of MuCell. Impressed by the unending interest from potential customers and the progression of the technology, investors agreed to the terms of Trexel's series C financing which raised $2.16 million in exchange for 13% of the company (see **Exhibit 8** for ownership positions). Bernstein described his investors' rationale:

> d'Arbeloff and the others continued to be enthusiastic about Trexel because of the strong interest from the world's plastics manufacturers. Typically, ventures develop a technology and hope they can find a market for it. In our case, the market was already screaming for our technology, so it was up to us to deliver.

With an infusion of capital and a modified business model, Bernstein again set out to get MuCell products into commercial production. To speed up commercialization, Bernstein refined the plan for the Fast Track Development program. Fast Track Development called for a smaller initial financial investment from partnering companies in exchange for a stronger commitment to the rapid initiation of actual production of simple products using the MuCell technology. Bernstein planned to roll out the Fast Track Product Development program at the National Plastics Exposition (NPE) in June of 1997.

Interest in the new program proved to be extremely strong. The NPE show generated interest from nearly 50 companies that manufactured products that seemed to be a good fit with the MuCell technology. **Table A** illustrates the range of products that companies were interested in producing in partnership with Trexel.

Bernstein was excited by the interest that Fast Track Development was generating:

> At this point, we didn't want money, we wanted answers. We intentionally marketed to organizations that had products that seemed to fit with what we had learned about the material and process characteristics of our technology. We quickly eliminated any customers who demanded that we develop samples before installing a production line in their facility. I knew that if we could get customers to commit to a MuCell production line in their plants early on, they would then have an incentive to really make the technology work. If you have space on your production floor being taken up by a MuCell line, your commitment to making the line work will be much stronger than if the equipment is sitting on Trexel's plant floor. In addition to our refusal to make samples, we also asked ourselves three questions. Is this a material and application that we understand and can transfer with little effort? Is the customer capable of

TABLE A Potential fast track development products from the NPE trade show.

Marine flotation	Seat cushions	Razors	Beverage containers
Refrigerator liners	Raincoats	Wine bottle corks	Gaskets
Label backing	Shoe soles	Power cables	Graphic arts production
Tape	Highway signs	Aircraft parts	Dashboard covers

working independently at his facility when the time comes? Does the customer have a target product that represents an interesting market opportunity? If and only if the answer to all of these questions was "yes" were we willing to undertake the project. In the end, we took less money up front in exchange for a commitment on the part of customers toward rapid development in their own facilities.

The Fast Track Development initiative changed the economics of Trexel's business model. **Table B** highlights the shift from the previous partnership model to the Fast Track Development model.

Under this new model, Trexel explored partnerships with 67 companies spanning a wide range of applications. **Table C** highlights the range of potential product partnerships.

In the end, Bernstein chose to engage in only the projects that he believed would be most likely to quickly produce successful commercial products. Many potential partnerships were not developed because the customers required product samples in advance of committing to a production line at their facility. He eventually signed nine Fast Track Development deals for products such as cushioned shoe inserts, strips to seal the doors of automobiles and disposable food service plates. Unfortunately, technological limitations, resource constraints

TABLE B Fast track development vs. traditional development contracts

	Traditional development contract	Fast track development
Payment:	$500,000	$50,000
Negotiation:	6-12 Months	1 Month
Term:	24 Months Plus	6 Months
Exclusivity:	At least 3 years subject to performance	1 Year

TABLE C Sample of potential fast track development partnerships

Partner	Product	Material
Rogers Corporation	Shoe inserts	Santoprene
Norton Plastics	Weather stripping	Polyvinyl chloride (PVC)
Blind Systems Inc.	Vertical blinds	Polyvinyl chloride (PVC)
Tray Form Plastics	Meat trays	Polystyrene
3M	Thin-film labels	Polypropylene (PP)
Owens-Illinois	Bottle cap liners	Polystyrene (PS)
Mattel	"Hot Wheels" tracks	Polyethylene (PE)
Anderson Windows	Window profiles	Polyvinyl chloride (PVC)

and a lack of follow-through on the part of development partners hindered the development of market-ready products. Bernstein described one failed Fast Track Development project:

> The shoe insert opportunity seemed very promising to us early on. Rogers Corporation was really committed to the technology and they were willing to do everything we asked of them. Unfortunately, as often happens with a developing technology, the customer kept getting more and more specific about the product requirements, and we couldn't get all of the aspects of the technology to match exactly to the customer's specifications. The material they needed to use for the inserts was simply not viscous enough for our process. The low viscosity prevented us from getting the proper cell structure in the foam because we couldn't apply the appropriate pressure to the material. The technology fell just short of meeting expectations.

In May of 1998, it became clear to Bernstein that Trexel's success would be dependent upon a more drastic change in the business model than ever before. Despite the evolution from long-term development contracts to the more streamlined Fast Track Development program, the actual results—measured by the success of commercial products—continued to disappoint Bernstein and Trexel's investors. Bernstein's view of Trexel's most likely path to success was becoming clearer in his own mind:

> Despite the popularity of the Fast Track Development program, it became very clear to me that success would never come if we continued to rely so heavily on our partners for development. Customers simply don't have the staying power to endure the process changes that MuCell required of them. Our partners found it difficult to dedicate the required resources to Trexel-sponsored projects because of the developmental nature of MuCell. Further complicating things was the fact that most marketing people want our products to be more than just cheaper than the existing products. For example, one customer was demanding that our drinking straws perform better than their current drinking straws. It didn't seem to be enough to simply save them money on a commodity-like product.
>
> In my mind, success would only come if we could control every step in the process. We needed to control the material, dimensional requirements, and tooling of equipment to make this technology ready for commercial production. In the end, this is not a turnkey technology and, therefore, it is extremely difficult to launch products in partnership with customer's marketing and engineering departments without constant support from Trexel.

Alex d'Arbeloff and the other investors agreed with Bernstein's assessment of the situation. It was Bernstein's responsibility to articulate Trexel's strategy including which specific application or applications the company would pursue. This impending meeting with the Board of Directors would set the stage for Trexel's future. Bernstein now turned his attention to deciding which project(s) he would recommend.

Potential Applications

Bernstein's task of choosing specific applications on which to focus future development efforts was a daunting one. The sheer number of potential applications required a great deal of market analysis and technological understanding on the part of Trexel's management team. Rigorous analysis and frequent debates eventually led to a list of five potential areas of focus for Trexel. Each of these five applications had several attractive characteristics, which only made it more difficult for Bernstein to eliminate options. While d'Arbeloff and the other investors wanted to pick a single application on which to focus, Bernstein was not yet convinced that eliminating some applications in order to concentrate on others was in the best interest of the company. Regardless, it was now up to Bernstein to select only the most promising application(s) for Trexel's MuCell technology.

Molded structural foam

Products made from this process included garbage cans, computer monitor and keyboard housings, beverage carriers and swimming pool panels. Bernstein and his management team estimated that the 100 worldwide structural foam molding companies collectively churned out over one billion pounds of material each year (at a cost of $.40 per pound) generating over $2 billion in revenue on an annual worldwide basis. Plastic products made using this process required enormous machines costing over $1 million each. Uniloy/Milacron was the dominant force in the manufacturing of this equipment, garnering over 80% of the world market share for such machinery. There were approximately 300 installed structural foam machines with approximately 20 to 30 new machines forecasted to come on line each year for the foreseeable future.

At the end-user level, the MuCell process resulted in a 30% material cost savings and an increase in the speed of manufacturing. In fact, Trexel's engineers had proven that a typical structural foam molding machine utilizing the MuCell process could operate at twice the speed of the same machine utilizing traditional process technology. Trexel engineers were confident in their ability to produce commercially viable products in the structural foam molding arena because the technological adaptation was nearly identical to what they had already been working on in other areas.

Bernstein's revenue model for structural foam molding combined a consulting contract, an equipment sale and a licensing agreement. First, Trexel would allow original equipment manufacturers (OEM's) to develop equipment specifically for the MuCell process. By giving the OEM's the know-how to develop MuCell equipment, Trexel would be providing the OEM's with an additional product line to sell to customers. In exchange for the opportunity to sell MuCell equipment, OEM's would agree to sell the equipment at a price comparable to other plastic manufacturing equipment, thus eliminating equipment cost as a barrier to end user adoption of the MuCell process. Once the equipment was purchased from the OEM's by plastic manufacturers, Trexel would negotiate a deal for a consulting contract, the sale of a proprietary Trexel-made supercritical fluid delivery system and a seven-year licensing agreement. The consulting contract called for Trexel to earn $15,000 for advising the plastic manufacturer on implementing the initial production line. This was expected to be a one-time fee. Trexel would manufacture, at a cost of $36,000, the crucial supercritical fluid delivery system, which would be sold to the plastic manufacturer for approximately $73,000. The seven-year licensing agreement called for the plastic manufacturer to pay Trexel either $25,000 or 20% of the total savings that the plastics company would realize by instituting the MuCell technology, whichever was greater. After the first year, the licensing agreement would revert to a fixed contract based on the licensing fees paid to Trexel in the initial 12 months, thus eliminating the need for Trexel to monitor the actual production of the plastic manufacturer after the first year.

Injection molding

This process was used in the manufacturing of a wide range of plastic products including buckets, ties for garbage bags, trays and nearly every plastic part under the hood of an automobile. Injection molding was an attractive segment from a technology perspective because the difficulties in controlling the MuCell "nucleation" were made easier by the use of physical molds. Since the plastic was foamed directly into a mold, it was significantly easier for Trexel engineers to control the "mini-explosion" that often plagued the MuCell process. The market for injection molding was estimated to be over $40 billion annually with over 25,000 injection molding machines expected to be sold annually over the coming years. While the significant installed base of over 100,000 machines was attractive, the opportunity for applying Trexel's microcellular process was more limiting. One of the inherent shortcomings of the MuCell process was the difficulty in making products that required a glossy finish. Trexel's marketing team estimated that there were nearly 5,000 potential customers using injection molding machines that could be adapted for MuCell production. Ultimately, 20,000 existing injection molding machines could be equipped to utilize the MuCell process while an additional 5,000 applicable machines would be purchased annually over the coming years. Trexel believed that 50% of the market would use the MuCell process for producing commodity resins while the remaining 50% would use Trexel's technology to produce more complex engineering resins. The manufacturers could expect to realize cost savings of nearly 25% by instituting a MuCell production line.

Bernstein believed that Trexel could enter into similar consulting arrangements as outlined in the structural foam plan, but would sell the supercritical fluid delivery system for $50,000, twice the cost of developing the delivery system for injection molding. Licensing agreements for injection molding would be fixed contracts based on the type of resin being used by the plastic manufacturer.

Commodity resin producers would pay $25,000 per year for seven years for each of the MuCell machines that the company utilized. Engineering resin producers would be charged $35,000 per machine per year for seven years.

Blow molding

Over five billion pounds of blow molded products were produced worldwide in 1997. The primary application for blow molding was the manufacturing of bottles for consumer packaged goods. Shampoo, motor oil, milk and a variety of household cleaning items were all bottled in blow molded plastic containers. With the exception of milk bottles, the majority of the MuCell-compatible blow molding was produced by 20 companies who collectively operated 200 high-volume machines that each generated close to eight million pounds of material annually. Milk bottles represented another relatively promising application. The three major milk bottle manufacturers operated nearly 1,500 blow molding machines accounting for over 1.5 billion pounds of blow molded plastic each year. Trexel's marketing team believed that an additional 300 milk bottle manufacturing machines would be sold each year over the next several years.

The technological feasibility of applying the MuCell process to blow molding appeared very promising. The use of molds made controlling the rapid cell nucleation process easier for Trexel's engineers. Additionally, the tubular shape of most blow molded products had proven to be relatively "MuCell-friendly" in the past. Trexel's engineers were confident that blow molded products could be produced using 25% less material than in traditional blow molding. The current cost of blow molding material averaged $.40 per pound. Bernstein believed that Trexel could command a $25,000 annual licensing fee per milk bottle machine for five years. In addition, he anticipated Trexel being able to garner 20% of the net cost savings as a royalty payment from the applicable non-milk bottle blow molding machines, which would convert to a fixed contract after the first year. Similar to the structural foam revenue model, Trexel expected to earn $15,000 per customer in consulting fees and an additional $73,000 per machine for the supercritical fluid delivery system.

PVC extrusions

Trexel's management believed that the PVC extrusion market held great promise for the MuCell technology. PVC extrusion, in a form suitable for MuCell, was applied almost exclusively to three product lines: exterior siding for houses, vertical window blinds and interior paneling. The market for PVC exterior siding was comprised of five companies who collectively represented 90% of the North American market and operated close to 100 machines producing nearly two billion pounds of material each year. Vertical window blinds and interior paneling were produced by 20 manufacturing companies located primarily in North America, Europe and South America who operated a total of 400 machines each producing one million pounds of material each year. Bernstein saw two key advantages to focusing on the PVC market. First, the thin shape and long production runs used in making PVC extrusion products fit well with the capabilities of MuCell at the current time. Second, Trexel's engineers believed they could produce the necessary PVC products using 25% less material with no loss of stiffness. Existing PVC material cost manufacturers nearly $.50 per pound. The risks in this application were two-fold. First, there was a risk in gaining commitment from partners because of the significant changes in both the equipment platform and manufacturing process required to produce PVC extrusion using the MuCell technology. Second, the production of exterior siding would require the approval of the industry building code committee. The interests of entrenched players in the industry might make the required approval difficult to receive.

The revenue model for PVC extrusions called for Trexel to receive $50,000 in a one time consulting fee from each customer who adopted the technology. Trexel would also charge $73,000 for the supercritical fluid delivery system required for each MuCell machine. The licensing arrangement would be structured such that Trexel would receive 20% of the cost savings realized by manufacturers in the first year, which would again convert to a fixed contract in subsequent years.

Meat trays and food packaging

Trexel had already demonstrated an ability to successfully manufacture meat trays on commercial extrusion lines in limited quantities. As a result, Bernstein and his managers were attracted to this potential application which utilized nearly two billion pounds of material each year. This market was comprised of 15 major companies operating a total of 400 machines. Trexel's engineers had already been successful in producing meat trays that offered both appearance and performance improvements over the traditional products in addition to eliminating the need for using harmful hydrocarbons in the manufacturing process. Experience had proven that Trexel could offer development partners a savings of $.035 per pound of material if they instituted the MuCell process. Revenue from the production of meat trays and food packaging would come from a $60,000 per machine annual fixed licensing agreement which would run for the life of Trexel's patents in addition to a one time $50,000 consulting fee and the $73,000 supercritical fluid delivery system sale.

DECISION TIME

Bernstein began to sort through the files he kept on each of the potential applications that Trexel was evaluating:

> We've tried to think through our choices in a systematic, disciplined way. Unfortunately, every time we think we have a plan, we learn something new about the technology, the market or the customer that forces us to rethink our strategy.

With the board meeting approaching, Bernstein knew it was time to focus on selecting the application(s) that would form the basis for Trexel's next round of development efforts.

EXHIBIT 1 Sample of plastic products using traditional foaming technology and other non-foamed plastic products with MuCell potential

Sample of products using traditional foaming process technology	Sample of non-foamed plastic products with MuCell potential
Meat trays	Bottles
Disposable flatware	Automotive parts
Weather stripping	Computer housings
Pallets	Television housings
Barrels	Phone housings
Art board	Buckets
Trash cans	Manifolds
Crates	Food trays
Wire insulation	Cutlery

EXHIBIT 2 MuCell process system diagram

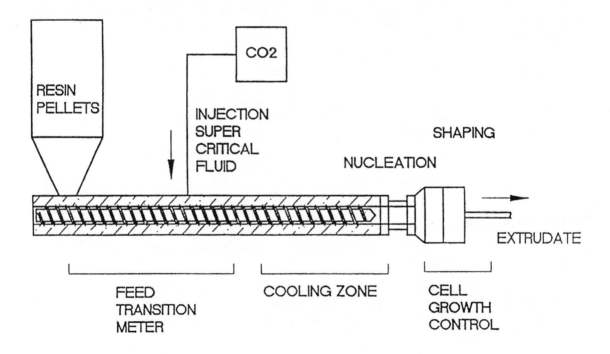

EXHIBIT 3 Trexel technology vs. traditional technology: nucleation comparison

MuCell Process
Large number of Nucleated clusters uniformly distributed. No competition with growth.

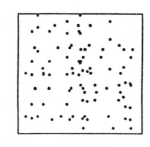 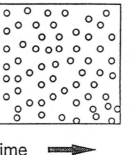 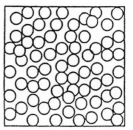

 time

Traditional Process
Nucleation clusters initially sparse. Cell growth by diffusion competes with nucleation for gas.

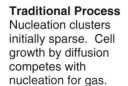 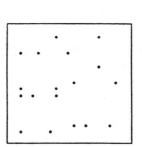 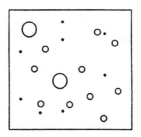 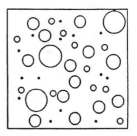

EXHIBIT 4 Trexel technology vs. traditional technology: mechanical property comparison.

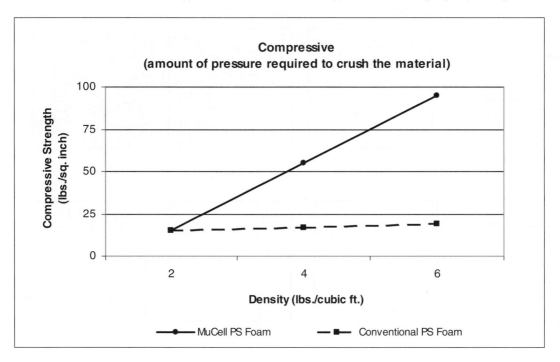

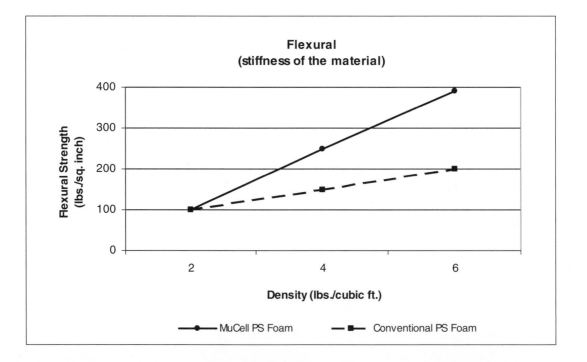

EXHIBIT 5 Trexel management team biographies

David P. Bernstein, President & CEO

Mr. Bernstein, who has held several executive positions in finance, general management, sales and marketing, focuses on the commercialization and marketing of advanced technologies. He spent nine years with Teradyne, a leading manufacturer of electronics testing equipment, where, as Vice President of Sales and Support, he built a 250-person worldwide organization to sell and service $150 million annually in capital equipment sales. While at Teradyne, he also negotiated a $250 million OEM agreement with the General Electric Company. As a Vice President of Thermedics Detection, a Thermo Electron Company, he built and managed a worldwide business selling operationally critical equipment to the Coca-Cola and Pepsi-Cola companies and establishing a worldwide support organization to service it. He also negotiated successful OEM and licensing relationships with leading European bottling equipment companies. Mr. Bernstein received a B.A. from Harvard College and an M.B.A. from Harvard Business School.

Matt Pallaver, Executive Vice President

Mr. Pallaver is responsible for the marketing development of Trexel projects in Asia, Europe and North America. He joined Trexel in 1993, after more than 15 years on marketing development, including seven years of management in new product development and commercialization with Siemens, Control Data, and Sperry Corporation. Mr. Pallaver received a B.S. in Mechanical Engineering from the Illinois Institute of Technology and an M.B.A. from the University of Oklahoma.

Dr. Richard Straff, Vice President, Research and Commercialization

Dr. Straff joined Trexel after 20 years with Hoechst Celanese. His experience includes diverse technical-management and research-management assignments in injection molding applications and new product developments for optical fibers, liquid-crystal polymers, polyester products, and other engineering plastics. He received a B.S. in Metallurgy, an M.S. in Polymer Materials, and a Ph.D. in Polymer Science from the Massachusetts Institute of Technology.

Daniel Szczurko, Vice President, Business Development

Mr. Szczurko is responsible for identifying and licensing product development programs to the plastics industry. He has 20 years of sales and marketing experience in new technologies, in addition to an extensive record of early technology concept sales to *Fortune* 100 companies in a variety of instrumentation and automation technologies. As Director of Strategic marketing for Thermedics Detection, a subsidiary of Thermo Electron, Mr. Szczurko was responsible for marketing analytical instrumentation in the areas of chromatographic, x-ray, and chemiluminescent, and fluorescent technology. He received a B.A. in Industrial Economics from Duquesne University.

Dr. Lee Chen, Research Director

Dr. Chen has more than 15 years of experience in polymer process development, screw and tooling design, and the application of computer modeling and simulation to polymer processes. He also has performed extensive research in extrusion, reactive extrusion, and polyurethane foam. Before joining Trexel, Dr. Chen was the Manager of Process Research and Development with BICC Cables Corporation. A recipient of the Shanghai Government Award for Outstanding Scientists and Technologists, Dr. Chen has authored many articles and studies in the area of mass flow effects, residence time distribution, and non-plug-flow solid conveying. He received a B.S. and an M.S. in Mechanical Engineering from Beijing Institute of Chemical Technology and a Ph.D. in Chemical Engineering from the University of Pittsburgh.

David Pierick, Vice President, Injection Molding Programs

Mr. Pierick has experience in injection molding technologies, polyolefin product development, and polyolefin structure/property relationships. Prior to joining Trexel, he acquired an international reputation as Manager of Product Development and Technical Sales at Montell Polyolefins. Mr. Pierick also has held the positions of Plant Engineer and Plant Manager for the Rehrig Pacific Company, a manufacturer of injection molded crates for the beverage industry. Mr. Pierick has authored publications on screw design and product performance. Mr. Pierick received a B.S. in Mechanical Engineering from the University of California at Los Angeles, an M.S. in Polymer Processing from the University of Lowell, and an M.S. in Polymer Science from University of Ferrara, Italy.

EXHIBIT 6 Trexel patent

US005158986A

United States Patent [19]

Cha et al.

[11]   **Patent Number:**   **5,158,986**

[45]   **Date of Patent:**   **Oct. 27, 1992**

[54]   **MICROCELLULAR THERMOPLASTIC FOAMED WITH SUPERCRITICAL FLUID**

[75]   Inventors: **Sung W. Cha**, Cambridge; **Nam P. Suh**, Sudbury; **Daniel F. Baldwin**, Medford; **Chul B. Park**, Cambridge, all of Mass.

[73]   Assignee: **Massachusetts Institute of Technology**, Cambridge, Mass.

[21]   Appl. No.:   **682,116**

[22]   Filed: **Apr. 5, 1991**

[51]   **Int. Cl.^{5}** **C08J 9/00**
[52]   **U.S. Cl.** **521/82;** 521/97; 264/50
[58]   **Field of Search** 521/82, 97; 264/50

[56]   **References Cited**

U.S. PATENT DOCUMENTS

3,796,779	3/1974	Greenberg	264/50
4,473,665	9/1984	Martini-Vvedensky et al.	521/79
4,719,246	1/1988	Murdoch et al.	521/64
4,728,559	3/1988	Hardenbrook et al.	428/35
4,761,256	8/1988	Hardenbrook et al.	264/50
4,832,881	5/1989	Arnold et al.	521/918
4,873,218	10/1989	Pekala et al.	521/64
4,906,672	3/1990	Stone et al.	521/130

OTHER PUBLICATIONS

LeMay, J. D. et al. "Low–Density Microcellular Materials", *MRS Bulletin,* Dec. 1990, pp. 19–45.

Primary Examiner—Maurice J. Welsh
Assistant Examiner—Rachel Johnson
Attorney, Agent, or Firm—Robert F. O'Connell

[57]   **ABSTRACT**

A supermicrocellular foamed material and a method for producing such material, the material to be foamed such as a polymerplastic material, having a supercritical fluid, such as carbon dioxide in its supercritical state, introduced into the material to form a foamed fluid/-material system having a plurality of cells distributed substantially throughout the material. Cell densities lying in a range from about 10^9 to about 10^{15} per cubic centimeter of the material can be achieved with the average cell sizes being at least less than 2.0 microns and preferably in a range from about 0.1 micron to about 1.0 micron.

11 Claims, 10 Drawing Sheets

EXHIBIT 7 Trexel Financial Statements

Income Statement	Fiscal Year 1996	Fiscal Year 1997
Net Revenue	$ 1,248,300	$ 2,338,000
R&D Project Expenses	$ 456,700	$ 861,000
Gross Profit	$ 791,600	$ 1,477,000
R&D Expenses (Internal)	$ 281,300	$ 818,300
SG&A	$ 541,300	$ 582,900
Total Operating Expenses	$ 822,600	$ 1,401,200
Income (Loss) from Operations	$ (31,000)	$ 75,800
Interest Income, net	$ 45,700	$ 95,300
Net Income	$ 14,700	$ 171,100

Balance Sheet	31-May-97 Fiscal Year 1996	31-May-98 Fiscal Year 1997
Assets		
Current Assets		
Cash	$ 1,491,700	$ 2,730,918
Accounts Receivable	$ 345,000	$ 338,861
Ppd. Exp. And Other Current Assets	$ 83,200	$ 173,235
Total Current Assets	$ 1,919,900	$ 3,243,014
Property, Plant & Equipment, net	$ 383,200	$ 941,054
Other Assets, net	$ 203,800	$ 140,809
Total Assets	$ 2,506,900	$ 4,324,877
Liabilities & Stockholder's Equity		
Current Liabilities		
Accounts Payable and Accrued Expenses	$ 234,800	$ 91,527
Deferred Revenue	$ 27,600	$ 200,830
Current Portion of Long Term Debt	$ 17,300	$ 136,128
Total Current Liabilities	$ 279,700	$ 428,485
Capital Lease Obligations	$ 34,400	$ 456,560
Stockholder's Equity		
Preferred Stock, Series C at liquidation value	$ –	$ 2,158,056
Preferred Stock, Series B at liquidation value	$ 2,200,000	$ 2,200,000
Preferred Stock, Series A at liquidation value	$ 550,000	$ 550,000
Common Stock, par value $.01	$ 16,400	$ 16,361
APIC	$ 1,018,500	$ 997,935
Accumulated Deficit	$ (1,592,100)	$ (2,482,520)
Total Stockholder's Equity	$ 2,192,800	$ 3,439,832
Total Liabilities and Equity	$ 2,506,900	$ 4,324,877

EXHIBIT 8 Trexel Shares and Options Outstanding

Shares	
Non-MIT Shares	1,581,525
MIT Shares	210,634
Series A Preferred (as converted)	118,945
Series B Preferred (as converted)	1,000,000
Series C Preferred (as converted)	599,460
Total	**3,510,564**
Options	
Outstanding Options	710,765
Ungranted Options	110,950
Total	**821,715**

Questions

1. Is the transfer of technology from universities to industry important? Why or why not?

2. What's different about "clean" technology ventures as compared to other technology-intensive enterprises?

3. What criteria should you use to evaluate the projects at Trexel? Which project (molded structural foam, injection molding, blow molding, PVC extrusions, or meat trays and food packaging) should Bernstein recommend to the board? Why?

BIODIESEL INCORPORATED

Joshua Maxwell shut down his laptop and looked out the window. From the second floor of the Graduate School of Management's new building, he could see a number of cars driving on the nearby freeway and sitting in the adjacent parking lot.

Josh was in his last term of the full-time MBA program at UC Davis. He would soon be graduating and entering a new chapter of his life. While he had the luxury of having several management-level job offers from which to choose, he was unsure whether he wanted to follow such a traditional route. There was one opportunity in particular that had recently come across his path which gave him pause.

Background

The previous term, Josh had been enrolled in Professor Dorf's class on Business and Sustainability. While the class was offered at the GSM, it was open to the entire university. In this class, he met Hannah Long, who was in her final year of her undergraduate studies in Agricultural Economics, and Matthew Hammond, who was a senior in the Mechanical Engineering department.

The three began working on a class project, which would ultimately turn into a formidable business opportunity. The impetus for their collaboration began with a lecture-discussion regarding the challenges and opportunities in the emerging renewable energy industry.

The Challenge

Dependence on energy is a worldwide reality. Energy powers the machines and equipment around us in order to make life more convenient and efficient. In our everyday lives, energy is synonymous with the forms that it can assume. The major generation sources—petroleum, coal, natural gas and nuclear—are non-renewable resources and have detrimental effects on the environment. In our daily lives, the two most common forms of this energy are liquid fuel (refined from petroleum) and electricity.[1]

Increasingly, developed and developing countries alike are consuming liquid fuel for the purposes of mobility, food production, and the facilitation of trade. All of these functions essentially provide a substitute for human effort. Due to the widespread consumption of petrol-based liquid fuel, an incredibly large global infrastructure and set of surrounding institutions have grown around the support of such consumption. The petrochemical fuel industry manifests itself in the form of oil fields and reserves, pipelines, transport ships, and fueling stations.

Prepared by MBA candidate Benjamin Finkelor; Assistance from MBA candidate Sonja Yates and Paul Yu-Yang under the supervision of Professor Richard C. Dorf, Graduate School of Management of UC Davis.

[1] Technically speaking, liquid fuel is a form of energy, and electricity is considered a carrier of energy. For the purposes of this proposal, the distinction is not significant.

The way energy is used worldwide is not sustainable. It is well-documented that the use of these fuels is depleting the world's natural resource reserves, harming communities in terms of health and displacement, and polluting the air and water in local environments. The drilling, refining, and transporting of oil leads to spills on land and in oceans, and when petrol-based fuels are used to power machines and automobiles, the air is polluted with greenhouse gases and particulate matter such as carbon dioxide, carbon monoxide, sulfur, and nitrous oxide emissions.

In spite of the drawbacks, the current energy industry is committed to the continuation of these ways, primarily because of considerable assets and investment in the existing form of infrastructure.

The challenge, which became clear to the team from class discussion and further brainstorming, is to find a form of fuel or technology that can mitigate the current negative affects on the environment of petrol-based fuel while utilizing the existing infrastructure. The urgency of this challenge is heightened by the astounding projected growth in the global population and per-capita consumption of liquid fuels.

The Concept

Matthew's coursework in engineering coupled with a bit of networking with fellow engineers suggested the emerging technology of biodiesel as a possible solution to this challenge. As the group explored the environmental benefits and the viability of the diesel fuel substitute, the three began to realize the potential of the biodiesel market.

Biodiesel is a vegetable- and/or animal-based product that serves as a substitute for traditional diesel fuel. Although its chemical composition is dissimilar from the petrol-based diesel, biodiesel will still work in diesel engines built in and after 1996 with no modification. For engines made before that time, modifications can be made to allow for the use of biodiesel fuel. The choice of biodiesel as a product of biomass is an intentional one. Producing a product that can be utilized by the existing infrastructure and social patterns of use[2] increases the likelihood of its adoption. "Entrepreneurs must locate their ideas within the set of existing understandings and actions that constitute the institutional environment yet set their innovations apart from what already exists."[3] This economic viability is coupled with a significant potential to the environment: biodiesel showcases an innovation that is a step in the right direction for air quality.

Biodiesel's greatest promise to sustainability as a renewable energy source is its lower emissions over conventional diesel. Compared to traditional diesel, biodiesel achieves significant reductions in harmful emissions. Additionally, the

---

[2] JoAnne Yates, "The Structuring of Early Computer Use in Life Insurance," *Journal of Design History,* 12(1999): 5–22.

[3] A. Hargadon and Y. Douglas, "When Innovations Meet Institutions," *Administrative Science Quarterly,* 46(2001): 476.

ozone-forming impact of biodiesel is nearly half of that of petroleum fuel. Further benefits can be counted when looking at lifecycle effects. If biodiesel is obtained using soybeans as an example, the amount of CO_2 taken up by soybeans and released upon burning the fuel, is a near zero sum balance. Contrast this with petroleum products where release of CO_2 is unidirectional into the atmosphere.

Because biodiesel is biodegradable and dependent on organic material as opposed to fossil fuels, the energy source is considered renewable. Production of biodiesel begins with feedstock, preferably in the form of oils or fats. Oils can be processed from oleic varieties of plants such as soy, canula, sunflower and safflower. Fats can come directly from grease such as and tallow/lard and recycled cooking grease from restaurants. The oils or fats are mixed with alcohol and a catalyst in a process that forms esters resulting in biodiesel defined as mono-alkyl esters of long chain fatty acids and glycerin.

Ultimately, the large-scale production of biodiesel would generate a dramatic impact on the economic value of the feedstocks involved. For example, according to one study, if biodiesel demand over the next ten years were to increase to 200 million gallons, a commensurate amount of soy oil would be required and net average farm income would increase by $300 million per year. A bushel of soybeans would increase by an average of 17 cents over the ten-year period.[4] The potential economic benefit to farmers seems considerable.

Even with such economies of scale, however, the wholesale price of 100 percent biodiesel would rarely be lower, and therefore cost-competitive, with traditional diesel fuel. Barring some crisis that would drive up the price of crude oil or reduce the capacity of diesel refineries, the current regulatory structure and assets devoted to petrol-diesel will more often than not yield a lower price with petrol-based diesel. Biodiesel as a fuel additive however, does provide a cost-competitive potential. Studies have shown that splash-mixing even 1 percent biodiesel with traditional diesel "can increase the lubricity of petroleum diesel by up to 65 percent."[5] This is not to mention the sulfur- and other emissions-reducing benefits that splash-mixing provides. As more consumer and regulatory pressure is placed on traditional diesel users, biodiesel producers will be able to charge the premium necessary to offset higher relative costs. Markets for 100 percent biodiesel will grow as well in such specialty markets as the marine industry, railroads, electricity generators, and even agriculture.

Biodiesel Incorporated

Josh, Hannah, and Matthew presented a compelling business case for their final class project: Biodiesel Incorporated. This new venture would enlist and

---

[4] National Biodiesel Board, "Benefits of Biodiesel," www.biodiesel.org.

[5] "Biodiesel Carries New Weight in Premium Diesel Market," *Biodiesel Bulletin,* Sept. 2002. http://www.biodiesel.org/news/bulletin/2003/080403.pdf.

develop a series of local producer's cooperatives in an effort to capitalize on the emerging biodiesel market as described in the following list:

■ Members would grow feedstock crops and gather crop residues with high fat content.

■ Capital equipment costs would be shared and spread over membership. Oils would be extracted from the collected biomass and biodiesel would be produced using these oils.

■ Biodiesel Incorporated would distribute the biodiesel locally using the existing petroleum-based infrastructure.

Advantages of the Cooperative Business Form

The cooperative model has been successfully used to allow small farmers to maintain a competitive edge against the larger corporate farming organizations. "Today, there are more than 4,000 agricultural cooperatives in the U.S., with a total net income of nearly $2 billion and net business volume of more than $89 billion."[6] A coop is owned and controlled by the members, with self-reliance and self-help being key characteristics—ideal for the implementation of emerging and disruptive innovations such as biodiesel.

Biodiesel Incorporated will:

■ Utilize the collective purchase power of the coop to obtain necessary capital-intensive equipment and to gain economies of scale.

■ Increase negotiating power, allowing it to
  ■ Stabilize crop prices and biodiesel output
  ■ Gain access to higher-volume contracts

■ Serve to unite rural communities and preserve agricultural economy

Biodiesel Incorporated offers the unique service of both the bargaining and manufacturing of biodiesel on behalf of its farmer members. It will serve to control the production of agricultural products (i.e., the biomass feedstock), the price and terms set for members' production, and price and terms for biodiesel output.

Questions

1. What are the key factors in determining if this is a viable business opportunity for Josh, Hannah, and Matthew?

2. What market drivers should they research and be aware of?

3. What are the flaws in the current business strategy?

4. What type of financing should they use if they choose to go forward with this?

5. What types of distribution channels should they go into?

6. How can they improve their chances for success?

7. What is the next step?

---

[6] http://co-operatives.ucdavis.edu/Agricultural Co-operatives.htm.

YAHOO! 1995

*I guess, three and a half years ago, if we were looking to start a business and
make a lot of money, we wouldn't have done this.*
—Jerry Yang, 1997

It was April of 1995—a key decision point for Jerry Yang and David Filo. These
two Stanford School of Engineering graduate students were the founders of
Yahoo!, the most popular Internet search site on the World Wide Web. Yang and
Filo had decided that they could transform their Internet hobby into a viable business. While trying to decide between several different financing and partnering
options that were available to them, they attended a meeting with Michael Moritz,
a partner at Sequoia Capital. Sequoia, one of the leading venture capital firms in
Silicon Valley, had been discussing the possibility of investing in Yahoo!.

Michael Moritz leaned forward in his chair. As he looked across the conference table at Jerry and Dave, he laid out Sequoia's offer to fund Yahoo!:

> As you know, we have been working together on this for some time now.
> We have done a lot of hard work and research to come up with a fair value
> for Yahoo!, and we have decided on a $4 million valuation. We at Sequoia
> Capital are prepared to offer you $1 million in venture funding in exchange
> for a 25 percent share in your company. We think that with our help, you
> have a real chance to make Yahoo! something special. Our first order of
> business will be to help you assemble a complete management team, after
> which we should be able to really start helping you to develop and man-
> age your site's vast amount of content.
>
> Right now, the biggest risk that you guys run is *not* making a decision.
> You *have* to make a decision, because if you don't, someone else is going

to run you over. You might get run over by Netscape. You might get run over by AOL. You might get run over by one of these other venture-backed start-ups. It is imperative that you make a decision now if you are going to survive. To help you make a decision, I am going to give you a deadline: tomorrow. If you don't want to do business with Sequoia, that's OK. I'll be disappointed, but that's OK. But you are going to have to call me by 10 A.M. tomorrow morning to tell me yes or no.[1]

Yang and Filo gazed around the Sequoia conference room and noticed the many posters of companies such as Cisco, Oracle, and Apple that were hung from the walls—all success stories from past Sequoia investments. They wondered if Yahoo!'s poster would someday join that group. The two were excited at the possibilities; however, they still had some decisions to make. There were several other financing options available, and they were still not sure if they wanted to accept Sequoia's funding. Yang responded:

That sounds like a pretty fair offer, Mike. Let us talk this over tonight, and we will get back to you by tomorrow after we weigh all of our options. However you have to realize that we're still grad students, and we don't even usually wake up by 10 A.M., so can you give us until noon?

Yahoo!

Yahoo! was an Internet site that provided a hierarchically organized list of links to sites on the World Wide Web. It offered a way for the general public to easily navigate and explore the Web. Users could click through multiple topic and category headings until they found a list of direct links to Web sites related to their interest. In addition, Yahoo! offered a central place where people could go to just to see what was out there. This made it easy for people with little previous exposure to the Web to start searching through Yahoo!'s lists of links, often just to see if they could find something of interest. In a little over a year since its inception, it had become one of the most heavily visited sites on the Web.

But Yang and Filo believed Yahoo! had the potential to be much more that a way for Web surfers to find what they were looking for. In 1995, John Taysom, a vice president of marketing of Reuters, a London-based provider of news and financial data, called Jerry Yang to explore the idea of a Yahoo!-Reuters partnership. It seemed to Taysom that affiliating with Yahoo! could help Reuters to build a distribution network on the Web.

"The first thing Jerry said to me," Taysom remembers, "was 'if you hadn't called me, I would have called you.'" Jerry *got* the news feed vision. He had been thinking about it for months. He further surprised Taysom by informing him that as far as he was concerned, Yahoo! was "not just a directory but a media property."[2]

---

[1] Michael Moritz, personal interview, November 10, 1998.

[2] Rob Reid (1997), Architects of the Web, John Wiley and Sons, New York, p. 253.

Yang further believed that: "Primarily we're a brand. We're trying to promote the brand and build the product so that it has reliability, pizzazz, and credibility. The focus of the business deals we are doing right now is not on revenues but on our brand."[3]

Dave and Jerry at Stanford

David Filo, a native of Moss Bluff, Louisiana, attended Tulane University's undergraduate program in computer engineering. In 1988, Filo finished his undergraduate work and enrolled in Stanford's master's program in electrical engineering. Completing his master's degree, he opted to stay at Stanford and try for his PhD in electrical engineering. Extremely competent in the technical arena, Filo had been described by many as a quiet and reserved individual.

Jerry Yang was a Taiwanese native who moved to California at the age of 10. Yang was raised by his widowed mother and grew up in San Jose with his younger brother, Ken. Yang was a member of the Stanford class of 1990 and completed both his bachelor's and master's degrees in electrical engineering. Yang also opted to stay at Stanford for a PhD in electrical engineering. Also technically competent, Yang was considered much more outgoing than Filo.

Yang and Filo met each other in the electrical engineering department at Stanford; Filo was Yang's teaching assistant for one of his classes. They also both worked in the same design automation software research group. They became close friends while teaching at the Stanford campus in Kyoto, Japan. Upon returning to the Stanford campus, they moved into adjacent cubicles in the same trailer to conduct their graduate research. They both enjoyed working together, as their individual personalities perfectly complemented each other, forming a unique combination.

Their office was not much to look at, but it served as a place for them to work on their research as well as a place from which they could run their website. "The launching pad (for Yahoo!) was an oxygen-depleted, double-wide trailer, stocked by the university with computer workstations and by the students with life's necessities… that prompted a friend to call the scene 'a cockroach's picture of Christmas'."[4] Michael Moritz remembered his early visits to Jerry and Dave's cube:

> With the shades drawn tight, the Sun servers generating a ferocious amount of heat, the answering machine going on and off every couple of minutes, golf clubs stashed against the walls, pizza cartons on the floor, and unwashed clothes strewn around . . . it was every mother's idea of the bedroom she wished her sons never had.[5]

---

[3] Jerry Yang (1995) interview in *Red Herring* (October), online back issue, p. 9.

[4] Randall E. Stross (1998) "How Yahoo! Won the Search War," *Fortune,* http://www.pathfinder.com.fortune/1998/980302/yah.html, p. 2.

[5] Rob Reid (1997), *Architects of the Web*, John Wiley and Sons, New York, p. 254.

Mosaic and the World Wide Web

In 1993, the University of Illinois-Urbana Champagne's National Center for Supercomputing Applications (NCSA) revolutionized the growth and popularity of the World Wide Web by introducing a Web browser they had developed called Mosaic. Mosaic made the Web "an ideal distribution vehicle for all kinds of information in the professional and academic circles in which it was known."[6] It provided an easy-to-use graphical interface that allowed users to travel from site to site simply by clicking on specified links. This led to the widespread practice of surfing the Web, as people spent hours trying to find new and interesting sites. This easy-to-use browser for navigating the Internet was estimated to have 2 million users worldwide in just over one year.

Creating Jerry's Guide to the World Wide Web

With Mosaic's introduction in late 1993, Filo and Yang, along with thousands of other students, began devoting large amounts of time to surfing the Web and exploring the vast content available. As they discovered interesting sites, they made bookmarks of the sites. The Mosaic Web browser had an option to store a bookmark list of your favorite sites. This feature allowed users to return directly to a page that they had visited, without having to navigate through several different links. As the popularity of the Web quickly increased, so did the total number of sites created, which in turn led to an increase in the number of interesting sites that Filo and Yang wanted to bookmark. Eventually, their personal list of favorite Web sites grew large and unwieldy, due to the fact that the earliest versions of Mosaic were unable to sort bookmarks in any convenient manner.

To address this problem, Filo and Yang wrote software using Tcl/TK and Perl scripts that allowed them to group their bookmarks into subject areas. They named their list of sites "Jerry's Guide to the World Wide Web" and developed a Web interface for their list. People from all over the world started sending Jerry and Dave e-mail, saying how much they appreciated the effort. Yang explained: "We just wanted to avoid doing our dissertations."[7]

The two set out to cover the entire Web. They tried to visit and categorize at least 1,000 sites a day. When a subject category grew too large, subcategories were created, and then sub-subcategories. The hierarchy made it easy for even novices to find websites quickly. "Jerry's Guide" was a labor of love— lots of labor, since no software program could evaluate and categorize sites. Filo persuaded Yang to resist the engineer's first impulse to try to automate the process. "No technology could beat human filtering," Filo argued.[8]

---

[6] Rob Reid (1997), *Architects of the Web,* John Wiley and Sons, New York, p. 11.

[7] Randall E. Stross (1998), "How Yahoo! Won the Search War," *Fortune,* http://www.pathfinder.com.fortune/1998/980302/yah.html, p. 2.

[8] Randall E. Stross (1998), "How Yahoo! Won the Search War," *Fortune,* http://www.pathfinder.com.fortune/1998/980302/yah.html, p. 3.

Though engineers, Yang and Filo had a great sense of what real people wanted. Consider their choice of name. Jerry hated "Jerry's Guide," so he and Filo opted for "Yahoo!," a memorable parody of the tech community's obsession with acronyms (this one stood for "Yet Another Hierarchical Officious Oracle"). Why the exclamation point? Said Yang: "Pure marketing hype."[9]

Yahoo!'s Growing Popularity

At first, Yahoo! was only accessible by the two engineering students. Eventually, they created a Web interface that allowed other people access to their guide. As knowledge of Yahoo!'s existence spread by word of mouth and e-mail, more people began using their site, and Yahoo!'s network resource requirements increased exponentially. Stanford provided them with sufficient bandwidth to the Internet, but bottlenecks came from limitations in the number of TCP/IP connections that could be made to the two students' workstations.[10] Additionally, the time required to maintain the site was becoming unmanageable, as Yang and Filo found themselves continually updating their Web site with new links. Classes and research fell behind as Yang and Filo devoted more and more time to their ever-expanding hobby.

Competing Services

A number of businesses already existed in the Internet search space. While none offered the same service that Yahoo! did, these companies could definitely provide potential competition to any new business that Yahoo! would start. Among the competitors were Architext, soon to be renamed Excite, Webcrawler at the University of Washington, Lycos at Carnegie Mellon, the World Wide Web Worm, and Infoseek, founded by Steven Kirsh. AOL and Microsoft in 1995 represented larger competitors who could enter the market either by building their own capability or acquiring one of the other start-ups.

Yahoo!'s human-crafted hierarchical approach to organizing the information for intuitive searches was a key component of its value proposition. Rob Reid, a Venture Capitalist with 21st Century Internet Venture Partners, explained how this made Yahoo! unique among Internet search providers.

> The Yahoo! hierarchy is a handcrafted tool in that all of its . . . categories were designated by people, not computers. The sites that they link to are likewise deliberately chosen, not assigned by software algorithms. In this, Yahoo! is a very labor intensive product. But it is also a guide with human discretion and judgment built into it—and this can at times make it almost uncannily effective. . . .

---

[9] Randall E. Stross (1998), "How Yahoo! Won the Search War," *Fortune,* http://www.pathfinder.com.fortune/1998/980302/yah.html, p. 3.

[10] Mark Holt and Marc Sacoolas (1995), "Chief Yahoos: David Filo and Jerry Yang," *Mark & Marc Interviews,* May, http://www.sun.com/950523/yahoostory.html.

This is the essence of Yahoo!'s uniqueness and (let's say it) genius. It isn't especially interesting to point to information that many people are known to find interesting. *TV Guide* does this. So do phone books, and countless Web sites that cater to well-defined interest groups. . . . But Yahoo! is able to build intuitive paths that might be singularly, or even temporarily important to the people seeking it. And it does this in a way that no other service has truly replicated.[11]

However, if Yahoo!, as a business, was to survive and flourish in the face of increasingly well-funded competition, it would quickly need to find some outside capital.

Leaving Stanford and Starting the Business

Yang and Filo had been in Silicon Valley long enough to realize that what they really wanted to do was to start their own business. They split much of their free time between their Internet hobby and sitting around thinking up possible business ideas.

"A considerable period of time passed before it occurred to them that the most promising idea was sitting under their noses, and some of the credit for their eventual illumination belongs to their PhD adviser, Giovanni De Micheli. Toward the end of 1994, De Micheli noted that inquiries to Yahoo! were rising at an alarming rate. In a single month, the number of hits jumped from thousands to hundreds of thousands daily. With their workstations maxed out, and the university's computer system beginning to feel the load, De Micheli told them that they would have to move their hobby off campus if they wanted to keep it going."[12]

By fall of 1994, the two received over two million hits a day on their site. It was then that Jerry and Dave commenced the search for outside backing to help them continue to build up Yahoo!, but with only modest hopes. Yang thought they might be able to bootstrap a workable system, using personal savings to buy a computer and negotiating the use of a network and a Web server in return for thank-you banners. Unexpected overtures from AOL and Netscape caused them to raise their sights, although both companies wanted to turn Filo and Yang into employees.

If they were going to abandon their academic careers (as they soon did, six months shy of their doctorates), they reasoned that they should hold out for some control. Filo and Yang had three main potential options to explore: (1) sell Yahoo! outright; (2) partner with a corporate sponsor; (3) start an independent business using venture capital financing.

---

[11] Rob Reid (1997), *Architects of the Web,* John Wiley and Sons, New York, pp. 243–244.

[12] James, Lardner, (1998), "Yahoo! Rising," *U.S. News,* May 18, http://www.usnews.com.usnews/issue/80518/18yaho.html, p. 3.

The Search for Funding

Looking to receive funding and create a credible business out of Yahoo!, Filo and Yang began preliminary discussions with potential partners in October 1994. One of the first people who contacted them was John Taysom, a vice-president of marketing at Reuters, the London-based media service. Taysom was interested in integrating Reuters' news service into Yahoo!'s Web pages. Yahoo! would gain the advantage of being able to provide news services from a well-known source, while Reuters would be able to begin developing its own presence on the Internet. Unfortunately, since Yahoo! did not generate revenues, it was in a poor negotiating position. Talks between the two were cordial, but they also progressed very slowly.

Yahoo! also talked to Randy Adams, founder of the Internet Shopping Network (ISN), a company that styled itself as "the first online retailer in the world." ISN, funded by Draper Fisher Jurvetson, was one of the first venture funded Internet companies. It had recently been purchased by the Home Shopping Network, in order to expand its possible exposure. ISN was interested in being a host site for Yahoo!, offering them the chance to finally generate some revenue. However, there were also definite possible disadvantages that came from being associated with a shopping network.

Another company that approached Yahoo! was Netscape Communications Corporation. Founded in April 1994 by Jim Clark, who also founded Silicon Graphics, and Marc Andressen, who created the NCSA Mosaic browser with a team of other UIUC students and staff, Netscape was a hot private company developing an improved browser based on the old Mosaic technology. Andressen contacted Yang and Filo over e-mail and, in Yang's words said, "Well, I heard you guys were looking for some space. Why don't you come on into the Netscape network? We'll host you for free and you can give us some recognition for it."[13] This was a fortuitous contact that allowed Yahoo! to move itself off of Stanford's campus. By early 1995, Yahoo! was running on four Netscape workstations.

Soon after, Netscape offered to purchase Yahoo! outright in exchange for Netscape stock. The advantage of this option was that Netscape was already planning its initial public offering and had tremendous publicity and momentum behind it. Coupled with high profile founders and backers like Clark, James Barksdale, former president and CEO of AT&T Wireless Services, and the venture capital firm Kleiner Perkins Caufield & Byers, this offer was a potentially lucrative one for the two Yahoo! founders. Additionally, Netscape's company culture was more in tune with what the two students were looking for, in comparison to some of the more established market players.

---

[13] Jeff Ubois (1996), "One Thing Leads to Another," *Internet World,* January, http://www.Internetworld.com/print/monthly/1996/01/yahoo.html, p. 1.

Corporate Partnerships

Yahoo! was also feeling tremendous pressure to partner or accept corporate sponsorship from other large content companies and online service providers like America Online (AOL), Prodigy, and Compuserve. These companies offered the carrot of money, stock, and/or possible management positions. They argued that if Yahoo! did not partner with them, as large players they could develop their own competing services that would cause Yahoo! to fail. One potential disadvantage with corporate funding was the potential taint that came with such sponsorship. Yahoo! had started as a grass-roots effort, free of commercialization. A second disadvantage was the lack of control that the two Yahoo! founders would have over their creation. "Building Yahoo! was fun, particularly without adult supervision. (Dave) and Jerry were also worried that selling to AOL would have 'most likely killed' Yahoo! in the end."[14]

With partner discussions beginning to heat up, Yang requested help from Tim Brady, a friend and second-year Harvard Business School student. As a class project, Brady generated a business plan for Yahoo! during the 1994–1995 Christmas vacation. (See the Appendix for excerpts of the business plan circa 1995.)

With Brady's business plan in hand, Filo and Yang began to approach different venture capital firms on nearby Sand Hill Road. Venture capital firms brought experience, valuable contacts in the Silicon Valley, and most importantly, money. However, they also required substantial ownership in return for their services. One venture firm that the Yahoo! founders approached was Kleiner Perkins Caufield & Byers. KPCB had an excellent reputation as one of the most prestigious VC firms in the Silicon Valley, and their list of successful investments included Sun Microsystems and Netscape. KPCB showed a definite interest in Yahoo!; however, Vinod Khosla of KPCB and Geoffrey Yang of Institutional Venture Partners had just invested $0.5M in Architext (later renamed Excite), another company started by Stanford engineering students that was developing a search-and-retrieval text engine. Architext was receiving increased press coverage, with a March 1995 *Red Herring* magazine spotlighting the company and its venture capital partners. KPCB proposed to fund Yahoo!, but only if they agreed to merge with Architext.

Sequoia Capital

Another venture capital firm that Yahoo! approached was Sequoia Capital. It was during partnership discussions with Adams at the Internet Shopping Network that Yang and Filo were first introduced to Michael Moritz, a partner at Sequoia Capital. Moritz went to visit Jerry and Dave, who were at the time still operating out of their tiny Stanford trailer. Said Yang, "The first time we sat down with Sequoia, Mike (Moritz) asked, 'So, how much are you going to charge subscribers?' Dave and I looked at each other and said, 'Well, it's

---

[14] Rob Reid (1997), *Architects of the Web,* John Wiley and Sons, New York, p. 256.

going to be a long conversation."[15] Fortunately, Moritz, who came from a journalistic background at *Time* was flexible in his thinking. Some of the major advantages that Moritz brought to the negotiating table were his contacts with publications and knowledge about how to manage content. Moritz talked about the roots of Sequoia's interest in working with Yang and Filo. "I think we are always enamored with people that seem to be on to something, even if they can't define that something. They had a real passion and a real spark."[16]

Sequoia Capital had a long tradition of success in the venture capital market, citing that the total market capitalization for Sequoia backed companies exceeded that of any other venture capital firm. Sequoia's trademark *modus operandi* was funding successful companies using only a small amount of capital. Its list of successful investments included Apple Computer, Oracle, Electronic Arts, Cisco Systems, Atari, and LSI Logic. Said Moritz, Sequoia preferred "to start wicked infernos with a single match rather than 10 million gallons of kerosene."[17]

In February 1995, Filo and Yang were weighing a number of possibilities and in no hurry to accept any of them, when Michael Moritz made them an offer. Sequoia Capital would fund Yahoo! for $1 million and would help them to assemble a top management team. In return, Sequoia would receive a 25 percent share of the company. Additionally, Moritz gave them only 24 hours to accept the deal before it was pulled off the table. "I felt a need to deliver them from the agony of indecision," claimed Moritz. With the deadline quickly approaching, Yang and Filo sat down to weigh their options. The decisions that they made that night would determine the direction of their careers as well and the future of Yahoo!

The Decision

Sitting in their tiny office on the Stanford campus, Jerry and Dave shared a late-night pepperoni and mushroom pizza as they explored their options and tried to come to a decision. It was already getting pretty late, and they only had until noon the next day to make their decision.

Yang took a bite from his pizza as he looked over the terms sheet that Sequoia had given them.

> We have some pretty tough decisions to make, and Michael has really forced the issue now with this 24-hour deadline. As I see it, we have a couple of options. The first is to accept Sequoia's offer and launch Yahoo! as our own company. We would be giving up a significant percentage of Yahoo!, but we really need the money if we are going to survive. Moritz and the rest of the resources at Sequoia could also prove to be invaluable as we try to assemble the rest of our management team.

---

[15] Jerry Yang, "Found You on Yahoo!" *Red Herring,* October 1995, p. 3.

[16] Michael, Moritz, personal interview, November 10, 1998.

[17] Anthony, Perkins, *Red Herring,* June 1996.

Our second option is to accept corporate sponsorship. This would allow us to get the funding we need and still retain 100 percent ownership of Yahoo!. However, I am worried about selling out to corporate America. We were fortunate to be able to develop our site in an educational setting as a noncommercial free site. I am afraid if we accept the corporate sponsorship, it will taint Yahoo!'s image.

Finally, we could agree to merge with an existing corporation. The word is that Netscape is pretty close to their IPO, and Architext has some really big time investors behind it. If we merge with Netscape or Architext in exchange for stock options, it could mean a lot of money for us in the next couple of years.

Filo got up from his seat and kicked aside some of the empty pizza boxes that had started to accumulate. He walked over to Yahoo!'s tiny office window and stared at Stanford's Hoover Tower, which was barely visible in the distance.

It's true that we could make some money if we sell to Netscape or Architext, but we would have to give up primary control of Yahoo! if we did. We would never know what we could have done if we would have maintained control of the site ourselves.

There is also a fourth option you forgot to mention. I'm excited by Sequoia's offer, but I'm wondering if maybe we are giving up too much of our company. A fourth option could be to not decide tonight and look for better terms with another VC firm. I know Michael said that we should decide quickly, but I would hate to give up 25 percent of our company, only to find out in a week that another firm would have offered us $3 million for the same percentage. I know that time is really important, and we like working with Michael Moritz. On the other hand, I don't want to be regretting our decision two months from now.

As they grappled with the alternatives facing them, Filo and Yang began to envision life outside of the Stanford trailer in which Yahoo! was born. It was well past 2 A.M., and they had to make a decision in less that ten hours. What should they do?

Questions

1. What makes Yahoo! an attractive opportunity (and not just a good idea)?
2. How will Yahoo! make money (i.e., business model)?
3. Identify the major risks in each of these categories: technology, market, team, and financial. Rank order them.
4. What are the advantages and disadvantages of each of the funding options they could pursue? Which one do you recommend?

Video Resources

Visit http://techventures.stanford.edu to view a video of the founders of Yahoo! and others discussing the outcome of the case.

EXHIBIT 1 Yahoo! Founders and Potential Investor

Jerry Yang

Jerry Yang was a Taiwanese native who was raised in San Jose, California. He co-created the Yahoo! online guide in April of 1994. Jerry took a leave of absence from Stanford University's electrical engineering PhD program after earning both his BS and MS degrees in electrical engineering from Stanford University.

David Filo

David Filo, a native from Moss Bluff, Louisiana, co-created the Yahoo! online guide in April 1994 and took a leave of absence from Stanford University's electrical engineering PhD program in April 1995 to co-found Yahoo!, Inc. Filo received a BS degree in computer engineering from Tulane University and a MS degree in electrical engineering from Stanford University.

Michael Moritz, Partner, Sequoia Capital

Moritz was a general partner at Sequoia Capital since 1988 and focused on information technology investments. Moritz served as a director of Flextronics International and Global Village Communication, as well as several private companies. Between 1979 and 1984, Moritz was employed in a variety of positions by Time, Inc. Moritz had an MA degree in history from Oxford University and an MBA from the Wharton School.

Appendix Selected Excerpts from the Yahoo! Business Plan.

Yahoo!'s first business plan was developed by Tim Brady as part of a course project at the Harvard Business School. The plan was continuing to evolve during discussions between Jerry Yang and David Filo at Yahoo! and Michael Moritz of Sequoia Capital. For this case, the company has provided excerpts of this business plan that are not proprietary.

The case writers thank Mr. J. J. Healy, director of corporate development, and others at Yahoo! for their efforts in providing this original archival information to enhance the learning experience of future entrepreneurs.

Business Strategy

Yahoo!'s goal is to remain the most popular and widely used guide to information on the Internet. The Internet is in a period of market development characterized by extremely high rates of both user traffic growth and entry of new companies focused on various products and services. By virtue of its early entry, Yahoo! has developed its current position as the leader in this segment. Yahoo!'s ability to expand its position and develop long-term, sustainable advantages will depend on a number of things. Some of these relate to its current position and others relate to its future strategy.

Today, Yahoo! solves the main problem facing all Internet users. It is next to impossible for users, faced with millions of pieces of information scattered globally on the Internet, to easily find that what is relevant to them without a guide like Yahoo! Not only is the amount of information huge, it is expanding almost exponentially.

All enhancements to Yahoo! will be governed by the goal of making useful information easy to find for individuals.

We believe that Yahoo's enormous following has been generated by the following list:

■   Yahoo! was first company to create a fast, comprehensive and enjoyable guide to the Internet, and in so doing, built a strong brand early and created momentum.

■   The unique interest-area based structure of Yahoo! makes it an easier and more enjoyable way for the user to find relevant information than the classic search engine approach where key words and phrases are used as the starting point.

■   Through its editorial efforts, Yahoo! has continually built a guide which is noticeably better than its competition through a combination of comprehensiveness and high quality.

The company will focus on the directory and the guide business and generate revenue from advertising and sponsorship.

Yahoo!'s strategy is to:

■   **Continue to build user traffic and brand strength** on the primary server site through product enhancements and extensions as well as through an aggressive marketing communications program.

■   **Develop and integrate the leading technology** required to maintain a leadership position. Underlying the extremely appealing guide is Yahoo!'s scaleable core technology in search engine, database structure, and communication software. These core technologies are relevant to the user's experience to the extent that it enables Yahoo! customers' access to a broader array of high quality information in an intuitive way, faster than any competitors product. Yahoo! is discussing a full license to advanced web-wide search engine technologies, web-wide index data, and crawler services with Open Text of Waterloo, Canada. Yahoo! will be the first guide with a seamless integrated directory/web-wide search product. The proposed agreement with Open Text also includes ongoing joint development of advanced search and database technologies leveraging the strengths of both companies. All jointly developed products will be distributed by Yahoo! allowing the company to continue to introduce advanced features on a regular and aggressive basis.

■   **Extend the reach to a broader audience** through establishment of contractual relationships with Internet access providers such as MSN, America Online, and Compuserve and very popular web sites.

- **Extend the reach and appeal to international users** through partnerships with international access providers who can operate foreign mirror sites for Yahoo and add localization in the form of foreign language, local advertisers, and local content.
- **Retain the users ("readership") of Yahoo!** through constant enhancements to the content and interface of the guide.
- **Rapidly extend the product line** by introducing regional guides, vertical market guides, and more importantly, individually personalizeable guides. Our intention is to be the first to market in all of most of these categories and outrun our competition by constantly "changing the competitive rules and targets." Our introduction of personalized guides will be a first in the market and will leverage core technology owned both internally as well as through our license with Open Text.

Market Analysis

The Internet, whose roots trace back almost 20 years, is experiencing a period of incredibly rapid growth in the area of online access base and user population. According to IDC and a recent report by Montgomery Securities, there are approximately 40 million users of the Internet, a majority using it only, for email. However, it is estimated that about 8 million people have access to the Internet and World Wide Web. Most of these access the Web from the workplace because of the availability of high bandwidth hardware and communications ports there. It is expected that over the next two to four years as higher bandwidth modems, home-based ISDN lines and cable modems are adopted, that both the growth and penetration of Web access into the home will increase dramatically. IDC estimates that by 2000, 40 percent of the homes and 70 percent of all businesses in the United States will have access to the Internet. In the Western European and Japan markets, the comparable penetration rates might be as high as 25 percent and 40 percent respectively. If this holds true, there will be as many as 200 million users on the Internet and Web by the year 2000.

Market Segmentation and Development

We believe that between now and the year 2000 there will be three principal user groups driving the growth of the Web:

- Large businesses using the Internet for both internal wide area information management and communication as well as intrabusiness communication and commerce.
- Small home based businesses using it for retrieval of information relevant to the business as well as for vendor communication and commerce.
- The individual user/consumer using it initially to find and access information which is relevant to their personal entertainment and learning and later to make purchases of products and services.

We also believe that the evolution of the Internet will include three stages of market development:

- Availability and proliferation of enabling technology.
- Establishment of widespread access and communication services.
- Widespread distribution of high value content.

We are currently in the first stage of market development consisting primarily of infrastructure building and including rapid growth in the adoption and sale of computer, network, and communication products and entering into the second stage involving the initial establishment of "access" service based businesses.

Internet Market Size

Estimates of the amount of current and projected revenue for Internet related business vary. However, primary research conducted by both Montgomery Securities as well as Goldman Sachs indicate that the total served market for Internet hardware, software, and services will total approximately $1B in 1995, up from approximately $300M in 1994. Projections are that these categories might grow to a total of $10B by the year 2000. Several research firms including Forrester and Alex Brown & Sons have estimated the revenues to be produced by Web-based advertising at approximately $20M in 1995, $200M in 1996, and over $2B by the year 2000.

Market Trends

During the current, rapidly expanding stages of market and industry development, the following trends are clear:

- There is large scale adoption of enabling technology in the areas of network hardware and software, as well as communication hardware and software. The World Wide Web with its inherent support of multimedia begs for the adoption of higher and higher bandwidth platform and communication hardware and software.
- Telecommunication companies and newly entering Internet access providers are rushing to put in place basic "hook-ups" in high bandwidth form.
- The price for high-speed computer and communication "port" hardware and software of adequate bandwidth to support acceptable levels of transport and display is still somewhat high. Partly for this reason, the adoption of fully capable ports onto the Web is still principally occurring at businesses.
- With the availability of 28.8K baud modems, ISDN lines and high performance/low price personal computers, home adoption of Internet access is on the rise and slated to have extremely high growth over the next five years. Adoption of cable modems could accelerate this trend.

- Formerly closed network online services such as America Online, Compuserve, and Prodigy are now offering Internet access and opening up their services. Other companies such as Microsoft as well as divisions of MCI, AT&T, and others are attempting to put in place Internet online services in which a range of programming content is presented.
- Companies such as Yahoo! which provide means to navigate the Web are growing rapidly as measured by amount of end user traffic.
- These high traffic sites already provide a high volume platform for delivering electronic advertising.

During this stage, and sustainably for all stages to come, there is one fundamental need which users have: The location of meaningful information easily and quickly on this large and exponentially growing source called the Internet.

Competition

Yahoo! intends to effectively beat any emerging competitors by:

- Establishing broader distribution earlier than any other competitor in order to maintain the Yahoo! guide as the most widely used in its class.
- Broadening the product line faster than the competition through the introduction of vertical market focused guides and personalizeable editions of the guide.
- Staying ahead of the competition with regular core product updates which continue to make it faster, easier to use, and more effective.
- Delivering high quality audiences and compelling results to advertisers.

Risks

The main risks facing Yahoo! are:

- *The ability to increase traffic and enhance the Yahoo! brand.* Management believes it can achieve both these goals.
- *Ability to introduce key new products faster and better than the competition.* We believe that our current core technologies and platform will allow us to do this if supplemented by funded expansion of product development and marketing functions.
- *Ability to develop an international presence and leading brand internationally before the competition.* At the present time, Yahoo! is being pursued by a number of very high visibility and capable international affiliates. The funded addition of limited marketing and business development resources will allow us to respond to these opportunities in a timely way.
- *The introduction of competitive products internally developed by access providers.* While there is no assurance that this will not happen, we have secured relationships with several of the leading providers already in which the Yahoo! product is featured and are in advanced discussions

with others. We believe that many of the access providers already respect Yahoo!'s strong brand, comprehensive guide and focus and are concluding that they will not be inclined to reinvent this late in lieu of a mutually favorable affiliate business relationship with Yahoo!.

■ *Ability to scale our support of both the traffic through our main site as well as mirror sites of our affiliates.* If the demands of traffic outgrow the bandwidth of servers we install, then response rates might go down and lead to customer dissatisfaction. Yahoo! has successfully scaled and operated its server site. We believe we will be able to support the needed growth.

■ *That the growth of the Internet industry as a whole slows significantly, or that the adoption of the Web as a significant platform for advertising does not grow as projected.* These are both out of Yahoo!'s control. However, the company believes that the industry is in a secure phase of adoption which should fuel growth.

Yahoo!'s sustainable advantages

The Internet is in a period of market development characterized by extremely high rates of both user traffic growth and entry of new companies focused on various products and services. By virtue of its early entry, Yahoo! has developed its current position as the leader in its segment. Yahoo!'s ability to sustain and grow its position will depend on a number of things. Some of these relate to its current core advantages and others relate to future execution of its strategy.

At present, Yahoo!'s core strategic advantages include:

■ *It's strong brand.* The company executed early and well with its unique, context focused, quick and intuitive guide and benefited from the widespread adoption of the Yahoo! product. The guide is the standard in the world of Web navigation.

■ *Yahoo!'s scalable core technology in search engine, database structure, and communication software.* These core technologies are relevant to the user's experience to the extent that it enables the Yahoo! customer's access to a broader array of high quality information in an intuitive way, faster than any competitor's product.

BARBARA'S OPTIONS

Introduction

Barbara Arneson strolled through the campus of the University of Maryland at College Park on a Spring evening in 2009. She often came to the quad at the end of a day for some quiet reflective time. Tonight she was mulling over her career options and the path her life would take in the next few years. Graduation was only five days away, and tomorrow Barbara would pick up her parents at the airport for a short visit and the ceremony. She hoped to be able to share her career decision with them and then relax over the next few days.

Having completed an undergraduate degree in biology a year prior, Barbara would soon be receiving her masters degree in computer science from Maryland before beginning a career in high technology. Barbara felt lucky that she had been offered a number of career options, thanks to the strong high-tech economy and growth of investment in the new field of bioinformatics, which applied software and Internet technology to the process of identifying and using genetic information for the life sciences industries. Barbara's personal objective was to work in product development and eventually move into management and maybe someday start her own company. As an interim step, she thought she might return to graduate school for an MBA in a few years.

On this beautiful evening, however, Barbara had to make a decision between two attractive job offers.

Barbara's Dilemma

Barbara had interviewed extensively with high-technology companies, and had decided BioGene Systems, Inc., and InterWeb Genetics Corp. were her top choices. She had hoped to receive an offer from at least one of them, but had received offers from both. Now she had a tough decision.

BioGene was a 7-year-old company that was successful and had grown rapidly. Its product-development group was highly regarded technically, and Rasha Motwani, to whom Barbara would report, had more than 10 years of development experience in several product areas in which Barbara was interested. She liked Rasha and felt she would learn a lot from her.

InterWeb was a start-up that had been funded about a year ago by two venture capital investors, both of whom had successfully funded technology companies. The InterWeb team was hard at work on its first product, which would launch in about a year. Barbara would be joining a team of about 10 engineers, most of whom had extensive experience in the areas relating to the product they were developing. The technical team leader was Robert Jackson, one of the company founders, who was only a few years older than Barbara and had a reputation as a technical "visionary."

Barbara had been trying to decide for several days. As she strolled toward her dorm, she reflected on her thoughts. "Okay, I've been trying to make a decision on the basis of my key priorities, namely the types of projects I would be working on, the quality of the people I would be working with, and the opportunities for personal growth. Although BioGene and InterWeb are not directly comparable—each has potential strengths and weaknesses—the fact is that I think I would be equally happy with either one. For me, the decision is a toss-up. I guess the only way to determine which is better is to evaluate the financial offers. Because both proposed similar salaries and benefits, that means analyzing the stock option offers."

The Stock Option Packages

Not all companies offer stock options to new college graduates. Because of Barbara's success at school and a hot job market in bioinformatics, both BioGene and InterWeb had included a stock option package in their offers.

A stock option gives an individual the right to purchase, during a fixed time period called the "term," a certain number of shares of stock from the company at a fixed price, called the "exercise price." The option expires at the end of the term, but it can be "exercised" (or bought) all or partially during the term, usually subject to certain conditions such as "vesting." Options have no financial risk to the employee—if the value of the stock remains below the exercise price, he or she need not ever exercise the option.

BioGene had offered Barbara options for 6,000 shares at an exercise price of $16.00 per share. BioGene had gone public in June 2008 at $10 per share, and the stock was currently selling for roughly $16. In extending the offer to Barbara, Karen Hershfield, manager of recruiting for BioGene, had said, "We have a proven record of rapid, profitable growth, and we expect that kind of growth to continue. You should receive a handsome return on this option package."

InterWeb had offered 60,000 shares at an exercise price of $0.10 per share. Because the company was private, this price reflected an arbitrary pricing decision at the time of the venture capital investment received by the company. Robert Jackson, in discussing the offer with Barbara, had commented, "The great thing about going with a start-up is that if it is successful, everybody gets rich. Our business plan shows that we should be in a position to do our IPO (initial public offering) in 3 or 4 years, and because companies usually go public at $10 to $15 per share, you can see that this option could be worth almost a million dollars!"

Both options had identical restrictions. Vesting was 25% per year with a term of 4 years. This meant that at the end of one year of employment, Barbara had the right to exercise 25% of the shares at her discretion any time. At the end of two years of employment, she would vest for another 25%, and so on. If at any time she left the company, she could, within 90 days, exercise any options for which she was vested, but any unvested options would terminate. Any unexercised portion of the option would expire at the end of 10 years from the start of employment.

Having decided that she would be equally satisfied with joining either company, Barbara was understandably excited by the prospect of big financial gains on stock options. She had even begun day-dreaming about what she could do with a financial windfall—travel abroad for a year, buy a new car for her parents, and pay for an MBA without worrying about 2 years of no salary and huge loan payments. Because she had learned a lot about financial analysis in her entrepreneurship courses, she had obtained and analyzed financial data from both companies, as shown in Exhibits 1 and 2. She also had analyzed the opportunity and strategies of both companies, and felt each had excellent prospects of achieving its objectives. She had examined stock market data for public high-tech companies and knew that the price/earnings (PE) ratios of bioinformatics companies averaged 25. InterWeb had offered a lot more shares than BioGene, but there was a higher element of risk. She knew that many start-ups failed to be successful.

She recalled that autumn day 5 years ago when her parents had dropped her off at school for the start of her freshman year. When she picked them up at the airport tomorrow, she wanted to share her career decision with them and make them proud of her.

Questions

1. What is the number of shares outstanding at BioGene as of May 31, 2009? What is its current PE ratio? Why do you think it is higher than the current average of other bioinformatics companies (Hint: consider the recent annual growth rates of revenues and profits)?

2. What is Barbara's potential percentage ownership in each firm?

3. Compare the firms in 4 years (i.e., 2013) when the stock options will be fully vested. Assuming Barbara remains employed until that time, which stock option offer is better? Make sure to include the cost of the stock options and state all critical assumptions.

4. In addition to compensation matters, what other factors would you suggest Barbara consider in making her decision?

EXHIBIT 1 BioGene Systems profit and loss history.

	FY05	FY06	FY07	FY08	FY09
Revenue	10.1	17.1	25.6	42.4	74.6
Cost of revenue	4.0	6.8	10.2	17.0	29.8
Gross margin	6.1	10.3	15.4	25.4	44.8
Expenses:					
Engineering	1.2	2.1	3.1	5.1	9.0
Marketing	2.5	4.3	6.4	10.6	18.7
G&A	0.6	1.0	1.5	2.5	4.5
Total expenses	4.3	7.4	11.0	18.2	32.2
Pretax profit	1.7	2.9	4.4	7.2	12.7
Taxes	0.7	1.2	1.7	2.9	5.1
After-tax profit	1.0	1.7	2.6	4.3	7.6
EPS (earnings per share)	$0.06	$0.08	$0.12	$0.19	$0.33

Note: (1) All numbers in millions except EPS. (2) Stock is traded on NASDAQ. Closing price on 5/31/09 was $16.25. (3) End of fiscal year for 2009 is June 30. FY09 numbers are estimates by stock market analysts and consistent with guidance by management.

EXHIBIT 2 InterWeb proforma profit and loss projections from business plan.

	FY08	FY09	FY10	FY11	FY12	FY13
Sales	0.0	0.0	5.0	20.0	41.0	62.0
Cost of sales	0.0	0.0	2.0	8.0	16.4	24.8
Engineering	0.7	1.0	1.5	2.4	4.9	7.4
Marketing	0.3	0.5	1.3	5.0	10.3	15.5
G&A	0.1	0.2	0.3	1.2	2.5	3.7
Total expenses	1.1	1.7	3.1	8.6	17.7	26.6
Pretax profit	−1.1	−1.7	0.0	3.4	7.0	10.5
Taxes	0.0	0.0	0.0	0.2	2.8	4.2
After-tax profit	−1.1	−1.7	0.0	3.2	4.2	6.3

Note: (1) 118.6 million shares outstanding as of 5/31/09. Management's business plan requires no additional venture capital or other funding. (2) End of fiscal year for 2009 is June 30.

SOLIDWORKS

In August, 1994, 12 months after Jon Hirschtick left a great job to found a new venture in the software industry, SolidWorks, the deal was looking good. The seed capital discussions had shifted into high gear as soon as Michael Payne joined the SolidWorks team. After working on the deal for nine months, Axel Bichara, the Atlas Venture vice president originating the project, finally got a syndicate excited about it: Atlas Venture, North Bridge Venture Capital Partners, and Burr, Egan, Deleage & Co. presented an offer sheet to SolidWorks two weeks after Michael was on board.

This process was particularly interesting because Jon and Axel had worked together for most of the past eight years. They met at MIT in 1986 and cofounded Premise, Inc., a computer aided design (CAD) software company, in 1987. After Premise was bought by Computervision, they joined that team as managers. Now, they sat on opposite sides of the table for Axel's first deal as the lead venture capitalist.

Jon and the other founders thought the valuation and terms were fair, but the post-money* equity issue was unresolved. They had to decide how much money to raise. Did they want enough capital to support SolidWorks until it achieved a positive cash flow, or should they take less money and attempt to increase the entrepreneurial team's post-money equity?

If they took less money now, they could raise funds later, when SolidWorks might have a higher valuation. But they would be gambling on the success of the development team and the investment climate. If their product was in beta testing with high customer acceptance, raising more money would probably be fast and fun, but if they hit any development snags, the process could take a lot of time and yield a poor result.

Jon Hirschtick: 1962–1987

Jon grew up in Chicago in an entrepreneurial family. He fondly remembers helping with his father's part-time business by traveling to stamp collectors' shows across the Midwest. In high school, he was self-employed as a magician.

The entrepreneurial impulse continued during his undergraduate years. Jon recalls the blackjack team he played with at MIT:

> We raised money to get started. At the same time, we developed a probabilistic system for winning at blackjack. The results were amazing! We tripled our money in the first six months, doubled it during the next six months, and doubled it again in the next six months. We produced a 900

This case was written by Dan D'Heilly and Tricia Jaekle under the direction of Professor William Bygrave. Funding provided by the Ewing Marion Kauffman Foundation and the Frederic C. Hamilton Chair for Free Enterprise Studies. © Copyright Babson College, 1995. All rights reserved.

---

*Post-money valuation: the value of a company's equity after additional money is invested.

percent annualized return. I learned a useful lesson: you really can know more than the next guy and make money by applying that knowledge. We tackled blackjack because people thought it was unbeatable; we studied it, and we won. The same principle applies to entrepreneurship. Opportunities often exist where popular opinion holds that they don't.

Jon's introduction to CAD came from a college internship with Computervision during the summer of 1981. Computervision was one of the most successful start-up companies to emerge during the 1970s. By the early 1980s, it dominated the CAD market.

After earning a master's degree in mechanical engineering at MIT, Jon managed the MIT CAD laboratory. He supervised student employees, coordinated research projects, and conducted tours for visitors.

Axel Bichara: 1963–1987

Axel was born in Berlin and attended a French high school. In 1986, while studying at the Technical University of Berlin for a master's degree in mechanical engineering, he won a scholarship to MIT. Axel had worked in a CAD research lab in Germany, so he selected the CAD laboratory for his work-study assignment at MIT.

Early CAD Software

CAD software traces its roots to 1969, when computers were first used by engineers to automate the production of drawings. CAD was used by architects, engineers, designers, and other planners to create various types of drawings and blueprints. Any company that designed and manufactured products (e.g., Ford, Sony, Black & Decker) was a prospective CAD software customer.

An Entrepreneurship Class: January 1987

Visitors to the MIT CAD lab often complained about problems that Jon knew he could solve. He enrolled in an entrepreneurship class to write a business plan for a CAD start-up company, Premise, Inc. Jon described the decision to quit his job and start a company:

> I once heard Mitch Kapor [founder of Lotus Development Company] use a game show metaphor to describe the entrepreneurial impulse. He said, "Part of the entrepreneurial instinct is to push the button before you know the answer and hope it will come to you before the buzzer." That's what happened for us: we didn't know how to start a company, or how to fund it, but Premise got rolling, and we came up with answers before we ran out of time.

Jon and Axel were surprised and delighted to find each other in the entrepreneurship class. They had worked together for the past month on a project at the CAD lab, and they decided to become partners in the first class session. Axel recalled: "It was a coincidence that we enrolled in the same class, but it was clear that we should work together. Jon had had the idea for a couple of months, and we started work on the product and the business plan immediately."

Axel took the master's exam at MIT in October 1987 and at Technical University of Berlin in July 1988. He was still a student at both universities when he and Jon started Premise. Axel graduated with highest honors from both institutions.

Premise, Inc.: 1987–1991

Premise went from concept to business plan to venture capitalist-backed start-up in less than six months. As Axel remembered:

> The class deadline for the business plan was May 14. On June 1, we had our first meeting with venture capitalists, and by June 22, we had a handshake deal with Harvard Management Company for $1.5 million. We actually received an advance that week. It was much easier than it should have been, but the story's 100 percent true.

In the first quarter of 1989, Premise raised its second round of capital. Harvard Management and Kleiner Perkins Caufield & Byers combined to finance the product launch. The product shipped in May to very positive industry reviews, but sales were slow. Premise's software didn't solve a large mass-market problem. As Jon later recalled: "I've seen successful companies get started without talent, time, or money—but I've never seen a successful company without a market. Premise targeted a small market. I had a professor who said it all, '"The only necessary and sufficient condition for a business is customers.'"

By the end of 1990, the partners had decided that the best way to harvest Premise was an industry buyout. They hired a Minneapolis investment banking firm to find a buyer. Wessels, Arnold & Henderson was considered one of the elite investment banking firms serving the CAD industry. Premise attracted top-level service providers because of the prestige of its venture capitalist partners. Jon explained: "Several bankers wanted to do the deal, and a big reason was because they wanted to work with our venture capitalists. We had top venture capitalists, and that opened all kinds of doors. This is often under-appreciated. I believe in shopping for venture capital partners."

Wessels, Arnold & Henderson were as good as their reputation. As Axel recalled: "We sold Premise to Computervision on 7 March 1991. Computervision bought us for our proprietary technology and engineering team. It was a good deal for both companies."

Computervision: 1991–1993

As part of the purchase agreement, Jon and Axel joined the management team at Computervision. They managed the integration of Premise's development team and product line for one year before Axel left to study business in Europe. Jon stayed on after Axel's departure.

Revenues for the Premise team's products grew 200 percent between 1991 and 1993, and perhaps as important as direct revenue, their technology was incorporated into some of Computervision's high-end products. In January 1993, Jon was promoted to director of product definition for another CAD product. He stayed in this position for eight months. After two years at Computervision, he was ready for new horizons. He resigned effective August 23, 1993. (See Exhibit 1 for excerpts from his letter of resignation.)

After a holiday in the Caribbean, Jon purchased new computer equipment, called business friends and associates, and began working on a business plan. He didn't have a clear product idea, but his market research suggested that the time was ripe for a new CAD start-up.

EXHIBIT 1 Excerpts from Jon Hirschtick's letter of resignation from Computervision (CV).

This is my explanation for wanting to leave CV. . . . The other day you asked me whether I was leaving because I was unhappy, or whether I really want to start another company. I strongly believe that it is because I really want to work on another entrepreneurial venture.* I want to try to build another company that achieves business value. . . .

I am interested in leaving CV to pursue another entrepreneurial opportunity because I seek to:

1. Be a part of business strategy decisions. I want to attend board meetings and create business plans, as I did at Premise.

2. Select, recruit, lead, and motivate a team of outstanding people. I believe that one of my strengths is the ability to select great people and form strong teams.

3. Represent a company with customers, press, investors, and analysts. I enjoy the challenge of selling and presenting to these groups.

4. Work on multidisciplinary problems: market analysis, strategy, product, funding, distribution, and marketing. I am good at cross-functional problem-solving and deal-making.

5. Work in a fast-moving environment. I like to be in a place where decisions can be made quickly, and individuals (not just me) are empowered to use their own judgment.

6. Work in a customer-driven and market-driven organization. I find technology and computer architecture interesting only as they directly relate to winning business. I want to focus on building products customers want to buy.

7. Have significant equity-based incentives. I thrive on calculated risks with large potential rewards.

8. Be recognized for having built business success. I measure "business success" by sales, profitability, and company valuation; I want to directly impact business success. Recognition will follow. I admit that this ego-need plays a part in my decision.

Summary

I've decided I want to work on an entrepreneurial venture. . . . This is more a function of what I do best than any problems at CV. . . . I don't have any delusions about an entrepreneurial company being any easier. I know first-hand that start-up companies have at least as many obstacles as large established companies—but they are the obstacles I want.

*Underlines in original.

CAD Software Market in the 1990s

By the 1990s, the hottest CAD software performed a function called solid modeling. Solid modeling produced three-dimensional computer objects that resembled the products being built in almost every detail. It was primarily used for designing manufacturing tools and parts. Solid modeling was Solid-Works' focus. The key benefits driving the boom in solid modeling were:

1. Relatively inexpensive CAD prototypes could be accurate enough to replace costly (labor, materials, tooling, etc.) physical prototypes.
2. The elimination of physical prototypes dramatically improved time-to-market.
3. More prototypes could be created and tested, so product quality was improved.

However, not all CAD software could manage solid modeling well enough to effectively replace physical prototypes.

Most vendors offered CAD software based on computer technology from the 1970s and 1980s. IBM, Computervision, Intergraph, and other traditional market leaders were losing market share because solid modeling required software architecture that worked poorly on older systems.

As one of the industry's newest competitors, Parametric Technology Corporation (PTC) was setting new benchmarks for state-of-the-art solid modeling software. (It was an eight-year-old company in 1994.) CAD was a mature and fragmented industry with many competitors, but PTC thrived because other companies tried to make older technology perform solid modeling functions.

Worldwide mechanical CAD software revenues were projected at $1.8 billion for 1995, with IBM expected to lead the category with sales of $388 million. PTC was growing over 50 percent annually and had the second highest sales, with $305 million in projected revenue. Industry analysts predicted 3 percent to 5 percent revenue growth per year, with annual unit volume projected to grow at 15 percent. The downward pressure on prices was squeezing margins, so many stock analysts thought that the market was becoming unattractive. However, PTC traded at a P/E between 21 and 40 in 1994.

Axel after Computervision: 1992–1994

After five years in the United States, Axel decided to attend an MBA program in Europe. From his experiences at Premise and Computervision, he had become intrigued with the art and science of business management, and he was ready for a geographic change.

INSEAD was his choice. Located in Fontainebleau, an hour south of Paris, INSEAD was considered one of the top three business schools in Europe. The application process included two alumni interviews, and one of Axel's interviewers was Christopher Spray, the founder of Atlas Venture's Boston office. Atlas Venture was a venture capital firm with offices in Europe and the United States. It had $250 million under management in 1994.

Since Axel had a three-month break before INSEAD started, Chris asked him to consult on a couple of Atlas Venture's projects. Axel found he enjoyed evaluating business proposals "from the other side of the table." He graduated in June 1993 and joined the Boston office of Atlas Venture as a vice president with responsibility for developing high-tech deals.

Axel reflected on the relationship between business school training and venture capital practice:

> I was qualified to become a venture capitalist because of my technical and entrepreneurial background; business school just rounded out my skills. You do not need a bunch of MBA courses to be a successful venture capitalist. Take finance, for example, I learned everything I needed from the core course. People without entrepreneurial experience who want to be venture capitalists should take as many entrepreneurship courses as possible.

Jon Founds SolidWorks: 1993–1994

Jon's business plan focused on CAD opportunities. He explained:

> I knew that this big market was going through major changes, with more changes to come. From an entrepreneur's perspective, I saw the right conditions for giving birth to a new business. I also knew I had the technical skills, industry credibility, and vision needed to make it happen. This was a pretty rare situation.

SolidWorks' product vision evolved slowly from Jon's personal research and from discussions with friends. He was careful to avoid using research that Computervision might claim as proprietary. He was concerned about legal issues, because he would be designing software similar to what Computervision was trying to produce. Axel explained:

> Both Computervision and SolidWorks wanted to produce a quality solid modeling product. Solid modeling technology was still too difficult to learn and use. Only PTC's solid modeling software really worked well enough. The rest made nice drawings but could not replace physical prototypes for testing purposes.

There were only 50,000 licensed solid modeling terminals in the United States, and most of them belonged to PTC, but there were over 500,000 CAD terminals. There were two main reasons PTC did not have a larger market: (1) its products required very powerful computers, and (2) it took up to nine months of daily use to become proficient with PTC software. Solidworks' goal was to create solid modeling software that was easier to learn and modeled real-world parts on less specialized hardware (see Table 1).

TABLE 1 Competitive positioning grid.

	Computer aided drafting and add-ons	Production solid modeling
Low-end:		
Windows	Autodesk	SolidWorks
~$5K per station	Bentley	
VAR channel	CADKEY	
High-end:		
UNIX	Applicon	PTC
~$20K per station	CADAM	
Direct sales		

This vision was not unique in the industry. Many CAD companies were developing solid modeling software, and the low-end market was wide open. SolidWorks's major advantage was its ability to use recent advances in software architecture and new hardware platforms—it wasn't tied to antique technology. Attracting talented developers was the top priority in this leading-edge strategy.

Team Building

Jon's wife, Melissa, enthusiastically supported his decision to resign from Computervision. Jon explained: "Some spouses couldn't deal with a husband who quits a secure job to start a new company. Melissa never gave me a hard time about being an entrepreneur."

Jon described his priorities in October 1993, when he decided to launch SolidWorks:

I knew I needed three things: good people, a good business plan, and a good proof-of-concept.* I needed a talented team that could set new industry benchmarks, but there was no way I could get those people without a persuasive prototype demonstration.

The venture capitalists wanted a solid business plan, but that wouldn't be enough. They wanted a strong team. I needed fundable people who were also CAD masters. Venture capitalists couldn't understand most complex technologies well enough to be confident a high-tech business plan was really sound, so they looked at the team and placed their bets largely on

---

*Proof-of-concept is a term that refers to a computer program designed to illustrate a proposed project. Also referred to as a prototype, it is used for demonstration purposes, and it is limited but functional in ideal circumstances.

that basis. If the proof-of-concept attracted the team, then the team and the business plan would attract the money. I needed a team that could create the vision and make venture capitalists believe it was real.

Jon worked on finding the team and developing the proof-of-concept concurrently; but the proof-of-concept was his first priority. He worked on it daily. In his search for cofounders, Jon talked to dozens of people; he even posted a notice on the Internet, but "none of those guys worked out."

Recruiting posed another dilemma—how to get people to work full time without pay, while the company retained the right to their output? He resolved this problem by creating consulting agreements that gave SolidWorks ownership of employees' work and made salaries payable at the time of funding. As it turned out, this arrangement only lasted nine months. Jon described his approach to recruiting:

> I always paid for the meal when I talked with someone about SolidWorks. I wanted them to feel confident about it, and that meant that I had to act with confidence. The deal I offered was: no salary, buy your own computer, work out of your house, and we're going to build a great company. I'd done it before, so people signed on.

Axel described Jon's management style as, "visionary, he's a talented motivator, and a strong leader."

Robert Zuffante: CAD Engineer/Consultant

A major development in 1993 was the addition of super-star consultant Bob Zuffante as manager of proof-of-concept development. Jon needed time to write the business plan and recruit his team. He had been working on the prototype for over a month when Bob took over development. Jon recalled the situation:

> I hadn't seen him since we were students together at MIT, but when I thought about the skills I needed, my mental Rolodex came up with his name. I always thought about working with him again. We talked in late November, and about a month later, he began work on the prototype.

Bob knew Jon and Axel from MIT, where he earned a master's degree in mechanical engineering. He had worked in the CAD industry for over 10 years and had managed a successful consulting business. His arrival at SolidWorks allowed Jon to focus on other pressing issues.

Scott Harris: CAD Marketer

Scott Harris worked at Computervision for 11 years, where he managed development and marketing activities. Most notably, Scott was the founder and manager of Computervision's product design and definition group. He also managed the 11-person solid modeling development group and acted as technical liaison between Computervision's customers and R&D engineers.

Scott was let go by Computervision during a large-scale layoff. He was skeptical when Jon first told him about the SolidWorks vision, but he became a believer after seeing a proof-of-concept demonstration. Scott stopped looking for a job and started working full time for SolidWorks almost immediately. Scott was impressed, "The prototype was the embodiment of a lot of the things I was thinking about. This was the way solid modeling should perform."

Scott started with SolidWorks about six weeks after Bob signed on. He became involved in the marketing sections of the business plan and in the product definition process. He ran focus groups, conducted demonstrations for potential customers, and analyzed the purchasing process. He kept the development team focused on customer needs—how did customers really use CAD software, and what did they need that current products lacked?

The Business Plan

When Bob came on in January, Jon turned to the business plan with a passion. The plan went through a number of versions as Jon and his advisors wrestled with key issues such as positioning, competitive strategy, and functionality. By the end of March, the plan was polished enough for Jon to show it to venture capitalists. Axel recalled:

> Jon and I decided that the business plan was ready to show in April, so I scheduled a presentation at Atlas. Jon gave the presentation to Barry [Fidelman, Atlas general partner] and myself—market, team, and concept. Overall, Barry was encouraging, but not excited. He thought Jon's story was not crisp enough; he was looking for money to take on some very large companies, and the CAD market was not that attractive. It was a rocky start.

Initial Financing Attempts

In addition to negotiating with Atlas Venture, Jon met with other venture capital firms and rewrote the business plan several times. Axel described the rationale behind this process:

> If you talk to too many people and you do not make a good impression, it will be much harder to get funding, because the word on the street will be, "this deal will not fly." Meet with four or five venture capitalists at most, then revise the plan if you are not getting the right response. After each major revision, show it again to the lead venture partner.

While there were promising discussions with several venture capitalists, Atlas did not want to be the sole investor, and SolidWorks did not win support from other venture capitalists during the spring or summer.

Jon was contacted by an established CAD software company in May 1994. It wanted to acquire SolidWorks—essentially the development team and the prototype. The proposal was attractive; it included signing bonuses and stock. Scott recalled his excitement: "This was a big shot in the arm. It meant that other industry insiders respected our vision and talent enough to put up their money and take the risk. This was like an cold bucket of Gatorade on a hot day."

Jon stopped seeking venture capital for about a month while he considered the buyout offer. If the offer was a boost to morale, the way the team rejected it was even more meaningful. Jon talked to each person (several other programmers had joined during the spring), and they were unanimous in wanting to continue toward their original goal. Affirming their commitment reinvigorated the team.

Turning Point: Michael Payne, CAD Company Founder

The most significant advance that summer began with a due diligence meeting set up by Atlas Venture. Atlas wanted the SolidWorks team to meet its agent, Michael Payne, who had recently resigned from PTC. Michael had cofounded PTC, the number one company in CAD software. He was one of the most influential people in the industry.

Michael had grown up in London. He earned his bachelor's degree in electrical engineering from Southampton University and his master's degree in solid-state physics from the University of London. He came to the United States and worked many years for RCA designing computer chips. Michael continued his education at Pace University, where he earned an MBA. His senior CAD development experience began in the 1970s, when he ran the CAD/CAM design lab at Prime Computer. He was subsequently recruited by Sam Geisberg, the visionary behind PTC. Michael recalled their first meeting in 1986: "Sam had some kind of crazy prototype, and I said, 'Hey, we can do something with that. This is what we should be working on.'"

PTC was founded in 1986 with Michael as vice president of development, and within five years the company had created a new set of CAD industry benchmarks. For FY 1993, PTC sales were $163 million, it earned a pretax profit margin over 40 percent, and it reached a market capitalization[*] of $1.9 billion. Michael's reputation as a development manager was outstanding. Remarkably, PTC had never missed a new product release date, and it released products every six months. This was considered a near-impossible feat in software development. He left PTC in April 1994 during a management dispute, about two months before the due diligence meeting with SolidWorks.

Jon had never met Michael but knew by reputation that he was a tough character. The SolidWorks team was worried about two possibilities: that Michael would say they were on the wrong track, or that he might take their ideas back to PTC. Jon recalled the meeting:

---

[*]Market capitalization is the value of the company established by the selling price of the stock times the number of shares outstanding.

Bob and I were on one side of the table and Michael and Axel on the other. I decided to gamble on a dramatic entrance. Before we told him anything about SolidWorks, I asked Michael to show his cards. I asked him to tell us what he thought were the greatest opportunities in the CAD market. Michael mentioned many of the things we were targeting. I couldn't imagine a better way to start the meeting.

We presented our plan and prototype. Michael asked us a lot of tough, confrontational questions. Afterwards, he told Atlas Venture, "These guys have a chance." Coming from him, that was high praise.

The due diligence meeting was also the beginning of a dialogue between Michael and Jon about joining SolidWorks. Over the next couple of months, Michael decided to join the team. Jon described the synergy between them:

You almost couldn't ask for two people with more different styles, but we got along well because we were united in our philosophy and vision. We found that our stylistic differences were assets; they created more options for solving problems.

Michael talked about his motivation for joining the SolidWorks team:

I couldn't go work for a big company because I didn't have any patience for petty politics. A start-up was my only option. The larger the company, the more focused it would be on internal issues rather than on making a product that customers would buy. Customers don't care about technique, they care about the benefits of the technology.

Jon focused on CAD features that he knew customers wanted, and he had a prototype demonstrating that he could do it. It was also quicker and easier than what was on the market. Being able to develop it was another matter. They still had to build it. Implementation, that's where he would be useful. He told them, "Give me whatever title you want; I just want to run development."

Team Adjustments

Michael's arrival created an imbalance in the SolidWorks team, and it took time to sort it out. In fact, Michael didn't join the team until the last week in August. Jon described his thoughts about team cohesion:

When I decided to start SolidWorks, I had three goals: (1) work with great people, (2) realize the vision of a new generation of software, and (3) make a lot of money. We didn't go looking for Michael Payne, but when he came along, it was an easy decision. It can be hard to bring in strong players, but if those are your three goals, the decision falls out of the analysis rather naturally.

Bob and I had to give up the reins in some areas so Michael could come on board. We weren't looking for a top development manager because we thought we already had two. The change took some getting used to, but it was clearly the right thing to do.

Jon focused on team building, and Michael became the development manager. There were still big talent gaps, especially in sales and finance, but those positions could be filled when they were closer to the product launch. Michael was satisfied, "We didn't have a vast team, but you don't start out with a vast team, and we had a terrific nucleus."

September 1994

Atlas arranged for Jon to talk with venture firms interested in joining the investment syndicate. The team met with Jon Flint of Burr, Egan, Deleage & Company and Rich D'Amore of North Bridge Venture Partners. After completing their due diligence investigations, both firms joined the syndicate. Jon Hirschtick recalled the situation:

> I was pleased that Jon Flint and Rich D'Amore decided to invest. I had met Jon many years earlier and thought very well of him. Rich also impressed me as a very knowledgeable investor. Both had excellent reputations and I looked forward to having them join our board.

An offer sheet was presented to SolidWorks two weeks after Michael officially joined the SolidWorks management team. Now the team had to decide how much money they really wanted. Michael's last venture, PTC, only needed one round of capital, and this team wanted to go for one round, too. SolidWorks' monthly cash burn rate was projected to average about $250,000 and they planned to launch the product in a year, so they needed $3 million for development. Sales and marketing would also need money; they decided that $1 million should be enough to take them through the product launch to generating positive cash flow. To that total, they added a $500,000 safety margin. SolidWorks asked Atlas to put together an offer sheet based on raising $4.5 million.

SolidWorks received the offer sheet during the first week of September. It gave a $2.5 million pre-money valuation with a 15 percent post-money stock option pool.* For SolidWorks' business plan projections, see Exhibit 2. These terms were fairly typical for a first round deal, but the SolidWorks team didn't like what happened to their post-money equity when they ran the numbers.

Questions

1. Why has this deal attracted venture capital?
2. Can the founders optimize their personal financial returns and simultaneously ensure that SolidWorks has sufficient capital to optimize its chance of succeeding? What factors should the founders consider?

---

*The pool of company stock reserved for rewarding employees in the future.

3. How can the venture capitalists optimize their return? What factors should they consider?

4. After you have answered questions 2 and 3, structure a deal that will serve the best interests of the founders, the company, and the venture capital firms.

EXHIBIT 2 Business plan projections.

	1994	1995	1996	1997	1998
Revenue	$ —	$ 175,000	$ 3,010,000	$8,225,000	$17,115,000
Cost of sales	$ —	$ 31,500	$ 541,800	$1,480,500	$ 3,080,700
Sales and marketing	$ 71,919	$ 765,920	$ 1,930,000	$3,030,000	$ 5,822,500
R and D	$ 605,544	$ 1,126,208	$ 1,350,000	$1,500,000	$ 2,050,000
G and A	$ 185,954	$ 445,175	$ 650,000	$ 800,000	$ 1,050,000
Total expenses	$ 863,417	$ 2,368,803	$ 4,471,800	$6,810,500	$12,003,200
Operating income	$ (863,417)	$(2,193,803)	$(1,461,800)	$1,414,500	$ 5,111,800
Margin analysis					
Cost of Sales		18.0%	18.0%	18.0%	18.0%
Gross Profit		82.0%	82.0%	82.0%	82.0%
Sales and marketing		437.7%	64.1%	36.8%	34.0%
R and D		643.5%	44.9%	18.2%	12.0%
G and A		254.4%	21.6%	9.7%	6.1%
Operating income (before taxes)		−1253.6%	−48.6%	17.2%	29.9%

ARTEMIS IMAGES

Christine Nazarenus tried to retain her optimism. Thirteen had always been a lucky number for her, but Friday, the thirteenth of July, 2001, had the earmarks of being the unluckiest day of her life. She was more than disappointed. She was shattered. Yet she knew that she had hard facts, not just gut feel, that offering images and products on the World Wide Web was the wave of the future. She was sure that the management team she had put together had the creativity and skills to turn her vision into reality. Managing her own company had seemed the obvious solution, but she hadn't counted on how overwhelming the start-up process would be. Now, two years later, she was trying to figure out what went wrong and if the company could survive.

It had been so clear on day one. Archived photographs and images had tremendous value if they could be efficiently digitized and catalogued. Sports promoters and publishers had stores of archived information, most of it inaccessible to those who wanted it. Owners and fans represented only part of the untapped markets that the Internet and digital technology could serve. She had conceived a simple business model: digitize documents using the latest technology, tag them with easy-to-read labels, and link them to search engines for easy retrieval and widespread use. But over the ensuing months, so many factors affected the look, feel and substance of the company that Artemis Images would become.

So many things seemed outside her control that she wondered how she could have been so sure of herself back in February of 1999. Enthusiastically, Chris had approached a number of friends and acquaintances to help in the formation of a new "dot.com" company that seemed a sure bet. Frank Costanzo, a former colleague from Applied Graphics Technologies (AGT), shared Chris's enthusiasm, as did long-time friend George Dickert. George, in turn, contacted Greg Hughes, who was enrolled in a Business Planning course. Grateful for the opportunity to help launch a real company, Greg took the idea and honed it as part of a class assignment. The plan was a confirmation of Chris's confidence in the venture. But as she looked over the original plan, she knew there was a lot of work yet to do. Greg understood the business idea, but he didn't understand the work involved to actually run a business. George and Frank understood digital technology and project management, but, like Chris, had never launched, much less worked for, a start-up company. Chris knew that she had the technology and talent she needed

© 2002 by Joseph R. Bell, University of Northern Colorado, and Joan Winn, University of Denver. Published in *Entrepreneurship Theory & Practice,* 28(2) Winter 2003. The authors wish to thank Chris Nazarenus and the staff of Artemis Images for their cooperation in the preparation of this case. Special thanks to Herbert Sherman, Southampton College, Long Island University, and Dan Rowley, University of Northern Colorado. This case is intended to stimulate class discussion rather than to illustrate the effective or ineffective handling of a managerial situation. *All events and individuals in this case are real, but some names may have been disguised.*

and felt confident that the four friends could construct a business model that would put Artemis ahead of the current image providers. Greg's business plan looked like the perfect vehicle to appeal to investors for the funds they needed to proceed.

The Business Idea

In 1999, Chris had been working for three years as VP-Sales out of the Colorado office of AGT, a media management company that provided digital imaging management and archiving services for some of the largest publishers and advertisers in the world. AGT had sent Chris to Indianapolis to present a content management technology solution to the Indianapolis Motor Speedway Corporation (IMSC) as it prepared marketing materials for the 2001 Indy 500. IMSC is the host of the 80-plus-year-old Indy 500, the largest single-day sporting event in the world, NASCAR's Brickyard 400, the second-largest single-day sporting event in the world, and other events staged at the track. Chris's original assignment was a clear one: IMSC needed to protect its archive of photographs, many of which had begun to decay with age. The archive included five million to seven million photographs and dynamically rich multimedia formats of video, audio, and in-car camera footage.

Chris discovered that the photo archives at IMSC were deluged with requests (personally or via letters) from fans requesting images. She was amazed that a relatively unknown archive had generated nearly $500,000 in revenues in 1999 alone. Further discussions with IMSC researchers revealed that requests often took up to two weeks to research and resulted in a sale of only $60 to $100. However, IMSC was not in a position, strategically or financially, to acquire a system to digitize and preserve these archives. Not willing to leave the opportunity on the table, Chris asked herself, "What is the value of these assets for e-commerce and retail opportunities?" Without a doubt, IMSC and some of her other clients (Conde Nast, BBC, National Motor Museum) would be prime customers for digitization and content management of their collections.

Chris knew that selling photos on the Internet could generate substantial revenue. She conceived of a business model where the system would be financed through revenue-sharing, rather than the standard model where the organization paid for the system up front. IMSC was interested in this arrangement, but it was outside the normal business practices of AGT. AGT wanted to sell systems, not give them away. They couldn't see the value of managing other organizations' content.

As Chris told the story, her visit to the archives at IMSC was her *Jerry Maguire* experience. In the movie, Jerry is sitting on the bed when everything suddenly becomes clear and now he must pursue his dream. Like Jerry, Chris believed so passionately that her idea would bear fruit that when AGT turned down Chris's request for the third time, she quit her job to start Artemis Images on her own.

Building a Team

When AGT was not interested in Chris's idea of on-site digitization and sale of IMSC's photo archives, Chris was not willing to walk away from what she saw as a gold mine. She contacted her friends and colleagues from AGT. Swept up in the dot.com mania, Chris named her company "e-Catalyst." e-Catalyst was incorporated as an S-corporation on May 3, 1999, by a team of four people: Christine Nazarenus, George Dickert, Frank Costanzo, and Greg Hughes. (See Exhibit 1 for profiles of these partners.) Expecting that they would each contribute equally, each partner was given a 25 percent interest in the company. Chris fully expected them to work as a team, so no formal titles were assigned, largely as a statement to investors that key additions to the team might be needed and welcomed. As another appeal to potential investors—and to broaden the team's expertise—Chris and George put together a roster of experts with content management, systems and technology experience as their first advisory board. Greg's professor and several local business professionals agreed to serve on the board of advisors, along with an Indy 500 winning driver-turned-entrepreneur, and Krista Elliott Riley, president of Elliott Riley, the marketing and public relations agency that represented Indy 500 and

EXHIBIT 1 Artemis Images management team 1999–2000.

Christine Nazarenus, 34, was formerly vice president of national accounts for AGT, one of the top three content management system providers in the world securing million dollar deals for this $500 million company. She is an expert in creating digital workflow strategies and has designed and implemented content management solutions for some of the largest corporations in the world including Sears, Conde Nast, Spiegel, Vio, State Farm, and Pillsbury. Ms. Nazarenus has extensive general management experience and has managed a division of over one hundred people. Chris holds a BA in communications from the University of Puget Sound.

George Dickert, 32, most recently worked as a project manager for the Hibbert Group, a marketing materials distribution company. He has experience with e-commerce, Web-enabled fulfillment, domestic and international shipping, call centers and CD-ROM. He has overseen the implementation of a million-dollar account, has managed over $20 million in sales, and has worked with large companies including Hitachi, Motorola, ON Semiconductor, and Lucent Technologies. Mr. Dickert has an MBA from the University of Colorado. George and Christine have been friends since high school.

Frank Costanzo, 40, is currently a senior vice president at Petersons.com. Petersons.com has consistently been ranked as one of the top one hundred sites worldwide. Mr. Costanzo is an expert in content management technology and strategy and was previously a vice president at AGT. Mr. Costanzo has done in-depth business analysis and created on-site service solutions in the content management industry. He has worked on content management solutions for the world's top corporations including General Motors, Hasbro, Bristol-Meyers Squibb, and Sears.

Greg Hughes, 32, is currently a senior sales executive with one of the largest commercial printers in the world. Mr. Hughes has 10 years' sales experience and has sold million-dollar projects to companies like US West, AT&T, R. R. Donnelly, and MCI. His functional expertise includes financial and operational analysis, strategic marketing, fulfillment strategies and the evaluation of start to finish marketing campaigns. Mr. Hughes has an MBA from the University of Colorado.

Le Mans Sports Car teams and drivers. Chris felt confident that her team had the expertise she needed to launch a truly world-class company.

Chris and George quit their jobs and took the challenge of building a company seriously. They contacted one of the Rocky Mountain region's oldest and most respected law firms for legal advice. They worked with two lawyers, one who specialized in representing Internet companies as general counsel and one who specialized in intellectual property rights. With leads from her many contacts at AGT, Chris contacted venture capitalists to raise money for the hardware, software licensing, and personnel costs of launching the business.

The dot.com bust of 2000 did not make things easy. Not wanting to look like "yet another dot.com" in search of money to throw to the wind, Chris and her team changed their name to Artemis Images. Artemis, the Greek goddess of the hunt, had been the name of Chris's first horse as well as her first company, Artemis Graphics Greeting Cards, her first entrepreneurial dabble at the age of 16. Chris had always been enthralled with beautiful images.

Artemis Images's Niche

In her work at AGT, Chris had observed that many organizations had vast stores of intellectual property (photos, videos, sounds and text), valuable assets often underutilized because they exist in analog form and may deteriorate over time. Chris's vision was to preserve and enable the past using digital technology and the transportability of the World Wide Web. Chris envisioned a company that would create a digitized collection of image, audio and video content that she could sell to companies interested in turning their intellectual property into a source of revenue.

Publishers and sports promoters were among the many organizations with large collections of archived photos and videos. Companies like Boeing, General Motors, and IMSC are in the business of producing planes, cars, or sporting events, not selling memorabilia. However, airplane, car, and sports fans are a ready market for photos of their favorite vehicle or videos of their favorite sports event.

Proper storage and categorization of archived photos and videos is complex and expensive. In 2000, the two common solutions were to sell the assets outright or to set up an in-house division devoted to managing and marketing them. Most organizations were unwilling to sell their assets, as they represented their priceless brand and heritage. Purchasing software and hiring specialized personnel to digitize and properly archive their assets was a costly proposition that lay beyond the core competence of most companies. Chris's work with AGT convinced her that there were literally thousands of companies with millions of assets that would be interested in a company that would digitize and manage their photo and video archives.

Chris understood a company's resistance to selling its archives, and the high cost of obtaining and scanning select images for sale. However, she also understood the value to an organization of having its entire inventory digitized, thus creating a permanent history for the organization. She proposed a revenue sharing model whereby Artemis Images would digitize a client's archives but would not take ownership. Instead, her company would secure exclusive license to the archive, with 85 percent of all revenue retained by Artemis Images and 15 percent paid to the archive owner. She expected that the presence of viewable archives on the Artemis Images website would lure buyers to the site for subsequent purchases.

The original business model was a "B2C" (business-to-consumer) model. Starting with the IMSC contract, Artemis Images would work with IMSC to promote the Indy 500 and draw the Indy race fans to the Artemis Images website. Photos of the current-year Indy 500 participants—and historical photos including past Indy participants, winners, entertainers, celebrities (e.g., Arnold Palmer on the Indy golf course)—would be added to IMSC's archived images and sold for $20 to $150 apiece to loyal fans. A customer could review a variety of photo options on the Artemis website, then select and order a high-resolution image. The order would be secured through the Web with a credit card, the image transferred to the fulfillment provider, and a hard copy mailed to the eager recipient. The website was sure to generate revenue easier than IMSC's traditional sales model of the past.

Having established the model with IMSC content in the auto racing market, Chris and George built the business plan around obvious market possibilities that might appeal to a wider range of consumers and create a comprehensive resource for stock photography. Since the Artemis Images team had prior business dealings with two of the three largest publishers in the world, publishing was the obvious target for future contracts. Future markets would be chosen similarly, where the Artemis team had established relationships. These markets would be able to build on the archive already created and would bring both consumer-oriented content and saleable stock images. Greg made a list of examples of some industries and the content that they owned:

- Sports: images of wrestling, soccer, basketball, bodybuilding, football, extreme sports
- Entertainment: recording artists, the art from their CDs, movie stars, pictures of events, pictures from movie sets
- Museums: paintings, images of sculpture, photos, events
- Corporations: images of food, fishing, planes, trains, automobiles
- Government: coins, stamps, galaxies, satellite imaging

As Chris and George worked with Greg to put together the business plan, they began to see other revenue-generating opportunities for their virtual-archive company. Customers going to IMSC or any other Artemis client's website would be linked to Artemis Images's website for purchase of photos or videos. Customer satisfaction with image sales would provide opportunities to sell merchandise targeted to specific markets and to syndicate content to other websites. For motor sports, obvious merchandise opportunities would include T-shirts, hats, and model cars. For landscapes, it might be travel packages or hiking gear. Corporate customers might be interested in software, design services, or office supplies. Unique content on Artemis Images's website could be used to draw traffic to other companies' sites. Chris and her team planned to license the content on an annual basis to these sites, creating reach and revenues for Artemis Images.

Another potential market for Artemis Images lay in the unrealized value of the billions of images kept by consumers worldwide in their closets and drawers. These images were treasured family heirlooms which typically sat unprotected and underutilized. Consumers could offer their photographs for sale or simply pay for digitization services for their own use. If just 10% of the U.S. population were to allow Artemis Images to digitize their archive and half of these people ordered just one 8" × 10" print, Artemis Images could create a list of 25 million consumers and generate revenues of approximately $250 million. Because images suffer no language barriers, the worldwide reach of the Internet and the popularity of photography suggested potential revenues in the billions.

Working together on the business plan, the Artemis team brainstormed ways they could attract customers to the Artemis Images site by providing unique content and customer experiences. A study by Forrester Research analyzed the key factors driving repeat site visits and found that high-quality content was cited by 75 percent of consumers as the number one reason they would return to a site. The Artemis team wanted to create a community of loyal customers through additional unique content created by the customers themselves. This would include the critical chats and bulletin boards that are the cornerstone of any community-building program. Artemis Images could continuously monitor this portion of the site to add new fan experiences to keep the experience "fresh." Communities would be developed based on customer interests.

As the company gained clients and rights to sell their archived photos and videos, Artemis would move toward a "B2B" (business-to-business) model. Chris and George knew marketing managers at *National Geographic,* CMG World Wide, the BBC, Haymarket Publishing (includes the Formula 1 archive), Conde Nast, and International Publishing Corporation. These large publishers controlled and solicited a wide range of subject matter (fashion, nature, travel, hobbies, etc.) yet often had little idea of what existed in their own archives or had difficulty in getting access to it. Finding new images was usually an expensive and time-consuming proposition. Artemis Images could provide the solution. For example, Conde Nast (publisher of *Vogue, Bon Appetit, Conde Nast Traveler, House & Garden,* and *Vanity Fair*) might like a photo for its travel magazine from the *National Geographic* archives. They would be willing to pay top dollar for classic stock images, given the number of viewers who would see the image. Price-per-image was typically calculated on circulation volume, much like royalty fees on copyrighted materials. Similarly, advertising agencies use hundreds of images in customer mockups. For example, an agency may desire an image of a Pacific island.

If Artemis Images held the rights to Conde Nast and *National Geographic,* there might be hundreds of Pacific island photos from which to choose. As with the B2C concept, a copy of the image would be transferred through the Web with a credit card or on account, if adequate bandwidth were available (only low-resolution images would be available to view initially), or via overnight mail in hard copy or on disk.

The transition from B2C to B2B seemed a logical progression, one that would amass a large inventory of saleable prints and, at the same time, draw in larger per-unit sales. The basic business model was the same. Artemis would archive photos and videos that could be sold to other companies for publication and promotion brochures. Chris and George expected that this model could be replicated for other vertical markets including other sports, nature, entertainment, and education.

While the refocus on the B2B market seemed a surer long-term revenue stream for the company, both B2B and B2C were losing favor with the investing community. Chris and George refocused the business plan as an application service provider (ASP). With the ASP designation, Artemis Images could position itself as a software company, generating revenue from the licensing of its software processes. In 2000, ASPs were still in favor with investors.

Artemis Images's revenue would come from three streams: (1) sales of images to businesses and consumers, (2) syndication of content, and (3) sales of merchandise. Projected sales were expected to exceed $100 million within the first four years, with breakeven occurring in year three. (See Exhibits 2, 3, and 4 for projected volume and revenues.)

To implement this strategy, Artemis Images, Inc., needed an initial investment of $500,000 to begin operations, hire the team, and sign four additional content agreements. A second round of $1.5 million and a third round of $3 million to $8 million (depending on number of contracts) were planned, to scale the concept to 28 archives and over $100 million in assets by 2004. (See Exhibit 5 for funding and ownership plan.)

The Content Management Industry

According to GISTICS, the trade organization for digital asset management, the content management market (including the labor, software, hardware, and physical assets necessary to manage the billions of digital images) was projected to be a $2 trillion market worldwide in the year 2000 (1999 Market Report). Content could include images, video, text and sound. Artemis Images intended to pursue two subsets of the content management market. The first was the existing stock photo market, a business-to-business market where rights to images were sold for limited use in publications such as magazines, books, and websites. Deutsche Bank's Alex Brown estimated this to be a $1.5 billion market in 2000. Corbis, one of the two major competitors in the digital imaging industry, estimated it to be a $5 billion market by 2000.

EXHIBIT 2 Anticipated sales volume and on-site operations.

Volumes 2001

	Jan-01	Feb-01	Mar-01	Apr-01	May-01	Jun-01	Jul-01	Aug-01	Sep-01	Oct-01	Nov-01	Dec-01	Total
Consumer Photos	0	0	7,500	7,500	27,000	9,000	9,000	22,500	18,000	9,000	4,500	22,500	136,500
Stock Photos	0	0	0	3,750	4,500	5,250	6,750	7,500	9,000	9,750	10,500	11,250	68,250
Subtotal	0	0	7,500	11,250	31,500	14,250	15,750	30,000	27,000	18,750	15,000	33,750	204,750
Licensing Deals	0	0	0	0	0	1	1	2	4	5	6	7	26
Merchandise Orders	0	0	6,000	6,000	21,600	7,200	7,200	18,000	14,400	7,200	3,600	18,000	109,200

Volumes 2002

	Jan-02	Feb-02	Mar-02	Apr-02	May-02	Jun-02	Jul-02	Aug-02	Sep-02	Oct-02	Nov-02	Dec-02	Total
Consumer Photos	12,000	15,000	15,000	15,000	54,000	18,000	18,000	45,000	36,000	18,000	9,000	45,000	300,000
Stock Photos	6,000	7,500	9,000	10,500	12,000	15,000	15,000	15,000	15,000	15,000	15,000	15,000	150,000
Subtotal	18,000	22,500	24,000	25,500	66,000	33,000	33,000	60,000	51,000	33,000	24,000	60,000	450,000
Licensing Deals	8	9	10	11	12	13	14	15	16	16	16	16	156
Merchandise Orders	9,600	12,000	12,000	12,000	43,200	14,400	14,400	36,000	28,800	14,400	7,200	36,000	240,000

Volumes 2003

	Jan-03	Feb-03	Mar-03	Apr-03	May-03	Jun-03	Jul-03	Aug-03	Sep-03	Oct-03	Nov-03	Dec-03	Total
Consumer Photos	24,800	31,000	31,000	31,000	111,600	37,200	37,200	93,000	74,400	37,200	18,600	93,000	620,000
Stock Photos	17,000	21,250	25,500	29,750	34,000	42,500	42,500	42,500	42,500	42,500	42,500	42,500	425,000
Subtotal	41,800	52,250	56,500	60,750	145,600	79,700	79,700	135,500	116,900	79,700	61,100	135,500	1,045,000
Licensing Deals	16	16	16	16	16	16	16	16	16	16	16	16	192
Merchandise Orders	15,600	19,500	19,500	19,500	70,200	23,400	23,400	58,500	46,800	23,400	11,700	58,500	390,000

ONSITE OPERATIONS (by quarters)

Year	2000	2001				2002				2003			
Quarter	Qtr 4	Qtr 1	Qtr 2	Qtr 3	Qtr 4	Qtr 1	Qtr 2	Qtr 3	Qtr 4	Qtr 1	Qtr 2	Qtr 3	Qtr 4
Onsites (cumulative)	1	4	4	7	10	13	13	16	19	22	25	28	28

Source: e-Catalyst Business Plan, February 28, 2000.

EXHIBIT 3 Projected monthly revenue stream

Revenues 2001

	Jan-01	Feb-01	Mar-01	Apr-01	May-01	Jun-01	Jul-01	Aug-01	Sep-01	Oct-01	Nov-01	Dec-01	Total
Consumer Photos	$ 0	$ 0	$ 149,925	$ 149,925	$ 539,730	$ 179,910	$ 179,910	$ 449,775	$ 359,820	$ 179,910	$ 89,955	$ 449,775	$ 2,728,635
Stock Photos	$ 0	$ 0	$ 0	$ 562,500	$ 675,000	$ 787,500	$1,012,500	$ 1,125,000	$ 1,350,000	$ 1,462,500	$1,575,000	$ 1,687,500	$ 10,237,500
Subtotal	$ 0	$ 0	$ 149,925	$ 712,425	$ 1,214,730	$ 967,410	$1,192,410	$ 1,574,775	$ 1,709,820	$ 1,642,410	$1,664,955	$ 2,137,275	$ 12,966,135
Syndication	$ 0	$ 0	$ 0	$ 0	$ 0	$ 8,333	$ 16,667	$ 33,333	$ 66,667	$ 108,333	$ 158,333	$ 216,667	$ 608,333
Merchandise	$ 0	$ 0	$ 45,000	$ 45,000	$ 162,000	$ 54,000	$ 54,000	$ 135,000	$ 108,000	$ 54,000	$ 27,000	$ 135,000	$ 819,000
Total	$ 0	$ 0	$ 194,925	$ 757,425	$ 1,376,730	$1,029,743	$1,263,077	$ 1,743,108	$ 1,884,487	$ 1,804,743	$1,850,288	$ 2,488,942	$ 14,393,468

Revenues 2002

	Jan-02	Feb-02	Mar-02	Apr-02	May-02	Jun-02	Jul-02	Aug-02	Sep-02	Oct-02	Nov-02	Dec-02	Total
Consumer Photos	$ 239,880	$ 299,850	$ 299,850	$ 299,850	$ 1,079,460	$ 359,820	$ 359,820	$ 899,550	$ 719,640	$ 359,820	$ 179,910	$ 899,550	$ 5,997,000
Stock Photos	$ 900,000	$1,125,000	$1,350,000	$1,575,000	$ 1,800,000	$2,250,000	$2,250,000	$ 2,250,000	$ 2,250,000	$ 2,250,000	$2,250,000	$ 2,250,000	$ 22,500,000
Subtotal	$1,139,880	$1,424,850	$1,649,850	$1,874,850	$ 2,879,460	$2,609,820	$2,609,820	$ 3,149,550	$ 2,969,640	$ 2,609,820	$2,429,910	$ 3,149,550	$ 28,497,000
Syndication	$ 283,333	$ 358,333	$ 441,667	$ 533,333	$ 633,333	$ 741,667	$ 858,333	$ 983,333	$ 1,116,667	$ 1,250,000	$1,383,333	$ 1,516,667	$ 10,100,000
Merchandise	$ 72,000	$ 90,000	$ 90,000	$ 90,000	$ 324,000	$ 108,000	$ 108,000	$ 270,000	$ 216,000	$ 108,000	$ 54,000	$ 270,000	$ 1,800,000
Total	$1,495,213	$1,873,183	$2,181,517	$2,498,183	$ 3,836,793	$3,459,487	$3,576,153	$ 4,402,883	$ 4,302,307	$ 3,967,820	$3,867,243	$ 4,936,217	$ 40,397,000

Revenues 2003

	Jan-03	Feb-03	Mar-03	Apr-03	May-03	Jun-03	Jul-03	Aug-03	Sep-03	Oct-03	Nov-03	Dec-03	Total
Consumer Photos	$ 495,752	$ 619,690	$ 619,690	$ 619,690	$ 2,230,884	$ 743,628	$ 743,628	$ 1,859,070	$ 1,487,256	$ 743,628	$ 371,814	$ 1,859,070	$ 12,393,80
Stock Photos	$2,550,000	$3,187,500	$3,825,000	$4,462,500	$ 5,100,000	$6,375,000	$6,375,000	$ 6,375,000	$ 6,375,000	$ 6,375,000	$6,375,000	$ 6,375,000	$ 63,750,000
Subtotal	$3,045,752	$3,807,190	$4,444,690	$5,082,190	$ 7,330,884	$7,118,628	$7,118,628	$ 8,234,070	$ 7,862,256	$ 7,118,628	$6,746,814	$ 8,234,070	$ 76,143,800
Syndication	$1,650,000	$1,783,333	$1,916,667	$2,050,000	$ 2,183,333	$2,316,667	$2,450,000	$ 2,583,333	$ 2,716,667	$ 2,850,000	$2,983,333	$ 3,116,667	$ 28,600,000
Merchandise	$ 117,000	$ 146,250	$ 146,250	$ 146,250	$ 526,500	$ 175,500	$ 175,500	$ 438,750	$ 351,000	$ 175,500	$ 87,750	$ 438,750	$ 2,925,000
Total	$4,812,752	$5,736,773	$6,507,607	$7,278,440	$10,040,717	$9,610,795	$9,744,128	$11,256,153	$10,929,923	$10,144,128	$9,817,897	$11,789,487	$107,668,800

Source: e-Catalyst Business Plan, February 28, 2000.

EXHIBIT 4 Pro forma financial summary 2000.

Summary profit and loss statement

	2000	2001	2002	2003	Total
Revenues	$ 0	$14,393,468	$40,397,000	$107,668,800	$162,459,268
Cost of sales	$ 0	$ 5,186,454	$11,398,800	$ 30,457,520	$ 47,042,774
Gross profit	$ 0	$ 9,207,014	$28,998,200	$ 77,211,280	$115,416,494
Operations	$ 439,847	$13,623,571	$27,109,143	$ 47,078,657	$ 88,251,217
Net income before tax	($ 439,847)	($ 4,416,556)	$ 1,889,057	$ 30,132,623	$ 27,165,277
Taxes (38%)	$ 0	$ 0	$ 0	$ 10,322,805	$ 10,322,805
Net income	($ 439,847)	($ 4,416,556)	$ 1,889,057	$ 19,809,818	$ 16,842,472

Summary balance sheet

Assets	2000	2001	2002	2003
Cash and equivalents	$ 428,020	$ 4,490,768	$ 4,958,270	$ 21,508,477
Accounts receivable	$ 0	$ 2,488,942	$ 4,936,217	$ 11,789,487
Inventories	$ 0	$ 0	$ 0	$ 0
Prepaid expenses	$ 0	$ 0	$ 0	$ 0
Depreciable assets	$ 0	$ 0	$ 0	$ 0
Other depreciable assets	$ 0	$ 0	$ 0	$ 0
Depreciation	$ 0	$ 0	$ 0	$ 0
Net depreciable assets	$ 0	$ 0	$ 0	$ 0
Total assets	$ 428,020	$ 6,979,710	$ 9,894,487	$ 33,297,964

Liabilities and capital				
Accounts payable	$ 367,867	$ 1,836,113	$ 2,861,833	$ 5,589,379
Accrued income taxes	$ 0	$ 0	$ 0	$ 866,113
Accrued payroll taxes	$ 0	$ 0	$ 0	$ 0
Total liabilities	$ 367,867	$ 1,836,113	$ 2,861,833	$ 6,455,492
Capital contribution	$ 500,000	$10,000,000	$10,000,000	$ 10,000,000
Stockholders' equity	$ 0	$ 0	$ 0	$ 0
Retained earnings	($ 439,847)	($ 4,856,403)	($ 2,967,346)	$ 16,842,472
Net capital	$ 60,153	$ 5,143,597	$ 7,032,654	$ 26,842,472
Total liabilities and capital	$ 428,020	$ 6,979,710	$ 9,894,487	$ 33,297,964

Source: e-Catalyst Business Plan, February 28, 2000.

EXHIBIT 5 Artemis images original funding plan.

Projected Plan	Round 1	Round 2	Round 3	Round 4	Exit
Financing assumptions:					
2003 Revenues	$110,000,000				
2003 EBITDA	$ 30,000,000				
2003 Revenue growth rate	40%				
2003 Valuation	$440,000,000				
Valuation/revenue	4				
Valuation/EBITDA	14.67				
Round 1 Financing	$ 500,000				
Round 2 Financing	$ 1,500,000				
Round 3 Financing	$ 3,000,000				
Round 4 Financing	$ 5,000,000				
	Round 1 Oct-00	Round 2 1-Jan	Round 3 1-Mar	Round 4 1-Jun	Exit 3-Dec
Number of shares outstanding					
Total number of shares outstanding prior to financing	6,000,000	7,200,000	9,000,000	11,250,000	11,250,000
Shares issues this round	1,200,000	1,800,000	2,250,000	1,406,250	1,406,250
Total number shares outstanding after financing	7,200,000	9,000,000	11,250,000	12,656,250	12,656,250
Valuations					
Premoney valuation	$2,500,000	$6,000,000	$12,000,000	$40,000,000	$440,000,000
Amount of financing	$ 500,000	$1,500,000	$ 3,000,000	$ 5,000,000	0
Postmoney valuation	$3,000,000	$7,500,000	$15,000,000	$45,000,000	$440,000,000
Price per share	$0.42	$0.83	$1.33	$3.56	$34.77
Resulting ownership					
Founders	83.33%	66.67%	53.33%	47.41%	47.41%
Round 1 investors	16.67%	13.33%	10.67%	9.48%	9.48%
Round 2 investors	0.00%	20.00%	16.00%	14.22%	14.22%
Round 3 investors	0.00%	0.00%	20.00%	17.78%	17.78%
Round 4 investors	0.00%	0.00%	0.00%	11.11%	11.11%
Total	100.00%	100.00%	100.00%	100.00%	100.00%
Value of ownership					
Founders	$2,500,000	$5,000,000	$ 8,000,000	$21,333,333	$208,592,593
Round 1 investors	$ 500,000	$1,000,000	$ 1,600,000	$ 4,266,667	$ 41,718,519
Round 2 investors	$ 0	$1,500,000	$ 2,400,000	$ 6,400,000	$ 62,577,778
Round 3 investors	$ 0	$ 0	$ 3,000,000	$ 8,000,000	$ 78,222,222
Round 4 investors	$ 0	$ 0	$ 0	$ 5,000,000	$ 48,888,889
Total	$3,000,000	$7,500,000	$15,000,000	$40,000,000	$440,000,000
Payback to investors	**Round 1**	**Round 2**	**Round 3**	**Round 4**	
Holding period (years)	3.25	3	2.75	2.5	
Times money back	83.44	41.72	26.07	9.78	
Internal rate of return (IRR)	290%	247%	227%	149%	

Source: e-Catalyst Business Plan, February 28, 2000.

Commercially produced images were also in demand by consumers. Industry insiders believed that this market was poised for explosive growth in 2000, as Web-enabled technology facilitated display and transmission of images directly from their owners to individual consumers. The archives from the Indianapolis Motor Speedway was an example of this business-to-consumer model. Historically, consumers who bought from the archive had to visit the museum at IMSC or write a letter to the staff. Retrieval and fulfillment of images then required a manual search of a physical inventory, a process which could take as long as two weeks. Web-based digitization and search engines would reduce the search time and personnel needed for order fulfillment and allow customers the convenience of selecting products and placing orders on-line. The *Daily Mirror,* a newspaper in London, had displayed its archived images on its own website and had generated over $30,000 in sales to consumers in its first month of availability. IMG, a sports marketing group, placed a value of $10 million on the IMSC contract.

Competition

There were a variety of stock and consumer photo sites ranging from those that served only the business-to-business stock photo market to amateur photographers posting their pictures. Most sites did not offer a "community," the Internet vehicle for consumer comments and discussion, a powerful search engine, and ways to repurpose the content (e-greeting cards, prints, photo mugs, calendars, etc.). In addition, the archives available in digital form were limited because other content providers worked from the virtual world to the physical world versus the Artemis Images model of working from the physical world to the virtual world. Competitors had problems with integrated digital workflows and knowing where the original asset resided due to the distributed nature of their archives. They scanned images on demand, which severely limited the content available to be searched on their websites.

Chris and Greg evaluated the five major competitors for their business plan:

www.corbis.com: Owned by Bill Gates with an archive of over 65 million images, only 650,000 were available on the Web to be accessed by consumers for Web distribution (e-greeting cards, screen savers, etc.). Only 350,000 images were available to be purchased as prints. The site was well designed and the search features were good, but there was no community on the site. The niche Corbis pursued was outright ownership of archives and scanning on demand. Corbis had recently acquired the Louvre archive, for a reported purchase price of over $30 million.

www.getty-images.com: An archive of over 70 million images. In 1999, this site was only a source to link to their other wholly owned subsidiaries, including art.com. There were no search capabilities, no community. This website functioned only as a brochure for the company. Like Corbis, Getty was focused on owning content and then scanning on demand.

www.art.com: A good site in design and navigation, this site was a wholly owned subsidiary of Getty and was positioned as the consumer window to a portion of the Getty archive. Similar to Corbis, customers were able to buy prints, send e-greeting cards, etc. Despite the breadth of the Getty archive, this site had a limited number of digitized images available.

www.mediaexchange.com: Strictly a stock photo site targeted toward news sources, the site was largely reliant on text. It was difficult to navigate and had an unattractive graphical user interface.

www.thepicturecollection.com: Strictly a stock site offering the *Time* photo archive, this site was well designed with good search capabilities. Searches yielded not only a thumbnail image but a display of the attached locator tags, or metadata.

www.ditto.com: The world's leading visual search engine, ditto.com enabled people to navigate the Web through pictures. The premise was two-fold: deliver highly relevant thumbnail images and link to relevant Web sites underlying these images. By 2000, they had developed the largest searchable index of visual content on the Internet.

Exhibit 6 shows a comparison of Artemis Images to the two major players in the stock photography market, Getty and Corbis. This table illustrates only revenues from stock photo sales and does not include potential revenue from consumer sales, merchandise, advertising or other potential revenue sources.

According to its marketing director, Corbis intended to digitize its entire archive, and was in the process of converting analog images into digital images, with 63 million images yet to be converted. While Getty and Corbis were established players in the content industry, they were just recently feeling the effects of e-commerce:

- In 1999, Corbis generated 80 percent of its revenues from the Web versus none in 1996.
- Getty's e-commerce sales were up 160 percent between 1998 and 1999.
- 34 percent of Getty's 1999 revenues came from e-commerce versus 17 percent in 1998.

Strategy

Artemis Images intended to provide digitization and archive management by employing a professional staff who would work within each client-company's organization, rather than in an off-site facility of its own. Chris's model was to provide digitized archive services in exchange for (1) exclusive rights to market the content on the Internet, (2) merchandising rights, and (3) promotion of Artemis Images's URL, effectively co-branding Artemis Images with each client-partner. Chris envisioned a software process that would be owned or licensed by Artemis, and which could be used for digitizing different archive media, such as photos, videos, and text.

Chris and George expected Artemis Images to partner with existing sellers of stock photography and trade digitizing services for promotion through their sales channels. Artemis Images would pursue these relationships with traditional

EXHIBIT 6 Anticipated sales volume comparisons.

Stock photo market						
	Artemis Images	Artemis Images	Artemis Images	Artemis Images	Getty	Corbis
	Indy Archive 2000*	2000	2001	2002	1999	1999
Archive size	5,000,000	5,000,000	50,000,000	95,000,000	70,000,000	65,000,000
Cumulative number of images digitized	345,600	345,600	6,796,800	21,542,400	1,200,000	2,100,000
% digitized**	7%	7%	14%	23%	1.71%	3.2%
# of image sales needed to hit revenue target	0	0	151,484	623,493	1,646,667	666,666
% of archive that must be sold to hit revenue target***	0	0	0.30%	0.16%	2.35%	1.00%
Revenues****	0	0	$22,722,600	$93,523,950	$247,000,000	$100,000,000
Revenue per image in archive	0	0	$0.45	$0.98	$ 3.53	$ 1.54
Revenue per digitized image	0	0	$3.25	$4.30	$205.83	$47.62

*Artemis Images had already secured an exclusive content agreement from the Indianapolis Motor Speedway Corporation.
**Estimates based on scanning 1,920 images a day per scanner, 2 scanners per archive. As scanning technologies improve, the throughput numbers were expected to go up.
***The percentage of the Artemis Images archive that needed to be sold to hit revenues projections varied between 0.03% and 0.22%, as compared to an actual 2.35% for Getty and to 0.6% for Corbis.
****The Artemis Images revenue numbers were based on selling a certain number of images at $150 per image; $150 was the minimum average price paid for stock photographs. Corbis was privately held; this figure was an estimate.

sales and marketing techniques. Sales people would call on the major players and targeted direct mail, trade magazine advertising and PR would be used to reach the huge audience of smaller players. In addition, content partners were expected to become customers, as they were all users of stock photography.

As Artemis Images gained clients, the company would have access to some of the finest and most desirable content in the world. Chris knew that the workflow expertise of the management team would put them in a good position to provide better quality more consistently than either Corbis or Getty. This same expertise would allow Artemis to have a much larger digital selection, with a website design that would be easily navigable for customers to find what they needed.

Using on-site equipment, the client's content would be digitized, annotated (by attaching digital information tags, or metadata) and uploaded to the corporate hub site. Metadata would allow the content to be located by the search engine and thus viewed by the consumer. For example, a photo of Eddie Cheever winning the Indy 500 would have tags like Indy 500, Eddie Cheever, win photo, 1998, etc. Therefore, a customer going to the website and searching for "Eddie Cheever" would find this specific photo, along with the hundreds of other photos associated with him. The Artemis corporate database was intended to serve as the repository for search and retrieval from the website.

The traditional content management strategy forced organizations to purchase technology and expertise. Artemis Images's model intended to alleviate this burden by exchanging technology and expertise for exclusive web distribution rights and a share of revenues. The operational strategy was to create an infrastructure based on installing and operating digital asset management systems at their customers' facilities to create a global digital archive of images, video, sound and text. This would serve to lock Artemis Images into long-term relationships with these organizations and ensure that Artemis Images would have both the historical and the most up-to-date content. Artemis Images would own and operate the content management technology, with all other operational needs outsourced including Web development, Web hosting, consumer data collection, and warehousing and fulfillment of merchandise (printing and mailing posters or prints). Artemis Images would scan thousands of images per day, driving down the cost per image to less than $2.00, versus the Corbis and Getty model of scan-on-demand, where the cost per image was approximately $40.00. The equipment needed for both the content management and photo production would be leased to minimize start-up costs and ensure greater flexibility in the system's configuration.

The original plan was to purchase and install software and hardware at their main office in Denver, Colorado, contract with a Web development partner, and set up the first on-site facility at Indianapolis Motor Speedway Corporation. The Denver facility would serve as a development lab, to create a standard set of metadata to be used by all of their partners' content. This consistency of annotation information was intended to allow for consistent search and retrieval of content. Artemis Images's goal was to build a world-class infrastructure to handle content management, consumer data collection, and e-commerce. This infrastructure would allow them to amass a large content and transaction volume by expanding to other market segments. Developing their own structure would ensure standardization of content and reduced implementation time. Outreach for news coverage and the development of community features would be negotiated concurrently. The time line in Exhibit 7 illustrates the Artemis Images development plan.

Financial Projections

Revenues were expected to come from four primary sources:

Consumer photos: IMSC's archive sold approximately 53,000 photos in 1999 to a market limited to consumers who visited the archive or wrote to its staff. Artemis Images based its projected sales on an average of

Three phases of development

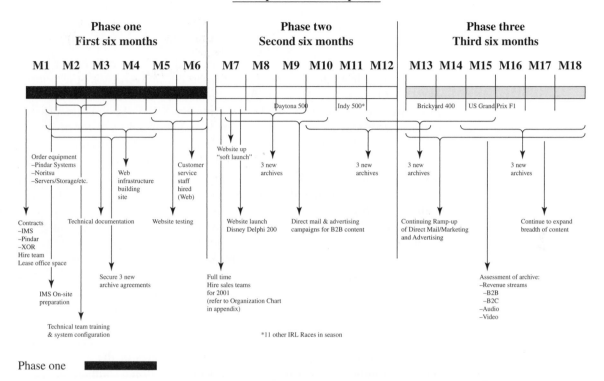

Phase one   ████████████

This phase was intended to take Artemis Images from initial funding to operationally being ready to sell images and take orders. The three main components included establishing the on-site facility at IMSC, construction and testing of the website and establishing the fulfillment operations. Phase one assumed money was in the bank.

Phase Two   ▭▭▭▭▭▭

This phase assumed that three additional archives had been secured and implemented, at least one of which would include breadth of content. Focus would be sales and ramping up revenues. B2B and B2C marketing strategies were to be executed and evaluated. Toward the end of Phase two three more archives would be secured.

Phase Three   ▭▭▭▭▭▭

Phase three continued to build more archives and breadth of content. Marketing and sales would continue to be core focus for revenue development. Audio and video content assessed based on the state of market and technology (e.g., bandwidths) and a decision would be made on timing to enter this market.

EXHIBIT 7 Artemis Images development time line.

Source: e-Catalyst Business Plan, February 28, 2000.

15,000 images sold per archive in 2001, increasing to 20,000 images per archive in 2003. Price: $19.99 (8" × 10").

Stock photos: Stock photos ranged in price from $150 to $100,000, depending on the uniqueness of the photo. Competitors Getty and Corbis, two of the leaders in this market, sold 2.35 percent and 0.6 percent of their archive, respectively. Based on an average selling price of $150, Getty generated approximately $6.00 in revenue for each image in its archive; Corbis generated approximately $1.85. Artemis Images constructed financial projections based on sales of 0.30 percent of its archive in 2001 and 0.16 percent of its archive in 2002. Artemis Images's margin was based on a return of $0.20 per image in its archive for 2001, increasing to $0.60 per image in 2003.

Syndication: The team's dot.com experience led them to believe that websites with exclusive content were able to syndicate their content to other websites. They anticipated that Artemis Images would generate revenues of $100,000 per year from each contract for content supplied as marketing tools on websites. Existing companies with strong content had been able to negotiate five new agreements per week for potential annual revenues of $5 million.

Merchandise: According to America Online/Roper Starch Worldwide, approximately 30 percent of Internet users regularly make purchases. Artemis Images used a more conservative assumption that only 1 percent of unique visitors would make a purchase. Estimates of the average purchase online varied widely, ranging from Wharton's estimate of $86.13 to eMarketers's estimate of $219. The Artemis Images team viewed $50 per purchase as a conservative figure.

Chris and George felt confident that Artemis Images would be able to reach the revenue projections for number of photos sold. IMSC's archive had sold approximately 53,000 photos in 1999, an increase of 33 percent over 1998. These sales had been generated solely by consumers who had visited the archive in person, estimated at 1 million people. In other words, one out of every 28 possible consumers actually purchased an image. Chris and George assumed that if even one out of 160 unique visitors to the website purchased a photo, the Artemis website would generate 42 percent more than IMSC's 1999 figures (see Exhibit 6 for projected sales volume). Chris and George believed that this projection was reasonable in light of the fact that IMSC did not market its archive and significant publicity and advertising would accompany Artemis Images's handling of the archive. As breadth of content and reach of the Web increased, 2002 revenues should easily be double those of 2001.

Since the team previously had configured and sold content management systems, they were familiar with the costs associated with this process, including both equipment and personnel. They carefully conducted research to stay abreast of recent improvements in technology and intended to be on the lookout for cost reductions and process improvements.

The Launch: Problems from the Start

Chris dove into the Artemis Images project with a vengeance. Having secured a five-year contract for exclusive rights and access to the IMSC archive, she found a dependable technician who was eager to relocate to Indianapolis to start the scanning and digitizing process. A reputable, independent photo lab agreed to handle printing and order fulfillment. Chris's visit to the Indy 500 in May 2000 was a wonderful networking opportunity. She met executives from large companies and got leads for investors and clients. She secured an agreement with a Web design company to build the Artemis Images site, careful to retain ownership of the design. She contacted over 100 potential venture capitalists and angel investors.

Personally, she was on a roll. Financially, she was rapidly going into debt. Frank and Greg, legal owners of the company, had long since contributed ideas, contacts, or legwork to the Artemis Images launch. While confident that his work on the business plan would appeal to investors, Greg viewed the start-up company as a risk to which he was unwilling to commit. Likewise, Frank decided to hold onto his job at Petersons.com, a unit of Thompson Learning, until the first round of investor funding had been secured. Frank continued to offer advice, but he had a wife and two preschool-age children to support.

Each meeting with a potential funder resulted in a suggestion on how to make the business more attractive for investment. Sometimes they helped, sometimes they just added to Chris's and George's frustration. Beating the bushes for money over two years was exhausting, to say the least. The lack of funds impacted the look and feel of the business and severely strained relationships among the founding partners. Heated discussions ensued as to the roles that each was expected to play, the reallocation of equity ownership in the company, and the immediate cash needed to maintain the Indianapolis apartment and pay the scanning technician and Web developers, not to mention out-of-pocket expenses needed to manage and market the business.

Chris and George appealed to their families for help. George's father contributed $5,000. Chris's mother tapped into her retirement, mostly to pay Chris's mortgage and to fund Chris's trips to potential clients and investors in London, New York, and Boston. By May 2001, Chris's mother's contribution had exceeded $200,000. A $50,000 loan from a supportive racing enthusiast provided the impetus for Artemis Images to reorganize as a C-corporation. All four original partners had stock in the new company, but Chris held the majority share (66 percent), George held 30 percent, and Frank and Greg's shares were each reduced to 2 percent. Financial projections were revised downward (see Exhibit 8).

The site was officially launched on May 18, 2001. It was beautiful. Chris held her breath as she put in her credit card late that evening when the site went live. The shopping cart failed and the order could not be processed. Chris knew she was in trouble.

EXHIBIT 8 Revised pro forma financial summary 2001

Summary Profit and Loss Statement					
	2001	2002	2003	2004	Total
Revenues	$ 5,312*	$373,779	$2,294,116	$4,735,400	$7,408,607
Cost of sales	$ 1,700	$ 43,368	$ 265,312	$ 564,480	$ 874,860
Gross profit	$ 3,612	$330,411	$2,028,804	$4,170,920	$6,533,747
Operations	$52,499	$328,550	$1,235,363	$2,035,430	$3,651,842
Net income before tax	($48,887)	$ 1,861	$ 793,441	$2,135,490	$2,881,905
Taxes (38%)	$ 0	$ 0	$ 283,638	$ 811,486	$1,095,124
Net income	($48,887)	$ 1,861	$ 509,803	$1,324,004	$1,786,781

Summary Balance Sheet

Assets	2001	2002	2003	2004
Cash and equivalents	$45,113	$ 78,260	$ 675.347	$2,615,573
Accounts receivable	$ 0	$ 13,610	$ 222,950	$ 462,200
Inventories	$ 0	$ 0	$ 0	$ 0
Prepaid expenses	$ 0	$ 0	$ 0	$ 0
Depreciable assets	$ 0	$ 0	$ 0	$ 0
Other depreciable assets	$ 0	$ 0	$ 0	$ 0
Depreciation	$ 0	$ 0	$ 0	$ 0
Net depreciable assets	$ 0	$ 0	$ 0	$ 0
Total assets	$45,113	$ 40,574	$ 898,297	$3,077,773

Liabilities and capital				
Accounts payable	$ 4,000	$ 12,355	$ 61,882	$ 105,868
Accrued income taxes	$ 0	$ 0	$ 283,638	$1,095,124
Accrued payroll taxes	$ 0	$ 0	$ 0	$ 0
Total liabilities	$ 4,000	$ 12,355	$ 345,520	$1,200,992
Capital contribution	$90,000†	$ 90,000	$ 90,000	$ 90,000
Stockholders' equity	$ 0	$ 0	$ 0	$ 0
Retained earnings	($48,887)	($ 61,781)	$ 462,777	$1,786,781
Net capital	$41,113	$ 28,219	$ 552,777	$1,876,781
Total liabilities and capital	$45,113	$ 40,574	$ 898,287	$3,077,773

Notes: *Approximately two-thirds of these transactions were executed by Artemis staff and friends to test the website.
† Chris's mother's contribution to her daughter for mortgage and living expenses is not included.

The Crash

From the first, the website had problems. The Web development contract stipulated that the website for the Indy 500 would go live by May 8, 2001, to coincide with the month-long series of events held at the Indianapolis Motor Speedway leading up to the Indy 500 on May 27. However, the Web development took longer than anticipated, and the site was first operational on May 18. Having neglected to test the Web interface properly, serious failures were encountered when the site was activated. The site went down for 24 hours, only to face similar problems throughout the following week, again shutting down on May 27. More technical difficulties delayed the reactivation of the site until May 31, after the Indy racing series had ended.

Throughout June, consumer traffic was far less than originally anticipated. The site was not easily navigable. The shopping cart didn't work. Yet the Web builder demanded more money. Fearful of a possible lawsuit, investors stayed away. The crash of the dot.coms added kindling to the woodpile. Chris and George started to rethink their original business model. They were held hostage, as they owned no tangible assets.

Website tracking data indicated that between May and July there had been at least $40,000 worth of attempted purchases. Chris read through hundreds of angry e-mails, and tried manually to process orders. Orders which were successfully executed resulted in spotty fulfillment. Many photos ordered were never shipped, were duplicated, or were incorrectly billed. At the same time, she tried to negotiate with the software developers' demand for payment and keep alive a $250,000 investment prospect.

On July 9, 2001, the Web development company threatened an all-or-nothing settlement. They wanted payment in full for the balance of the contract even though the sites didn't work. Absent full payment, they would shut down the sites within the week. The investor offered to put up 80 percent of the balance owed on the full contract to acquire the code to fix it. The company refused. On Friday, July 13, Chris had to tell IMSC that in less than 48 hours the sites would be shut down. The investor took his $250,000 elsewhere.

On Tuesday, July 17, Chris called an emergency meeting with George. George had had enough. The stress was affecting his health, his relationships, and his lifestyle. He believed that his family had already contributed more money than he had a right to ask. He was putting in long hours with no money to show for his efforts. His girlfriend had been putting pressure on George to quit for some time. Now he had run out of reasons to stay.

Chris was devastated. How could she face the people in Indianapolis? It was hard for her to come to grips with having let them down. Having put so much of herself into this venture, she wasn't sure she could let go. At the same time, she wasn't sure how to go on.

Chris reflected, "At one time, I defined success by my title, my salary, and my possessions. Working for AGT, I had it all. I started Artemis Images because I really cared about IMSC and making the Indy motorsports images available to its fans. Now, I realize that there is a profound satisfaction in building a company. I can see my future so clearly, but living day to day now is so hard. And I'm still enthralled with beautiful images."

Questions

1. Discuss why Chris started her company. What was the opportunity?
2. What is your evaluation of the team's qualifications for this business?
3. Discuss the division of ownership among the team.
4. Evaluate the business model for Artemis. Is it strong and will the firm be profitable?

SIRTRIS PHARMACEUTICALS: LIVING HEALTHIER, LONGER

"You can live to be a hundred if you give up all the things that make you want to live to be a hundred."
Woody Allen

One Saturday in February 2007, Dr. David Sinclair and Dr. Christoph Westphal co-founders of Sirtris Pharmaceuticals, a Cambridge, MA-based life sciences firm, navigated the company's narrow hallways and cramped offices to a conference room for their regular weekend strategy planning session.

When they reached the conference room, Sinclair and Westphal reviewed their activities during the past week. Sinclair, who was an associate professor of pathology at Harvard Medical School and co-chair of Sirtris's Scientific Advisory Board, had had interviews with Charlie Rose, the *Wall Street Journal,* and *Newsweek*. Westphal, who was Sirtris's CEO and vice chairman, had closed a $39 million round of financing, bringing the total amount of invested capital in the company to $103 million.

Sinclair and Westphal were riding a wave of interest generated, in part, by their company's promising research into age-related diseases, such as diabetes, cancer, and Alzheimer's. The company's research into disease, however, only partly explained its appearance on the covers of *Scientific American, Fortune,* and the *Wall Street Journal.* According to their suggestive headlines—"Can DNA Stop Time: Unlocking the Secrets of Longevity Genes" (*Scientific American*), "Drink wine and live longer: The exclusive story of the biotech startup searching for anti-aging miracle drugs" (*Fortune*) and "Youthful Pursuit: Researchers seek key to Antiaging in Calorie Cutback" (*Wall Street Journal*)__ Sirtris was hoping to develop drugs that could treat diseases of aging, and in so doing had the potential to extend the lifespan of human beings[1].

---

[1] Leonard Guarente and David Sinclair, "Can DNA Stop Time: Unlocking the Secrets of Longevity Genes," *Scientific American,* March 2006; David Stipp, "Researchers seek key to antiaging in calorie cutback," *Wall Street Journal,* October 30, 2006. David Stipp, "Drink Wine and Live Longer: The exclusive story of the biotech startup searching for anti-aging miracle drugs," *Fortune,* February 12, 2007. See Appendix for cover of the *Wall Street Journal* article.

Professor Toby Stuart and Senior Researcher David Kiron, Global Research Group, prepared this case, with advice and contributions from Alexander Crisses (MBA 2008). HBS cases are developed solely as the basis for class discussion. Cases are not intended to serve as endorsements, sources of primary data, or illustrations of effective or ineffective management.

The Sirtris team had, in fact, established a link between resveratrol, a compound found in red wine-producing grapes, and sirtuins, a newly discovered family of enzymes with links to improved longevity, metabolism and health in living things as diverse as yeast, mice and humans. Sinclair and Westphal were building Sirtris around the development of sirtuin-activating drugs for the diabetes market. The Sirtris team had developed its own proprietary formulation of resveratrol, called SRT501, and was developing new chemical entities (NCEs) that were up to 1000x more potent activators of sirtuins than resveratrol.

In today's strategy session, which included Dr. Michelle Dipp, Sirtris's senior director of corporate development and Garen Bohlin, the company's chief operating officer, the team was discussing their upcoming move to new laboratory space in another part of Cambridge, and three of the more pressing strategic issues facing the firm.

- ■ **In-licensing.** One issue was whether to in-license compounds from a biotech company to diversify Sirtris's drug development platform beyond its narrow focus on SIRT1, one of seven sirtuin variants in the human body. Several members of the Sirtris executive team were advocating a more balanced risk portfolio as the company started to increase investment in its drug development efforts.

- ■ **Partnership with Pharma.** As is almost always the case in biotech, the team was in discussions about a partnership with a few large pharmaceutical firms. They were considering (a) what it would mean for the organization to become tied to a pharmaceutical company at this stage of its development and (b) whether to postpone a deal until Sirtris had more clinical data. Was this the right point in the company's history to do a deal?

- ■ **Nutraceuticals.** Sinclair received a near-constant stream of emails and phone calls from the public requesting Sirtris's proprietary version of resveratrol, SRT501. For some time, he had contemplated selling SRT501 as a nutraceutical, an off-the-shelf health supplement that would not require FDA approval. This idea raised many questions about market opportunity, commercialization strategies, and the potential impact of a nutraceuticals offering on the Sirtris brand and the all-star group of scientists that had allied themselves with the organization.

Anti-Aging Science

The quest for long life has spurred the imagination of people in virtually every era in human history. Ancient Greeks imagined immortal gods, the sixteenth century Spanish adventurer Ponce de Leon searched for the fountain of youth, and twenty-first-century scientists test rodents for life-extending biological compounds.

Until the 1990s, the only proven means of increasing life-span in any animal was to reduce its calorie intake. In 1935, Cornell University researchers discovered that reducing calorie intake in rodents by 40% increased their lifespan by an average of 30-40%. Conventional wisdom became that calorie reduction (CR) activates an evolutionary adaptive process that lowers metabolism to help animals through periods of food shortages or droughts.

Decades passed before scientists could shed light on the biological mechanism triggered by CR. When new information arrived in the 1980s and early 1990s, it contradicted what had become the conventional wisdom. The new work indicated that instead of lowering metabolism, calorie reduction is a biological stressor that activates a defensive metabolic response. Few scientists paid much attention to the shift in view, as few serious scientists focused their academic careers on anti-aging studies.

Longevity research began to gain traction as an academically credible field of study in the early 1990s, after MIT professor Leonard Guarente and his laboratory traced the molecular pathway of calorie reduction in yeast to sirtuins (silent information regulators), which are proteins (enzymes) that are found in all cells. (See **Exhibit 1** for timeline of scientific milestones in longevity research.) In humans, there are seven types of sirtuins in different parts of cells and in different parts of the body. Sirtris was focusing 90% of its R&D on one sirtuin, called SIRT1 in humans. For simplicity, any reference to sirtuins, unless otherwise noted, is to the family of sirtuins or SIRT1.

David Sinclair

In 1993 while sightseeing in Sydney, Australia, Guarente was taken to dinner by some local yeast researchers, a group that included David Sinclair, then a young doctoral candidate at the University of New South Wales. During the dinner, Guarente mentioned that he had two students working on aging in yeast. "I was incredibly excited by Lenny's work," said Sinclair. "I asked him why the longevity field was so pre-occupied with looking for genes that ended life, rather than genes that could extend it. By the time we finished dinner, I told him I was going to work with him at MIT."

Sinclair grew up in St. Ives, near Sydney Australia, the eldest son of parents who both worked in the medical diagnostics industry. In high school, he was known as a talented class clown and risk-taker, a young man who aced science classes but could not resist setting off minor explosions in chemistry lab.[2]

Two years after their first meeting, Sinclair made good on his promise and joined Guarente's MIT lab as a postdoctoral fellow. Sinclair quickly established himself as a creative and productive researcher, publishing a 1997 article in *Cell* with Guarente that described how the yeast equivalent of SIRT1 increased the longevity of yeast. When yeast cells divide (a sign of aging in yeast cells), they spin off extra copies of genetic material called extrachromosomal rDNA circles (or ERCs). With each successive cell division, ERC copies accumulate in the nucleus. The original cell, faced with copying both its original genetic material and an increasing number of ERCs, soon runs out

---

[2] David Stipp, "Drink Wine and Live Longer," *Fortune*, February 12, 2007.

of energy and eventually dies. But when an extra copy of the sirtuin gene was added to the cell nucleus, the formation of ERCs was repressed and the cell's life span was extended by 30 percent.

In 1999, Sinclair left Guarente's lab for a tenure track position at Harvard Medical School, but continued to collaborate with Guarente.[3] They found that extra copies of the sirtuin gene extended the life span of roundworms by as much as 50 percent. "We were surprised not only by this commonality in organisms separated by a vast evolutionary distance but by the fact that the adult worm body contains only non-dividing cells," wrote Guarente and Sinclair in their 2006 *Scientific American* article.[4]

The search was on for sirtuin activating compounds, or STACs. This was a high-stakes search. "No chemical or drug had ever increased the activity of sirtruins" said Dr. Dipp, affectionately known within the firm as "The General". "A compound that could activate sirtuins would increase the speed of cellular metabolism. It could have far reaching implications for human healthcare."

In 2003, Sinclair discovered that resveratrol, a compound found in red wine, activated sirtuins in yeast cells, a discovery which indicated that in fact it might be possible to develop a drug that could activate the sirtuin enzyme. One way to activate the sirtuin enzyme was to optimize the effects of resveratrol by giving it in highly purified form. Another was to mimic the effects of resveratrol using an entirely new, more potent chemical structure. Sirtris was pursuing both approaches through its SRT501 and new chemical entity drug development projects.

When Sinclair's *Nature* article was published September 11, 2003, it was hailed by many scientists as a seminal paper, but it also drew criticism from members of the scientific community, including former colleagues from Guarente's MIT laboratory. The article also drew the attention of Dr. Christoph Westphal, who had recently been promoted to general partner at Polaris Venture Partners, one of the larger Boston-area venture capital funds.

Christoph Westphal

In his four years at Polaris, Westphal had had several successful investments and stints as founding CEO. Between 2000 and 2004, Westphal co-founded five companies and served as the original CEO at four of them. In all cases,

---

[3] Dr. Sinclair obtained a BS with first-class honors at the University of New South Wales, Sydney, and received the Commonwealth Prize for his research. In 1995, he received a Ph.D. in Molecular Genetics and was awarded the Thompson Prize for best thesis work. He worked as a postdoctoral researcher with Dr. Leonard Guarente at M.I.T. until being recruited to Harvard Medical School at the age of 29.

Dr. Sinclair has received several additional awards including a Helen Hay Whitney Postdoctoral Award, and a Special Fellowship from the Leukemia Society, a Ludwig Scholarship, a Harvard-Armenise Fellowship, an American Association for Aging Research Fellowship, and is currently a New Scholar of the Ellison Medical Foundation. He won the Genzyme Outstanding Achievement in Biomedical Science Award for 2004. http://medapps.med.harvard.edu/agingresearch/pages/faculty.htm, Accessed 1.4.07.

[4] Leonard Guarente and David Sinclair, "Can DNA Stop Time: Unlocking the Secrets of Longevity Genes," *Scientific American,* March 2006.

Westphal recruited a CEO to replace him and remained on the board as lead investor once he got the company off the starting blocks. Two went public— Alnylam and Momenta—and had a combined market value of $1.4 billion in early 2007. Philip Sharp, a Nobel Prize winning biologist and Sirtris advisor, described Westphal's business and science acumen: "Christoph's combination of skills is very rare. I haven't seen his equivalent in 30 years of working in biotech."[5] In 2002, MIT's *Technology Review* recognized Westphal as one of the country's top 100 Young Innovators under 35.

The son of two doctors, Westphal was a former McKinsey consultant and physician, who sped through an MD/Ph.D. program at the Harvard Medical School in less than six years. A polyglot (English, French, German, and Spanish) and accomplished musician (cello), Westphal was described by several Sirtris board members as having "extraordinary energy" and a "rock star" reputation in the biotechnology world. "He has an unusual combination of abilities— to understand the science and its commercial potential, and explain it all clearly in an understated way that resonates with investors," said John Freund, Managing Director and cofounder of Skyline Ventures as well as a Sirtris director at the time of the case.

Westphal had a distinctive approach to building biotech companies—his own mode of operation. Westphal's major successes, Alnylam and Momenta, both went public before many market watchers believed them to be ready for an IPO. In both cases, Westphal teamed with world-renowned authorities (Alnylam with Paul Schimmel, a prominent scientist at the Scripps Institute and biologist Philip Sharp; Momenta with Robert Langer, an MIT Institute Professor and one of the world's most prolific scientist/entrepreneurs). Westphal described the elements he looked for and the approach he took in starting and building companies:

> You need fantastic science. Second, you need a great story. Third, you need great venture capital support and lots of money. I am a big believer in raising as much money up front as possible.

Applying this model and exploiting an ever-growing network among the biotech industry's prominent players, Westphal had clearly developed a successful approach to launching early stage companies and then passing the CEO's baton to a different leader. In 2003, Westphal was looking for his next investment opportunity, when he encountered Sinclair's paper in *Nature*. Westphal quickly realized that this was a novel and possibly watershed discovery *if* it could be extended to humans. Westphal phoned Sinclair to discuss the prospects of commercializing his discovery.

Launching Sirtris

At the time, Sinclair had been thinking of commercializing his work for many years. In 1999, he almost joined his mentor, Guarente, in launching Elixir Pharmaceuticals, a longevity-oriented biotech company. Several years later, as

---

[5] David Stipp, "Drink Wine and Live Longer," *Fortune,* February 12, 2007.

Sinclair finalized the 2003 *Nature* paper, he began exploring opportunities to form a company of his own.

Sinclair described his initial meeting with Westphal, "He came in and refused to sign a nondisclosure agreement. So, I told him I wouldn't talk to him. And he said, 'David, if I walk out of this office, I'm not coming back. So I suggest you tell me as much as you can.'" I wound up telling him more than I normally would have. It soon became apparent that he's one of the smartest people I'd ever met. But it took me months to realize that he's also a nice guy."[6]

After their initial meeting, Westphal expressed an interest in starting a company with Sinclair, but could not do so until he found someone to replace him as CEO of Acceleron, one of the companies he had launched while at Polaris. Meanwhile, Sinclair continued discussions with other investors. Westphal reentered the picture six months later, expressing a readiness to invest and pull the venture together. Sinclair and Harvard (the owner of several pieces of intellectual property that would be licensed by Sirtris) decided to move forward with Westphal as the founding CEO.

After an agonizing decision process, Westphal chose to join Sirtris as its full-time CEO. Unlike his other start-ups, his plan this time was to remain with the company, which meant that he would be leaving behind the venture capital life and a high six-figure salary. Westphal explained his decision:

> Many people thought it was too risky to leave a successful VC career. I was taking a $500,000 paycut and my wife and I had just purchased an expensive house in Brookline close to Fenway Park. At the time, David had no data that showed resveratrol activated sirtuins in mammals or could affect mammalian glucose levels or insulin, although we hoped all that would prove true. My VC friends were telling me that I was not being rational. In some ways they were right, but I was excited about Sirtris in a way I had not yet been at my other companies.

Scientific Advisory Board

Westphal set out to attract a world-class Scientific Advisory Board (see **Exhibit 2** for details on the SAB). Virtually all early-stage biomedical companies create boards of scientific advisors. Among other roles, SAB members advise the company on matters related to scientific and experimental strategies; they sometimes assist in securing access to intellectual property produced by SAB members; and they serve as portals that keep the company abreast of developments in the broader scientific community.

Sirtris's goal was to collect the brightest scientists in the field of sirtuin research, including those who would generate IP for Sirtris and be the "eyes and ears" of the company. Sinclair explained the formation of the Sirtris SAB: "Christoph said, 'Give me a list of the top 10 people in your field.' Within a

---

[6] David Stipp, "Drink Wine and Live Longer," *Fortune,* February 12, 2007.

week or two, we were having conference calls with all of these people. In one case, an academic was going to start a rival company, and Christoph flew out to St. Louis and convinced him to join us instead." One observer described the Sirtris SAB and board of directors in the following terms (see **Exhibit 3** for Sirtris Board of Directors):

> Since combining forces with Sinclair, Westphal has organized what is arguably the most pedigree-rich scientific advisory board in biotech, including MIT's Sharp; Robert Langer, one of medicine's most prolific inventors, also of MIT; Harvard gene-cloning pioneer Thomas Maniatis; and Thomas Salzmann, formerly executive vice president of Merck's research arm. The group now numbers 27, among them many of the world's leading experts on sirtuins. Westphal also assembled an impressive list of directors—they include Alkermes's Pops; Aldrich, the Boston hedge fund manager and biotech veteran; and Paul Schimmel, a prominent scientist at the Scripps Research Institute in La Jolla, Calif., who has co-founded half-a-dozen biotech concerns. Westphal's right arm at Sirtris is chief operating officer Garen Bohlin, formerly a senior executive at Genetics Institute, a biotech now owned by Wyeth.[7]

Growing Sirtris

During the spring and summer of 2004, Westphal and Sinclair went on a road show to local Boston-area venture capital groups. In August, they obtained a $5 million seed (Series A) round of financing from Polaris Ventures, Cardinal Partners, Skyline Ventures and Techno Venture Management ("TVM"). Sinclair described their early efforts to raise capital:

> Christoph was very good at getting us in to talk with the majority of VCs in the Boston area over a short period of time. Although a lot of people said "no" to us, Christoph set a small window of time to invest and more than a few people started getting nervous about being left out. The short timeframe built its own momentum and helped drive interest in the company.

In November 2004, Westphal and Sinclair secured another $13 million in a Series A-prime round. (See **Exhibit 4,** which details five investment rounds; including investors, dates and investment amounts. See also **Exhibit 5,** which details pre-IPO financing at three comparable biotechnology firms.) Regarding the first two funding rounds, Westphal explained their ability to raise funds before the firm had any mammalian data:

> We were very early in terms of the science. We raised $18 million without any mammalian data, something that is almost unheard of in today's world of biotechnology. Part of our success was getting people bit by the

---

[7] David Stipp, "Live Forever?," *Fortune,* February 5, 2007.

Sirtris bug. We had a long-term vision of where we could go with the biology and the anti-aging message is extremely powerful, especially when you are talking to a bunch of aging, overweight guys who are prime targets for the drugs you want to develop.

Sirtris had several decisions to make about how to focus its drug development efforts. Sinclair had theorized that sirtuins played a role in diabetes, cancer, heart disease, neurodegenerative diseases and diseases related to mitochondrial disorders. One thing was clear: there would be no effort to claim anti-aging effects from any drug the firm produced, since the FDA did not recognize aging as a disease. The firm decided to develop a drug for diabetes, a large market with epidemic numbers of type II diabetes in developed and developing countries, and an orphan drug[8] for the mitochondrial disease MELAS (mitochondrial encephalopathy, lactic acidosis, and stroke-like symptoms syndrome), a rare genetically inherited disorder. (See **Exhibit 6** for details on the global diabetes market.)

Between 2004 and 2005, Sinclair began conducting experiments that sought a connection between resveratrol and sirtuins in mice. "By early 2005, David started getting data in his lab that showed resveratrol was going to extend lifespan in a mammal," Westphal said, "At Sirtris, we had evidence you could lower glucose and insulin in mice. All of a sudden, we were getting real proof that this actually could be a drug; that this could actually be a very valuable company." Sirtris began hiring its R&D team, successfully staffing leadership positions with executives who had had long stints at large pharmaceutical and biotech companies – including Millenium Pharmaceuticals, Alkermes, and GlaxoSmithKline – identifying small molecules, developing drugs and advancing drugs through clinical trials.

With their mammalian data in hand, Sinclair and Westphal went back on the road seeking additional capital to finance Sirtris's R&D efforts. In March 2005, Sirtris closed on a $27 million Series B-round of financing.

In the next year, Sirtris made three significant advances. (See **Exhibit 7** for a timeline of Sirtris scientific proofs of principle.) First, the company created a formulation for SRT501 that kept resveratrol in its active form and increased its absorption into the bloodstream. The result was that SRT501 could get five times more resveratrol into the blood stream than the best other preparations currently available. Second, Sirtris began conducting clinical trials in India, assessing SRT501 as a diabetes therapy. The transition from an R&D-only company to a clinical-stage company was, as Westphal remarked in a press release, "an important milestone in our plan to develop a rich

---

[8] Orphan drug status was a special designation by the FDA that granted drug makers a seven-year marketing monopoly along with tax reductions for an approved drug for diseases afflicting less than 200,000 people. The purpose of the orphan drug classification was to provide an incentive to drug makers to focus some of their R&D on smaller, less profitable markets. Some of the more well-known targets of orphan drugs include cystic fibrosis and snake venom.

pipeline of therapeutics to treat diseases of aging." Third, and perhaps most significantly, Sinclair completed several experiments examining the effects of high doses of resveratrol on obese mice. In one experiment, middle-age mice fed high calorie diets with resveratrol were compared to a control group fed similar diets but without resveratrol. Remarkably, the mice fed resveratrol could run further; were leaner; and, lived 30% longer than the high fat, no-resveratrol mice in the control group. (See **Exhibit 8** for a picture of mice on diets with and without the drug.)

Sirtris researchers were exhilarated by these findings in part because the data suggested that sirtuins could play a role in managing or even delaying the onset of type-II diabetes. They were also excited because in the past experimental data with the rat model (Zucker-fa/fa) had tended to foreshadow a higher probability of success for drugs in human trials.

In April 2006, Sirtris closed on a $22 million Series C round of financing, and obtained a $15 million line of venture debt from Hercules Technology Growth Capital.

Throughout 2006, Sirtris continued to gain momentum as Sinclair's research made its way into high profile academic journals, newspapers, and other media. In June 2006, Sirtris announced that it had successfully completed dosing eighty-five human subjects in a Phase 1 safety and pharmacokinetic trial of SRT501. In October 2006, an article about Sinclair's work appeared on the front page of the *Wall Street Journal.* The following month, papers published in *Cell* and *Nature* demonstrated that resveratrol increased the stamina of mice two-fold and significantly extended their lifespans. The 2006 *Nature* article also demonstrated that sirtuin activation increased within the cell the number of functional mitochondria, the powerhouses that sustain a range of cellular activities including glucose metabolism.

The discovery that sirtuin activation increased the number of functional mitochondria was, Sinclair suggested, "quite intriguing" since the number of functional mitochondria was known to decline with age. Moreover, it was well-known in the scientific community that people with above average numbers of functional mitochondria, such as famed cyclist Lance Armstrong, had above average stamina levels.

In late fall of 2006, Sinclair received an unsolicited email from Red Sox owner John Henry requesting a meeting. Westphal described the meeting with Henry:

> He visits David and me here at Sirtris. And he's a very shy, wonderful gentleman. After we present him the company, he says, "How can I be helpful to you?" And I look at him, and I say, "I think you could invest $50 million in the company." And he says, "I don't think I can do $50 million, but I think I can do $20 million." And I said, "Can we close in two weeks?" John teamed up with Peter Lynch, the legendary former fund manager of Fidelity's Magellan Fund, who did extensive due diligence, and they said, "OK." Everyone wanted in after that.

By February 2007, Westphal closed on another round of financing, raising $35.9 million, from company executives, venture firms that had contributed to previous rounds, as well as John Henry and Peter Lynch. Westphal explained their fundraising success:

> We've always been able to raise money, I think because we had money. We weren't desperate. They [potential investors] knew that they couldn't get away with trying to hammer us on the valuation, because if we don't get the valuation we want we just won't raise the money.

Moving Forward

Nutraceuticals

Westphal and Sinclair had a long-running debate over the commercial opportunity represented by their proprietary formulation of resveratrol (SRT501). Sinclair had a strong belief that there would be great public demand for this formulation long before the results of the clinical studies on SRT501 were completed, and that Sirtris, one way or another, should start selling SRT501 to the public as a nutraceutical. Sinclair received an average of 30-50 emails a day from people requesting information on how to obtain resveratrol.

For a biotech company focused on drug development, a Sirtris foray into nutraceuticals would not be unprecedented. For instance, Sigma-Tau Pharmaceuticals, a 50-year old Italy-based drug company specializing in rare diseases, had developed a nutraceuticals business around its FDA-approved drugs for metabolic and renal conditions (carnitine deficiency) as well as cancer (antineoplastic therapy). Sigma-Tau had two nutraceutical divisions and sold physician-recommended and clinically tested dietary supplements for patients with ulcerative colitis or irritable bowel syndrome.

The opportunity in nutraceuticals might be substantial. The global nutraceuticals market, which included sales of health and nutritional supplements, such as vitamin C and fish oil, had estimated annual sales of $120 billion, and was growing at a compound annual growth rate of 7%.[9] For example, the market for glucosamine chondroitin, a joint-health supplement most commonly taken by the elderly, exceeded $1 billion, and had been expanding at a double-digit rate.[10] The economics of the nutraceuticals marketplace were compelling. Total manufacturing costs for small molecule drugs were typically low, often in the range of 25 cents per pill or less. Other costs would depend on the commercialization strategy, of which there were numerous options. Current vendors of formulations of resveratrol that were far less bioavailable than SRT501 were charging prices in the vicinity of $1 per capsule.

---

[9] http://biz.yahoo.com/bw/080220/20080220005585.html?.v=1. Accessed February 22, 2008.

[10] Eric Nagourney, "Study Sees Little Benefit in Chondroitin for Arthritis," *New York Times,* April 17, 2007.

Westphal, however, had doubts about whether the timing was right for such a venture, and whether the firm should even market a nutraceutical under its corporate brand. There were several issues:

The nutraceuticals market was unpredictable. It was, by and large, unregulated. No FDA approval was necessary for selling supplements. No evidence was necessary to prove a product's effectiveness or even its composition. Sinclair had tested resveratrol pills and found that some brands on the market had no active resveratrol at all. Other brands had only trace amounts of active resveratrol, far too little to have any meaningful effects on humans. Even if Sirtris had a scientifically proven product, it was not clear that science alone would be enough to differentiate Sirtris's offering from the dozen or so resveratrol supplement providers. How would the company distinguish itself? A final concern was the potential for consumer allegations about resveratrol's safety. "The potential that someone could attribute a death to SRT501 consumption could easily derail a nutraceuticals business," said Sinclair.

Another issue concerned Sirtris's identity. Was the company a scientifically rigorous enterprise focused on developing FDA-approved medicines that physicians would prescribe to improve the health of patients with aging-related disorders? Would the corporate brand suffer if it started selling 501 as a nutraceutical, especially if it became, in part, a nutraceutical retailer like the makers of glucosamine? On one hand, the Sirtris office walls were plastered with pictures of Sinclair and Westphal with Nobel prize winning biologists and luminaries from the venture capital world. On the other hand, there were the pictures of Sinclair and Westphal with celebrities, including Barbara Walters and Mel Gibson. Rich Aldrich, one of the company's original investors and current board member, summarized the issues this way: "The Sirtris story is a balance. It's carefully constructed from the core science and Christoph and David's public outreach. It's not clear if a nutraceutical approach will taint that story or extend it."

Even if they did consider the nutraceuticals business, what was the business case for market entry? How much of a return would warrant their participation in the market? And how would market entry be achieved? Should the company set up a subsidiary or create a spin-off company devoted to SRT501? Would the subsidiary have the Sirtris name, e.g., Sirtris Health, or would the subsidiary have a different name, as Lexus is to Toyota? Some Sirtris executives were favoring a wholly owned nutraceutical subsidiary with a different name, in order to make clear that the Sirtris brand was focused on drugs based on their NCE development platform. "Our long-term investors like this option because they are in this company for the NCEs. They're not in it for the 501 data," said Westphal.

Yet another issue was retail format. How would a nutraceutical be sold: its own retail stores, the Internet, supplement stores, such as GNC? Would developing its own retail distribution take away from its drug development? Would partnering with a GNC retailer reduce control over the brand?

Pharmaceutical Deal

Throughout 2006, new evidence emerged from laboratories around the world confirming Sinclair's hypothesis of a connection between resveratrol, sirtuins and metabolic activity in mammals. Several pharmaceutical companies began talking with Westphal about a possible drug development partnership. He anticipated that the terms of a deal could include a significant upfront cash payment, an equity purchase agreement, non-milestone-based (i.e., guaranteed) R&D support that would be likely to step up over a four to five year period, and payments tied to clinical development milestones. Sirtris would also receive royalties on sales of any drugs resulting from its SIRT activation program (See **Exhibit 9** for more details on the terms of a deal that might be possible and **Exhibit 10** for details of three deals between big pharma and other private biotechnology firms.) As an alternative, the team also believed that they could ink a deal with terms similar to those in **Exhibit 9** but in which the pharmaceutical partner would take a 51% equity stake in the company.

As Westphal, Sinclair, and Dipp resumed the conversation they had been having in early 2007 about the reasons for and against a deal, one positive aspect that continued to come up was that having a drug development deal was often a condition of having a successful initial public offering. Sirtris executives were hoping to take the company public at some point in the not-too-distant future. A successful IPO could deliver some of the financial resources a company needs to move drug development from the laboratory through clinical trials. "The typical biotech playbook says to get a partnership deal done, then file for an IPO. Public investors are often reassured by the prior involvement of a pharmaceutical company," said Westphal. Several board members intent on satisfying the company's capital needs had already voiced their interest in exploring whether a deal could be completed on attractive terms. Jeff Leiden, a life sciences venture capitalist and the former president and chief operating officer of Abbott Pharmaceuticals and a friend of Westphal's, noted that a typical biotechnology company would not be a strong IPO candidate until it had developed some clinical data, successfully made headway on two different research programs, established intellectual protection around its discovery of one or more new chemical entities, and signed at least one significant deal with a pharmaceutical company.

Another reason to contemplate a deal was that it might be a relatively inexpensive source of additional capital. Pharmaceutical partners were often willing to purchase an equity stake at a premium to VC investors. As well, they might be willing to finance Sirtris's R&D programs.

There were reasons not to do a deal. Westphal and Sinclair were building a company that they hoped would have great medical and investment impact. Even though Westphal had been the founding CEO of several companies, this was the first company that he had actively managed beyond the first two years.

This was also Sinclair's first company. How would a deal affect the founders' control over the company and its future?

Another reason to consider deferring a deal was Sirtris's promising development of new chemical entities (NCEs). Resveratrol is a naturally occurring substance and because it was already on the market as a nutraceutical, intellectual property protection related to it would be limited to "methods of use" patents, which would cover the use of resveratrol to treat particular diseases, such as diabetes, as well as formulation work. Sirtris had recently made strides in synthesizing new compounds that could be as much as 1000 times more potent than resveratrol at activating SIRT1 and that would be eligible for NCE composition of matter patents. Partnering with a large pharmaceutical company would require out-licensing these new compounds without knowing their full value. If Sirtris waited for new NCE data to arrive, it might be able to arrange a more lucrative deal than what might have been doable in early 2007. Of course, if the NCE data came in and it did not produce the results Sirtris was expecting, the pharma deal terms would be substantially worse, assuming that the pharmaceutical companies remained interested at that point.

In-Licensing or Expanding the Scientific Base

From the very beginning of the company, Sirtris executives had had an internal debate over how much of the company's resources to focus on SIRT1 versus alternative sirtuin targets. The biotechnology industry was littered with companies and drug development projects that had stalled in moving from mouse studies and toxicity screens to human trials. Several board members were advocating that the firm diversify its product development platform. There were two main alternatives. One was to investigate the six other human sirtuins. The other was to in-license from another biotechnology firm a compound that had a better known mechanism of operation.

The study of sirtuins was still in its infancy, so investigating the clinical possibilities of other sirtuins would require a great deal of basic research, financing and time. (See **Exhibit 11** for Sirtris financial data, 2004-2006.) Even so, the clinical role of the other human sirtuins offered tantalizing commercial options. SIRT3, for instance, was found in mitochondria and was considered a potential drug target, but little was currently known about its functional role. Developing a research platform based on SIRT3 might complement the company's own drug development efforts in other mitochondrial disorders, including diabetes and MELAS.

Several members of the SAB and the board of directors considered the in-licensing option to be an appealing alternative, although others disagreed. It would balance their investment in SIRT1, which was absorbing 90% of the firm's R&D expenditures. Dr. Dipp explained that, after investigating more than one hundred potential compounds to in-license, Sirtris had found a few anti-diabetic compounds that had better characterized effects than sirtuin-activating compounds. "It would be what we call "me-too" drugs. We know how they work, and if we could get them on the market they would get at least a small percentage of market share. It's something to have in your back pocket."

When the firm launched in 2004, Sinclair and Westphal debated whether to in-license a compound and decided against doing so. "I was the only person at the company," said Sinclair, "who thought that SIRT1 activation was the right bet to make. I told Christoph, 'don't stop it until you know it's wrong.' If I'm wrong, find out sooner than later, and then in-license something." Westphal offered another view, "For the first eight months of this company, I was sitting there like a venture guy thinking that resveratrol will not be a great drug. It's a great story, but we'll have to bring in other stuff to build the company around."

Even though new experimental data seemed to confirm that resveratrol activated SIRT1, and that SIRT1 activation could be clinically important, Sinclair, Westphal, Bohlin and Dipp continued to debate whether to focus the firm's time, money, and other resources on that one target or divert more of the firm's resources to additional targets, including non-sirtuins. They still did not have evidence that SIRT1 had the effects in humans that Sinclair believed they would one day see.

Conclusion

After discussing these three issues for several hours, Sinclair and Westphal decided to summarize their views on the decisions they needed to make.

At a general level, they remained convinced that sirtuin-activating drugs, if they could be successfully developed, would have a revolutionary impact on human disease. However, they recognized that Sirtris was still many years from completing development of these drugs, much less manufacturing and selling them. To reach that distant point would require successfully navigating technical and regulatory hurdles that had stymied the majority of other biotechnology companies at similar points in their history. According to a pharmaceutical industry association report, only one in five compounds entering clinical trials gained FDA approval.[11] And, only 30% of approved drugs recovered the average development cost of a new medicine.[12] Given all of the unknowns about what could happen Sinclair and Westphal described several options for addressing the risks they faced:

One approach would be to fully "hedge their bets:" Sirtris could try to complete a pharmaceutical deal, in-license a compound with better known effects, and consider a nutraceuticals business around SRT501.

Another approach would be to "swing for the fences" (or, in a frequently used metaphor in the industry, "one shot at gold"). This would continue the

---

[11] PhRMA Pharmaceutical Profile Industry Report, 2007.
http://www.phrma.org/files/Profile%202007.pdf, Accessed February 28, 2008.

[12] PhRMA Pharmaceutical Profile Industry Report, 2007.
http://www.phrma.org/files/Profile%202007.pdf, Accessed February 28, 2008.

firm's focus on a sirtuin-activating drug development platform. If successful, Sirtris could become, as one pharmaceutical executive suggested, "the Google of biotech." However, an IPO would be less likely in the interim, since markets often prefer that biotech firms have a validating deal with a large pharmaceutical company.

A third approach would be a "middle of the road" path that incorporated some hedging, such as pursuing an in-licensing deal, but also accepted some risk, e.g., deferring a potential pharmaceutical deal. Alternatively, Sirtris could try to complete a pharmaceutical deal now, but forego in-licensing and the nutraceuticals project.

The company seemed to have arrived at a critical juncture in its development. What approach should Sinclair and Westphal take? And why?

EXHIBIT 1 Anti-Aging Science Timeline

1884:	William Jones MD reports that a fasting spider lives unusually long, 204 days. [Longevity in a fasting spider, *Science* 3: 4, Jan. 4, 1884]
1934	Clive McCay and Mary Crowell of Cornell University report delayed aging in rats on limited calorie diets. [McCay, C. M., and M. F. Crowell. 1934. Prolonging the life span. *Science Monthly* 39:405–414]
1986	Roy Walford and Richard Weindruch show limited-calorie diet produces *"youthful"* old mice. [Walford R. Weindruch R, *Journal Nutrition* 116: 641, 1986; The Retardation of Aging and Disease by Dietary Restriction, 1988]
2000:	Leonard Guarente of MIT mimics calorie restriction's prolonged lifespan effects in yeast cells. [*Science* 289: 2126, 2000]
2003	David Sinclair of Harvard extends life of yeast cells with resveratrol. [*Nature* 425: 191, Sept. 11, 2003]
2004	Marc Tatar of Brown University and David Sinclair of Harvard report resveratrol extends life of roundworms. [*Nature* 430: 686, July 14, 2004]
2006:	Italian researchers are first to report resveratrol prolongs the lifespan of a vertebral animal (cold-water fish). [*Current Biology* 16: 296, Feb. 7, 2006]
2006	David Sinclair and colleagues demonstrate resveratrol extends life of warm-blood mammal (mice) and overcomes deleterious effects of a high-fat diet. [*Nature* 444: 337, Nov. 1, 2006]
2006	*JAMA*. 2006 Apr 5;295(13):1539-48.
2006	Lagouge, M. et al., Resveratrol improves mitochondrial function and protects against metabolic disease by activating SIRT1 and PGC-1alpha. *Cell*. 2006 Dec 15;127(6):1109-22.
2007	Civitarese, A.E. et al., Calorie Restriction Increases Muscle Mitochondrial Biogenesis in Healthy Humans. *PLoS Med*. 2007 Mar 6;4(3).

Source: Adapted from http://www.longevinex.com/article.asp?story=Time%20Of%20Research%20Involving%20Calorie%20Restriction%20and%20Longevity, accessed February 25, 2008; and Sirtris documents.

EXHIBIT 2 Sirtris Scientific Advisory Board

BioPharma expertise	Sirtuin expertise

Tom Salzmann, Co-Chair SAB
    Former EVP Merck
	Biochemistry
	    John Denu, Wisconsin
	    Anthony Sauve, Cornell

Julian Adams
    President and CSO, Infinity
	Vern Schramm, AECOM
	    David Sinclair, Co-Chair SAB, HMS
	    Eric Verdin, UCSF

Peter Hutt Former FDA General Counsel	Mouse Genetics, Phenotypes

Bob Langer
    MIT, NAS, NAE, IOM, Co-Founder Momenta, AIR
	Fred Alt, HMS, NAS, HHMI
	    Johan Auwerx, ICB

Tom Maniatis
    Harvard, NAS, Co-Founder GI, Acceleron
	Structural Biology
	    Cynthia Wolberger, JHU, HHMI

Phil Sharp
    Co-Founder Biogen and Alnylam, NAS, Nobel Prize
	Links to Disease
	    Ron Kahn, Joslin, NAS
	    Jeff Milbrandt, Washington University

Ted Sybertz
    SVP Genzyme, Schering (Zetia)
	Jerrold Olefsky, UCSD
	    Pere Pulgserver, HMS

Eric Gordon
    Former head of Medicinal Chemistry at Bristol-
    Myers, Founder of Vicuron Pharmaceuticals
	Eric Ravussin, Pennington Institute
	    Li-Huei Tsai, MIT, HHMI

Chris Walsh Harvard, NAS, IOM, Genzyme, Vicuron	

Source: Company.

HHMI = Howard Hughes Medical Institute
ICB = Mouse Clinical Institute in Strasburg, France
NAS = National Academy of Sciences
NAE = National Academy of Engineering
AECOM = Albert Einstein College of Medicine

EXHIBIT 3 Sirtris Board of Directors (as of February 2007)

Name	Background
Richard Aldrich (F)	Founder, RA Capital
John Clarke	Managing General Partner, Cardinal Partners
Alan Crane	Momenta
John Freund	Managing Director, Skyline Ventures
Stephen Hoffman, Ph.D., M.D.	Managing Director, TVM Capital
Wilf Jaeger	Three Arch Partners
Stephen Kraus	Bessemer Venture Partners
Richard Pops (F)	Chairman, Alkermes
Paul Schimmel, Ph.D. (F)	Professor, The Scripps Research Institute
David Sinclair, Ph.D. (F)	Associate Professor, Harvard Medical School
Christoph Westphal, M.D., Ph.D. (F)	CEO and Vice Chair, Sirtris Pharmaceuticals, Inc.

Source: Company.

F = Founder

EXHIBIT 4 Sirtris Investors and Investment Rounds

Round	Date	Investors	Investment	Pre-$ Valuation
A	Aug. 2004	Polaris Venture Partners, Cardinal Partners, Skyline Ventures and TVM	$5 M	$2.8 M
A1	Oct. 2004	Polaris Venture Partners, Cardinal Partners, Red Abbey, Skyline Ventures, TVM and The Wellcome Trust	$13M	$10.9M
B	Mar. 2005	Three Arch Partners, Cargill Ventures, Novartis Bioventures Fund, Polaris Venture Partners, TVM, Red Abbey, Cardinal Partners, Skyline Ventures, and The Wellcome Trust	$27 M	$32.5M
C	Apr. 2006	Bessemer Venture Partners, Genzyme Ventures, QVT Fund LP, Alexandria Real Estate Equities, Inc. Polaris Venture Partners, TVM Capital, Cardinal Partners, Skyline Ventures, Three Arch Partners, The Wellcome Trust, Novartis Bioventures Fund, Cargill Ventures, , Cycad Group, Hunt Ventures, and Red Abbey	$22 M ($22 million in Series C-equity placement, plus $15 million in venture debt financing from *Hercules Technology Growth Capital*)	$95.4M
C1	Feb. 2007	John Henry Trust, Peter Lynch and several investors from previous rounds	$35.9 million	$184M

Source: Company.

EXHIBIT 5 Pre-IPO Financing of Comparable Biotechnology Companies

Company	Round/Date	Investors	Investment
Anesiva (IPO 2004)	2Q2001	InterWest Partners, Alta Partners, J.P. Morgan Partners	$13 million
	2Q/2002	Bear Stearns Health Innoventures, Alta Partners, HBM BioVentures, J.P. Morgan Partners, InterWest Partners, MIC Capital, China Development Industrial Bank (CDIB Ventures)	$50 million
	4Q/2003	Bristol Myers Squibb	$45 million
Cytokinetics (IPO 2004)	2Q/1998	Mayfield, Sevin Rosen Funds, Individual Investors	$5.3 million
	3Q/1999	International BM Biomedicine Holdings, Vulcan Capital, Mayfield, Sevin Rosen Funds, NMT New Medical Technologies, Duke University, Individual Investors	$20 million
	4Q/2000	Credit Suisse, Alta Partners, Lombard Odier Darier Hentsch, Mayfield, Sevin Rosen Funds, Vulcan Capital, NMT New Medical Technologies, Duke University, Individual Investors	$50 million
	2Q/2001	GlaxoSmithKline	$14 million
Momenta (IPO 2004)	2Q/2002	Cardinal Partners, Polaris Venture Partners	$4.4 million
	2Q2003	Atlas Venture, MVM Limited, Polaris Venture Partners, Cardinal Partners	$19 million
	1Q2004	Mithra Group, Polaris Venture Partners, Atlas Venture, MVM Limited, Cardinal Partners	$20 million

Source: Adapted from Growthink Research.

EXHIBIT 6 Diabetes Market Data

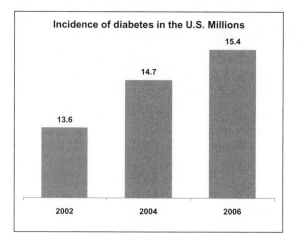

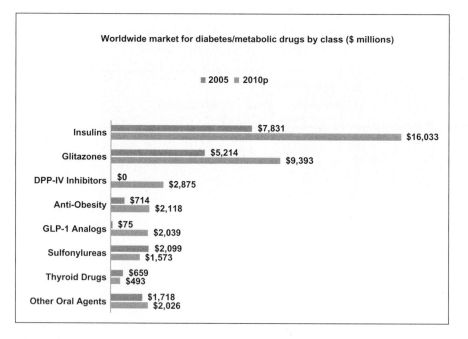

- 90% of the total diabetic population has Type 2 Diabetes
- Total direct and indirect costs associated with diabetes were over $100 billion in 2002

Source: Company.

EXHIBIT 7 Proof of Principle for SIRT1 Activators

Pre-clinical Therapeutic	*Nature* 2006 Nov 16:444(7117):337-42.
	Cell. 2006 Dec. 15; 127(6):1109-22.
	Unpublished
Human Physiological	*PLoS Medicine.* 2007 Mar 6:4(3).
	JAMA. 2006 Apr 5;295(13) 1539-48.
Human Genetic	*Cell.* 2006 Dec 15;127(6):1109-22.

Source: Company.

EXHIBIT 8 Effects of High Doses of Resveratrol in Mice

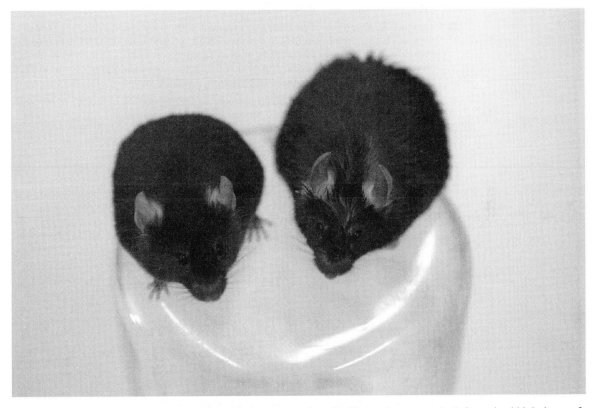

Two mice fed the same high-calorie diet in a Sirtris-sponsored study: The svelte one on the left received high doses of resveratrol.

Source: Company.

EXHIBIT 9 Potential Sirtris Pharma Deal Terms

- Five year term.

- 19.9% OR 51% equity stake at a 50% premium to most recent share price.[a]

- $75 million upfront for an exclusive option to join Sirtris in developing and marketing compounds from its SIRT1 Activator Program.

- 5 years of guaranteed R&D support totaling $100 million. Payments to step up over time. $10 million in year one.

- A combination of royalties, possibly manufacturing profits, and co-promotion fees that equate to approximately a 50/50 split of profits in the U.S. This is a significant point for Sirtris since the SIRT 1 activator program is a core program, and the one that represents 90% plus of the firm's value.

- The pharmaceutical company will lead marketing and country specific development ex-U.S., Sirtris to receive substantial, double digit royalties on ex-U.S. sales.

- Roughly $200 million in milestones concurrent with risk reducing progress. Roughly 15% upon successful completion of safety/PK of a SIRT 1 activator NCE in humans; Roughly 25% based on observation of glucose effects in phase 1b of NCE in man; Roughly 30% upon successful completion of a phase 2a efficacy study for an NCE in man; and roughly 30% upon completion of phase 2b studies.

a: Two different equity stakes were under discussion—19.9% or 51% of Sirtris's equity. All other terms would remain the same in both scenarios.

Source: Company.

EXHIBIT 10 Comparison Agreements between Comparable Biotechnology Start-ups and Large Pharmaceutical Companies

Company	Deal Partner	Agreement	Terms
Anesiva is a late-stage biopharmaceutical company that seeks to be the leader in the development and commercialization of novel therapeutic treatments for pain. The Company has four drug candidates in clinical development for multiple potential indications. IPO in 2004 (formerly called Corgentech—this was the name when it went public).	Bristol-Myers Squibb Company	Jointly develop and commercialize Corgentech's E2F Decoy (edifoligide sodium), a first-of-its-kind E2F Decoy treatment currently in Phase III development for the prevention of vein graft failure following coronary artery bypass graft (CABG) and peripheral artery (i.e., leg) bypass graft surgery.	Bristol-Myers Squibb will make an initial payment to Corgentech of $45 million comprising cash and an equity investment in Corgentech, with the potential for an additional $205 million in clinical and regulatory milestone payments. Bristol-Myers Squibb and Corgentech will share development costs in the U.S. and Europe based on a pre-agreed percentage allocation.
Cytokinetics is a biopharmaceutical company dedicated to the discovery, development and commercialization of novel classes of small-molecule therapeutics, particularly in the field of cytoskeletal pharmacology. IPO in April 2004	GlaxoSmithKline	A broad strategic collaboration to discover, develop and commercialize novel small-molecule therapeutics targeting mitotic kinesins for applications in the treatment of cancer and other diseases.	GSK has committed funding of approximately $50 million over the minimum 5-year research term, including a $14 million upfront cash payment and a $14 million purchase of Cytokinetics preferred stock. In addition, GSK could make milestone payments to Cytokinetics ranging from $30-50 million per target for products directed to each of over 10 mitotic kinesins that will be the subject of collaborative activities.
Momenta Pharmaceuticals was founded based on proprietary technology developed at MIT that enables high throughput, detailed characterization and engineering of sugars. IPO in June 2004	Sandoz/Novartis	A strategic alliance covering joint product development and commercialization in the area of complex pharmaceutical products. The collaboration will apply Momenta's novel technological capabilities related to complex sugars and the leadership of Sandoz in the generic pharmaceuticals industry to pursue the joint goal of commercializing products. Under the terms of the agreement, Sandoz and Momenta will jointly manage product development and commercialization.	Sandoz has committed $600,000 in an upfront cash payment and has the right to purchase $5-10 million in Momenta equity. R&D payments estimated to be $12 million and up to $50 million in contingent payments to accompany development milestones.

Source: Adapted from Recombinant Capital.
NOTE: All deals inked when biotech was still private.

EXHIBIT 11 Sirtris Financial data 2004–2006 (in thousands, except share and per share amounts)

	Period from March 28, 2004 (date of inception) through December 31, 2004	Year Ended December 31,		Period from March 25, 2004 (date of inception) through December 31, 2006
		2005	2006	
Statement of operations data:				
Revenue	$ —	$ 68	$ —	$ 68
Operating expenses:				
Research and development	1,190	7,062	14,242	22,494
General and administrative	699	3,865	4,340	8,904
Total operating expenses	1,889	10,927	18,582	31,398
Loss from operations	(1,889)	(10,859)	(18,582)	(31,330)
Interest income	45	1,143	2,447	3,635
Interest expense	—	—	(878)	(878)
Net loss	$ (1,844)	$ (9,716)	$ (17,013)	$(28,573)
Net loss per share—basic and diluted	$ (0.93)	$ (3.16)	$ (3.52)	
Weighted average number of common shares used in net loss per share—basic and diluted	1,995,468	3,087,716	4,854,646	
Pro forma net loss per share—basic and diluted			$ (0.20)	
Shares used in computing pro forma net loss per share—basic and diluted			85,603,228	

Source: Company.

Appendix Wall Street Journal Cover

THE WALL STREET JOURNAL.

MONDAY, OCTOBER 30, 2006 | *© 2006 Dow Jones & Company, Inc. All Rights Reserved.*

Youthful Pursuit

Researchers Seek Key to Antiaging In Calorie Cutback

A Controversial Hypothesis
Draws Scientists, Investors;
Will It Work in Humans?

Fighting Fat in Lab Mice

By David Stipp

In the 1930s, researchers stumbled onto a surprisingly simple way to slow the biological forces of aging: cutting normal calorie intake by about a third. Scientists found it boosts animals' life spans by 30% to 40%, and considerable evidence suggests that calorie restriction, or CR, would slow human aging too.

But only steely ascetics could hack its hunger pangs. So the finding remained a little-known curiosity in the back halls of science.

David Sinclair

Now a coterie of scientists and biotech ventures are rekindling interest in CR as they try to mimic its antiaging effects with medicines. It is still a highly speculative quest, and many researchers fret that it hasn't completely shaken its association with centuries of dubious nostrums to slow aging, from inhaling virgins' breath to eating gold to implanting monkey glands.

Much of the new focus is on a substance in red wine called resveratrol. The interest in it started three years ago when a group led by Harvard Medical School biologist David Sinclair reported that it boosted yeast cells' life span by 70% via a mechanism resembling CR. He later co-authored a study showing that it also boosts life span in fruit flies and round-worms. But his tendency to make bold leaps based on tentative data has also sparked intense controversy. One big question: Does he really understand the workings of CR well enough to mimic them in a drug?

Last spring, Italian scientists reported that resveratrol boosted life span more than 50% in a kind of short-lived fish. Intriguingly, fish on resveratrol had much faster swimming speeds as they aged, and spent far more time moving around, than did undosed control fish.

At least two groups of researchers are now testing whether resveratrol can extend life span in mice—the first such studies in mammals. At a meeting of the American Aging Association in June, Dr. Sinclair and colleagues presented preliminary results from a study showing that resveratrol had "CR-like protective effects" against the buildup of fatty deposits in the livers of mice on high-calorie diets. That suggests that resveratrol could lead to new drugs for diseases of aging associated with rich diets, such as adult-onset diabetes.

A company that Dr. Sinclair co-founded in 2004, Sirtris Pharmaceuticals Inc., of Cambridge, Mass., has begun testing a resveratrol-based drug in diabetic patients. It has raised $82 million from venture capitalists, a hefty sum for an early-stage biotech. (Sirtris's chief executive, Christoph Westphal, is married to a reporter for this newspaper.)

Leonard Guarente

It faces competition from Elixir Pharmaceuticals Inc., also based in Cambridge, which Dr. Sinclair's former mentor, Massachusetts Institute of Technology biologist Leonard Guarente, co-founded in 1999 to develop drugs based on gene variants that slow aging. The niche also includes BioMarker Pharmaceuticals of Campbell, Calif., and LifeGen Technologies of Madison, Wis., both of which focus on mimicking CR with drugs.

The companies hope to develop therapies for diseases, not antiaging pills. One reason is that the Food and Drug Administration doesn't recognize aging as a problem warranting treatment. But if a drug could retard aging, it might delay the onset and possibly the progression of age-related diseases. "When you slow aging," says University of Illinois epidemiologist

S. Jay Olshansky, "you push a host of diseases to later ages at one fell swoop—cancer, heart disease, Alzheimer's, diabetes, as well as everything else that's negative about growing older."

Some researchers believe antiaging drugs could also improve health in late life—rather than prolong misery—letting people stay in relatively good shape until a swift demise. Their case rests partly on the svelte, energetic look of old animals on CR. "Often it's hard to identify the cause of death" in post-mortem studies on such animals, says Richard Weindruch, a University of Wisconsin CR researcher. "The only apparent problem is that they died."

Still, some experts on aging doubt that enough is known about CR to guide the development of drugs that mimic its effects. "We know a lot about CR's effects," says Edward Masoro, a leading gerontologist. "But what bothers me is that I don't think we've figured out CR's basic mechanism yet."

Dr. Sinclair's idea that resveratrol mimics CR has come under heavy fire. His main adversaries are two researchers who used to rub elbows with him when they all studied together with MIT's Dr. Guarente. The skeptics maintain that resveratrol's mode of action is still murky; instead, they are looking at other mechanisms that may account for how CR works.

The resveratrol doses used in the lifespan-extension studies in animals were far higher than the amount people can get by drinking wine—they were roughly equivalent to hundreds of glasses a day. Resveratrol is available as a dietary supplement, but to replicate the doses used in the studies, a person would need to take scores of pills a day. (Sirtris says it is developing prescription drugs that work like resveratrol but are hundreds of times more potent.) The dietary supplements haven't been tested in clinical trials, so their efficacy isn't proven, nor is it clear what dose might make people live healthier or longer. And although they seem safe at modest doses, megadoses may not be.

Nevertheless Dr. Sinclair, a 37-year-old Australia native, thinks taking small doses over time may yield health benefits and has been taking the supplements for three years.

The story of resveratrol has its roots in scientists' increased understanding of CR. In 1989 researchers theorized that it acti-

(over please)

DOWJONES

Source: David Stipp, "Researchers seek key to antiaging in calorie cutback," *Wall Street Journal,* October 30, 2006.

Questions

1. What made Sirtris an attractive opportunity (and not just a good idea) at the time that Christoph Westphal joined as CEO? How will Sirtris make money?

2. Identify the major risks in each of these categories: technology, market, team, and financial. What is the most important category and risk?

3. Should Sirtris do the deal with the pharmaceutical company? Why or why not? Should Sirtris launch a neutraceuticals business? Why or why not?

4. Assume that Christoph Westphal and his team have just been told that J.P. Morgan Securities, which is a reputable investment bank, is eager to help Sirtris go public soon by filing for an Initial Public Offering (IPO). Furthermore, assume that Glaxo Smith Kline (GSK) has also just contacted them to discuss an offer to buy Sirtris entirely. What are the advantages and disadvantages of each of these three options that they could pursue in order to finance and operate the venture (e.g., stay private for now and fund the company with more venture capital and corporate partnerships, take the company public via an IPO, or agree to be acquired to become an operating division of an established company)?

COOLIRIS: BUILDING AN A+ TEAM

Introduction

It was well past 2AM on a warm evening in July 2007 at the Kleiner Perkins Caufield Byers (KPCB) incubator in California. Josh Schwarzapel, the young and energetic co-founder of Cooliris, checked his email again to see if their top recruit had accepted the Cooliris employment offer. Across the gray cubicle wall, the technical team, Austin and Kyan, were coding steadily on the next product release. Although Austin and Kyan had been working feverishly for months, they urgently needed more engineers if they were to make the next release deadline.

Three months earlier, when the Cooliris co-founders received their first round of funding from KPCB, Josh had accepted the challenge to help Cooliris expand by building a world-class technical team. At the time Josh had felt full of confidence: how difficult could it be for a young company to attract great talent when it had the backing of one of the world's most successful venture capital firms, an incredible technical vision, an early product with great traction, and another product in the pipeline?

Much to Josh's surprise, however, it had been challenging to build the team. Josh understood the importance of an outstanding team to the success of a new venture. As a student at Stanford studying entrepreneurship, Josh had heard luminaries in the field highlight the necessity of a world-class team for eventual success. But, as he struggled to find and attract such a team, Josh began to realize that his courses had not prepared him well for *how* to build such a team. Nonetheless, in Cooliris' high-accountability culture, pointing the finger at an issue from his formal education would not excuse a failure. Soujanya Bhumkar, the company's CEO, had been supportive and helpful along the way, but Josh could sense the pressure: from the investors, from Soujanya, and worst of all from the team who had been forced to work long hours as they waited for the much needed new hires.

Now, in the early hours of the morning, Josh once again began to ask himself difficult questions. Why had he struggled to recruit a great team? In the past he had always been successful, both as a student at Stanford and a collegiate volleyball player. What had gone wrong? Inevitably, Josh began to ask himself . . . "Is it me? Have I failed?"

This case was prepared by Nathan Furr via the Stanford Technology Ventures Program at Stanford University with assistance from Josh Schwarzapel, Soujanya Bhumkar, and Professor Thomas Byers, as the basis for class discussion rather than to illustrate either effective or ineffective handling of an administrative situation. Some facts have been disguised.

Revised October 2009.

COOLIRIS BACKGROUND AND HISTORY

At the time of Cooliris' founding, the Internet underwent a transformation frequently referred to as Web 2.0. During the first major Internet wave spanning the late 1990s, entrepreneurs and corporations focused primarily on establishing a presence on the Internet using standard interface design to communicate with their audience. However, as the Internet evolved, a second wave emerged during the early 2000s that shifted the focus from corporate-generated content to user-generated content and from proprietary interfaces to modular interfaces that could be recombined (sometimes referred to as "mashable"). As part of this movement, a host of new websites, many of them funded by venture capitalists, sprung up to empower everyday users to become content publishers by posting blogs, making profiles, publishing videos, and sharing photos. Social networks such as MySpace and Facebook became household names and social networking, or the connecting of users to each other in online communities, became a common buzzword for many online ventures.

With the democratization of content creation and publishing, however, the sheer volume of content became a very real problem for most web users. As millions of participants posted billions of new photos, videos, and blogs, the quantity and type of information expanded exponentially. Narrow search terms returned hundreds of thousands of results, and information coming from one's social graph on sites like Facebook and MySpace was becoming too lengthy to consume. In short, a growing challenge for Internet users was how to interact with and process the exploding volume of new content. It was this challenge that led to the founding of Cooliris.

Soujanya, Josh and Austin Shoemaker founded Cooliris in January 2006 around the idea that the Internet had indeed become a fundamental element in the lives of billions of people, but the user interaction metaphors had changed very little since the first browsers. In their view, the Internet had always been characterized by a clunky, non-intuitive navigation experience based on the table format inherent in HTML code. Although flash animations and video content had enriched basic text, the way people interacted with the Internet was still constrained within the 2D framework of the original browsers.

The initial idea for Cooliris had emerged earlier in 2005 during a conversation between Soujanya and a friend, Mayank Mehta. As the two talked about how to make the Internet a more rich experience, Mayank suggested the idea of creating a mouse-over preview of embedded links to other web pages. The preview would allow the user to see the content underneath the link in a contextual window without leaving the original page, thereby creating a more multi-dimensional media experience. Soujanya, a former engineer turned serial entrepreneur, was struck by the insight and began discussing it with colleagues to get their feedback.

During this period, Josh and Soujanya met for coffee to discuss ideas. Josh caught the vision of a better Internet experience and connected Soujanya with Austin, a fellow student with exceptional technical capability. Together they laid out a plan for Cooliris and their first product: Cooliris Previews. The team began development right away, self-funded on Soujanya's credit card and working part time while Josh and Austin finished school. By September of 2006, the team had released Cooliris Previews as a free Internet browser plug-in.

Over the course of a few months, the product began to get significant traction, becoming a featured plug-in for Firefox and attracting thousands of users who desired a richer online experience. The success of Previews validated the primary Cooliris hypothesis that users desire a richer interaction with content, and so the team developed the concept for a second product: PicLens. The new PicLens product would allow users to view online photos from photo sharing sites or online image searches as full-screen slideshows rather than as low-quality thumbnail images or larger files that had to be downloaded individually. PicLens felt like a natural next step for the young company, but as the team built the product, their vision of the company began to evolve into something bigger. The team realized that the challenge of improving web navigation extended beyond a browser add-in to the fundamental way in which users access, discover, and navigate information. The team began soliciting feedback from investors and industry friends on how to take their ideas to the next level. In one such meeting, Randy Komisar, a partner at KPCB, suggested that his firm might be able to provide some funding and incubate the company in their adjacent offices. After several weeks of follow-on meetings with the KPCB partnership, Cooliris received and signed a term sheet for investment, taking up residence in the KPCB incubator next door to offices of such famous venture capitalists as John Doerr and Brook Byers.

Cooliris Hiring Process

After receiving funding and moving into their new offices on Sand Hill Road, the Cooliris team identified several critical next steps for the company. The team and investors decided that one of the most critical action items should be to hire a top-notch technical team to execute on the Cooliris vision. As Josh and Soujanya defined their recruiting strategy, they both agreed that they needed to recruit people who were both entrepreneurial and technically brilliant. Furthermore, they firmly believed in the wisdom of hiring great team members right from the beginning. As Guy Kawasaki of Garage Technology Ventures said, "A players hire A players, B players hire C players, and C players hire bozos."

As the team discussed recruiting, hiring great team members seemed a comparatively easy task given their recent round of prestigious funding and their exciting technical vision. In Josh's eyes, new recruits would get in on the ground floor of an exciting opportunity, operate in a highly supportive setting, and be mentored by industry-leading venture capitalists. In this positive light, the recruiting task seemed easy.

Excited about his new role and freshly graduated from college, Josh took the lead on recruiting the new team. Initially, Josh anticipated that recruiting should take about half of his time, leaving the other half of his time open for business development. Conventional entrepreneurial wisdom suggested that the Cooliris team should start the recruiting process by tapping their social networks for potential hires. Aware of the competition for great technical talent, however, Josh developed an incentive to motivate his extended social network to proffer the best technical candidates: Cooliris would pay $1,000 for a candidate recommendation that led to a hire. Josh soon flooded his own extensive social network with the news about the exciting opportunity to join the Cooliris team. Similarly, KPCB and the entire Cooliris team reached out to their social networks to find the best potential recruits available. At the same time, Josh recognized that the Cooliris team might not know or have links to all the best technical people. To fill this gap, Josh also searched online databases such as LinkedIn and Google, looking for technical talent at similar companies using search terms such as "3D graphics engineer." Finally, Josh posted advertisements on LinkedIn and the San Francisco Bay Area edition of Craigslist.

Over the course of the next few weeks, Josh's search effort yielded mixed results. Not surprisingly, tapping the team's social networks produced the best leads. Searching for candidates on LinkedIn and Google also produced candidates, but not at the caliber the team desired. In contrast to searches, advertisements on LinkedIn and Craigslist typically produced subpar candidates. Later, Josh discovered that the reason the ads proved so disappointing was because candidates with significant talent ignored ads in general; these candidates were already receiving good offers through other channels.

In the end, after an exhaustive initial search, Josh reviewed over 1,200 resumes in search of the ideal team. Of the resumes browsed, Josh reached out to potential hires who he estimated had at least a 50% chance of being an "A player." Once Josh had filtered through the initial list of resumes, he then reached out to candidates via an introductory email explaining the Cooliris opportunity and the team's interest in the candidate (see EXHIBIT 1 for the text of a sample email). In total, Josh contacted 400 candidates via email to invite them to talk more with Cooliris.

In the end, the Cooliris team brought in 50 candidates for a first round interview. Because the incubator was located behind the main KPCB office, it was a little difficult to find. Furthermore, the doors to the incubator were always locked. To solve this problem, Josh gave candidates instructions on how to drive around to the back of the building to where the incubator was located. He told candidates to call once they arrived so that he could personally meet them and show them into the Cooliris offices.

At the beginning of the interview, a candidate would sign a non-disclosure agreement after which either Josh or Soujanya would give the candidate a ten-minute outline of Cooliris's vision and the products that had already been developed. At the same time, because the company was in stealth mode and had some very high potential ideas, Josh and Soujanya were careful not to reveal too much about the future direction of the company or some of their upcoming products. Therefore, in a typical interview, after a brief presentation about the company, the interviewer spent the rest of the hour screening the candidate for his or her technical ability. Josh and Soujanya also experimented with interviewing candidates by themselves or with the entire team.

After a thorough examination of the candidates' abilities during the first round, the Cooliris team decided to invite nine candidates back for a final round interview. The final interview lasted at least two hours and focused on a deeper technical discussion. Although the Cooliris team was still evaluating the candidate for fit and talent, in the final round interview the Cooliris team revealed a little more about the exciting future of the company. Lastly, the interview always included a long chat with Soujanya about expectations. In particular, Soujanya believed that it was important to have open and clear communication about the potential upsides as well as the risks involved; otherwise, both the candidate and the Cooliris team would be entering into a relationship under false pretences—a bad start to any relationship.

Of the nine candidates who received final round interviews, the team decided to extend offers to five high-potential candidates. Josh and Soujanya carefully crafted the offers to be as financially competitive as possible, benchmarking against what Google might pay for a similar position as well as giving candidates potential upside through equity in the company. It now seemed that Cooliris could finally add needed resources to the skeleton technical team who had already been stretched to the maximum.

Two Unanswered Questions

Then came the surprise. Despite the apparent excitement that candidates exhibited during the interviews, of the five offers extended, four offers were turned down. Fortunately, the fifth candidate verbally accepted the offer before departing for a long-planned trip to Europe. Although the yield for his efforts seemed slim, Josh felt that if they could at least hire one candidate then the last few months would not be wasted effort. Indeed, recruiting had taken an immense amount of time—much more than the 50% of his time that he expected. He had worked late nights, weekends, and holidays, all in an effort to succeed in building a technical team.

In the end, though, even the fifth candidate decided not to join Cooliris. Shortly after returning from Europe, the candidate emailed to say he had second thoughts and decided he would pass on the opportunity. The candidate's retraction came as a shock and led Josh to reflect seriously on the failure of the recruiting process. What had the last two and a half months been about? Why had he failed to build a great technical team? What could he have changed to make the process successful? And finally, the most challenging question of all, could it be him? Was the problem that he lacked what was needed to be an entrepreneur? These questions plagued him, but, on another level, Josh realized their challenges boiled down to two key questions. First, *who* is actually an "A player" and second, *how* do you attract those people to join your team? Soujanya peeked over the rim of Josh's cubicle and with his usual earnestness suggested that Josh get some rest: "Hey Josh, it's okay. Let's talk about it in the morning." Josh nodded in agreement, grabbed his bag and headed for the door. The entire weekend was blocked out for the team to meet and discuss what had gone wrong in the recruiting process and what, if anything, could be changed. As he headed home, Josh wondered what he should suggest at the meeting. Were the team's standards too high? Should they hire whomever they could find to help? What was the problem? Why couldn't they recruit world-class team members?

Questions

1. You have just raised your first round of financing and want to build a team that can innovate in a completely new area: how would you go about identifying the right candidates and striking a balance between ingenuity and experience? Where exactly would you search for them? How do you successfully attract your top choices to join your venture?

2. Make a list of what Cooliris is doing right and doing wrong, if anything, with its current recruiting process? How should they improve it?

3. How should recruiting processes differ for hiring various functional positions in this venture? For example, do the same rules apply to hiring engineers as sales and business development talent?

EXHIBIT 1: Email to Potential Candidates

Hi (candidate name here),

I took a look at your profile and you are definitely the kind of guy that we would like to work with for our startup, Cooliris. To give you context, we're leveraging 3D graphics to build an immersive media environment for browsing web content (check out our downloadable app at www.cooliris.com). We've recently raised Series A investment from Kleiner Perkins (the same investors as Google, Amazon, Intuit etc.) and are working with people like Bill Joy (Chief Scientist at Sun Microsystems) and Randy Komisar (former CEO of Lucas Arts).

We'd be willing to explore both full time and contracting options with you, although we would greatly prefer people open to full time. Would you be interested in chatting further?

Sincerely,

Josh Schwarzapel
Cofounder and VP of Business Development
www.cooliris.com

Information Sources
on the Internet

The complete contents of appendix C are maintained on the textbook website at http://techventures.stanford.edu. The directory includes over 150 websites arranged by category, ranging from business plan resources to industry information to sources of capital. A sampling of the resources is listed below. Note that the standard prefix "www" is omitted on all web addresses, except as shown.

C1 General Information on Entrepreneurship

AlwaysOn	alwayson.goingon.com
Entrepreneur America	entrepreneur-america.com
Entrepreneurship.org	entrepreneurship.org
Fast Company	fastcompany.com
Global Entrepreneurship Monitor	gemconsortium.org
Inc.	inc.com
Kauffman Foundation	kauffman.org
Service Corps of Retired Executives (SCORE)	score.org
Wall Street Journal	startupjournal.com

C2 Student Organizations and Venture Forums

Collegiate Entrepreneurs Organization	c-e-o.org
Forum for Women Entrepreneurs	fwe.org
MIT Enterprise Forum	http://enterpriseforum.mit.edu
Students in Free Enterprise	sife.org
SVASE	svase.org
Venture clubs	venturea.com/clubs2.htm
Women entrepreneurs	springboardenterprises.org

absorptive capacity   A firm's ability to assimilate new knowledge for the production of innovations. The more related knowledge a firm has, the easier it is for it to assimilate the new knowledge.

acquisition   When one firm purchases another, the acquired company gives up its independence and the surviving firm assumes all assets and liabilities.

adaptive enterprise   An enterprise that changes its strategy or business model, as the conditions of the marketplace require.

advertising revenue model   Selling firm, usually media companies, provides space or time for advertisements and collects revenues for each use.

affiliate revenue model   A model based on steering business to an affiliate firm and receiving a referral fee or percentage of revenues.

alliance   See *partnership*.

angels   Wealthy individuals, usually experienced entrepreneurs, who invest in business start-ups in exchange for equity in the new ventures.

architectural innovation   A change in how components of a product are linked together while core design concepts are left untouched.

asset   Something of monetary value that is owned by a firm or an individual.

asset velocity   Ratio of sales to net assets.

balanced scorecard   A strategy formulation device as well as a report of performance.

barriers to entry   Whatever keeps a firm from entering an industry or market.

base case   The calculation of cash flows based on a set of assumptions that portray outcomes that are most likely to happen.

best customer   One who values your brand, buys it regularly whether your product is on sale or not, tells his or her friends about your product, and will not readily switch to a competitor.

board of advisors   A group constituted to provide advice and contacts to a venture. The members have extensive skills and knowledge and provide good advice.

board of directors   A group composed of key officers of a corporation and outside members responsible for the general oversight of the affairs of the entity.

book value   The net worth (equity) of the firm, calculated by total assets minus intangible assets (patents, goodwill) and liabilities.

bootstrap financing   Launching a start-up with modest funds from the entrepreneurial team, friends, and family.

brand   A combination of name, sign, or symbol that identifies the goods sold by a firm.

brand equity   The brand assets linked to a brand's name and symbol that add value to a product.

breakeven   The point at which the total sales equal the total costs.

burn rate   Defined as cash in minus cash out on a monthly basis.

business design   A design that incorporates the venture's selection of customers, its offerings, the tasks it will do itself and those it will outsource, and how it will capture profits.

business-format franchise   Involves the provision of a complete method including a license for the trade name and logo, the products and methods, the form of the physical facility, the strategy, and the purchasing system.

business method patent   A type of a utility patent that involves the classification of a process.

business model   A set of planned assumptions about how a firm will create value for all its stakeholders; the resulting outcome of the business design process.

business plan   A document that describes the opportunity, product, context, strategy, team, required resources, financial return, and harvest of a business venture.

cannibalization   The act of introducing products that compete with a firm's existing product line.

capacity   The ability to act or do something. A firm has processes and assets that need to be expanded as the firm grows its sales volume.

cash flow   The sum of retained earnings plus the depreciation provision made by a firm. The cash coming into a firm minus the cash going out over a predetermined time period.

C-corporation   A business that provides limited liability, unlimited life, the ability to accept investments from other corporations, and the ability to merge with other corporations.

certain   An outcome resulting from an action in that it will definitely happen.

challenge   A call to respond to a difficult task and the commitment to undertake the required enterprise.

champion   An executive or leader in the parent company who advocates or provides support and resources as well as protection of the venture when parent company routines are breached. The champion helps describe and defend the venture and secure the necessary resources.

chasm   A large gap between visionaries and pragmatists in the adoption process.

cluster   A geographic concentration of interconnected companies in a particular field. Clusters can include companies, suppliers, trade associations, financial institutions, and universities active in a field.

collaborative structure   Primarily consists of teams with few underlying functional departments. In a collaborative structure, the operating unit is a team that may consist of 5 to 10 members.

competitive advantage   A firm's distinctive factors that give it a superior or favorable position in relation to its competitors.

competitive intelligence   The process of legally gathering data about competitors.

complement   A product that improves or perfects another product.

complementors   Companies that sell complements to another enterprise's product offerings.

concept summary   A simple statement of the problem the new venture is solving and how the new venture will act to solve it.

conjoint analysis   A quantitative measure of the relative importance of one attribute as opposed to another.

convergence   The coming together or merging of several technologies or industries thought to be different or separate.

copyright   An exclusive right granted by the federal government to the owner to publish and sell literary, musical, or other artistic materials. A copyright is honored for 70 years after the death of the author.

core competencies   The unique skills and capabilities of a firm.

corporate culture   The basic style of a company and how people work with each other.

corporate new venture (or **corporate venture**)   A new venture started by an existing corporation for the purpose of initiating and building an important new business unit or organization, solely owned subsidiary, or spin-off as a new public company.

corporate venture capital   An initiative by a corporation to invest in either young firms outside the corporation or units formerly part of the corporation. These are often organized as corporate subsidiaries, not as limited partnerships.

corporation   A legal entity separate from its owners. A body of owners granted a charter to act as a separate entity distinct from its owners.

cost driver   Any expense that impacts total firm costs: fixed, variable, semi-variable, and nonrecurring.

creative destruction   The creation of new industrial structures and companies and the destruction of older structures.

creativity   The ability to use the imagination to develop new ideas, new things, or new solutions.

customer relationship management   A set of conversations that consist of (1) economic exchanges, (2) the product offering that is the subject of the exchange, (3) the space in which the exchange takes place, and (4) the context of the exchange with the customers.

customization   Provision of a product designed to meet a user's preferences.

debt capital   Money that a business has borrowed and must repay in a specified time with interest.

design   The activity leading to the arrangement of concrete details that embodies a new product idea or concept.

design patents   Grants of exclusive right of use for new original, ornamental, and nonobvious designs for articles of manufacture for a period of 14 years.

diffusion of innovations   The process by which innovations spread through a population of potential adopters.

diffusion period   The time required to move from 10 percent to 90 percent of the potential adopters.

discount rate   The rate (r) at which future earnings or cash flow is discounted because of the time value of money.

disruptive innovation   An innovation that uses new modules and new architecture to create new products.

dominant design   A design whose major components and underlying core concepts do not vary substantially from one product model to the other and that commands a high percentage of the market share for the product.

due diligence   Gathering and verifying facts and data provided in a business plan before making a commitment to the terms of an investment deal.

dynamic capitalism   The process of wealth creation characterized by the dynamics of new, creative firms forming and growing and old, large firms declining and failing.

dynamic disequilibrium   The constant change of factors in an economy.

e-commerce   Digitally enabled commercial transactions between and among organizations and individuals.

economic capital   The value of an economy and the associated standard of living.

economics   The study of humans in the ordinary business of life. Economics can also be defined as the study of how society manages its scarce resources.

economies of scale   The concept that larger volumes sold reduce per-unit costs.

economies of scope   Economies obtained by sharing of resources by multiple products or business units.

elevator pitch   A short version of the venture story that quickly demonstrates that the entrepreneurs know their business and can communicate it effectively.

emergent industries   Newly created or re-created industries formed by product, customer, or context changes.

emotional intelligence   A bundle of four psychological capabilities that leaders exhibit: self-awareness, self-management, social awareness, and relationship management.

entrepreneur   (1) A person who undertakes an enterprise or business with the chance of profit or loss (or success or failure); or (2) a person or group that engages in the initiation and growth of a purposeful enterprise for the production of goods and services.

entrepreneurial capital   A combination of entrepreneurial competence and entrepreneurial commitment.

entrepreneurial commitment   A dedication of the time and energy necessary to bring the enterprise to initiation and fruition.

entrepreneurial competence   The ability (1) to recognize and envision taking advantage of opportunity and (2) to access and manage the necessary resources to take advantage of the opportunity.

entrepreneurial intensity   The degree of commitment of the entrepreneurial team to the growth of the firm.

entrepreneurship   Focused on the identification and exploitation of previously unexploited opportunities.

equity capital   The investment by a person in ownership through purchase of the stock of the firm.

ergonomics   The science of making a physical task easier and less stressful to accomplish.

ethics   A set of moral principles for good human behavior. Ethics provides the rules for conducting activities in a manner acceptable to society.

execution   A discipline for meshing strategy with reality, aligning the firm's people with goals, and achieving the results promised.

exit strategy   The way entrepreneurs or investors get their money out of a venture.

family-owned business   A firm that includes two or more members of a family who hold control of the firm.

financial capital   Financial assets such as money, bonds, securities, land, patents, and trademarks.

first-mover advantage   Gain accruing to the first to enter a market.

flexibility   A measure of a firm's ability to react to a customer's needs quickly.

flow-through entities   Firms where all profits flow to the owners free of prior taxation.

focus group   A small group of people who are brought together to discuss a product or service.

follower strategy   An entrant to a market that follows the initial one and attempts an improvement of the first mover's strategy.

founders   The people responsible for starting a firm, usually all the members of the initial leadership team.

franchise   A legal arrangement in which the owner of a business format has licensed it to an individual or a local firm.

franchisee   An individual or a local firm that receives the right through a contract to use the franchisor's business format, brand, and logo in a specific geographic region.

franchisor   The organization that owns and operates a firm that controls the business format and its associated trademarks and logo.

global   A strategy that emphasizes worldwide creation of new products, sales, and marketing.

globalization   The integration of markets, nation-states, and technologies enabling people and companies to offer and sell their products in any country in the world.

growing industries   Industries that exhibit moderate revenue growth and have moderate stability and uncertainty.

harvest plan   A plan that defines how and when the owners and investors expect to realize or attain an actual cash return on their investment.

high growth business   Characterizes a business corporation that aims to build an important new business and requires a significant initial investment to start up.

horizontal merger   A merger between firms that make and sell similar products in a similar market.

human capital   Combined knowledge, skill, and ability of the people in the enterprise.

hybrid model   This model, sometimes called "bricks and clicks," utilizes the best of the Internet as well as other channels. A hybrid model can extend a company's reach to new market segments as well as globally.

implementation   Putting into effect the elements of the business plan.

increasing returns   When the marginal benefits of a good or of an activity are growing with the total quantity of the good or the activity consumed or produced.

incremental innovation   An innovation that is a faster, better, and/or cheaper version of an existing product.

independent venture   A new venture that is not owned or controlled by an established corporation. An independent venture is typically unconstrained in its choice of a potential opportunity yet is usually constrained by limited resources.

industry   A group of firms producing products that are close substitutes for each other and serve the same customers.

initial public offering   The first public equity issue of stock by a company.

innovation   Invention that has produced economic value in the marketplace. It is the commercialization of new technology.

installed base profit model   One of the most powerful profit models; the supplier builds a large installed base of users who then buy the supplier's consumable products.

integrity   Truthfulness, wholeness, and soundness; the consistency of our words and our actions or our character and our conduct.

intellectual capital   The sum of knowledge assets of an organization.

intellectual property   The valuable intangible property owned by persons or companies.

international   A strategy that aims to create value by transferring products and capabilities from the home market to other nations using export or licensing arrangements.

internet   A worldwide network of computer networks linking businesses, organizations, and individuals.

intrepreneurship   The entrepreneurial process within the confines of an established corporation.

IPO   See *initial public offering*.

joint venture   A short-lived partnership with each partner sharing in the costs and rewards of the project; common in research, investment banking, and the health care industry.

just-in-time   A method that focuses on reducing unnecessary inventory and removing non-value-added activities by receiving items only when needed.

killer app   See *disruptive application.*

knowledge   The awareness and possession of information, facts, ideas, truths, and principles in an area of expertise.

knowledge management   The practice of collecting, organizing, and disseminating the intellectual knowledge of a firm for the purpose of enhancing its competitive advantages.

layout   The arrangement of a facility to provide a productive workplace. This can be accomplished by aligning the form of the space with its use or function.

leadership   The ability to create change or transform organizations. A real measure of leadership is the ability to acquire needed new skills as the situation changes.

lead users   People who have an advanced understanding of a product and are experts in its use.

lean systems   Operations systems that are designed to create efficient processes by using a total systems perspective.

learning organization   A firm that captures, generates, shares, acts, and uses its corporate experiences to improve and adapt.

license   A grant to another firm to make use of the rights of the licensor's intellectual property.

licensing   Occurs when a firm (the licensor) grants the right to produce its product, use its production processes, or use its brand name or trademark to another firm (the licensee). In return for giving the licensee these rights, the licensor collects a royalty fee on every unit the licensee sells.

limited liability company   A way of legally organizing a firm to limit liability of the participants.

local   A strategy focusing all efforts locally (or regionally) since that is the venture's pathway to a competitive advantage.

logistics   The organization of moving, storing, and tracking parts, materials, and equipment. Logistic systems usually are based on electronic networks such as a supply-chain intranet.

loyalty   A measure of a customer's commitment to a company's product or product line.

management   A set of processes such as planning, budgeting, organizing, staffing, and controlling that keep the organization running well.

marketing   A set of activities with the objective of securing, serving, and retaining customers for the product offerings of the firm. Marketing is getting the right message to the right customer segment via the appropriate media and methods.

marketing objectives statement   A clear description of the key objectives of the marketing program.

marketing plan   A written document serving as a section of the business plan and containing the necessary steps required to achieve the marketing objective.

market potential   A prospective of the maximum sales under expected conditions.

market research   The process of gathering the information that serves as the basis for a sound marketing plan.

market segment   A group with similar needs or wants and may include geographical location, purchasing power, and buying attitudes.

market segmentation   The division of the market into segments that have different buying needs, wants, and habits.

mature industries   Industries that have slow revenue growth, high stability, and intense competitiveness.

merger   The combining together of two companies.

metanational   A company that possesses three core capabilities: (1) being the first to identify and capture new knowledge emerging all over the world; (2) mobilizing this globally scattered knowledge to out-innovate competitors; and (3) turning this innovation into value by producing, marketing, and delivering efficiently on a global scale.

modular innovation   An innovation that uses new components and modules, but does not disrupt the linkages between modules.

module   An independent, interchangeable unit that can be combined with others to form a larger system.

moral principles   Tenets concerned with goodness (or badness) of human behavior and usually provided as rules and standards of human behavior.

multidomestic   A strategy that calls for a presence in more than one nation as resources permit.

natural capital   Those features of nature, such as minerals, fuels, energy, biological yield, or pollution absorption capacity, that are directly or indirectly utilized or potentially utilizable in human social and economic systems.

negotiation   A decision-making process among interdependent parties who do not share identical preferences.

net present value (NPV)   The present value of the future cash flow of a venture discounted at an appropriate rate (r).

network economies   Observed effects in industries where a network of complementary products is a determinant of demand.

new venture team   A small group of individuals who possess expertise, management, and leadership skills in the requisite areas.

new venture valuation rule   Uses the projected sales, profit, and cash flow in a target year (N) and the projected earning growth rate (g) for five years after year N to calculate value of a firm.

niche business   A firm that seeks to exploit a limited opportunity or market to provide the entrepreneurs with independence and a slow-growth buildup of the business.

nonprofit organization   A corporation or a member association initiated to serve a social or charitable purpose.

oligopoly   An industry characterized by just a few seller firms.

on-time speed   A measure of lead-time, on-time delivery, and product development speed.

open source innovation   An innovation that is the product of many firms and individuals working together under a common goal and an agreed-to governance system.

operation   A series of actions.

operations management   The supervising, monitoring, and coordinating of the activities of a firm carried out along the value chain. Operations management deals with processes that produce goods and services.

opportunity   A timely and favorable juncture of circumstances providing a good chance for a successful venture or progress; an auspicious chance of an action occurring at a favorable time.

opportunity cost   The value (cost) of the forgone action.

option   The right to purchase an asset at some future date and at a predetermined price.

organic growth   Growth enabled by internally generated funds.

organic organizations   Organizations that are flexible and effectively adapt to change.

organizational capital   An enterprise's management processes, work procedures, information technologies, and communication methods.

organizational culture   The bundle of values, norms, and rituals that are shared by people in an organization and govern the way they interact with each other and with other stakeholders.

organizational design   The design of an organization in terms of its leadership and management arrangements, its selection, training, and compensation of its talent (people), its shared values and culture, and its structure and style.

organizational norms   The guidelines and expectations that impose appropriate kinds of behavior for members of the organization.

organizational rituals   The rites, ceremonies, and observances that serve to bind together members of the organization.

organizational values   The beliefs and ideas of an organization about what goals should be pursued and what behavior standards should be used to achieve these goals. Values include entrepreneurship, creativity, honesty and openness.

outsourcing   Purchasing services or goods from suppliers rather than doing or producing them within the firm.

partnership   Business association of two or more people or firms who agree to cooperate with one another to achieve mutually compatible goals that would be difficult for each to accomplish alone. There are two types of partnerships: the general partnership and the limited partnership.

patent   A grant by the U.S. government to an inventor giving exclusive rights to an invention or process for 20 years. A U.S. patent does not always grant rights in foreign countries.

PE ratio   The ratio of the price of a stock to the company's earnings.

personalization   The provision of content specific to a user's preferences and interests.

pessimistic case   When the outcome of the calculation of cash flows, based on a set of assumptions, is less than expected.

place   The channels for distribution of the product and, when appropriate, the physical location of the stores.

plant patent   A grant of exclusive right of use for a term of 20 years for certain new varieties of plants that have been asexually reproduced.

positioning   The act of designing the product offering and image to occupy a distinctive place in the target customer's mind.

post-money value   The valuation accorded a company after investment by venture capitalists or angels.

preferred stock   Stock with preferences or claims on dividends and assets before common stock owners.

pre-money value   The value accorded a company before investment from venture capitalists or angels.

pricing policies   Methods for setting prices for various customer categories and volume discount plans.

private placements   The sale of stocks or bonds to wealthy individuals, pension funds, insurance companies, or other investors without a public offering or any oversight from the Securities and Exchange Commission.

process   Any activity or set of activities that takes one or more inputs, transforms and adds value to them, and provides one or more outputs.

product   The item or service that serves the needs of the customer.

product-distribution franchise   A license to sell specific products under a manufacturer's trademark and brand.

productivity   The quantity of goods and services produced from the sum of all inputs, such as hours worked and fuels used.

product offering   Communicates the key values of the product and describes the benefits to the customer.

product platform   A set of modules and interfaces that form a common architecture from which a stream of derivative products can be efficiently developed and produced.

product sales model   Sales of a product in units to a customer. Selling items for a price.

profit   The net return after subtracting the costs from the revenues.

profit margin   The ratio of profit divided by sales revenues.

profit model   The mechanism a firm uses to reap profits from its revenues.

pro forma   Provided in advance of actual data. Pro forma statements are forecasts of financial outcomes.

promotion   The communication of an initial product message using public relations, advertising, and sales methods to attract customers.

proprietary   That which is owned, such as a patent, formula, brand name, or trademark associated with the product or service, and not usable by another without permission.

prototype   A model that has the essential features of the proposed product or service but remains open to modification.

public offering   The sale of a company's shares of stock to the public by the company or its major stockholders.

quality   A measure of a product that usually includes performance and reliability.

radical innovation   An important new development that leads to a new industry or way of operating.

rapid prototyping   The fast development of a useful prototype that can be used for collaborative review and modification.

real option   The right to invest in (or purchase) a real asset (the start-up firm) at a future date.

regional   See *local*.

regret   The amount of loss that a person can tolerate.

regular taxable corporation   An enterprise subject to taxes on its reported profits.

relational coordination   Describes how people act as well as how they see themselves in relationship to one another.

reliability   A measure of how long a product performs before it fails.

resilience   The ability to recover quickly from setbacks. It is a skill that can be learned and increased.

restricted stock   Stock issued in an employee's name and reserved for his or her purchase at a specified price after a period of time.

return on capital   The ratio of profit to the total invested capital of a firm.

return on equity   The ratio of net income divided by owner's equity.

return on invested capital   The ratio of net income to investment.

return on investment (ROI)   The ratio of net income divided by invested capital.

revenue model   Describes how the firm will generate revenue.

revenues   A firm's revenues are its sales in dollars expressed after deducting for all returns, rebates, and discounts.

risk   The chance or possibility of loss, which could pertain to finance, physicality, or reputation. Risk is a measure of the potential variability that will be experienced in the future.

robust product   A product that is relatively insensitive to aging, deterioration, component variations, and environmental conditions.

sales cycle   The length from the first contact with a customer until a sale transaction is made.

sales forecast   An estimate of the amount of sales to be achieved under a set of assumed conditions within a specified period of time.

scalability   The extent to which a firm can grow in various dimensions to provide more service.

scale of a firm   The extent of the activity of a firm as described by its size. The scale of a firm's activity can be described by its revenues, units sold, or some other measure of size.

scenario   An imagined sequence of possible events or outcomes, sometimes called a mental model.

scope of a firm   The range of products offered or distribution channels utilized.

S-corporation   A firm that has elected to be taxed as a partnership under the subchapter S provision of the Internal Revenue Code.

seed capital   The first funds used to launch the new firm.

self-organizing organization   Teams of individuals that benefit from the diversity of the individuals and the robustness of their network of interactions.

selling   The transfer of products from one person or entity to another through an exchange mechanism.

six forces model   A model for evaluating the competitive forces in an industry: (1) firm rivalry, (2) threat of entry by new competitors, (3) threat of substitute products, (4) bargaining power of customers, (5) bargaining power of complementors, and (6) bargaining power of suppliers.

small business   A firm with fewer than 50 employees operating as a sole proprietorship, a partnership, or a corporation owned by a few people.

social capital   The accumulation of active connections among people in a network. Social capital refers to the resources available in and through personal and organizational networks.

social entrepreneur   A person or team that acts to form a new venture in response to an opportunity to deliver social benefits while satisfying environmental and economic values.

sole proprietorship   A firm owned by only one person and operated for his or her profit.

sources of innovation   For new ventures, these include universities, research laboratories, and independent inventors.

spin-off unit   An organization that is first established within an existing company and then sent off on its own.

staged financings   The provision of capital to entrepreneurs in multiple installments, with each financing conditional on meeting particular business targets. This helps ensure that the money is not squandered on unprofitable projects.

stock   The owner's shares of the corporation.

stock options An offer in a plan under which employees can purchase shares of the company at a fixed price. Stock options take on value once the market price of a company's stock exceeds the exercise price. A stock option gives employees the right to buy the company's stock in the future at a preset price.

story A narrative of factual or imagined events. The story tells the goal, the challenge, and the response of the new firm.

strategic control The process used by firms to monitor their activities and evaluate the efficiency and performance of these activities and to take corrective action to improve performance, if necessary.

strategic learning A cyclical process of adaptive learning using four steps: learn, focus, align, and execute.

strategy A plan or road map of the actions that a firm or organization will take to achieve its mission and goals.

subscription revenue model A type of business that offers content or a membership to its customers or members and charges a fee permitting access to the information or participation for a certain period of time.

sunk costs A cost that has already incurred cannot be affected by any present or future decisions. In other words, funds and time invested on a new venture are already spent, regardless of any action taken today or later.

supply-chain management A firm's processes and those of its suppliers that enable the flow of materials, resources, and information to meet customer demand.

sustainable competitive advantage A competitive advantage that can be maintained over a period of time.

switching costs The costs to the customer to switch from the product of an incumbent company to the product of the new entrant.

synergy The increased effectiveness and achievement produced as a result of the combined action of two or more firms.

talent The people, often called employees, of an organization.

team A small number of people with complementary capabilities and skills who are committed to a common objective, goals, and tasks for which they hold themselves mutually accountable.

technology Devices, artifacts, processes, tools, methods, and materials applied to industrial and commercial purposes.

telework All kinds of remote work from home, satellite offices, and on the road.

term sheet A summary of the principal conditions for a proposed investment by a venture capital firm in a company.

theory of a business How a firm comprehends its total activities, resources, and relationships.

throughput The amount of units processed within a given time.

throughput efficiency The ratio of value-adding time and the sum of value-adding time and non-value-adding time. The goal is to reduce non-value-adding time.

tipping point The moment of critical mass or threshold that results in a jump in adoption of a product or service.

trademark Any distinctive word, name, symbol, slogan, shape, sound, or logo a firm uses to designate its product.

trade-name franchise A franchise name that primarily involves a brand name, such as Western Auto or ACE Hardware.

trade secret An intellectual asset protected by confidentiality, nondisclosure, and assignment of inventions agreements as well as physical barriers such as safes and limited access.

transaction fee revenue model A type of business that provides a transaction source or activity for a fee.

transnational A strategy resting on a flow of product offerings created in any one of the countries of operation and transferred between countries.

triple bottom line The three factors of a product or business: economic, environmental, and social equity.

trust   A firm belief in the reliability or truth of a person or an organization.

uncertain   An outcome resulting from an action in that the outcome is not known or is likely to be variable.

unique selling proposition   A short version of a firm's value proposition often used as a slogan or summary phrase to explain the key benefits of the firm's offering versus that of a key competitor.

usability   A measure of the quality of a user's experience when interacting with a product.

utility patents   Rights of exclusive use issued for the protection of new, useful, nonobvious, and adequately specified processes, machines, and manufacturing processes for a period of 20 years.

valuation rule   The algorithm by which an investor, such as an angel or venture capitalist, assigns a monetary value to a new venture.

value   The worth, importance, or usefulness to the customer. In business terms, value is the worth in monetary terms of the social and economic benefits a customer pays for a product or service.

value chain   The sequence of steps or subprocesses that a firm uses to produce its product or service.

value proposition   Summarizes the values offered to the customer.

value web   Consists of the extended enterprise within a network of interrelated stakeholders that create, sustain, and enhance its value-creating capacity. It is usually based on an Internet infrastructure to manage operations dispersed in many firms.

venture capital   A source of funds for new ventures that is managed by investment professionals on behalf of the investors in the venture capital fund.

venture capitalists   Professional managers of investment funds.

versioning   The creation of multiple versions of a products and selling their modified versions to different market segments at different prices.

vertical integration   The extension of a firm's activities into adjacent stages of productions (i.e., those providing the firm's inputs or those that purchase the firm's outputs).

vertical merger   The merger of two firms at different places on the value chain.

viral marketing   Building knowledge of a product through word of month.

virtual organization   A venture that manages a set of partners and suppliers linked by the Internet, fax, and telephone to provide a source or product.

vision   An informed and forward-looking statement of purpose in response to an opportunity.

working capital   The amount of funds available to support a firm's normal operations, such as unexpected or out-of-the-ordinary, one-time-only expenses. Working capital is a firm's current assets minus its current liabilities.

INDEX

Note: Page numbers followed by *f* and/or *t* refer to figures and tables, respectively.